W9-CNM-543

# GUIDING CHILDREN TO MATHEMATICAL DISCOVERY

## THIRD EDITION

LEONARD M. KENNEDY

California State University
Sacramento

Wadsworth Publishing Company
Belmont, California
A Division of Wadsworth, Inc.

Education Editor: Roger Peterson
Production Editor: Carolyn Tanner
Designer: Robert Hu
Copy Editor: Charles D. Cox
Technical Illustrator: John Foster
Cover Design: Tony Naganuma



Printed in the United States of America

3 4 5 6 7 8 9 10—84 83 82

---

Library of Congress Cataloging in Publication Data

Kennedy, Leonard M.
Guiding children to mathematical discovery.

Bibliography: p.
Includes index.
1. Mathematics—Study and teaching (Elementary)
I. Title
QA135.5.K43 1980 372.7 79-20904
ISBN 0-534-00757-0

---

# CONTENTS

# PREFACE

This book is for preservice and inservice teachers of contemporary mathematics in the elementary school. It presents a program of instruction that is based on the work of such learning theorists as Piaget, Bruner, Gagne, and Skemp and on successful classroom practice. The book emphasizes the need for presenting mathematics through carefully sequenced activities in which the structure of mathematics is highlighted. Through such a program children first develop an understanding of the underlying concepts and principles by engaging in activities with appropriate models of the mathematics they are learning. Upon achieving this understanding, children formulate the generalizations that make for more permanent learning and learn the computational and other skills they need to use mathematics effectively now and in the future.

The material in this text will help the teacher present mathematical concepts in ways that lead each child to see—in his or her own way—their relevance in a wide variety of situations. Numerous examples show a teacher how a classroom can be organized to meet the varying needs of children; the roles of mathematics laboratories, classroom learning centers, individually prescribed instruction, and teacher-directed lessons are clearly explained. Each topic of contemporary elementary school mathematics programs is discussed. The background a child needs for each topic is considered, then activities and materials for introducing the topic to children and developing their understanding of it are presented. A wide variety of materials and procedures for dealing with each topic is featured.

This third edition of *Guiding Children to Mathematical Discovery* reflects current thinking about the content of elementary school mathematics and the ways it should be presented to children. It retains the emphasis on structure and the guided-discovery approach featured in the earlier editions. It also retains the easy-to-read style of writing and plentiful illustrations that made the first two editions so popular with their readers.

While retaining the best features of the previous editions, the author has changed the text to present a current approach to the teaching of elementary school mathematics:

1. Chapter 1 discusses the forces that affect the content and procedures of mathematics, and the topics commonly found in today's elementary school program.
2. Chapter 2 emphasizes the impact of Piaget, Bruner, Gagne, and Skemp on the way mathematics is presented to children, and includes a *model of the*

*learning process* that guides the organization of instruction around mathematical goals, modes of learning, and the hierarchy of concepts and skills for a topic.

3. Chapter 3 discusses ways to diagnose children's backgrounds and how to organize the classroom for teaching. New to the book is extensive information about *minicalculators;* their role in a program and guidelines for selecting a calculator are discussed in Chapter 3.

4. Greater emphasis has been placed on mathematical problem solving, with both new material and a new position in the text—it is now discussed in Chapter 4.

5. Beginning with Chapter 4, and continuing in each of the remaining chapters, is a section entitled *Common Pitfalls and Trouble Spots.* These sections discuss some of the problems teachers face while teaching mathematics; they also discuss trouble spots for children. Each discussion contains one or more suggestions for avoiding or minimizing both pitfalls and trouble spots. These pitfall and trouble spot sections should help teachers diagnose and remedy children's mathematical misunderstandings, as well as revise lesson plans accordingly. In the last chapter, we will summarize the many pitfalls and trouble spots into two mistakes that teachers frequently make.

6. Chapters 5 through 10 present updated information about teaching whole numbers and their characteristics and principles, the Hindu-Arabic numeration system, and operations. The low-stress algorithms for addition, subtraction, and multiplication developed by Barton Hutchings are included in Chapters 8 and 10. Common fractions, decimal fractions, and percent are discussed in Chapters 11, 12, and 13. A rationale for continuing to teach common fractions is given along with materials and procedures for teaching their meanings and operations. The role of decimal fractions and percent and instruction about them are the subjects of Chapter 13.

7. Chapter 14 contains a completely revised discussion of measures and the processes of measuring. It is devoted almost exclusively to *the metric system.* Background information about the system is given for readers who are unfamiliar with it. A scope-and-sequence chart suggests an order for presenting the system to children, and activities appropriate for developing children's understanding of measures and the uses of nonstandard and standard units of measure are presented. All commonly used units of measure in the metric system are included, along with time and money.

8. The chapter on number theory has been reorganized to include information about statistics and graphs and probability. (These were formerly in another chapter.)

9. Altogether, Chapters 3, 8, 10, 11, and 13 present more than twenty activities using a minicalculator. Many of these pertain to work in basic areas of mathematics, while others are enrichment-type activities that will extend children's interest in and understanding of mathematics. Some are general in nature and can be used as the springboard for teacher- and pupil-developed activities.

10. The other chapters have been revised to reflect current thinking about their content. The study questions and activities and the list of further reading at the close of each chapter have been updated.

11. Three appendices have been added. Appendix A contains prefixes and suffixes that are part of many of the mathematical terms used in this book. The derivation and meaning of each one is identified. Appendix B contains the titles of books that can be used to extend the reader's understanding of the content and procedures of elementary school mathematics. Appendix C contains the names and addresses of companies that produce materials pertaining to the metric system and minicalculators. These companies offer a variety of materials; you can become familiar with them by writing for catalogs.

The author is indebted to many persons who have helped in this revision of *Guiding Children to Mathematical Discovery.* I am grateful to users who have given suggestions for strengthening the book: Lois Lackner of Loyola University of Chicago, June D. Oxstein of California State University at Los Angeles, M. A. Fowler of Northeastern Illinois University, Joy McDaniel of Freed Hardeman College, C. M. Berberick of Trinity College, Sister Catherine Norton of Edgewood College and Merle Grady of the University of Dallas. I especially thank Paul C. Burns of the University of Tennessee, Hiram D. Johnston of Georgia State University, and Edward Silver of Northern Illinois University. It is my hope that they will find that the new book reflects their recommendations in satisfactory ways. However, I alone accept responsibility for any deficiencies a reader may find. The continuing support of my wife, Mary, is deeply appreciated.

# HOW TO USE THIS BOOK

Each chapter of this book begins with a list of performance objectives that explain what you may expect to gain from the chapter and a list of key terms you will find in it. The objectives and terms for each chapter should be read before reading the chapter itself. If there are terms with which you are not familiar, you should first look them up in the glossary/index and/or Appendix A, which includes prefixes and suffixes of mathematical terms included in this book. Within the chapters is a *self-check* for each objective. Each self-check should be read and responded to to determine how successfully you understand the mathematical concepts and the procedures for using teaching materials and learning aids.

If you perform the self-checks successfully, you will acquire the understanding and skills required to become a successful teacher of elementary school mathematics. Each time that a self-check is unsuccessful, you should review the materials upon which it is based, or read other material dealing with the objective. Then repeat the self-check.

At the end of each chapter are Study Questions and Activities. These will add depth and breadth to the understanding of the concepts and procedures included in the chapter. Complete as many of these as time, energy, and inclination dictate. There are also annotated further reading lists at the end of each chapter that offer a selection of additional material.

There is an unannotated bibliography in Appendix B. It includes books on special mathematical topics, books for teachers of particular age-groups of children, books about mathematics laboratories and learning centers, books about mathematical games, and selected National Council of Teachers of Mathematics yearbooks. You should examine as many of these books as possible and choose the ones that seem the most relevant to your objectives. This bibliography should help you to build a personal mathematics library.

I hope you will find this book interesting and helpful.

Leonard M. Kennedy

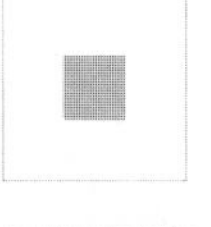

our symbol for Common Pitfalls and Trouble Spots

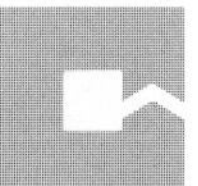

our symbol for Self-check

# GUIDING CHILDREN TO MATHEMATICAL DISCOVERY

THIRD EDITION

# 1 Mathematics in the Elementary School

Upon completion of Chapter 1, you will be able to:

1. State orally or in writing five different features of a mathematics program that can lead children to mathematical literacy.
2. Identify three factors that have influenced current mathematics programs, and explain orally or in writing at least one influence of each factor.
3. List five characteristics of successful contemporary mathematics programs.
4. Name orally or in writing eight skill areas commonly included in today's mathematics programs.
5. Explain why classification and sorting activities are an important part of mathematics in today's schools.
6. Distinguish between a number system and a numeration system, and list the number systems that elementary school children study.
7. Identify the characteristics of the Hindu-Arabic numeration system.
8. Name and give examples of the two basic number operations and their inverses.
9. Name the basic properties included in elementary programs, and identify the number systems in which each property is applicable.
10. Distinguish between discrete and continuous objects, and give examples of continuous objects.
11. Name at least two reasons for including geometry in a mathematics program, and cite examples that illustrate the validity of the reasons.
12. Name at least three reasons each for including number theory and probability activities in a program.
13. Cite examples of sources of statistics for children to use.

Key Terms you will encounter in Chapter 1:

automation
computer
minicalculator
National Science Foundation
School Mathematics Study Group
professional mathematicians
learning theories
problem solving
classification
sorting
attribute
number
operation
number system
numeration system
base number
whole number
fractional number
natural number
real number
rational number
algorithm

sum
addend
factor
product
closure property
commutative property
associative property
distributive property
identity element
inverse property
reciprocal
number sentence
discrete material
continuous material
measurement
metric system
topological relationship
transformation
number theory
prime number
composite number
least common multiple
greatest common factor
probability
statistics
graph

The challenge facing American education has never been greater. Today's technological society requires that schools help children prepare for a life of continuous learning. New jobs are appearing regularly, and the requirements for old ones are changing at an ever-increasing pace. Our schools must design programs that will enable graduates to cope with the unknown changes to come.

The mathematics program of the elementary school must help meet the challenge. It must provide children with the knowledge, attitudes, and skills they will need to be mathematically literate. Through an inductive approach to learning, teachers can show children the excitement in discovering how to perform operations on numbers. Children can search for patterns among numbers and can develop an awareness of how patterns are useful in organizing and synthesizing ideas about numbers. Rather than the teacher or a textbook presenting generalizations of mathematics, children can be guided to formulate them, using their own thought patterns and expressing generalizations in their own words. Later, the generalizations can be stated in the words of the mathematician. Through processes of inquiry, discovery, pattern searching, and generalizing, children can learn to think creatively and use mathematics as a means of solving everyday problems. Instead of memorizing a given approach to a problem situation, children can develop a grasp of the principles involved, thus enabling them to find alternate approaches to the problems.

Computers and minicalculators are commonplace and are replacing paper and pencil and other slower means of computing. Even so, children must realize that they need to master certain mathematical skills. All children should understand the concepts involved in the operations of addition, subtraction, multiplication, and division. As their understanding develops, they will master the basic facts for these operations. They must also have an understanding of the algorithms for the various operations.

During their elementary school years, children must learn mathematics in a manner that will:

1. Allow them opportunities to manipulate concrete objects and models to develop their understanding of the characteristics of number and numeration systems.
2. Lead them through varied experiences and dialogue to discover mathematical concepts and develop their understanding of computational and measurement processes.
3. Let them work according to their individual learning styles and at rates that are appropriate for each individual.
4. Inspire them to enjoy the study of mathematics and develop positive attitudes toward the subject.
5. Lead them to recognize the importance of mathematics in an increasingly technological society.

---

## Self-check of Objective 1

You have just read five features of a program to develop children's mathematical literacy. Can you name these five features? Also, can you give examples of why it is important that children learn mathematics in the manner described by these features?

---

## FACTORS INFLUENCING ELEMENTARY SCHOOL MATHEMATICS

Nearly every aspect of the elementary school mathematics curriculum has been under review in recent years. The increasing use of computers, minicalculators, and automation techniques, the interest of professional mathematicians, and investigations into the learning process have all had an impact on the content of mathematics and procedures for teaching it.

The computer industry, minicalculators, and automation are three developments that have placed a great premium on mathematical theory and understanding. Electronic change-making cash registers and miniaturized calculating machines are reducing the need for individuals with the skills required for performing complex calculations. There is an increasing need for individuals with the skills and knowledge to program computers and operate automated electronic devices. With the advent of adaptable computers and printing devices, and with the decreasing costs involved, the typical data-processing job is daily coming closer to being within reach of every business and industry.

(1) Automation, often described as the process of operating machines by machines, is a direct outgrowth of the development of computers. Today's space flights are possible because of the immediacy with which computers can operate devices controlling the entire operation. In less spectacular situations, type for newspapers can be set, air traffic controlled, electricity distributed, and dresses knitted by machines controlled by computers. Automation is rapidly eliminating many jobs and changing the requirements for others. At the same time, it is making new jobs available in many different industries and occupational fields. Many of these jobs require persons who have a much deeper understanding of mathematics than has been true of jobs in the past.

Following World War II and the Korean War the United States government funded the National Science Foundation (NSF) as a means of developing a national policy to promote science research and education. By 1958 the NSF was funding the work of the School Mathematics Study Group (SMSG), which was centered first at Yale and then at Stanford University. A number of professional mathematicians and mathematics educators, headed by E. G. Begle, developed mathematics materials for secondary schools. Later, their interests turned to the junior high and elementary schools, and materials were developed for these levels also. By 1960 a number of elementary school mathematics projects were under way. Notable among them were the University of Illinois Mathematics Project, the Greater Cleveland Mathematics Program, the Madison Project, and the Minnesota Mathematics and Science Teaching Program. An English counterpart of the American programs was the Nuffield Project, which also had an impact on mathematics teaching in this country. The outgrowth of this interest by professional mathematicians was the "New Math" movement of the 1960s with the introduction of such new topics as geometry, number theory, probability, and sets, emphasizing number properties and the structure of mathematics.

(2) Another influence on elementary school mathematics is research into how children learn. The studies of the learning process by William Brownell, Jean Piaget, Robert Gagne, Jerome Bruner, Richard Skemp, and others are receiving continuing attention from curriculum developers and educators at all levels.

As early as the 1930s Brownell stressed the need for helping children to see the relationships of the parts to the whole and the whole to the parts.[1] This was the beginning of the meaning theory, which emphasized that children should have a chance to use manipulative and exploratory materials as they explore the meanings of numbers. The research of Brownell and others showed that children *can* develop an understanding of the meaning of what they are doing as they work with numbers—without any loss of speed in learning the basic facts and in developing skill in performing operations on numbers.

Studies by Piaget indicate the importance of considering children's levels of cognitive development as learning activities are planned. Bruner, Gagne, and Skemp emphasize the importance of considering the structure of mathematics as programs are developed and activities are planned. (Further attention is given to research into the learning process in later chapters.)

While there is not full agreement on the content, instructional procedures, learning materials, and goals of mathematics today, certain characteristics are common to successful elementary school programs:

1. The content is presented in a sequence that takes into account the structure of mathematics.
2. Both the level of cognitive development and the mathematical background of each child are considered as activities are planned.
3. New topics are introduced in a concrete-manipulative mode. Learning then proceeds to the semiconcrete-representational and, finally, the abstract mode.
4. The content includes geometry and other topics as well as the traditional arithmetic.
5. The language and symbolism of mathematics are systematically developed.

---

## Self-check of Objectives 2 and 3

Three factors that have influenced current mathematics programs are given. Can you identify each one and state at least one way each has influenced the program?

List five characteristics of successful contemporary mathematics programs.

---

[1] See, for example, William A. Brownell, "Psychological Considerations in the Learning and the Teaching of Arithmetic," in National Council of Teachers of Mathematics, *The Teaching of Arithmetic*, Tenth Yearbook (New York: Columbia University, Bureau of Publications, Teachers College, 1935), pp. 1–32.

## SKILL AREAS INCLUDED IN ELEMENTARY SCHOOL PROGRAMS

Educators do not fully agree on the content of elementary school programs: not all skill areas are considered basic by everyone. In 1977 the National Council of Supervisors of Mathematics presented a position paper identifying ten basic skill areas that should be developed by students before they complete high school: problem solving; applying mathematics in everyday situations; alertness to the reasonableness of results; estimation and approximation; appropriate computational skills; geometry; measurement; reading, interpreting, and constructing tables, charts, and graphs; using mathematics to predict; and computer literacy.[2] With the exception of computer literacy, these skill areas are included in most elementary school programs and are discussed in this book under these headings: problem solving, application, and estimation; classifying and sorting; numbers and operations; measurement; geometry; number theory; probability; and statistics and graphing.

---

### Self-check of Objective 4

Eight skill areas included in most elementary school mathematics programs are discussed in this book. Name them before you continue reading.

---

### 1. Problem Solving, Application, and Estimation

Problem solving, application, and estimation are not mathematical topics such as those that follow. Nevertheless, it is important that children develop the skills needed to solve problems involving mathematics and to apply mathematical procedures to everyday situations. Children need to be able to make reasonable estimates in consumer situations and to check the accuracy of minicalculator computations and paper-and-pencil calculations. You should take advantage of in- and out-of-school situations to give your children real problems to solve and opportunities to apply their developing computational, statistical, graphing, measurement, and other mathematical skills in meaningful ways. Throughout this book you will read about problem situations you can use to introduce new topics and to which children can apply the mathematical knowledge and skills they are learning. Chapter 4 contains specific suggestions for helping children become skillful users of mathematics.

[2] National Council of Supervisors of Mathematics, "Position Paper on Basic Skills," *The Arithmetic Teacher*, XXV, No. 1 (November 1977), pp. 19–22.

### 2. Classifying and Sorting

Activities that involve children in classifying and sorting objects should begin as early as kindergarten to provide a foundation upon which many numerical, geometric, and measurement concepts are built. Jean Piaget is a Swiss developmental psychologist whose work encompasses more than mathematics learning alone. His findings have important implications for the way mathematics should be presented to children. He has stressed the importance of classification activities. His research has shown that children do not develop a mature concept of number until they understand clearly the properties of objects and relationships between collections of objects. Particularly, the concept of set inclusion must be well understood before counting and addition will make sense. For example, before a child can determine the number of objects in a set, he or she must realize that all of the objects being considered are actually included in the set. Likewise, before addition is understood, a child must recognize that it is reasonable for the objects in the separate sets to be combined to form one set.

Young children's classifying and sorting activities give them the background they need for understanding more advanced concepts. Their first experiences should be with collections of objects that have readily distinguishable attributes, such as color or shape, that can be easily sorted into smaller groupings. As children mature, they will use other attributes—size, thickness, length, width, and so on—as they sort their collections. Later, upper-grade children will deal with sets that are abstract rather than concrete, such as a set of numbers that can be separated into prime and composite numbers and sets of geometric figures that can be classified according to the number of sides each one has.

---

## Self-check of Objective 5

Explain orally or in writing why it is important for children to have classification and sorting experiences.

---

### 3. Numbers and Operations

**a. Number Systems.** The major portion of the time spent on mathematics in the elementary school is devoted to numbers and operations. By the time children complete the sixth grade, most of them have become acquainted with the rational number system. The elementary school program also includes work with the less extensive systems of whole numbers and integers.

The set of whole numbers consists of the numbers {0, 1, 2, 3, . . .}. A subset of the whole numbers is the set of natural numbers {1, 2, 3, . . .}. The only difference between the two number systems is the absence of zero in the set of natural numbers. The natural numbers were the first numbers used by man; zero was added later. Today we use the whole numbers for counting, with zero being the number associated with a set that has no members.

The set of integers includes numbers equivalent to all of the whole numbers and the numbers that are the negatives of the nonzero whole numbers: {. . . , −3, −2, −1, 0, 1, 2, 3, . . .}. The negative integers provide answers for subtraction in the form $b - a = c$, where $a > b$, and $a$ and $b$ are integers.

Rational numbers are those that can be expressed in the form $a/b$, where $a$ and $b$ are integers, and $b \neq 0$. Included are numbers equivalent to all of the integers and numbers not equivalent to the integers expressed in the form $a/b$. Rational numbers can be expressed as common fractions (½, ⅓, ²⁄₁); decimal fractions, both terminating and repeating (0.5, 0.333 . . . , 2.0); and percent (50%, 33⅓%, 200%). Rational numbers are also called fractional numbers. In the elementary school, work with fractional numbers is usually confined to those that are greater than zero.

Elementary school children have contact with real numbers as they work with *pi* ($\pi$) expressed as 3⅐ or 3.1416. . . . Usually, the fact that pi is an irrational number and a part of the real number system is not brought out at the elementary school level.

**b. The Hindu-Arabic Numeration System.** Children must learn to express numbers in both numeral and word form. They study the Hindu-Arabic numeration system thoroughly to learn its numerals and how they are used to express numbers. During their study, children learn that the Hindu-Arabic system has certain characteristics:

1. There is a *base number*, which is ten.
2. There is a symbol for zero. It is a place-holder in a numeral like 302, and the numeral that indicates the number of a set having no members.
3. There are as many symbols, including zero, as the number indicating the base. The symbols of the Hindu-Arabic system are 0, 1, 2, 3, 4, 5, 6, 7, 8, and 9.
4. The place value scheme has a ones place on the right, a base position to the left of the ones place, a base times base $(b^2)$ position next, a base times base times base $(b^3)$ position next, and so on.
5. The system makes it possible to use algorithms for doing computation. (An algorithm is a rule or procedure for finding the

solution for a particular problem. An example is the addition algorithm used for finding answers for addition problems.)

---

## Self-check of Objectives 6 and 7

Give an example of a number system; of a numeration system.

Several characteristics of the Hindu-Arabic numeration system have been identified. Name at least four of them.

---

**c. Operations.** The basic number operations are *addition* and *multiplication* and their inverses, *subtraction* and *division*. Addition is defined as the operation that assigns to an ordered pair of numbers, called addends, another number called their sum. Subtraction can be defined in terms of addition. When it is defined this way, it is the operation used to find a missing addend when a sum and the other addend are known. Multiplication is the operation that assigns to an ordered pair of numbers, called factors, another number, called their product. Division can be defined as the operation used to find a missing factor when a product and the other factor are known. These operations are extended to the nonnegative rational numbers during elementary school years. Some children also learn about addition and subtraction of negative integers, and are introduced to multiplication and division of these numbers.

---

## Self-check of Objective 8

You have just read about the basic operations of addition and multiplication. Name the inverse operation for each basic operation. Can you think of "real-world" situations where each of these operations is used?

---

**d. Basic Properties.** There are basic properties, or laws, that hold the various number systems together as *systems*. These properties, along with the operations to which they apply, are developed in an orderly fashion to reveal the structure of each system.

The closure property applies to addition and multiplication in all sets of numbers. In each set there is a sum for every ordered pair of addends and a product for every ordered pair of factors. The closure property does not apply to subtraction or division in the set of whole numbers. Not until integers are used is it possible to subtract every ordered pair of numbers. There is closure for division when the set of rational numbers is used.

The commutative property applies to addition and multiplication in all number systems. In the set of whole numbers, $3 + 4$ equals $4 + 3$. The general notation for any pair of numbers is $a + b = b + a$. For multiplication in any set of numbers, the product of $a \times b$ equals the product of $b \times a$. The commutative property does not apply to subtraction and division.

The associative property applies to addition and multiplication in each number system. In the set of whole numbers, $(3 + 4) + 5 = 3 + (4 + 5)$. In the first part of the equation, 3 and 4 are added, then 7 and 5 are added to make 12. In the second part, the 4 and 5 are added, then 3 and 9 are added to make 12. In either half of the equation, the sum is 12. This is expressed generally as $(a + b) + c = a + (b + c)$. For multiplication the associative property is expressed as $(a \times b) \times c = a \times (b \times c)$. This property does not apply to subtraction and division.

The distributive property applies to multiplication over addition. It makes it possible to find a product of one number and two or more numbers by either multiplying each of the numbers by the single number and adding the products, or by adding the numbers and multiplying their sum by the number. With whole numbers this is demonstrated by:

$$3 \times (4 + 5) = 3 \times 9 = 27$$

or

$$3 \times (4 + 5) = (3 \times 4) + (3 \times 5) = 12 + 15 = 27.$$

It is shown generally as

$$a \times (b + c) = (a \times b) + (a \times c).$$

The distributive property is also used when performing division. The division of 36 by 3 can be explained as $36 \div 3 = (30 + 6) \div 3 = (30 \div 3) + (6 \div 3) = 10 + 2 = 12$.

It can also be shown that multiplication and division are distributive over subtraction. However, this application is seldom used in the elementary school.

e. **Identity Elements.** Identity elements occur in all number systems. An identity element is a number which, when operated on with another number, results in the other number. Zero is the identity for addition. In the set of whole numbers, $0 + 3 = 3 + 0 = 3$. The general notation for the addition identity is $0 + a = a + 0 = a$. The identity for multiplication is one. This is expressed generally as $1 \times a = a \times 1 = a$.

f. **Inverse Property.** An inverse is defined as a number that operates on another to yield the identity. For addition, the inverse of a number is the number which, when added to the first, results in zero. An example is $3 + (-3) = 0$. This is shown generally as $a + (-a) = 0$. The inverse for multiplication is called the reciprocal. It is a number which, when multiplied by another, results in the product one. The reciprocal for 4 is $1/4$, and for $2/3$ it is $3/2$. In general terms it is expressed as $a/b \times b/a = 1$.

g. **Number Sentences.** Number sentences are used to record and communicate mathematical ideas. As children work, they learn that number sentences are a form of "shorthand" used to make statements about number situations. Sentences indicate equalities (for example, $4 + 2 = 6$) and inequalities (for example, $4 + 2 \neq 7$), and relations such as $6 > 3$ and $3 < 5$.

It is important that children develop their knowledge of number and the Hindu-Arabic numeration system carefully and systematically. As a teacher you must plan activities that take into account each child's level of maturity with respect to developmental stages and degree of understanding of mathematical concepts. By doing this you will increase the probability that each child will develop the mental images required for understanding number systems and their operations and properties.

## Self-check of Objective 9

Several basic properties are discussed. Name three or more. In which number systems is each property applicable?

### 4. Measurement

Children's activities in mathematics bring them into contact with both separate, or discrete, and continuous materials. Discrete materials are things that can be counted, such as marbles in a bag. Continuous materials are things that cannot be counted, such as a package of meat. Continuous materials are measured rather than counted. The types of measures treated in the elementary school are linear, area, volumes (both cubic units and liquid measure), angles, weight (or mass), time, and temperature. Money is also included and treated as a part of measurement, even though it is unlike any of the other measures.

The mathematics framework for California schools lists four general concepts to be developed by children as they study measures and processes of measuring:[3]

1. Measurement is a comparison of the object being measured with a "unit" and yields a number to be attached to the object as the mea-

[3] Statewide Mathematics Advisory Committee, *Mathematics Framework for California Public Schools Kindergarten through Grade Eight* (Sacramento: California State Department of Education, 1972), pp. 51–52. (Used by permission.)

sure of the object. (Measurement may thus be conceived as resulting in a pairing of objects with numbers. This pairing is of a type that leads to the notion of a function. Therefore, measurement may be treated as a special case of a function.)

2. The choice of a measurement "unit" is arbitrary, but standard units are agreed upon for accurate communication and simplified computation.

3. Measurement is approximate, and the precision of the measurement depends upon the measurement unit employed.

4. Any process of measurement has the following basic properties:

   a. If object A is part of object B, the measure of A is less than or equal to the measure of B.

   b. If objects A and B are congruent, their measures are equal.

   c. If objects A and B do not overlap, then the measure of the object consisting of the union of A and B is the sum of the measures of A and B.

Passage of the Metric Conversion Act of 1975 resulted in an ever-increasing use of the metric system in the United States. Consequently, much emphasis is placed on instruction in the metric system in many programs. However, until conversion is complete, children will need to be familiar with the essential features of both the metric and English systems of measure. Once children are acquainted with the English and metric systems, they should have frequent and varied activities in using both.

---

## Self-check of Objective 10

Give an example of both a discrete and a continuous object. Give examples of continuous objects you think should be available to children.

---

### 5. Geometry

Historically, geometry has been studied mainly after completion of elementary school. Elementary school children were exposed to geometry only briefly; they learned to recognize some geometric shapes and became acquainted with formulas for determining areas and perimeters of simple plane figures. Today, geometry has assumed a much more prominent role in the program. Through inductive, discovery-type activities, children develop an intuitive understanding of some of its basic concepts.

Geometry is now included in the elementary school because we recognize that children can learn and use many geometric concepts, that its study encourages creativity and inquiry, and that it provides children with a break from the study of numbers and computation.

Geometric shapes and forms surround us. We daily see geometric forms in both natural and man-made objects. In man-made objects they are usually a necessary part of the object's structure, and are frequently included for aesthetic reasons as well. For children, the study of geometry begins with an examination of natural and man-made objects so that they can develop an intuitive understanding of two- and three-dimensional figures. Later, carefully selected models are used to introduce certain specific geometric concepts. For example, models of cubes, prisms, and pyramids give opportunities for classifying figures according to the shapes of their faces or the number of their edges or vertices. As children mature, they refine the language they use to describe each figure.

By experimenting with models, children will also develop an intuitive understanding of certain topological relationships, such as the inside and outside of a simple closed curve, or a closed space figure, such as a cube; congruence and similarity; and some geometric transformations.

---

## Self-check of Objective 11

Several reasons for including geometry in the elementary school program are given. Can you cite at least two? Can you also give examples that illustrate the validity of each of these reasons?

---

### 6. Number Theory

The study of number theory introduces children to odd and even numbers, prime and composite numbers, and concepts such as least common multiple (LCM) and greatest common factor (GCF). There are several reasons for including number theory in a program. Children who are familiar with the concept of least common multiple can use LCMs to determine the lowest common denominators when adding and subtracting unlike common fractions. Knowledge of greatest common factor is helpful when simplifying common fractions (reducing to lowest terms).

Another reason for including number theory is that it is interesting to many children. A properly taught unit gives a program a motivating factor that might otherwise be missing. The study of number theory also provides many opportunities for children to practice mathematical skills, such as addition or multiplication involving the basic facts for these operations.

## 7. Probability

An intensive study of probability theory is not made in the elementary school. The approach throughout is intuitive as children experiment with probability devices such as spinners, dice, coins, and colored marbles. The writers of the School Mathematics Study Group (SMSG) unit on probability for the intermediate grades give these reasons for having children learn about probability.[4]

> Children, as well as adults, are often confronted with situations which involve probability. While these confrontations may not be crucial to children, they are important and they are frequent, so a fundamental knowledge of probability is useful.
>
> There is a need on the part of both children and adults to be better informed in this field. It is evident that both have many misconceptions and a lack of background in probability. Many people have little intuition about experience involving chance.
>
> The practical uses of probability are many, and they are increasing almost daily. Probability is used in making decisions in such diverse fields as military operations, scientific research, design and quality control in manufactured products, insurance calculations, business predictions, weather forecasting, and governmental operations.

The writers of the SMSG book on probability for the primary grades list the following fringe benefits for work in this area:[5]

> (1) First, it should be fun for children, and making mathematics fun is one way to improve instruction. Perhaps, because games are used as an integral part of the development of the unit [on probability], some children who have heretofore *not* enjoyed mathematics or who have found the work too hard will become more interested or will experience an unusual success.
>
> (2) Second, one of the goals of mathematics instruction is to promote systematic thinking rather than a hit-or-miss approach or jumping to conclusions. Especially the lessons on combinations and permutations (a word not used in the unit, by the way) demonstrate the advantage of a systematic approach and provide many opportunities for practice.
>
> (3) Third, certain arithmetic skills may be practiced and reinforced. Addition facts are used in many lessons, and subtraction and multiplication are required in some of the supplementary materials. For those classes in

[4] David W. Blakeslee, et al., *Probability for Intermediate Grades, Teacher's Commentary,* rev. ed. (Stanford, Calif.: School Mathematics Study Group, Leland Stanford Junior University, 1966), p. 1. (Used by permission.)

[5] David W. Blakeslee, et al., *Probability for Primary Grades, Teacher's Commentary,* rev. ed. (Stanford, Calif.: School Mathematics Study Group, Leland Stanford Junior University, 1966), p. 2. (Emphasis in original. Used by permission.)

which practice in the use of rational numbers (comparison and addition) is needed, there are opportunities galore. Children who have had difficulty in forming or interpreting addition and multiplication tables may find them easier to use after this unit.

Fourth, there are opportunities for independent investigation—one might almost say "research"—by individual children. The teacher, too, may seize opportunities to undertake more difficult concepts if it seems desirable with some groups of children.

Whether you use the SMSG units, other commercial programs, or develop your own, the same fringe benefits apply.

---

## Self-check of Objective 12

Both number theory and probability are relatively new additions to mathematics programs in the elementary school. Name several reasons for their inclusion.

---

### 8. Statistics and Graphing

Newspapers and magazines are filled with advertisements in which the claim for the superiority of one product over another is based on statistics. Radio and television advertisements contain similar claims. Reports of medical research use statistics to support the claim that smoking is harmful to a person's health. We are bombarded daily with statistics about fuel shortages, endangered animal species, and the effect of automobile exhaust emissions on the quality of the air we breathe. Every citizen must learn to interpret statistical data and to read tables and graphs. Children's ability to handle data meaningfully grows over time as they collect, organize, display, and interpret information throughout their elementary school years.

Work with statistics can begin as early as kindergarten. Children can make simple block graphs showing the colors of shoes they have worn to school on a particular day. All through the primary years, children can organize and interpret information about themselves: for example, their pets, families' sizes, color of clothing, favorite TV shows.

Intermediate-grade children work with data from a broader range of sources. In addition to information about themselves, they will gather data from science and social studies projects and from the community in which they live.

## Self-check of Objective 13

Sources of statistics should be from the child's world. Can you name possible sources of data for collecting and graphing activities?

## SUMMARY

A mathematics program for elementary school children provides opportunities for varied experiences involving manipulative materials. Such programs allow children to proceed at individual rates according to each child's background and learning style, and are designed to enhance their interest in mathematics and awaken their appreciation of the subject's importance. Three factors have influenced the content and procedures for helping children learn mathematics. These are (1) the prevalence of automation, minicalculators, and computers, (2) the interest of mathematicians in the teaching of mathematics, and (3) research into learning. Automation, minicalculators, and computers have created new jobs requiring a mathematical background that includes more than skill in computation. Mathematicians have introduced new topics and emphasized the importance of organizing instruction so that it develops children's understanding of the properties of numbers and numeration systems and the language and symbols of mathematics. The research of such learning theorists as William Brownell, Jean Piaget, Jerome Bruner, Robert Gagne, and Richard Skemp focuses attention on the need for allowing children opportunities to use manipulative materials as they investigate mathematical concepts, and for taking into account children's levels of intellectual development and background for learning new concepts and skills.

Educators and learning theorists do not fully agree on the content of elementary school mathematics, but eight skill areas are included in most programs. Problem solving, application, and estimation skills are important and are included in all programs. Classification and sorting activities provide a basis for understanding certain number, geometric, and measurement concepts. Number and numeration systems and operations on numbers receive major attention. The whole numbers, integers, and positive rational numbers are studied along with the Hindu-Arabic numeration system. The operations of addition and multiplication and their inverses, subtraction and division, are applied in each number system. Commutative, associative, distributive, and other properties contribute to children's understanding of the structure of mathematics.

Measurement of continuous objects is emphasized through study of both the metric and customary systems of measure. The study of geometry encourages creativity and inquiry in children. Number theory and probability help develop skills that are useful in other areas of mathematics, serve to motivate many children, and provide opportunities for children to practice basic operations. Statistics and graphs serve to help children develop thinking skills and improve their ability to solve problems.

## STUDY QUESTIONS AND ACTIVITIES

1. Ask a number of elementary school children to tell you about the mathematical skills they are learning. How many of the skills identified in this chapter do they mention?

2. Compare an arithmetic book published before 1960 and a contemporary mathematics book for the same grade level. What skill areas are in the contemporary book that are not contained in the older one? Are there skill areas in the old book not found in the new one? What skill areas are common to both books?

3. Examine the teacher's manual for each book (see question 2) to find out how the objectives for teaching mathematics in the new book differ from those in the older one.

4. Look through a contemporary elementary textbook series, beginning with the first-grade book. Identify the grade level at which each of these properties is introduced (if indeed, they are introduced at all): (1) commutative property for addition, (2) associative property for addition, (3) identity element for addition, (4) commutative property for multiplication, (5) associative property for multiplication, (6) distributive property of multiplication over addition, and (7) identity element for multiplication. Is each property identified by name when it is introduced? If not, what name is used in connection with each one when it is introduced? Do the textbooks and/or teacher's manual indicate that children are expected to use and identify each property by name before they complete the elementary school?

5. Several authors listed in this chapter's reading list have written about the dangers they believe are associated with a "back-to-basics" movement. Read one of these articles, summarize the author's concerns, and share your findings with your classmates. Talk to teachers, administrators, parents, and school board members in a district with which you are acquainted to find out if a "back-to-basics" movement has been or is presently in it. If so, what has

been the effect of this movement on the mathematics program in that district?

## FOR FURTHER READING

*The Arithmetic Teacher.* The February 1979 issue has the theme "The Case for the Comprehensive Mathematics Curriculum" and identifies topics other than "basics" that are important. Articles deal with probability, statistics, geometry, estimation, computers, and other topics.

Brodinsky, Ben. "Back to the Basics: The Movement and Its Meaning," *Phi Delta Kappan*, LVVIII, No. 7 (March 1977), 522–527. Discusses the forces behind "back to basics," its strengths and weaknesses, its supporters, educators' responses, and other points associated with the movement. While not devoted strictly to mathematics, the article clearly identifies the problems teachers face as a result of one of today's major influences on what is taking place in classrooms across the country.

Brownell, William A. "Psychological Considerations in the Learning and the Teaching of Arithmetic," *The Teaching of Arithmetic*, Tenth Yearbook of the National Council of Teachers of Mathematics. New York: Bureau of Publications, Teachers College, Columbia University, 1935, pp. 1–32. Brownell discusses three theories of teaching arithmetic: "drill," "incidental," and "meaning," with the last emphasized as the most fruitful.

Bruner, Jerome S., et al. *Studies in Cognitive Growth.* New York: John Wiley & Sons, Inc., 1966. Theories about cognitive growth and the learning process appear in chaps. 1 and 2.

———. *Toward a Theory of Instruction.* Cambridge, Mass.: Harvard University Press, 1966. How a theory of instruction should be developed. Discusses cognitive growth, the role of visual images, motivation, individual differences, and evaluation.

Davis, Robert B. "New Math: Success/Failure?" *Instructor*, LXXXIII, No. 6 (February 1974), 53–55. A review of the historical origins and shifts in content in mathematics. In addition to a change in emphasis and some change of content, there has also been a change in the manner of instruction. A contention is that in the rush to teach pure mathematics, applied mathematics, and projects, the computational skills of some children have been neglected.

Glennon, Vincent J. "Mathematics: How Firm the Foundations?" *Phi Delta Kappan*, LVII, No. 5 (January 1976), 302–305. Glennon reviews a number of influences on mathematics—the "new math," "back-to-basics" movement, Piaget and other theorists, and individualization and grouping trends. Then he pleads for a continued emphasis on sequential, systematic, and structured teaching using appropriate techniques for individualizing and teaching.

National Council of Supervisors of Mathematics. "Position Paper on Basic Goals," *The Arithmetic Teacher*, XXV, No. 1 (October 1977), 19–21. There is a need for a clear definition of the goals of school mathematics. The goals must be broader than computational skills alone. The Council's position paper identifies ten basic skills to be developed by completion of high school, defines them, and gives a rationale for a broad definition of basic skills.

Rappaport, David. "Historical Factors that Have Influenced the Mathematics Program for the Primary Grades," *School Science and Mathematics*, LXV (January 1965), 25–33. The stimulus-response theory, the meaningful theory, Piaget's work, and experimental programs are presented.

———. "The New Math and Its Aftermath," *School Science and Mathematics*, LXXVI, No. 7 (November 1976), 563–570. Rappaport makes a case for a two-track mathematics program. One track would be for the approximately 40 percent of the students who are capable of grasping mathematics concepts as presented through the "new math" movement and another for the remaining students whose contributions lie in fields other than mathematics and who should be taught only the mathematics required for effective consumerism and to solve basic problems.

Reys, Robert E. "Stop! Look! Think!" *The Arithmetic Teacher,* XXV, No. 1 (October 1977), 8–9. Presents the dangers and misunderstandings associated with the establishment of minimum competencies in mathematics and the development of means to test them. The best that can come from the competency movement is a benefit that will accrue from the exercise of getting parents, teachers, and laymen together to try to reach agreement on the competencies.

Smith, William D. "Minimal Competencies—A Position Paper," *The Arithmetic Teacher,* XXVI, No. 3 (November 1978), 25–26. The author accepts the National Council of Supervisors of Mathematics 1977 position paper as the best collection of skill areas constructed to date. He contends that it is necessary to go beyond a statement of minimal competencies to specify when competencies should be attained and how they should be measured. He lists seven recommendations pertaining to the competencies and their measurement and use.

Suydam, Marilyn. "The Case for a Comprehensive Mathematics Curriculum," *The Arithmetic Teacher,* XXVI, No. 6 (February 1979), 10–11. Children who are taught only the basic computational skills are cheated. Their program must include a variety of other topics and skills if they are to develop the understanding of mathematics they need.

# 2 Foundations for Effective Teaching

Upon completion of Chapter 2, you will be able to:

1. Trace orally or in writing the historical development of mathematics instruction, from the copying method of colonial schools to present day guided-discovery activities.

2. Name four stages of mental growth through which Piaget believes each person progresses, and give examples of how each of the stages following the first influences the mathematics program.

3. State two reasons why it is important to keep the hierarchy of concepts and skills for a topic in mind while planning learning activities.

4. Identify the three dimensions of a model designed to help teachers make decisions about teaching mathematics, and explain the significance of each dimension to the planning of activities.

5. State orally or in writing the importance of each of the following ideas when planning and teaching mathematics:
   a. Appropriate and adequate motivation.
   b. Children's varying manners and rates of learning.
   c. Practicing mathematical skills.

Key Terms you will encounter in Chapter 2:

mental-discipline theory
stimulus-response theory
meaning theory
sensorimotor stage
preoperational stage
reversibility
concrete operations stage
operations
formal operations stage
cognitive level
hierarchy of skills
structure
model
cognitive goals
facts
skills
comprehension
understanding
application
analysis
affective domain
teaching-learning sequence
concrete-manipulative mode
semiconcrete-representational mode
abstract mode
motivation
individual differences
practice

Picture a classroom in which children are working in pairs doing investigations dealing with weight or volume measures, while in another part of the room a group of eight or ten children is seated at a large table working with geoboards under their teacher's guidance. At a round table, several children are using cordless earphones to listen to the "My, My—A Fly" lesson from the *Computapes* program.[1] In a carpeted corner small groups of children are playing mathematics games. Still other boys and girls are outdoors measuring the playground with metric trundle wheels.

Perhaps you have never observed such a classroom. More likely, the classes you have been in are ones in which traditional teacher-directed, textbook-oriented lessons are featured. Fortunately for children, teachers are becoming aware of the need for changes in the ways boys and girls learn mathematics and are developing more child-centered programs based on sound learning theories and successful classroom practices.

Your effectiveness with children will depend in part upon how well you understand the mathematics you help them learn and the theoretical base upon which you build their learning experiences. If you do not understand all of the mathematics topics and concepts discussed in Chapter 1, you should plan now to overcome your weaknesses before you begin working with children. You may do this by studying mathematics-content books for teachers or by enrolling in an appropriate course in a junior college or four-year institution.

This chapter will help you understand and apply the learning theories that provide the bases for mathematics instruction in today's schools. The first section contains a brief discussion of early learning theories. More recently developed studies of learning by Jean Piaget, Jerome Bruner, Robert Gagne, and Richard Skemp are discussed next. Finally, a model of the factors considered by effective teachers as they plan mathematics teaching-learning strategies is discussed.

## EARLY THEORIES OF LEARNING

Mathematics as part of the elementary school curriculum in the United States has a long history. Computation, arithmetic for business, commerce, and certain other occupations, was commonly taught in colonial

[1]*Computapes*, a 112-lesson, 56-cassette tape program, is available from Science Research Associates, Inc., 259 East Erie Street, Chicago, Ill. 60611.

schools. Instruction usually took the form of dictation from a book by the schoolmaster and copying by the students. Students seldom had books.

During the nineteenth century the mental-discipline theory of learning came to be practiced in the schools. According to this theory, the brain was understood to be similar to a muscle which could be exercised in much the same way as any other muscle. It was believed that arithmetic provided good subject matter for such exercise and consequent development of the mind. By the middle of the century, arithmetic books claimed to provide courses in mental discipline as well as instruction in mathematics. Examples of addition involving 7 or 8 addends as large as 999,999 and with 32 addends as large as $9999.99 were common. The multiplication of 82,030,405 by 23,456 and division of 843,000,329,058 by 203,963,428 appear in a mathematics book published in 1874.[2]

The twentieth century can be divided into three periods during which different psychological theories influenced instruction in mathematics. The first two are dealt with briefly in this section; more recent theories are covered more extensively in the following section.

In the early part of the century, Edward Thorndike's stimulus-response theory replaced the mental-discipline theory. This theory was based on the assumption that learning is accomplished when a student forms bonds, or connections, between a given stimulus and a response to that stimulus. It was believed that by establishing the correct bonds between each stimulus and response, a child developed the "tools" for further learning. The role of the teacher became one of providing children with ample opportunities to establish the correct response for a given stimulus. Children were continually drilled on the basic facts and operations of arithmetic, and they were retained in a grade for failing to attain a given standard in arithmetic as frequently as for failing in any other area of the curriculum.

The meaning theory of William Brownell began to replace the stimulus-response theory in the 1930s. This theory is built on the belief that children must understand what they are learning if learning is to be permanent. Instruction that allows children opportunities to use manipulative materials as they investigate the meanings of mathematical concepts is advocated by proponents of the meaning theory. This theory is still valid as the basis for many classroom activities.

## Self-check of Objective 1

Three different approaches to teaching mathematics are discussed in the historical order in which they were featured in schools. Discuss each approach in chronological order and in terms of the way children were instructed.

[2] See James B. Thompson, *New Practical Arithmetic for Grammar Departments* (New York: Clark & Maynard, Publishers, 1874), for examples similar to those mentioned here.

## RECENT STUDIES OF THE LEARNING PROCESS

More recently, the research of Piaget and other contemporary theorists has provided guidance in planning children's learning experiences. Piaget's research has led him to conclude that children progress through four stages of mental growth as they mature:

1. The sensorimotor stage (zero to two years).
2. The preoperational stage (two to seven years).
3. The concrete operations stage (seven to eleven years).
4. The formal operations stage (eleven years and older).

The *sensorimotor stage* occurs during the time between birth and about two years of age. Teachers are not directly concerned with children this young, but it should be noted that even at this early age the foundations for later mental growth and mathematical understanding are being developed. Initially, events occur in disconnected sequences. Persons and objects cease to exist when they are out of sight. Later, children are able to recognize and hold mental images of persons and objects even when they can no longer see them. This is a prerequisite for thinking and developing the ability to connect present events with past experiences.

The beginnings of mathematical understanding emerge during the *preoperational stage*. During this period children use objects to stand for other things. A doll may stand for an adult, such as a parent or teacher, or for another child. Blocks become buildings, boats, and roadways. The ability to represent ideas with objects precedes the ability to represent them with pictures and, later, symbols. Piaget shows that during this period children go through three stages of growth in developing a concept of number.[3]

Initially, children make what Piaget calls a "global comparison" of quantities or amounts. For example, in one experiment children were shown a container of small wooden beads. Those still in the global comparison stage asserted that the quantity of beads changed when they were poured from the original container to a container of different size or shape; the quantity seemed to increase or decrease depending on how it appeared in the new container. In another series of experiments, Piaget had children put ten flowers in ten vases, one flower in each vase. The children then removed them from the vases and placed them in a bowl, bunched together. Then they took ten more flowers and again put one flower in each of the ten vases. The flowers were removed and placed in a second bowl, spread out. Children in the global comparison stage of operation were unable to ascertain that there was an equal number of

[3] Jean Piaget, *The Child's Conception of Number* (New York: The Humanities Press, 1952). See Part I, Sec. II, for a description of experiments with beads, and Part III, Sec. IX, for a description of experiments with flowers.

flowers in each of the two sets. They believed the number of flowers in the spread-out set was greater than the number in the bunched set, even though they had performed the operations of placing the flowers in the vases and then of removing them and placing them in the bowls.

Later, children establish what Piaget calls "intuitive correspondence without lasting equivalence." In the experiments, children at this stage knew that the quantity of beads should remain the same when poured from one container to another; they tended to think of conservation of quantity. Yet, when the shape of the second container was substantially different from the first, children were again unable to conserve quantity, and they said that the number of beads had changed. With the flowers, children at this level of understanding were able to establish and maintain a one-to-one correspondence between the flowers and vases and the two sets of flowers so long as the bunches were not spread too much. The correspondence was lost when the flowers in one group were spread more widely than those in the other or when one bunch was compacted more than the other.

Finally, children reach a third stage of thinking about numbers when they establish a lasting correspondence between the elements of two sets, even though configurations of the sets are changed. In the experiments they were able to maintain the idea that the number of beads did not change when poured from one container to another. When working with the flowers, the children understood the correspondence between the sets and did not lose it as the flowers were rearranged.

In the preoperational stage children cannot mentally reverse actions to the point of the beginning of action on an object. This lack of what Piaget calls *reversibility* accounts for children's inability to recognize that the quantity of beads or number of flowers does not change. They are still using what he calls *intuitive thinking*. Intuitively they believe what they see because they cannot yet think logically.

Piaget has not established any school grade levels at which these stages occur. It is accurate to say, however, that children in Piaget's experiments were of the ages of children in preschool and kindergarten classes and first and second grades. Teachers in these grades should keep in mind these stages of growth in children's concepts of numbers.

At about the age of seven children move into the *concrete operations stage*. Now they begin to form mental pictures of objects and think in terms of the whole rather than parts. As they move these mental images about in their minds, children achieve reversibility. Piaget calls such mental activities *operations*. According to Piaget, children must internalize these mental operations before they can think logically.

During this period children need frequent experiences with objects they can manipulate so they can refine their thoughts about numbers, time, space, and other concepts. Since the mental images children create and move about in their minds are a product of their experiences, those who see and manipulate an abundance of objects, shapes, and symbols have clearer mental images than those whose experiences are meager. By the time this period comes to a close, children who have had a variety of meaningful experiences will have mastered many operations. Toward the end, children will think effectively without concrete materials.

The *formal operations stage* is reached when children begin to form hypotheses, analyze situations to consider all factors that bear upon them, draw conclusions, and test them against reality.[4] This stage occurs at about twelve, when children generally move into the adult stage of thinking.

Piaget contends that all children pass through each of these four stages and that no stage is ever skipped. To the question of whether it is possible to accelerate children through the stages, Piaget replies: "It's probably possible to accelerate, but maximal acceleration is not desirable. There seems to be an optimal time. What this optimal time is will surely depend upon each individual and on the subject matter. We still need a great deal of research to know what the optimal time will be."[5]

---

## Self-check of Objective 2

Piaget's research is influencing the way children are being taught mathematics in today's schools. Identify Piaget's four stages of mental growth and give an example of how stages 2, 3, and 4 influence the mathematics program.

---

Jerome Bruner and Robert Gagne, American psychologists, and Richard Skemp, an English mathematician-psychologist, have investigated learning in mathematics and have theories that have implications for teaching the subject.[6] According to them, teachers must be aware of the hierarchy of concepts and skills for a topic. It is important that the essential steps in a process, such as subtraction of whole numbers, be developed sequentially and without omissions if learning is to be effective. Thus you must plan instruction carefully in order to guide children through the steps of a process in proper progression.

In *The Process of Education,* Bruner stresses the role of structure this way: "Grasping the structure of a subject is understanding it in a way

[4] Irving Adler, "Mental growth and the art of teaching," *The Arithmetic Teacher,* XIII, No. 7 (November 1966), p. 579.

[5] Jean Piaget, quoted in Frank G. Jennings, "Jean Piaget: Notes on Learning," *Saturday Review* (May 20, 1967), p. 82.

[6] Jerome Bruner, *The Process of Education* (New York: Vintage Books, Random House, Inc., 1960); Robert M. Gagne, *The Conditions of Learning,* 3rd ed. (New York: Holt, Rinehart and Winston, Inc., 1977); and Richard R. Skemp, *The Psychology of Learning Mathematics* (Hammondsworth, England: Penguin Books, Ltd., 1971).

that permits many other things to be related to it meaningfully. To learn structure, in short, is to learn how things are related."[7]

Skemp says this about the importance of structure: "The study of the structures themselves is an important part of mathematics; and the study of the ways in which they are built up, and function, is at the very core of the psychology of learning mathematics."[8] According to Skemp, the learner acquires an understanding of the structure of mathematics by developing a mental *schema*. He says that a schema has two functions: to integrate existing knowledge, and to serve as a mental tool for the acquisition of new knowledge.[9]

**a. The Integrative Function.** Mathematics learning progresses in stages from lower-level to higher-level concepts. Concepts of a higher level are tied to those of lower levels by a well-developed mental schema. Skemp says that when learners see something as an example of a concept, they become aware of it at two levels: as itself, and as a member of a class.[10] In mathematics, children recognize $40 + 76 + 60 = \square$ as an example of addition with three addends. Upon closer examination, they will see that it is a member of a broader set of examples in which the addition of three or more numbers can be simplified by using the associative and commutative properties of addition. Thus, they can add the 40 and 60, and then add 76 to 100 to get the total sum of 176.

A schema involving the commutative and associative properties applied to addition of whole numbers can be extended to other number systems. Carefully designed activities will make the extension from the whole numbers to the fractional numbers easier. For example, children in a fourth-grade class knew that the associative and commutative properties apply to addition of whole numbers. They had used the properties when adding three or more addends. When they began adding rational numbers expressed as common fractions, they added numbers in sentences like $\frac{1}{4} + \frac{3}{4} = \square$, $\frac{2}{3} + \frac{1}{3} = \square$, and $\frac{1}{2} + \frac{1}{2} = \square$. One day Mr. Dean, their teacher, gave them this sentence:

$$\frac{1}{4} + \frac{1}{2} + \frac{2}{3} + \frac{1}{6} + \frac{1}{2} + \frac{1}{3} + \frac{3}{4} + \frac{5}{6} = \square.$$

At first the children were uncertain about how to proceed. Mr. Dean prompted them to study the sentence carefully and to recall what they

[7] Jerome S. Bruner, *The Process of Education* (New York: Vintage Books, Random House, Inc., 1960), p. 7.

[8] Richard R. Skemp, *The Psychology of Learning Mathematics* (Hammondsworth, England: Penguin Books, Ltd. A Pelican Original. Copyright © Richard R. Skemp, 1971), p. 39.

[9] Skemp, p. 39

[10] Skemp, p. 39.

knew about addition of whole numbers that might help them with the new sentence. Different children offered suggestions. Finally, the children recognized that by using the commutative property and associating pairs of like fractions to get a sum of one for each pair, it was easy to get the total sum.

**b. The Schema as a Mental Tool.** Skemp believes that schemas are indispensable tools for learning. He describes an experiment designed to test this belief.[11] A set of symbols was created and presented to two groups of learners. For one group, the symbols were organized in such a way that the learning of subsequent symbols was enhanced by the manner in which they were related to the symbols presented earlier. The second group was presented the same symbols, but they learned them in a rote way, with no meaningful structures. The results of tests given immediately after study of the symbols, a day later, and four weeks later were dramatically in favor of the group that worked with the well-organized schema. This group recalled 69% of the symbols, while the comparison group recalled 32%. A day later the percentages were 69% and 23%, respectively. After four weeks the first group still recalled 58%, while the comparison group dropped to only 8%. This experiment shows clearly how a well-organized schema will help children learn and retain new concepts and skills.

You must keep structures for different topics and concepts in mind as you teach. Figure 2-1 shows the hierarchy of steps for subtraction with whole numbers. Beginning with the simplest skills at the bottom, each step becomes progressively more complex. Because each successive step requires knowledge of the lower-level concepts and skills, it is important that no steps be bypassed as children learn to subtract. You must remember, too, that these steps are spread over several years, and are not dealt with completely at any one grade level. This hierarchy is discussed more fully in the next section where it is related to the learner's behavior and a teaching-learning sequence.

---

## Self-check of Objective 3

Bruner, Gagne, and Skemp give reasons for the importance of the hierarchy of concepts and skills for a mathematical topic. Can you identify these reasons? Can you think of an example of a topic other than the one shown in the chart that will trouble a child who lacks the prerequisite concepts and/or skills? What are the necessary concepts and/or skills for that topic?

---

[11] Skemp, pp. 40–42.

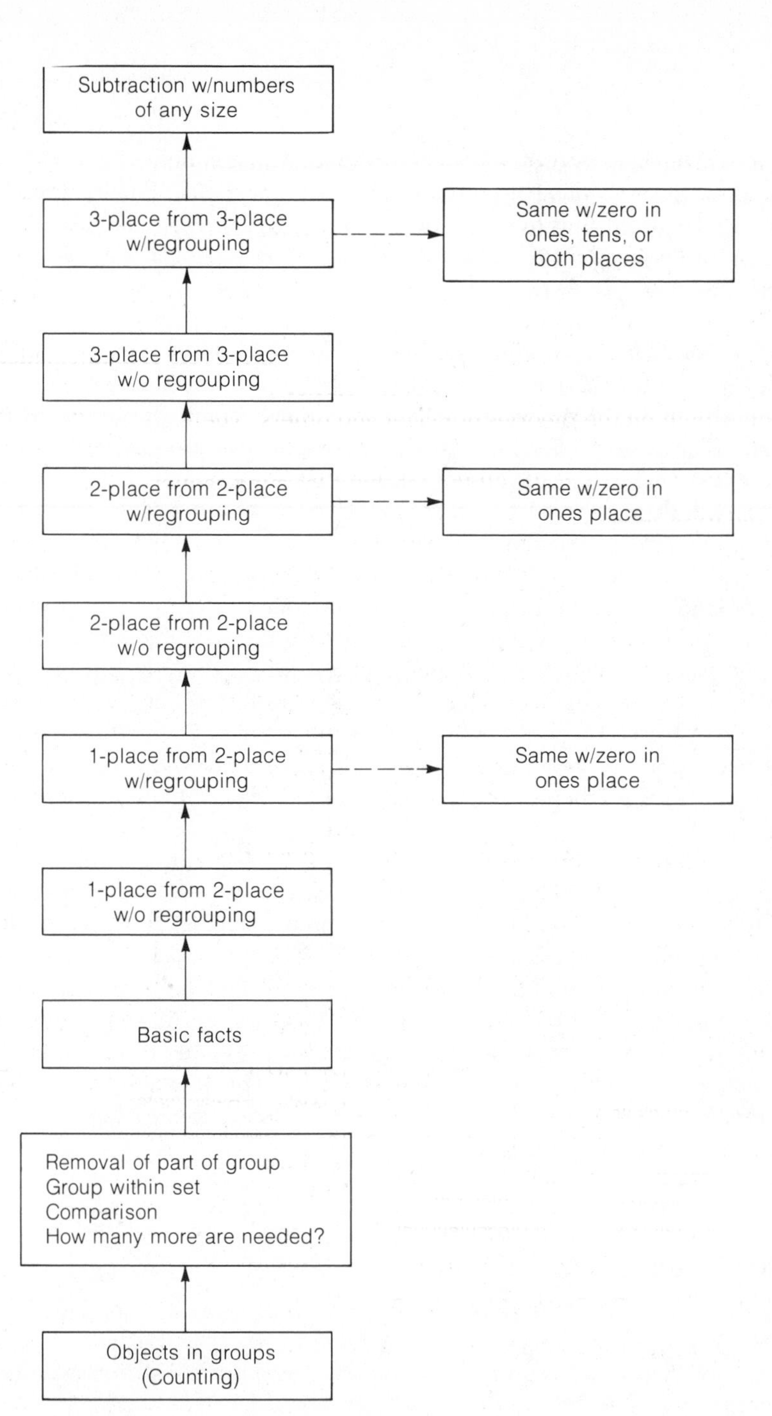


**Figure 2-1** A hierarchy of steps for learning to subtract whole numbers

## A MODEL FOR DECISION MAKING

While the theories of Piaget and others provide a theoretical basis for organizing children's experiences, they are of little value unless you are systematic in making decisions about your mathematics units and daily lessons. One way to systematize your decision-making process is to consider a model. "A model is a way of representing and testing an idea. Ideas are sometimes difficult to communicate, so models are used to help in the communication. Models can be actual constructions, like wind tunnels and space capsules, or they can be more symbolic representations."[12]

The model that follows illustrates how instruction about subtraction with whole numbers can be organized to relate the levels of mathematical behavior we should strive for in each learner to a teaching-learning sequence built on the theories of Piaget and others. There are three dimensions to the model, each of which is equally important (Figure 2-2). These are (1) the learner's goals, (2) the teaching-learning sequence, and (3) the hierarchy of steps.

objective 4

## 1. The Learner's Goals

Consider first the children you teach. In the model, cognitive goals are represented by four broadly stated types of mathematical behavior: (1) facts and skills, (2) understanding and comprehension, (3) application, and (4) analysis.[13] These four categories are represented by the model's layers, with the simplest at the bottom.

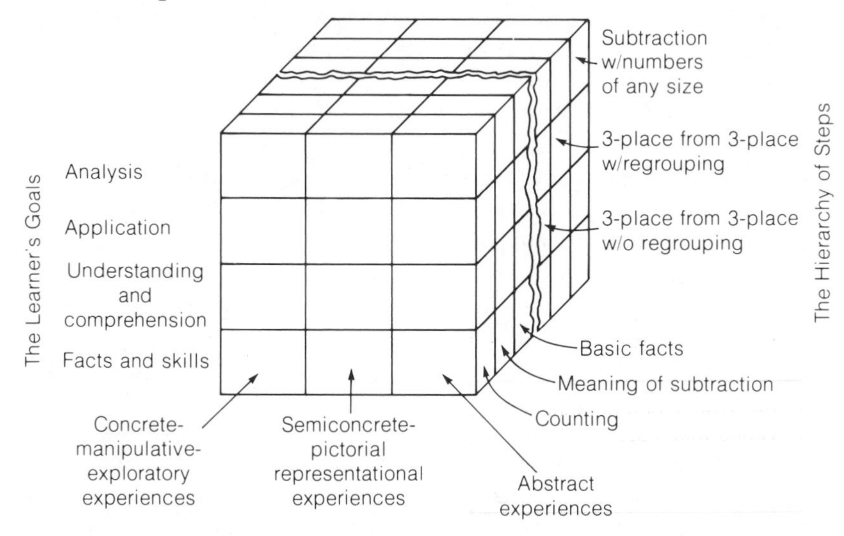


**Figure 2-2** A model for decision making in mathematics teaching

*Source:* Based on material in Edward G. Begle and James W. Wilson, "Evaluation of Mathematics Programs," *Mathematics Education* (Chicago: National Society for the Study of Education, 1970).

[12] Len Pikaart and Kenneth J. Travers, "Teaching Elementary School Mathematics: A Simplified Model," *The Arithmetic Teacher*, XX, No. 5 (May 1973), p. 32.

[13] These categories are adapted from Edward G. Begle and James W. Wilson, "Evaluation of Mathematics Programs," *Mathematics Education* (Chicago: National Society for the Study of Education, 1970), pp. 367–404. Other categories are found in Benjamin S. Bloom, ed., *Taxonomy of Educational Objectives, The Classification of Educational Goals, Handbook I; Cognitive Domain* (New York: Longmans, Green and Co., 1956); and Len Pikaart and Kenneth J. Travers, "Teaching Elementary School Mathematics: A Simplified Model," *The Arithmetic Teacher*, XX, No. 5 (May 1973), pp. 332–342.

Facts and skills are the least cognitively complex types of mathematical behavior, and the easiest to select, teach, and evaluate. Facts about subtraction consist of basic combinations like $8 - 6 = 2$ and $6 - 6 = 0$. One skill involves computation of answers for examples like $69 - 13 = \square$ and $45 - 22 = \square$.

The second category—understanding and comprehension—is more complex than simple knowledge of facts and skills. Principles and concepts are involved at this level. Thus a child who understands that the operation of subtraction does not have closure in the set of whole numbers will recognize that $64 - 78 = \square$ has no whole-number answer. A child operating at this level not only knows that he or she must regroup to complete the example $4003 - 629 = \square$, but will explain fully each step of the regrouping process.

The next cognitive level is application. In order to apply the facts and skills, concepts, and properties already learned, a child must recall the relevant knowledge, select an appropriate operation, and perform the necessary steps to complete an answer. The work is of a routine nature, and the student works in a familiar context and with operations already practiced. Examples of application involving subtraction are represented by a student who recognizes that subtraction is used to determine the difference in the sizes of crowds attending football games on two consecutive Saturdays and by one who composes a reasonable story problem for a sentence like $4003 - 629 = \square$.

Analysis, the highest level, requires the student to make ". . . a nonroutine application of concepts. The items may require the detection of relationships, the finding of patterns, and the organization and use of concepts and operations in a nonpractical situation."[14] Analysis is demonstrated by a student who proves that the difference between two odd numbers is always an even number. It is not until upper grades of the elementary school that most children are able to work at this level. Even then, children will operate at a lower cognitive level with most of the mathematics they are learning.

Teachers who are unaware of the goal dimension of mathematical behavior tend to limit children's work to the lowest level; that is, they concentrate on the development of factual knowledge and computational skills. It is important that the higher goals be considered, too, so that elementary school children will learn to use mathematics in everyday situations and build a foundation for more advanced work in the subject in later years.

When goals of mathematics education are considered, it is also important to give attention to goals in the affective domain. The affective domain includes outcomes related to attitudes, interests, appreciations, and anxieties. Unfortunately, little research has been devoted to the affec-

[14] Begle and Wilson, p. 394.

tive domain, and less is known about how to deal with it. Research does offer some clues, and a few are considered here.

It has been commonly assumed that mathematics is disliked by most children. Recent surveys indicate that about as many children rate it as the subject they like best or second best as those who indicate that it is the most disliked subject.[15] Earlier studies consistently showed that girls disliked mathematics more than boys. However, a 1975 study indicated that mathematics was the only school subject without a sex difference in preference. Suydam and Weaver also report that children's attitudes toward mathematics tend to become increasingly less positive as they progress through school, that there is a positive relationship between a teacher's attitude toward mathematics and the attitudes of his or her children, and that the methods and materials a teacher uses, as well as his or her manner, probably affect pupils' attitudes.

The studies reported by Suydam and Weaver suggest that you have a very direct role to play in determining children's feelings about mathematics. If you like mathematics and show children that you do, have realistic goals, show mathematics to be useful and interesting, and make learning meaningful, enjoyable, and successful, you will be more likely to create positive feelings in your children than if you do not.[16]

## 2. The Teaching-Learning Sequence

The second dimension in the model is the teaching-learning sequence. The characteristics of successful elementary school mathematics programs are listed in Chapter 1. There it is stated that new topics should be presented in a concrete-manipulative mode, followed by a semiconcrete-pictorial mode, and, finally, an abstract mode. In the model the concrete mode is shown at the left, the semiconcrete in the middle, and the abstract at the right.

The meaning of this part of the model is illustrated with three activities using subtraction to compare the sizes of groups as an example. The first activity involves objects, such as colored plastic airplanes. Children might make a group of blue planes, say eight, and a group of red planes, say three. They would match the blue planes one to one with the red planes and determine that there are more blue planes. If the matched pairs are set aside, the unmatched blue planes are isolated. The number of isolated planes—five—is the difference in the sizes of the two groups. Many activities similar to this provide the concrete experiences children need to grasp this meaning of subtraction.

[15] Marilyn N. Suydam and J. Fred Weaver, *Using Research: A Key to Elementary School Mathematics* (Columbus, Ohio: ERIC Information Analysis Center to Science, Mathematics, and Environmental Education, 1975), p. 1–1.

[16] Suydam and Weaver, p. 1–6.

Later, children will use pictures of real objects and representations of objects, such as colored squares, circles, and other shapes organized in groups on pages of their workbook and other practice materials. They may draw lines to match pictures in one group with those of another. They can count the unmatched pictures to determine the differences in the sizes of any two collections, but they cannot physically remove any of them.

Eventually, subtraction sentences like $8 - 3 = 5$ will be introduced to represent actions done with objects (concrete mode) and pictures (semiconcrete mode). The children's activities will then involve them in writing sentences and answers for situations such as these: "Joanie has 17 baseball cards, while Johnny has 8. Joanie has how many more cards than Johnny?" "Juan made 8 wreaths for the classroom; Tricia made 6. How many fewer wreaths did Tricia make than Juan?"

Each time you plan a lesson, you need to decide which of the three modes is the appropriate one to use. Teachers tend to concentrate on the semiconcrete and abstract modes because textbooks contain materials for them. However, the concrete mode should be used frequently, particularly as new topics and skills are introduced. Commercial and teacher-made materials needed for the concrete-manipulative mode and ways to use them are discussed throughout this book, beginning with Chapter 4.

### 3. The Hierarchy of Steps

The model's third dimension is the hierarchy of steps for a topic. The steps for subtraction of whole numbers serve as an example. Counting, the prerequisite skill for all operations, including subtraction, is at the front, while subtraction of numbers of any size is shown at the back. (The intervening steps that are omitted in the model are contained in the flow chart of Figure 2-1.) As you plan activities—subtraction in our example—you must be aware of the place of each of your children in the hierarchy so that you avoid having children spend time with concepts and skills they already understand or having them try to cope with those for which they lack the necessary background. Chapter 3 discusses materials you can use to learn about the hierarchy of steps for each of the topics included in an elementary school program.

### 4. Integrating the Three Dimensions

Viewing the model as a whole, rather than its parts, will help you visualize how the three dimensions fit together. Goals, subject matter, and the teaching-learning sequence come into sharper focus to help you develop a systematic way of dealing with your long-range and daily planning. Each activity you plan should be based on your knowledge of the children's goals, where they are in relation to the hierarchy of skills for a topic, and what their level of cognitive development indicates about the materials and procedures you should use. In the chapters that follow, you will read about the skill areas of mathematics included in contemporary programs and materials and procedures that will help children operate at the higher levels of the hierarchy of goals. Chapter 3 deals with ways of assessing children's levels of cognitive development and knowledge and background and ways to organize learning activities.

---

## Self-check of Objective 4

Identify the three dimensions of the model for decision making in teaching mathematics. Explain orally or in writing the significance of each dimension to the planning of mathematics units and daily lessons.

---

## FURTHER IMPLICATIONS OF RESEARCH ABOUT LEARNING

It is seldom that a model will include everything that relates to the subject it illustrates. The model just discussed is no exception, because there is no place in it to illustrate the role of motivation, children's varying manners and rates of learning, and practice in mathematics teaching. These topics are discussed here.

### 1. Appropriate and Adequate Motivation

Mathematics' importance in the elementary school curriculum reflects its position in our technological society. Some children will become specialists in mathematics when they reach adulthood; others must develop competence in mathematics at a nonspecialist level. In either case, the importance of the subject is not always readily recognized by children in elementary school. Instructional procedures that will motivate children are therefore especially important.

Most children exhibit a natural curiosity which prompts them to learn by questioning, observing, analyzing, listening, and investigating. You can capitalize on this curiosity to motivate children's interest in mathematics, regardless of their ages, if you allow them opportunities to engage in activities appropriate to their stages of mental growth. Commercial and teacher-made learning aids and games and puzzles can be used in all grades to help children explore basic mathematical concepts. Dienes and Golding say the following about classrooms in which children are allowed to explore mathematical concepts alone or as a part of a small group:[17]

[17] Z. P. Dienes and E. W. Golding, *Learning Logic, Logical Games* (New York: Webster/McGraw-Hill, 1971), pp. 10–11. (Used by permission.)

> Drastic changes in the mathematics curriculum would not be possible if we were to retain the traditional classroom procedures and atmosphere at the same time. In fact, we hope that teachers will endeavor to change "a teaching situation" into "a learning situation." It must be emphasized that, with the kind of approach suggested here, far less "whole class teaching" will take place. Much of the work will be done by children working in small groups, or even individually. These groups can be formed by the teacher, or she may allow the children to form themselves into groups. They will work together quite happily, especially if the work is not spoiled for them by the creation of a reward-punishment system. Children are essentially interested in finding out new things about their world, and we do not have to spoil this interest by introducing compulsions or rewards for work well done. A smile from the teacher, or a pat on the back, is quite sufficient reward for a task well completed.
>
> If we work in this way, the children will be encouraged to learn mathematics for its own sake, and not in order to excel or outdo their classmates in competition. Groups will form, will change in composition, and re-form, as some children learn more quickly than others. There will be a place for individual work, too, and there will be times when it will be more profitable for the whole class to work together.

You should also take advantage of every situation to help children see a need for learning mathematics. It is necessary on the playground as children measure lines for a playing field or keep score for a game. Boy Scout and Girl Scout or simple construction activities often require mathematical skills. Relating mathematics to other subject areas can help achieve this as well. The importance of mathematics in a science investigation or a social studies project can be stressed. All activities in which children need to use mathematics for solving problems help to maintain or to recreate their interest in learning it by making its importance evident.

The importance of carefully organized activities has already been discussed. Such activities do more than help children learn mathematics in an orderly fashion. They also help maintain children's interest. Students involved in a well-organized and properly sequenced "hands-on" program usually have positive attitudes toward mathematics because it is understandable and interesting.

## 2. Children's Varying Manners and Rates of Learning

Children exhibit different levels of mathematical understanding as soon as they begin school. Some children have skill with counting but little other understanding. Others begin school with practically no knowledge of numbers, and a few have advanced beyond their years. Children also vary in their interest in mathematics. Knowledge of these differences should help you determine which early number experiences would be most meaningful and most needed.

Abilities to grasp abstract ideas also vary greatly. Some children readily grasp the meanings of counting, addition, subtraction, and other mathematical topics at an early age. In the early grades, they move rapidly from concrete situations to abstract processes. Later, these children will probably spend little time on concrete representations of mathematical processes, but will operate efficiently at the semiconcrete or abstract level. On the other hand, the opposite is true for many children. These children need many concrete experiences. They should not be hurried into using semiconcrete and abstract processes. You should recognize that some of these children may never learn to work with numbers at sophisticated levels.

Differences in the ways children learn will determine in part the mode by which you present topics. For example, when children are introduced to a new concept, some of them will need more teacher guidance than others. The questions you ask these children will help them see important features of the work, such as number patterns or the way addition can be illustrated with objects and number sentences. Children who grasp concepts readily can be given less structured directions. Open-ended questions can always be closed, if necessary, when children do not understand them. The opposite is not true. Once you ask a direct question leading to a specific answer, you cannot open it to encourage children to broaden their thinking about it.

Children in the kindergarten and first grade need to work individually or in small groups in order that everyone's needs will be provided for adequately. Children must have activities suited to their stage of mental growth, learning style, and level of mathematical understanding if they are to succeed.

## 3. Practicing Mathematical Skills

Once children understand the meaning of addition, subtraction, multiplication, and division, and have been introduced to the basic facts for these operations, practice is essential. Otherwise, many children will habitually use immature processes, such as finger counting, to obtain answers. As children learn the algorithms for the four operations on whole numbers, practice is also necessary. Children who learn the operations and their algorithms in meaningful ways usually require less time for practicing them than when rote procedures are used, so the extensive drill formerly required is no longer necessary for many children.

Children need to practice skills learned in connection with rational numbers expressed as common and decimal fractions, percent, and measurement and geometry, too. For example, after they have learned to use a centimeter rule and meterstick to measure objects, they need frequent opportunities to use the instruments to maintain and extend their skills in using them accurately.

These considerations will promote successful practice sessions:

1. Practice and understanding go hand in hand.

2. The reasons for practice must be clear to children.
3. The kind and amount of practice are not likely to be the same for all children at the same time.
4. The practice sessions should be brief and occur often.
5. A variety of materials and procedures should be used.

---

## Self-check of Objective 5

State orally or in writing reasons why it is important to keep each of these ideas in mind while you plan and teach mathematics: (a) appropriate and adequate motivation, (b) children's varying manners and rates of learning, and (c) practicing mathematical skills.

---

## SUMMARY

Procedures for presenting mathematics to children have changed as new theories of learning and instruction have been introduced. In colonial times instructors presented information orally and students recorded it in their copy books. During the nineteenth century the mental-discipline theory of learning influenced instruction; exercises involving very large numbers were common. In the early 1900s Thorndike's stimulus-response theory provided the basis for instruction. Brownell introduced the meaning theory in 1935. This theory still has many implications for how mathematics should be presented to children.

More recently, the theories of Piaget, Bruner, Gagne, and Skemp have influenced instruction. Piaget has emphasized the importance of considering children's levels of cognitive development, while the others have stressed the need for considering structure and sequence in presenting topics to children. A model illustrates how cognitive goals, a teaching-learning sequence, and the structure of a topic are interrelated. It is also important that you consider motivation, children's varying manners and rates of learning, and ways to organize practice exercises as you plan activities.

## STUDY QUESTIONS AND ACTIVITIES

1. Read from books and articles that interpret Piaget's theories of mental development to learn more about the four stages of cognitive growth and the characteristics of children at each stage. The articles and books on this subject in this chapter's reading list will be useful.

2. Select a topic such as addition or multiplication of whole numbers. Examine the table of contents of each book in a mathematics series or the series' scope-and-sequence chart and list the major headings it shows for your topic. Do these headings show that the hierarchy of steps for the topic have been well accounted for in the series? Why, or why not?

3. Read Gersting's article reviewed in the reading list for this chapter. Design a bulletin board, either real or a model, that you believe children will respond to actively rather than passively. Bring it to class to be evaluated by classmates.

4. Davis discusses "moves" in "A Model for Understanding Understanding in Mathematics" (see the reading list). What does he mean by a "move"? Describe uses of "moves" in teaching mathematics.

5. Read Skemp's article on instrumental and relational understanding (see the reading list). Distinguish between the two types of understanding. Based on your experiences with mathematics, which of these understandings have you developed? Give examples of either or both types of understanding you presently possess. Can you think of any situations in which instrumental understanding might be preferred over relational understanding?

## FOR FURTHER READING

Adler, Irving. "Mental Growth and the Art of Teaching," *The Arithmetic Teacher*, XIII, No. 7 (November 1966), 576–584. A discussion of a dozen facets of Piaget's theory of mental growth, and thirteen implications of the theory for teaching mathematics.

Copeland, Richard W. *Diagnostic and Learning Activities in Mathematics for Children*. New York: Macmillan Publishing Company, Inc., 1974. Presents performance tests to assess children's levels of understanding of different mathematical concepts and discusses implications of the results.

Davis, Edward J. "A Model for Understanding Understanding in Mathematics," *The Arithmetic Teacher*, XXVI, No. 1 (September 1978), 13–17. Citing recent research, the author indicates that the kinds of logical things (moves) teachers do in the classroom affect the quality of teaching. One study indicates that teachers judged superior by their peers made approximately twice as many moves as inexperienced teachers and those rated lowest by their peers. Moves associated with teaching concepts, generalizations, facts, and procedures are illustrated along with questions to assess understanding of the moves.

Duckworth, Eleanor. "Piaget Rediscovered," *The Arithmetic Teacher*, XI, No. 7 (November 1964), 496–499. A discussion of Piaget's theories on intellectual development, teaching, and teacher training.

Dyrli, Odvard E. "Programmed Piaget," *Learning*, V, No. 2 (October 1976), 78–80, 82, 84, 86. A 22-unit "program" helps the reader understand Piaget's four levels of cognitive development and their implications for the teacher. This article is recommended as good for the novice with regard to Piaget's theories.

Elkind, David. "What Does Piaget Say to the Teacher?" *Today's Education*, LXI, No. 8 (November 1972), 46–48. Piaget's interview technique is used to identify some of the implications of his research for teachers. Three aspects of the technique and how they might affect teaching are reviewed.

Farnham-Diggory, Sylvia. "On Readiness and Remedy in Mathematics Instruction," *The Arithmetic Teacher*, XV, No. 7 (November 1968), 614–622. The author emphasizes the importance of developing coordinators between mathematical symbols and other information so comprehension will occur. She cites the work of Piaget and presents an alternative to Piaget's view regarding preoperational responses of children.

Fennema, Elizabeth. "Manipulatives in the Classroom," *The Arithmetic Teacher*, XX, No. 5 (May 1973), 350–352. Seven reasons for using manipulatives, or learning aids, while teaching mathematics are discussed. The author also cautions against the use of manipulatives in a random way.

Gagne, Robert M. *The Conditions of Learning*, 3rd ed. New York: Holt, Rinehart and Winston, 1977. A presentation of Gagne's latest views on how learning takes place. While this book deals with learning in general, many of the examples pertain to mathematics.

Gersting, Judith L., Joseph E. Kuczkowski, and Elaine V. Alton. "Banish the Boring Bulletin Board," *The Arithmetic Teacher*, XXV, No. 4 (January 1978), 44–46. Describes reasons for and ways of changing bulletin boards from those that elicit passive responses to ones which will elicit active responses. Two bulletin boards are illustrated.

Getzels, J. W. "Creative Thinking, Problem-solving, and Instruction," *Theories of Learning and Instruction*, ed. Ernest R. Hilgard. Chicago: National Society for the Study of Education, The University of Chicago Press, 1964, pp. 240–267. A discussion of Piaget's work, including his four stages of mental development of children.

Johnson, David R. "If I Could Only Make a Decree," *The Arithmetic Teacher*, XVIII, No. 6 (October 1971), 147–149. The decree Johnson would make is: "Let it be known that as of . . . all mathematics educators must begin to improve the 'art of questioning' in their classrooms." A three-step process for improving questions is given.

Lankford, Francis G., Jr. "Implications of the Psychology of Learning for the Teaching of Mathematics," *The Growth of Mathematical Ideas—Grades K–12*, Twenty-fourth Yearbook of National Council of Teachers of Mathematics. Washington, D.C.: The Council, 1959, pp. 405–430. Eleven points from the study of psychology of learning are listed and discussed to guide the teaching of mathematics.

Lovell, Kenneth R. "Intellectual Growth and Understanding Mathematics: Implications for Teaching," *The Arithmetic Teacher*, XIX, No. 4 (April 1972), 277–282. Lovell discusses the implications of Piaget's research for the mathematics educator.

Masalski, William J. "Mathematics and the Active Learning Approach," *The Arithmetic Teacher*, XXVI, No. 1 (September 1978), 10–12. Active learning, which involves students in physical interaction with materials and/or the environment, can be promoted through use of organizing centers, open-ended tasks, and the solution of good problems. Defines each means of active learning and gives examples.

Michaels, Linda A., and Robert A. Forsyth. "Measuring Attitudes Toward Mathematics? Some Questions to Consider," *The Arithmetic Teacher*, XXVI, No. 4 (December 1978), 22–25. The authors identify a number of factors to consider when preparing to test children's attitudes toward mathematics. They discuss points related to how to collect data, the attitudes to assess, types of items and their uses, length of scale, and validity.

Nasca, Donald. "Math Concepts in the Learner Centered Room," *The Arithmetic Teacher*, XXVI, No. 4 (December 1978), 48–52. Nasco identifies three stages in the acquisition of mathematical concepts. Premature emphasis on symbolism interferes with concept learning. Examples of topics presented in ways that stress understanding are illustrated.

O'Hara, Ethel. "Piaget, the Six-Year-Old, and Modern Math," *Today's Education*, LXIV, No. 3 (September/October 1975), 32–36. Piaget's four stages of cognitive growth are briefly reviewed. The focus is on preoperative first graders and what their level of cognitive development means for the mathematics program.

Piaget, Jean. *The Child's Conception of Number.* New York: The Humanities Press, 1952. Piaget describes, with many anecdotes regarding children's reactions, his research into how children think about numbers.

Sears, Pauline S., and Ernest Hilgard. "The Teacher's Role in the Motivation of the Learner," *Theories of Learning and Instruction*, ed. Ernest R. Hilgard. Chicago: National Society for the Study of Education, The University of Chicago Press, 1964, pp. 182–209. The authors compare using promise of reward or threat of punishment as a motive for learning versus using interest, curiosity, and self-selected goals as motives. They favor the second type of motivation and show how it can be used.

Skemp, Richard R. *The Psychology of Learning Mathematics.* Hammondsworth, England: Penguin Books, Ltd., 1971. This book deals with how mathematics is learned rather than how it is taught. The first half looks at the thought patterns people adopt when they do mathematics, and analyzes them psychologically. The second half applies the psychology of the first half to the learning of mathematics, beginning with the child's first concepts of number.

———. "Relational Understanding and Instrumental Understanding," *The Arithmetic Teacher,* XXVI, No. 3 (November 1978), 9–15. Relational understanding indicates that the learner knows the relationships among parts—knows the schema—of a topic or concept, while instrumental understanding relies on rules only. The advantages of relational understanding outweigh those of instrumental understanding, according to Skemp, even though it is a harder type of understanding to develop.

Weaver, J. Fred. "Some Concerns about the Application of Piaget's Theory and Research in Mathematics Learning and Instruction," *The Arithmetic Teacher,* XIX, No. 4 (April 1972), 263–270. Discusses cases where Piaget's evidence has been disregarded in teaching, and also where his research has been unnecessarily invoked.

# 3 Planning for Effective Teaching

Upon completion of Chapter 3, you will be able to:

1. Describe the information contained in a textbook series' scope-and-sequence chart.
2. Explain what performance objectives are and name reasons for using them.
3. Describe the kind of information revealed by Piagetian-type tests, and indicate the importance of such information.
4. Distinguish between achievement and diagnostic tests, and describe the values of analytical diagnostic tests for program planning.
5. List steps in preparing a criterion-referenced diagnostic test.
6. Identify at least six sources of information about children's mathematical backgrounds besides tests.
7. Describe a mathematics laboratory and some materials commonly found in one.
8. Describe the features of individually prescribed instructional units.
9. Describe why certain topics should be presented through teacher-directed lessons.
10. Describe different ways of beginning mathematics laboratories or classroom mathematics centers, and name the advantages of each type of beginning.
11. Prepare materials that can be used to give children positive reinforcement for accomplishments in mathematics.
12. Explain the role of the minicalculator in the elementary school mathematics program.

Key Terms you will encounter in Chapter 3:

scope-and-sequence chart
performance objective
Piagetian-type test
achievement test
diagnostic test
criterion-referenced test
analytical diagnosis
cumulative record
mathematics laboratory
learning center
individually prescribed unit
teacher-directed lesson
contract
minicalculator

The model in Chapter 2 identifies three important dimensions of the planning process to consider while making decisions about teaching mathematics. In it, the hierarchy of steps for a topic, pupils' goals, and the teaching-learning sequence are highlighted. The present chapter expands on these dimensions to give you specific information about planning children's activities. First, sources of information about the content of mathematics and children's goals are described. Tests and other assessment procedures and ways to organize children for teaching are then discussed. Finally, the role of the minicalculator in a contemporary program is considered.

## INFORMATION ABOUT CONTENT

One of the first steps in planning is to become aware of the mathematics your children will be learning. Where you turn for information will depend upon your district's program. In many districts a textbook series serves as the basis for the program. In others there are district-made programs, with teacher's guides, kits of instructional aids, practice pages, and other materials.

Determining the content for a textbook series is easy when there is a scope-and-sequence chart, which gives detailed information about how the series is organized. Publishers frequently include a chart in each teacher's manual to show the content and organization of all of the books in their series. A part of the scope-and-sequence chart for the Scott, Foresman and Company *Mathematics Around Us* series is shown in Table 3-1. By reading the chart, you can learn about the geometry content for each grade level, beginning with kindergarten.

If you use a textbook, you should make frequent use of the scope-and-sequence chart to be sure you know about each topic in the series. The hierarchy of steps for a topic is carefully developed so that gaps are eliminated, and the steps are spread over a period of time so that children are not overwhelmed with information for which they are not ready. In addition to knowing about the content for each topic for the grade level you teach, it is helpful to know what is presented at earlier levels and what follows in subsequent years. This information will help you determine the background your children should possess before beginning new work and what their current work leads to in the future. If a textbook series has no

**Table 3-1** Geometry Scope-and-Sequence Chart for *Mathematics Around Us,* K–6*

| K | 1 | 2 | 3 |
|---|---|---|---|
| Shapes, 15–16<br>Shapes with corners, 51–52<br>Preparation for congruence, 17–20<br>Mirror pictures, 73 | Three-dimensional shapes, 85–86<br>Congruent figures, 86<br>Inside of closed figures, 87<br>Triangle, square, rectangle, 87–90<br>Circle, 87–90 | Triangle, square, rectangle, 71–72, 96, 133–134<br>Circle, 72, 96<br>Open and closed figures, 69–70<br>Sides and corners of polygons, 73–74<br>Measuring sides of polygons, 75–76, 137–138<br>Readiness for perimeter, 159–162<br>Congruent figures, 131–132<br>Symmetry, 229–232<br>Coordinate geometry<br>• Giving ordered pairs for points, 163–164<br>• Locating points for ordered pairs, 165–166 | Triangle, square, rectangle, 46–47<br>Pentagon, octagon, 46<br>Segment, 52–53<br>Diagonal, 54–56<br>Angle, right angle, 332–333<br>Circle, 46–47<br>• Radius, 328–329<br>• Diameter, 329<br>Congruence<br>• Figures, 48–51, 231<br>• Segments, 52–53<br>Cube, 342<br>Similar objects, 334–335<br>Coordinate geometry<br>• Giving ordered pairs for points, 184<br>• Locating points for ordered pairs, 182–183, 185, 186 |

| 4 | 5 | 6 |
| --- | --- | --- |
| Polygon, regular polygon, quadrilateral, parallelogram, 72–75, 78–80, 82, 148, 235, 299 | Polygons, 54–55, 296–297 • Scalene, isosceles, equilateral triangles, 293 | Polygons, 143, 156–169, 190–192, 296 |
| Segment, 66 | Point, line, segment, ray, parellel lines, intersecting lines, 282–283 • Perpendicular lines, 290 | Circles, 288–294 |
| Diagonal, 278 | Angle, 285, 287–288 • Acute and obtuse, 285 | Angles, 295–299 |
| Point, endpoint, line, ray, parallel lines, intersecting lines, 66–69, 71 | Sum of angle measures, 294–298 | Central angle, 296–299 |
| Angle, right angle, 70–71 | Constructions, 220, 291 | Constructions, 82, 298–299 |
| Circle, diameter, radius, 81 | Congruence, 348–353 • Angles, 292–293 | Cube, prism, pyramid, 220–223 |
| Constructions, 103, 316, 327 | Prism, pyramid, 112–115 • Faces, vertices, edges, 111–112 | Polyhedron, 218–223 |
| Congruence, 76–77 | Similar figures, 358–361 | Faces, vertices, edges, 219–223 |
| Cube, 283 | Symmetry, 354–356 | Cylinder and sphere, 223, 289–293 |
| Symmetry, 224–227 | Coordinate geometry, 226–228 | Similar figures, 318–321, 352–353 |
| Coordinate geometry, 212–217, 222–223 • Graphing function rules, 218–219 • Finding function rules, 220–221 | | Coordinate geometry, 341–342 • Moving and enlarging figures on a grid, 350–353, 356–359 • Four quadrants, 354–359 |

scope-and-sequence chart, you can obtain information about the series' content from the table of contents for each book.

## PERFORMANCE OBJECTIVES

District-produced programs do not usually have scope-and-sequence charts. More often, information about content is contained in each program's performance objectives. Performance, or behavioral, objectives have been used since 1931, when Ralph Tyler recommended them in an article in the April 15 *Educational Research Bulletin.*[1] Tyler believed it was important that teachers consider the kinds of behaviors students would acquire as a result of their study of a subject. A quarter of a century passed before performance objectives were widely accepted. Now they are used in many schools to give direction to programs in all subject areas. These objectives are written so that the ways children show that learning is taking place are stated clearly.

Here are performance objectives for weight measurement from one district's program:

1. Given two objects of different weights, each of which can be lifted by a child with one hand, the student will identify the heavier or lighter object.

2. Given a balance with a pan or platform at the end of each arm and a collection of objects having slightly different weights, the student will use the balance to compare the weights of any two objects.

3. Given a balance with a pan or platform at the end of each arm and a collection of objects, such as different-sized wooden blocks, the student will use nonstandard units, such as dried beans or marbles, to weigh each object.

4. Given a balance scale and a set of gram weights to 1000 grams, the student will weigh objects weighing less than 1 kilogram to the nearest gram.

5. Given a metric bathroom scale, the student will weigh himself or herself to the nearest kilogram.

6. Given an incomplete table of metric units of weight, the student will complete the table to show the number of milligrams, centigrams, and decigrams in a gram, and the number of grams in a dekagram, hectogram, and kilogram.

[1] See Justin M. Fishbein, "The Father of Behavioral Objectives Criticizes Them: An Interview with Ralph Tyler," *Phi Delta Kappan*, LV, No. 1 (September 1973), pp. 55–57.

7. Given a metric balance scale, pound and ounce weights, and a kilogram weight, the student will show that there are a little more than two pounds in a kilogram.

These performance objectives present the hierarchy of concepts and skills involved in learning the metric system of weight. They tell what children should know and be able to do by the time they complete elementary school. They also give the sequence in which the concepts and skills are presented.

This set of objectives and those for other mathematics topics serve several useful purposes:

1. They give you knowledge of the purposes of instruction to use as you plan activities.
2. They give you information to use to plan assessment activities to determine each child's level of understanding and skills before you place him or her in a program of study.
3. They inform learners of the purposes of their activities.
4. They give you, administrators, and parents a basis for determining the success or failure of your program.

Once you begin working with children, you should find out if there are performance objectives for the school's mathematics program. Commercially prepared lists are appearing with increasing frequency. They are often included with textbook series, kits of mathematics laboratory materials, and cassette, filmstrip, and film-loop programs. Many teachers and mathematics specialists have written performance objectives for their districts' programs. You should use these prepared lists whenever possible.

If you do not have a prepared list, you can write your own. A number of books designed to help teachers learn to write objectives have been published in recent years.[2]

Some educators object to performance objectives on the grounds that they limit the program to a set number of objectives, that they are prepared in advance and therefore do not take into account incidental classroom happenings that can lead to significant learning, and that they omit forms of behavior that cannot be measured. If a teacher limits children's experiences to those based only on performance objectives, these are legitimate objections. However, there is no reason why a teacher must

[2] See, for example, Larry S. Hannah and John U. Michaelis, *A Comprehensive Framework for Instructional Objectives* (Menlo Park, Calif.: Addison-Wesley Publishing Company, 1977); H. H. McAshan, *Writing Behavioral Objectives, A New Approach* (New York: Harper & Row, Publishers, Inc., 1970); and W. James Popham, *Criterion-referenced Instruction* (Belmont, Calif.: Fearon Publishers, Inc., 1973).

limit a program to predetermined objectives and to outcomes that can be directly measured. Nor is there any reason why teachers using performance objectives cannot capitalize on incidental happenings to improve children's learning experiences.

---

## Self-check of Objectives 1 and 2

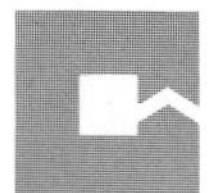

Scope-and-sequence charts are discussed and an example is given. Can you describe the type of information contained in a typical scope-and-sequence chart?

Explain what a performance objective is and name at least two reasons for using such objectives.

---

## ASSESSING CHILDREN'S LEVELS OF UNDERSTANDING AND GROWTH

The success of any mathematics program depends on how closely it is related to the needs and abilities of the children in it. You must know what these are for children in general, as well as for each of your children if the program is to be successful. Therefore assessment processes that help determine these specific needs and abilities must be understood. In the past many teachers have interpreted assessment to mean the evaluation of children's progress for the purpose of assigning grades at the end of a term. Today most teachers recognize that other purposes are served.

1. You should use varied assessment procedures and base a program of instruction on their results. Since most activities are designed to progress sequentially, gaps in children's backgrounds must be avoided. On the other hand, the instructional program should not require children to dwell too long on topics that they already understand adequately.

2. You should use varied assessment procedures to learn the specific needs of each child in your class. Children's needs with respect to mathematics will vary over time; also, their needs will vary as different topics are studied. A child who is proficient in one aspect of mathematics will not necessarily be proficient in all others. You should recognize these differences within individuals as well as those among different children.

3. You should help each child to use the results of assessment procedures to establish reasonable goals for himself or herself. As

children begin to study a mathematical topic, they must know their general goals. In addition, each child should know the specific goals of each lesson.

A teacher who knows the purposes of assessment will use varied procedures to meet each purpose.

### 1. Tests

**a. Piagetian-type Tests.** The Nuffield project in Great Britain relied heavily on Piaget's learning theories to develop an effective mathematics program. The project has prepared a series of guides covering various topics for teachers of children in British infant and junior schools. (These schools correspond to our elementary schools.) Many of the guides cover materials and procedures for teachers of four-, five-, and six-year-old children. To help teachers match activities with each child's level of cognitive development, checkups were written for the project by a team from Piaget's Geneva institute.

Three guides, *Checking Up I*, *Checking Up II*, and *Checking Up III*, were written to give information about the checkups.[3] Procedures for assessing children's cognitive development in topics such as one-to-one correspondence, conservation, geometry, and measurement are included in the books. A part of the checkup for one-to-one correspondence is reproduced in Table 3-2. The materials and procedures used for the test, along with typical children's replies, are given. Note that replies are organized in groups that correspond with the three levels of development Piaget found within the preoperational stage. Responses like those under c indicate that children have reached the level of lasting correspondence. Teachers of young children need information about the cognitive development of their children that tests like these reveal. The tests in *Checking Up I*, *Checking Up II*, and *Checking Up III*, and in Richard Copeland's two books, *How Children Learn Mathematics* and *Diagnostic and Learning Activities in Mathematics for Children*,[4] are good models to follow as you develop your procedures for getting this information.

---

## Self-check of Objective 3

The sample page from *Checking Up I* shows a Piagetian-type test. Describe the kind of information about children revealed by such a test. Why is this information important to teachers of young children?

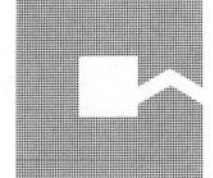

---

**Table 3-2** Sample of a Checkup Test Used to Determine Children's Levels of Cognitive Growth*

| **Summary Checkup OC** | **Typical Replies** |
|---|---|
| *One-to-one correspondence and transitivity*<br>Objective<br>*Transitivity: if* $A = B$ *and* $B = C$, *then* $A = C$.<br>Material<br>*A collection of counters, say twelve (A)*<br>*A collection of small bricks (B)*<br>*A collection of small miscellaneous objects (C)* | a<br>*The child may argue that because the line now looks different there are not as many bricks as counters. He may say:*<br>*'There are fewer counters because the line is shorter,'*<br>*or*<br>*'There are more because the line is longer.'*<br>*Children may also say:*<br>*'There are more counters because they are close together.'* |
| Part 1<br>Procedure<br>*The teacher should spread out nine counters on the table, and ask the child: "Can you put on the table as many bricks as there are counters?"* | *b*<br>*The children will say that there are as many bricks as counters, but will not be able to justify their reply.*<br>*c*<br>*Children will say that there are as many bricks as counters. They will be able to justify their reply in the following manner:* |
| Note<br>*Should the child not be able to establish this initial correspondence, then the teacher should let him have further practice.*<br>*Once the child has established a one-to-one correspondence between the two collections the teacher should ask the child if he needs the remaining bricks.* | *i*<br>*'It doesn't matter if they are in a long line because they are exactly the same bricks as before when they were in front of the counters,'*<br>*or*<br>*'You haven't put any more, and you haven't taken any away.'*<br>*ii*<br>*'You can put them back as they were before, and you'll see that they are the same.'* |

**Table 3-2**
(continued)

| Summary Checkup OC | Typical Replies |
|---|---|
| Note<br>*Should the child, with the remaining bricks, fill in a space in the line of bricks already arranged, the teacher should remove the remaining bricks and ask the child:*<br>*'Are there as many bricks as counters?'*<br>*'Why?'*<br>*Once the child's reply has been given, the teacher should then tell the child:*<br>*'We are now going to spread out the bricks.'*<br>*The teacher should do this in such a way that the line of bricks is longer than the line of counters, and then ask:*<br>*'Are there as many bricks as counters?'*<br>*'Why?'* | *iii*<br>*'The line is longer but there's more space between the bricks,'*<br>*or*<br>*'The line is shorter and there's less space between the counters.'*<br>*Only those children giving replies b or c should go on to Part 2.* |

*The Nuffield Foundation, *Checking Up I* [London: John Murray (Publishers), Ltd., 1970], p. 34. (Used by permission.)

**b. Achievement Tests.** Achievement tests are used to determine the success of a mathematics program and to assist teachers in placing children in the program by assessing each child's general level of achievement. Two types of achievement tests are available—standardized and program. Standardized tests dealing only with mathematics, such as the

[3]These three and other Nuffield books are available in the United States from John Wiley & Sons, Inc.

[4]Richard W. Copeland, *How Children Learn Mathematics*, 2nd ed. (New York: Macmillan Publishing Company, Inc., 1974), and *Diagnostic and Learning Activities in Mathematics for Children* (New York: Macmillan Publishing Company, Inc., 1974).

*Contemporary Mathematics Test*,[5] are available from some publishers, while some publishers include a mathematics test as part of their multisubject test batteries. The mathematics portions of the *SRA Achievement Series* and the *Stanford Achievement Tests*[6] are examples of mathematics tests that are part of a battery of tests.

Standardized achievement tests include items that test children's understanding of a variety of mathematical concepts and skills. A single test is usually designed to cover mathematics learned over a period of years, rather than in one year alone. These tests are frequently given to all of the children in a school or to children in selected grades as a part of a school, district, or state testing program. The manuals that come with standardized tests include the information needed to administer and score them.

Program achievement tests accompany many textbook series and laboratory kits. Program tests serve the same purposes as standardized tests, but are designed to test children's knowledge of the concepts and skills included in a specific program rather than mathematics in general. Examples of program tests are those that accompany the *Laidlaw Mathematics Program.*[7]

c. **Diagnostic Tests.** Achievement tests serve useful purposes, but they seldom yield enough information to determine a child's specific place in a program. A more comprehensive, or analytical, diagnosis of each child's level of achievement is usually required before the right placement can be made. An analytical diagnosis frequently includes a criterion-referenced test as one source of information. A criterion-referenced test is developed from a list of performance objectives, with items to test a child's performance for each objective. A test may contain test items dealing with only one objective, such as a test that includes the 100 basic addition facts, or several objectives, such as those dealing with adding two numbers less than 100 with zeroes in the ones place, with numbers other than zero in the ones place and no regrouping (carrying), and with numbers other than zero in the ones place and with regrouping.

Before a test is administered, a criterion is established to determine whether or not a child has mastered an objective. A commonly used criterion is a given number of items correct out of the total number of items from an objective. For example, for the 100 addition facts, a criterion of 95 out of 100—95%—might be established as evidence of mastery.

[5] *Contemporary Mathematics Test* (Del Monte Research Park, Monterey, Calif.: California Test Bureau).

[6] *SRA Achievement Series* (Chicago: Science Research Associates, Inc.); and *Stanford Achievement Tests* (New York: Harcourt Brace Jovanovich, Inc.).

[7] Andria Troutman, et al., *Laidlaw Mathematics Program* (River Forest, Ill.: Laidlaw Brothers, 1978).

For each of the other addition skills identified above, the criterion might be two out of three correct items for each type.

A criterion-referenced test is frequently used at the beginning of a unit as a pretest to determine which children need to do the work and at the unit's end to determine which children have mastered the objectives. When a pretest shows that a child has already mastered the unit's objectives, the child should not be required to do the work for that unit. When a posttest reveals that a child has not mastered some or all of the objectives, that child should do further work with those objectives, then or in the future. Criterion-referenced tests have been developed to replace achievement tests in some statewide testing programs.

Diagnostic tests covering concepts and skills in specific areas of mathematics, such as addition and subtraction of whole numbers, are available from some publishers. The *Stanford Diagnostic Arithmetic Tests*[8] are an example of a commercial diagnostic test. Some school districts have produced their own tests, which may be criterion-referenced. If commercial or district tests are unavailable, you should make your own to properly place children in your program. These guidelines will be helpful:

1. Determine the type of test you are going to make.
2. Choose the content. Use the texts for the present and previous grades or a list of performance objectives as a source of information. Determine the hierarchy of skills for each topic included. Use at least three items for testing each concept or skill. Children's responses may be by demonstration with concrete materials or by written or oral answers. (The younger the children to be tested, the more demonstration-response and oral items there will be. Older children will have more written-response items.)
3. Include all of the basic facts for addition, subtraction, multiplication, and division in tests for children in grades four through six. (See page 162 for a definition of the basic facts for addition and page 166 for a table that contains these facts. Definitions of basic facts for the other operations and a table of the 100 multiplication facts are given in later chapters.)
4. Organize the test items. If basic facts are to be tested, they should come first in random order. Some teachers prefer to put all items for each of the other objectives together in a simple to complex order. Others mix the items so a child must think about each item separately rather than being able to rely on a pattern to complete all of one type before going on to the next type.
5. Separate the test into subunits, if necessary. Primary-grade children

[8]*Stanford Diagnostic Arithmetic Tests* (New York: Harcourt Brace Jovanovich, Inc.)

should be able to complete each part in 15 to 20 minutes, while intermediate-grade children should work no more than 30 minutes.

6. Duplicate the written parts. Collect and organize the concrete materials the children will need for their demonstrations.

Children should be made to feel as much at ease as possible when they take a test. One way to put them at ease is to use an analogy between having to take a mathematics test and going to a doctor. The purpose of going to a doctor is to find out what health problems, if any, a person has. Mathematics tests are used to find out what mathematics problems, if any, a person has. Just as a doctor will prescribe medicine, exercise, and even surgery to overcome health problems, you will prescribe activities to overcome mathematical problems.

Not all children will complete all of the analytical tests that are given early in a school year. For example, children who have difficulty with addition and subtraction of whole numbers should not go on with the multiplication and division tests. Rather they should be given activities that will help them overcome their problems before they are given other tests and activities.

The items on completed tests should be carefully analyzed to determine the nature of any errors they contain. Not only should you mark each item right or wrong, but you should also note what each child does that is right or wrong. Correct responses often reveal the mature procedures a child uses, while incorrect responses may reveal immature procedures. In particular, each incorrect item should be examined to determine, if possible, why it was missed.[9] When only one or two items are incorrectly answered for no apparent reason, it is likely that the child was careless in solving them. Sometimes analysis reveals that errors are the result of nothing more than a child's use of a wrong number for a particular combination, such as 54 as the product of 8 and 7. At other times, analysis will reveal more complex problems, such as misunderstanding of the regrouping process used during addition.

**d. Individual and Class Profiles.** Individual and class profiles often accompany diagnostic tests to simplify record keeping. The measurement portion of the individual profile for the fourth level of one school district's program is shown in Figure 3-1. It contains information dealing with seven areas of measurement. The task(s) under each area is an abbreviated statement of one of the performance objectives for the fourth level, or third grade. A criterion-referenced test is given to determine each child's level of understanding of measurement concepts and skills.

The test results are recorded on the profile to give the teacher

[9] See Robert B. Ashlock, *Error Patterns in Computation*, 2nd ed. (Columbus, Ohio: Charles E. Merrill Books, Inc., 1977) for a systematic coverage of children's computational errors and ways to analyze and remediate them.

information for planning activities. Thus children with common needs can be identified and grouped to work together on suitable activities.

Class profiles are similar to individual profiles. They contain topics along one axis and children's names along the other. A summary record of each child's achievement is recorded following each name. These profiles provide an easy way to view the progress being made by a class of children. If profiles are maintained and go to the children's new teacher each year, the new teacher has an easy-to-use source of information about the children's earlier work.

---

## Self-check of Objectives 4 and 5

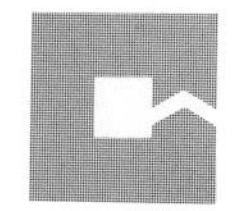

Criterion-referenced tests are used to diagnose children's understanding of mathematics objectives. Describe a criterion-referenced test. Tell why diagnostic tests are important in program planning.

You may have to prepare a diagnostic test one day. List the steps you will follow when you do.

---

### 2. Evaluation of Daily Work

During the course of a school year, children complete many activities—some introduce new concepts and skills, others provide practice in performing the four operations with different types of numbers or develop other skills, while still others are for enrichment. The ways children work with manipulative materials, their work habits, and the activities they select when they are given choices should all be noted. Their written responses on practice pages should receive special attention. Incorrect responses should be analyzed carefully to determine the nature of any errors. It is much easier to correct errors when they are first made than when they go undetected and are repeated over time.

When analysis of a child's written work reveals many errors, the topic may be too difficult for that child. If this is so, an adjustment should be made in the child's work. To constantly ask the child to perform tasks for which he or she is not ready will lead to discouragement, possibly so much that the child will quit trying to learn.

### 3. Observation of Children

You will find that observation of each child during class discussions and work periods pays big dividends: information is revealed about a child's interests in and attitudes toward mathematics, knowledge and understanding of the subject, and applications of mathematics in problem-solving situations. As children discuss a topic, note each child's contributions. Do some children tend to dominate the discussion? Are there some who never or infrequently contribute? What is the quality of each child's

The Measurement Portion ot a Student's Profile

| 5.0 Measurements | Not introduced | Progressing | Mastery |
|---|---|---|---|
| Money | | | |
| *5.1.1.E value of all coins & $1.00 bill* | | | |
| *5.1.2.E value of $'s & coins < $5.* | | | |
| Liquid Capacity | | | |
| *5.2.1.E use all liq. measures to gal.* | | | |
| Weight | | | |
| *5.3.1.E weigh objects to 24 lbs.* | | | |
| Time | | | |
| *5.4.1.E tell time, five minutes* | | | |
| *5.4.2.E write time, five minutes* | | | |
| *5.4.3.E read calendar* | | | |
| *5.4.4.E match event with time* | | | |
| *5.4.5.E arrange hands, five minutes* | | | |
| Linear | | | |
| *5.5.1.E measure, nearest quarter inch* | | | |
| *5.5.2.E measure, nearest centimetre* | | | |
| *5.5.3.E foot ruler to measure objects* | | | |
| *5.5.4.E yardstick to measure* | | | |
| *5.5.5.E in. in ft., yd.; ft. in yd.* | | | |
| Area | | | |
| *5.6.1.E count sq. units in region* | | | |
| Volume | | | |
| *5.7.1.E count cubic units* | | | |
| *5.7.2.E volume of rect. solid* | | | |

**Figure 3-1** The measurement portion of a student's profile

*Source: Arithmetic Instructional Management System,* San Juan Unified School District, Carmichael, Calif. (Used by permission.)

contribution? Does a child reveal that he or she is developing insights into the topic as it is discussed?

While children use learning aids, you should observe them carefully for clues that indicate how well they are developing an understanding of the aids and their uses. Are there children who use a device in a mechanical manner, applying it in an imitative way? Do others reveal that they understand the meaning of the device by applying it to new situations or in original ways? Do some children use several different devices to represent or illustrate a given process? Do some rely on learning aids more than they should?

You should observe each child during study periods to note work habits, use of materials, and how he or she performs operations on numbers. When children begin work immediately and apply themselves consistently, you can usually conclude that they understand what they are doing. On the other hand, children who are slow getting to work or cannot keep their minds on it show that they may not understand the work. By circulating among the children as they work, you can observe those who are having difficulty and give them immediate assistance. Often a child can overcome difficulty in a short time with immediate help. If several children have a common problem, they should form a small group for special assistance.

You should observe children from time to time as they use measuring instruments such as rulers, scales, and thermometers. If they use them incorrectly, they should get immediate help from you or another child.

Children should receive encouragement from you as they work. A soft-spoken, "That's the way," or "You're on the right track" is often the only praise needed by a child who is uncertain about whether he or she is working correctly.

### 4. Interviews with Children

A child who is having difficulty with a topic in mathematics is usually unable to determine the nature of his or her problem without help. Often an interview is the quickest way to help. A child can be asked to "think aloud" as he or she works an example, perhaps one involving addition with regrouping. Have the child explain each step, to reveal any faulty thinking. Note any deficiencies in the child's understanding of subordinate concepts involved in the operation. Interviews need not be lengthy; 2 or 3 minutes are usually sufficient.

### 5. Evaluation of Children's Uses of Mathematics

How children use mathematics in other curricular areas will reveal much about their attitude toward and understanding of the subject. When children use charts or graphs to organize and report data during a science or social studies project, it indicates they can use mathematics and recognize its value. A child who does not should be helped to see its uses and value through experiences that offer opportunities to use mathematics in many ways. Once a project has been completed, you should help the child evaluate his or her work to determine how adequately it served its intended purpose.

### 6. Parent-Teacher Conferences

Parents of a child can provide valuable information often unavailable from any other source. A child's attitude toward mathematics may be revealed by actions and statements at home. A child who has a high interest in mathematics will often read books about mathematics, work number puzzles, or pursue other mathematical recreations. A child who does not like mathematics may hide this dislike in the classroom but reveal it to his or her parents by word or action. When a child needs special help to overcome learning problems, parent consultation is often needed to plan remedial action. The child may have to be helped, in part, by his or her parents at home.

Sometimes children's problems may partly result from the conflict between their teacher's methods and the way parents think they should learn mathematics. A child may be capable of learning an operation—division with whole numbers, for example. However, if he or she is learning the process by a discovery approach in school and at home is taught a specific procedure familiar to his or her parents, he or she is sure to be confused and may not learn to perform the operation at all. Parents should be kept informed of how the mathematics curriculum is proceeding in the classroom and the reasons behind the way mathematics is taught. This will avoid misunderstanding and confusion.

### 7. Cumulative Records

A cumulative folder started for each child when he or she begins school and maintained to provide a record of progress through the grades will contain useful information for mathematics teaching. A record of achievement and program test scores is usually kept in the folder. The scores of several tests administered periodically during preceding years indicate an achievement pattern established over time. These scores generally indicate whether a child has achieved above, at, or below the norm.

Copies of student profiles kept in the cumulative folders are also a source of useful information. In addition to having data from tests, profiles frequently contain other information, such as teachers' notations about daily work, observations of children's work habits, notes recorded during interviews, and even information from parent conferences.

### 8. Help of Specialists

Occasionally you will have children with such severe problems that it is difficult for them to learn mathematics no matter what kind of activities you have for them. Usually children with severe learning problems have difficulty in all subject areas. However, you may have a child who has difficulty with mathematics alone. In either event, you should seek help as soon as possible.

A district mathematics specialist may be able to assist you and the child by providing materials specially suited for children with specific learning problems. The specialist may also be able to help you devise new materials if ready-made items are not available. A school psychologist may also be of assistance. A psychologist's help is especially valuable when a child's learning difficulties are the result of deep-seated problems. In some cases the need for a psychologist's assistance will be short-lived because the problem is quickly solved. When problems are persistent, assistance may extend over a period of months or years.

---

## Self-check of Objective 6

There are many sources of information about children's mathematical backgrounds besides tests. Name at least six of them, and tell why each one is an important part of an effective assessment program.

---

## ORGANIZING THE LEARNING ENVIRONMENT

The classroom described at the beginning of Chapter 2 is characteristic of a child-centered learning environment. More and more, mathematics laboratories or classrooms equipped with an array of mathematics learning aids are replacing more traditional classrooms as the setting in which children learn mathematics.

### 1. Mathematics Laboratories

A mathematics laboratory, as defined in this book, is a special room in a school in which children learn mathematics. It is equipped with a wide variety of learning aids for investigations by children of all ages. There are commercial aids such as Cuisenaire rods, abacuses, and geoboards, and others made by teachers, such as cut-out fraction "pies" and fraction strips. There are also viewing centers for slides, motion pictures, filmstrips, and film loops, and listening centers for cassette and reel-to-reel tapes. Games, puzzles, mathematics books, and files of problem cards are also available.

The laboratory is operated by a teacher with special training in teaching mathematics. During the course of a day all of the children in a school will work in the laboratory. Or, children who need remedial help or enrichment experiences will work there, while others remain in their classrooms to do their work. Teachers from the school and neighboring schools may come to the laboratory for workshops or to make things for the laboratory or their own classrooms.

---

## Self-check of Objective 7

Can you describe a mathematics laboratory, and name some of the materials contained in one?

---

### 2. Classroom Mathematics Centers

Another suitable environment is a classroom mathematics center. A well-equipped center contains most of the elements of a mathematics laboratory. It serves only one class, however, so it contains materials suitable for the needs of a particular group of children. A center frequently occupies one corner of the classroom, although children take materials from it to use in other parts of the room.

In addition to meeting the needs of each child, a well-functioning center will help children learn to manage their own activities and develop self-direction, independence, and sensitivity for the rights of others.

### 3. Individualized Learning Units

Individualized learning units are used in some schools to account for individual differences. Self-paced learning units have been developed commercially and by school districts and individual teachers. Good units have well-stated objectives, appropriate pre- and posttests for placing individuals in the unit and testing their progress, carefully sequenced activities, and materials at varying levels of difficulty.

Individually prescribed instructional (IPI) units can be useful. They provide one means for varying children's experiences in the classroom. They also help children develop independent study habits and offer opportunities for them to investigate topics not found in their textbooks or district program. However, rather than serving as the sole means of organizing learning, IPI units should be used selectively. You must guard against the practice of having a child use a textbook as an individualized learning program by having him or her proceed page by page at a self-determined pace. Textbooks are not designed to be used in this way. The 1977 Yearbook of the National Council of Teachers of Mathematics contains chapters which offer useful information for teachers who prepare their own units.[10]

### 4. Teacher-Directed Lessons

Even though a school may have a mathematics laboratory or each classroom learning centers and IPI units, there are times when teacher-directed lessons are the most effective way to present a concept or process to children. A child's first work with many topics, such as cardinal numbers, place value, regrouping in addition and subtraction, the meaning of decimal fractions, and number theory, needs to be guided carefully as you use a series of well-sequenced teacher-directed lessons. You can help children avoid faulty thinking and improper use of learning aids during their early work with a new topic. Children's questions, comments, and responses to tasks during these lessons will help you direct their attention to particular aspects of the work they might otherwise overlook. You can also guide the pace of the work when you are in direct control of a lesson.

You may begin teacher-directed lessons with a large group—perhaps an entire class. Later, you will find that it is probably necessary to subdivide the group. Some children's quick grasp of a concept will make them ready for independent work at the abstract level; others will need further work with concrete and semiconcrete materials but still be able to work independently or in small groups; a third group may need to continue working under your close direction. Lessons organized in this manner offer a way to account for individual differences in rates and styles of learning while still allowing you to maintain close supervision over their learning.

---

## Self-check of Objectives 8 and 9

Identify the features of good individually prescribed mathematics units.

Reasons for using teacher-directed lessons are given. Identify at least three reasons why some topics should be introduced through teacher-directed lessons.

---

### 5. Starting a Laboratory or Center

When you establish a laboratory or classroom learning centers, you must decide how to begin. Since each school is unique, there is no one way to start. There are, however, methods that have been used successfully by other teachers which will help you choose the way that is best for you.

[10] F. Joe Crosswhite, ed., *Organizing for Mathematics Instruction* (Reston, Va.: National Council of Teachers of Mathematics, 1977).

a. **Instant Beginning.** An instant beginning works best when it is initiated at the start of a school year. During the summer the laboratory or center can be equipped and the materials put in order for the children when they begin school in the fall.

During the first days of school, children's understandings and skills should be assessed so they can be organized in groups according to their needs. Groups of three to six children work well, although you may choose to have more or fewer children work together. Groups can be combined for student-teacher discussions and other activities that involve larger numbers of children. A flexible grouping plan is necessary; otherwise, children can get "locked" into a group that no longer meets their needs.

Once groups are organized, most children will need help getting started. Children who have learned self-management through independent learning activities in an earlier grade or other subject area will need less help than those whose experiences with self-learning have been meager.

The first days must be planned carefully, with more emphasis on procedures than mathematics activities. Once children learn to cope with procedures, their attention will shift to their own activities, and they will disregard what others are doing. Children differ in their ability to adjust to new learning situations, so it is likely you will have some children who need specific instructions for a period of weeks, while others will become independent quite soon.

All children need to know the purposes of the activities you give them. Certainly, each child's activities need to be related to his or her needs as revealed by tests and other assessment procedures. One way to give children information about activities and purposes is with oral or written contracts. A contract is an agreement between teacher and child in which the work to be completed in a given amount of time is identified. Written contracts are preferred for children just beginning work in a laboratory or center.

The work covered by a contract may be selected by the child or specified by the teacher, and may require a day, week, or longer period of time to complete. If problem cards, games, work sheets, film loops, and other learning aids are keyed to performance objectives by a coding system, it is easy for a teacher or child to match activities with objectives on a contract. A special contract form can be used, or the contract can be written on a plain sheet of paper. A contract should contain one or more objectives and their related activities. Eventually, most children can prepare their own contracts, filling in activities, organizing groups, and arranging appropriate time schedules. However, you will still be responsible for approving each contract before children begin work.

Some teachers prefer or are required to be more direct in selecting and organizing activities. Learning centers in the laboratory or classroom make this possible while still providing for individual differences. A learning center can be as simple as a problem card directing children to estimate the number of centimeters in the length and width of a book, desk, table, and other objects and then directing them to determine each object's measure with a centimeter ruler. A problem card response sheet and ruler for each child are then provided at the center.

Or, a center can contain a series of sequentially arranged activities dealing with addition and subtraction of fractional numbers represented by common fractions having like denominators. Such a center will have several learning stations, each self-contained with its own problem cards and learning aids, filmstrips or film loops, cassette tape programs, games, and puzzles.

A sample contract governing children's work at such a center is illustrated in Figure 3-2. The contract contains information about five stations at a center called "Adding and Subtracting Rational Numbers." It includes a performance objective, instructions for each station, and a place for a child to check off activities as they are completed. When a contract has been completed, it should be returned to you or an aide to be discussed and evaluated. You or your aide can meet with children in groups rather than individually to discuss their work. A new contract can be given the children at the conclusion of the discussion.

The role of the teacher changes frequently as children work in a laboratory or mathematics center. For example, Mr. Jackson was helping some children work with decimal fractions. The children had already used cardboard squares cut in ten or a hundred parts to learn about tenths and hundredths of a whole. Mr. Jackson wanted them to extend their understanding of place value to include the positions to the right of the ones place. He had the children bring their squares to a table where a large demonstration abacus had been placed. As children showed him sets consisting of whole squares and tenths of squares, he used the abacus to demonstrate how to represent each set. Then he wrote the decimal fraction numeral for each one. Later, Mr. Jackson demonstrated how to represent decimal fractions with hundredths. Before the lesson was completed, different children were demonstrating how to represent the sets on the abacus and how to write the decimal fraction for each one.

The teacher also has the role of guiding children's laboratory activities. The evaluation, discussion-leader, and class-management roles have already been mentioned.

**b. Gradual Beginning.** Some teachers make the changeover gradually rather than beginning all at once. A gradual beginning is particularly good for teachers in schools where traditional instructional procedures have been used in the past and where administrators are reluctant to make a change.

Ms. Spence, a third-grade teacher, attended two learning center

**Figure 3-2** A sample contract

Adding and Subtracting Rational Numbers

Name ______________
Date to be completed ________
Date returned to teacher ______

*Performance Objective:* When you have satisfactorily completed the activities at this center, you will be able to add and subtract fractional numbers represented by fractions having like denominators.

______ *Station 1.* There are two envelopes containing materials for activities dealing with addition of fractional numbers. Read the directions on the envelopes, then complete the activities on each problem card. Write your answers on a response card. Return the materials to the right envelopes.

______ *Station 2.* There are two envelopes containing materials for activities dealing with subtracting fractional numbers. Complete the activities and return the materials to the right envelopes. When you have finished all of the activities, staple your response card to your contract.

______ *Station 3.* Choose a partner to play "Common Fraction Addition."[11] The directions are on the cover of the box holding the game. Be sure to put the board and playing pieces back in the box when you are finished.

______ *Station 4.* Watch the film loop *Fractions: Adding and Subtracting (Common Denominators).*[12] You can watch the film more than once, if you wish.

______ *Station 5.* The listening center has a cassette called *Adding Fractional Numbers.*[13] Get one of the work sheets and put your name on it before you listen to the tape. When you have completed the work sheet, staple it to your contract.

workshops at her state mathematics conference and made plans to use the new ideas and materials she brought back.

First, she checked through the children's profiles to see what help they needed and formed a group of seven boys and girls who needed to

[11]This game is from Leonard M. Kennedy and Ruth L. Michon, *Games for Individualizing Mathematics Learning* (Columbus, Ohio: Charles E. Merrill Books, Inc., 1973), pp. 88–90.

[12]This film is from Jott Films, P.O. Box 745, Belmont, Calif. 94002.

[13]This cassette is from *Mathematics Teaching Tape Program, Primary Level,* Houghton Mifflin Company, 110 Tremont Street, Boston, Mass. 02107.

improve their understanding of linear measure. She decided the work should be with both the English and metric systems.

Ms. Spence looked over the teacher's manual for her textbook to see what it had to say about measurement. The manual suggested several teacher-directed activities to use before children worked the textbook exercises. She made problem cards for some of these activities. But one activity seemed too complicated for a problem card, so she recorded its directions on a cassette tape so the children could listen to an explanation of what they were to do. She also used two cards dealing with English measures from a packet she bought at the conference.[14] In addition, she made two similar cards dealing with metric measures. She also ordered two measurement filmstrips listed in her district's multimedia catalog and made a game called "grand prix."[15]

Ms. Spence organized these materials into a learning center with six stations: two with problem cards made from the teacher's manual, one with cards from the conference, one with the cassette recording, one with the filmstrips, and one with the game. She also made a contract for each child.

One idea Ms. Spence learned about at the conference was "pat-on-the-back" forms. The teacher of one of her workshops stressed that these forms give a visual way to congratulate a child for a particular achievement. The teacher may give a certificate which the child can take home to show his or her family. Or, the child may be awarded a bonus card good for free time on the playground or time to read a favorite book.[16] Two of the forms are shown in Figure 3-3.

Another suggestion from the leader was to use badges for recognition of good work (Figure 3-4). Children can proudly wear these badges and receive the plaudits of classmates, adults at school, and parents or guardians at home. Ms. Spence prepared pat-on-the-back forms and badges for her children.

When the learning center was ready for the children, Ms. Spence took one group to the center and explained how they would work there. She gave each child a copy of the contract and answered all questions about the new way of working and learning. While this group worked at the center, Ms. Spence was free to be with the other children to work with them in the more traditional way.

You can begin using a learning center in the same way Ms. Spence did by preparing and organizing materials for one center at a time and moving children into each one over a period of several weeks.

[14]*Open-Ended Task Cards.* Teachers Exchange of San Francisco, 600 35th Avenue, San Francisco, Calif. 94121.

[15]"Grand prix" is from Kennedy and Michon, *Games for Individualizing Mathematics Learning*, pp. 98–99.

[16]These forms were developed by the San Francisco Unified School District under Research Teaching Project Title VI-G/VI-B. (Used by permission.)

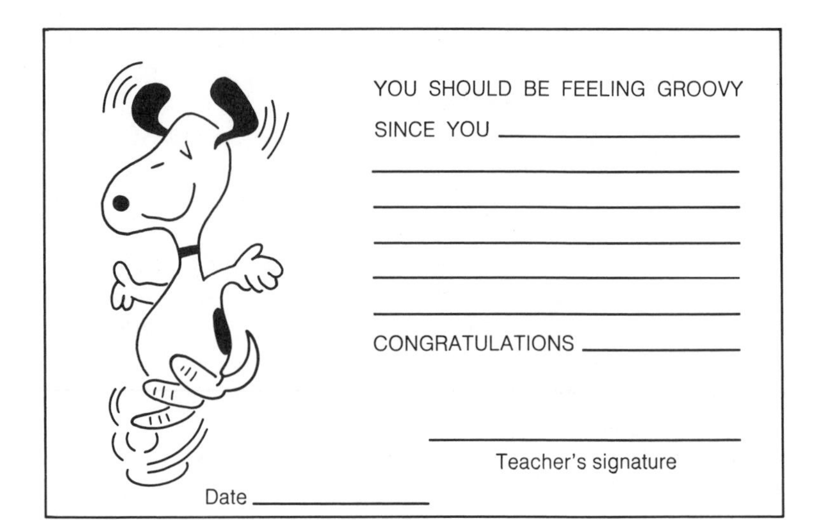

YOU SHOULD BE FEELING GROOVY

SINCE YOU ______________________

CONGRATULATIONS ______________

______________________
Teacher's signature

Date ______________

**Figure 3-3** Samples of forms used to reward children's good efforts

BONUS CARD

YOU HAVE EARNED ______________
POINTS BECAUSE OF THE EXTRA EFFORT
YOU HAVE PUT FORTH IN COMPLETING
YOUR WORK FOR THE DAY.

congratulations!!!!!

There are other ways you can make a gradual beginning. One is to set up centers for all of the children to use one day a week. Children work at the center on the one day, perhaps Friday, and work in the traditional way on the other days. Gradually, you have them spend more time on center activities and less on traditional ones.

Another gradual beginning can be made by letting children work at the centers after they complete their traditional assignments. Over a period of time, you can reduce the traditional assignments and lengthen the time spent at centers. If activities at the centers are closely related to children's daily assignments, the children will see the values of center activities more easily than when activities are unrelated to their work.

The role of the mathematics textbook diminishes when laboratories or learning centers are used. Instead of being the primary source of activities, it takes its place among other equally important materials. The text supplies the sequence for many activities and serves as the basis for many directed lessons. It also provides many pages of practice materials

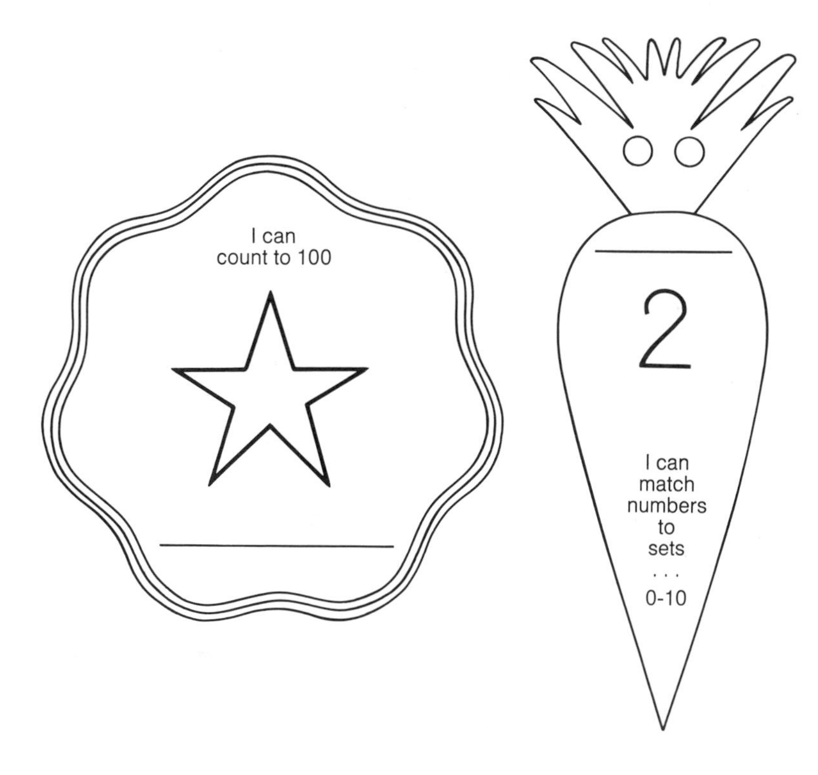
I can
count to 100
2
I can
match
numbers
to
sets
. . .
0-10

**Figure 3-4** Examples of badges awarded children for achievement

for children to complete on a selective basis. Not all children will complete the same exercises, nor will all children complete the same number of pages.

---

## Self-check of Objectives 10 and 11

One way to begin a mathematics laboratory or classroom mathematics center is to set one up and have it ready to go when children arrive for school in the fall. Another is to begin gradually. List some of the benefits of each type of beginning, and identify some factors to consider as you start either a laboratory or classroom center.

Design a pat-on-the-back form and a badge that you can use as a means of recognizing a child's accomplishment in some area of mathematics.

---

## USING MINICALCULATORS IN A PROGRAM

The minicalculator is an instrument that is too new to have an established place in an elementary school mathematics program. At the present it has strong advocates who believe it should be used extensively by children at all grade levels, and it has opponents who believe it should never be allowed across the threshold of the classroom. There has been little organized research to determine specific benefits and give clear direction to those planning to use it, so present day uses are still tentative. Even so, students have used calculators long enough to show that the device is a useful aid to learning and that it does have a place in the classroom.

The following are some of the ways a calculator is used:

a. **Check Written Computation.** Children use the calculator to check their answers to practice exercises. For example, after completing a set of examples using the subtraction algorithm, a child can do the same examples on a calculator and compare its answers with the written work. Whenever there is a discrepancy, the calculator should be used a second time. If the discrepancy persists, the handwritten work should be done again to correct it. The positive effect of immediate feedback on children's learning is well documented; the minicalculator is one means of providing such feedback.

b. **Practice Basic Facts.** Children can play a game in which they "race" the calculator. One person shows a card containing a basic fact, such as $7 \times 8 = \square$, or gives the fact orally, and then uses the calculator to determine the answer. Children attempt to name the answer orally or in writing before it is displayed on the calculator. Each child tries to get more points—one for each faster answer—than the calculator.

c. **Solve Problems.** One major goal of school mathematics is to help children become skillful problem solvers. Minicalculators contribute to this by giving children opportunities to solve realistic problems which are too complex and/or tiresome to solve by written computation alone. For example, children can learn to do comparison shopping. Have them bring in containers and packages for different food, grooming, and laundry items, with prices still attached. Children can determine the price per unit—gram, kilogram, ounce, pound, dozen, and so forth—to see which container or package for each similar item has the best unit price.

$$\begin{array}{r} 362 \\ \times 489 \\ \hline 3258 \\ 2896\phantom{0} \\ 1448\phantom{00} \\ \hline 177{,}018 \end{array} \qquad \begin{array}{r} 3258 \\ 28960 \\ +144800 \\ \hline 177{,}018 \end{array}$$

d. **Gain Insight into Operations.** Calculators can be used to give meaning to the operations and show the rationale behind the algorithms. Children who determine the answer to $384 \div 64 = \square$ by using continuous subtraction of 64 on a calculator will have a better understanding of the repeated subtraction concept of division. The multiplication algorithm shown in the margin can be completed in stages on a calculator. That is, by finding $9 \times 362$, $80 \times 362$, and $400 \times 362$ and recording the separate products and then determining their sum, a child can gain insight into how the algorithm works.

**e. Extend Children's Mathematical Experiences.** The minicalculator opens the door to many interesting experiences that might otherwise be bypassed because the necessary computations are too tedious and time consuming. For example, children can study the pattern that develops as these multiplications are completed.

$$1 \times 1$$
$$11 \times 11$$
$$111 \times 111$$
$$1111 \times 1111$$

Once these are completed, the child predicts the product of 11,111 × 11,111. Does the calculator show that the prediction is accurate? (Most calculators will not have a display large enough to show the complete product, but will show enough to check the prediction's accuracy.) There are many similar patterns children can study.

Another interesting activity is to make a calculator "speak." When a calculator is turned upside down, some numerals look like letters—8 → B, 3 → E, 6 → G, 4 → H, 1 → I, 7 → L, 0 → O, 5 → S. The quotient 10,033,582 ÷ 13 equals a greeting you might give a male friend when you meet. Children can make up their own puzzles for classmates to solve.

Before you use calculators in your classroom, you should answer questions like these:

1. Does your district have a policy on minicalculators? If it has, make certain your plans are consistent with the policy. If there is none, perhaps you can help establish one. In the absence of a district policy, you should discuss your plans with your principal.

2. What kind of calculator should you select? There are several kinds, so it is important that you choose one that is appropriate for children. Most persons who have used calculators in the elementary school recommend that children use one with algebraic rather than arithmetic logic. The algebraic calculator has a separate key for each operation — [+], [−], [×], [÷] —and one for [=]. On an algebraic calculator the order of pressing each key follows the same order as numerals and symbols in a number sentence. To add 7 + 3 = □, the [7] key, the [+] key, the [3] key, and the [=] key are pressed in that order.

   Don't consider a calculator with less than an eight-digit display. The calculator should have a floating decimal. Another feature to consider, at least for calculators for children in grades four through six, is a constant, either automatic or keyed. A constant saves considerable key punching, particularly when the number being used is large. Percent and square root keys are also useful for older children. An AC adapter makes it possible to plug the calculator into any convenient outlet in the room. An adapter also keeps batteries charged, thus reducing the need for frequent battery changes.

There are several special-purpose calculators you may want to consider. There are at least three kinds that permit children in a group to see the readout. One is used on an overhead projector (Figure 3-5); children see the readout on a screen.[17] Another has a large display unit with neon lights that shows the audience the same readout as the operator sees on the calculator.[18] The PCS 4650 is a gigantic calculator with a readout visible from 60 feet. All operations are visible as the calculator's uses are demonstrated so that children can follow along on their own machines.[19]

There is a calculator that is especially helpful for blind or partially sighted children.[20] It has a spoken output with a 24-word vocabulary that allows the user to hear each key as it is pressed and the answer as it appears on the readout. It even says "low" when the battery needs charging.

3. How do you respond to criticism from parents, administrators, and teachers who object to calculators in the classroom? The best way to respond is to have your reasons for bringing calculators into your program clearly in mind. Be able to demonstrate how you will have children use them, and point out their benefits to children. Have a plan for using the calculators. Decide whether children will go to a calculator center or will use them at their desks. Also, decide whether a child will be permitted to take a calculator home. If you do permit this, have reasons and a checkout plan.

4. How many calculators should you have? The answer to this question depends upon how you will have children use them. If they are only for checking computation, three or four will be enough. As you add uses, you will need to increase the number of calculators. At most, you will need no more than one for every two children. Once you introduce calculators into your program, there will probably be children who will want to bring their own. The rules for using personal calculators should be the same as for classroom machines.

[17] The Educator Overhead Calculator is available from Stokes Publishing Company, P.O. Box 415, Palo Alto, Calif. 94302.

[18] The EduCALC is available from Educational Calculator Devices, Inc., P.O. Box 974, Laguna Beach, Calif. 92652.

[19] The PCS 4650 is available from Ju-Rav Equipment, P.O. Box 1145, Pleasanton, Calif. 94566.

[20] The Speech PLUS Calculator is available from Telesensory Systems, Inc., 3408 Hillview Avenue, P.O. Box 10099, Palo Alto, Calif. 94304.

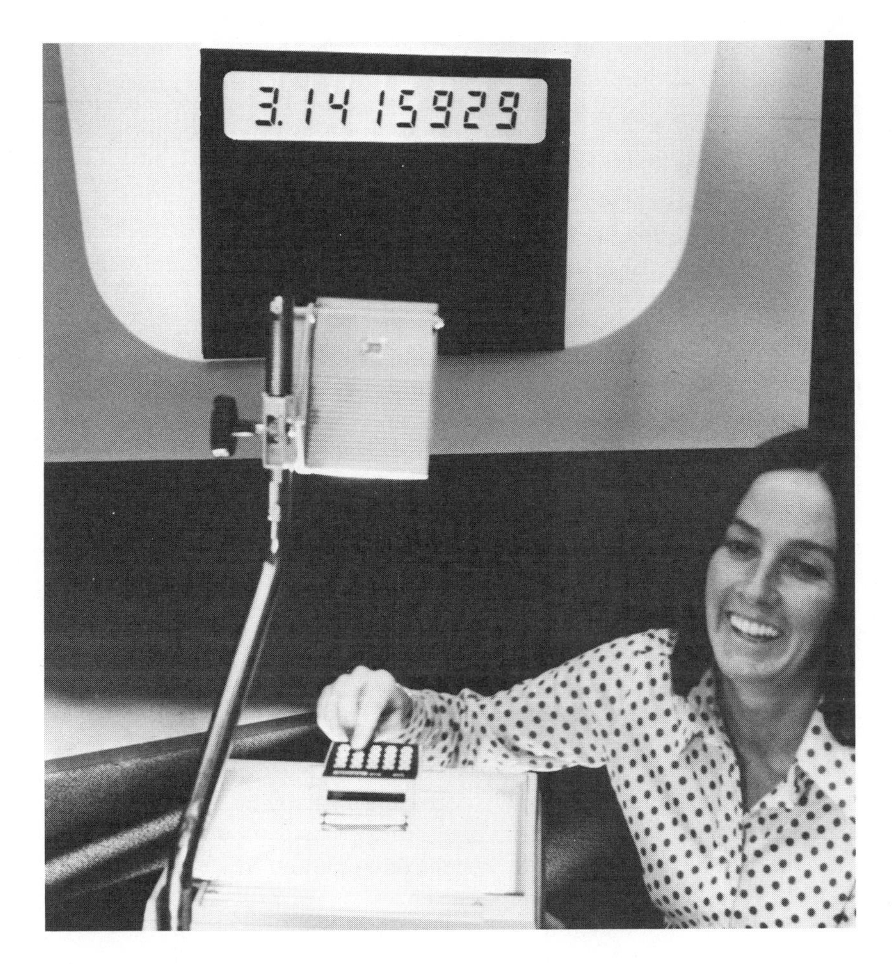


**Figure 3-5** An overhead calculator

Photo courtesy of Stokes Publishing Company, Palo Alto, California.

It is clear that you will need to use care as you plan activities with the calculator. You are advised to keep abreast of developments by reading current articles about calculators in journals such as *The Arithmetic Teacher*[21] and attending sessions dealing with them at mathematics conferences and district workshops. Further suggestions for using minicalculators appear in several chapters of this book.

---

## Self-check of Objective 12

Describe four of the uses of a minicalculator in an elementary school program.

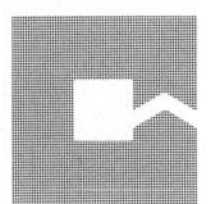

---

[21]*The Arithmetic Teacher* is published eight times each year by the National Council of Teachers of Mathematics, 1906 Association Drive, Reston, Va. 22091.

## SUMMARY

Content, objectives, and children's mathematical backgrounds must be considered carefully as activities are planned. A scope-and-sequence chart is a useful source of information about the content of a program based on a textbook series. Lists of behavioral objectives provide information about the content and objectives of many district-prepared and commercial programs. Children's levels of skills and understanding must be assessed periodically so that activities will be suitable for them. Piagetian-type tests provide information about children's levels of cognitive development. Achievement tests give useful information about children's general levels of accomplishment, while criterion-referenced diagnostic tests give specific information that is helpful in placing children in a program. Other assessment techniques supplement test information. These include analysis of daily work, observation of children at work, interviews with children, observation of their uses of mathematics, parent-teacher conferences, and consultation with specialists.

Children's activities must be carefully planned and organized to maximize learning experiences. Mathematics laboratories, classroom learning centers, individually prescribed learning units, and teacher-directed lessons are ways to organize activities. Each procedure has strengths and weaknesses, and the role of each needs to be considered. The textbook can provide the sequence of topics and many useful activities, but it should not be used as the exclusive learning aid.

Minicalculators are gradually becoming a part of the learning aids in elementary classrooms. They can be used to check computation, practice basic facts, solve problems, gain insight into operations, and extend children's experiences into fields of mathematics they might otherwise not consider. Calculators should be selected with care, with children's ages and uses of the machines taken into account. You should have a plan for their use fully developed before you introduce calculators into your classroom.

## STUDY QUESTIONS AND ACTIVITIES

1. Use the materials from *Checking Up I* quoted in this chapter, from one of the *Checking Up* books itself, or from another source to conduct a brief study of some aspect of kindergarten or first-grade children's concepts of mathematics. Which level of understanding has each child attained? How can teachers at the primary level use the ideas and investigative techniques developed by Piaget to improve their teaching?

2. Look at a textbook's teacher's manual; a filmstrip, film loop, or cassette program's manual; or a mathematics laboratory kit to see

which, if any, have performance objectives. When you find one that has performance objectives, examine the material carefully to see how it is designed to meet the objectives. Critique the textbook series, program, or kit in terms of how well you think it meets the objectives.

3. Examine a newly published commercial diagnostic test and its manual. What specific mathematical skills and understandings does it diagnose? Compare it with a contemporary textbook designed for children of the same age as the test is meant for. Are the same skills and understandings included in both test and textbook? How closely do the vocabulary and mathematical notation of the two agree? What difficulties might arise for children when a test and a book do not use the same vocabulary and notation?

4. Study carefully the written computation of children's practice exercises. As you study each paper, see if you can determine patterns in any mistakes that appear on it. Does the child show an evident lack of understanding of the work? Are there certain number facts or steps in a computational process that the child misses consistently? Or, are the errors apparently careless ones?

5. Examine the cumulative records of an elementary school class. (Make it one beyond first grade.) How do the records indicate that not all children have attained the same level of achievement in mathematics? Observe children during a mathematics lesson. What indications of individual differences do you observe?

6. Locate and visit a mathematics laboratory. Discuss its uses with the teacher in charge. In the teacher's view, what are the values of the laboratory? If possible, talk to several children to get their reactions to the laboratory.

7. Observe a teacher and children working in a classroom where learning centers are used. What learning aids are in the centers? How has the teacher organized the children to work in them? What different roles did the teacher fulfill during the time you observed the work?

8. Arnold believes that a system of management by learning activities is superior to a system of management by objectives (see the reading list for this chapter). What are the differences between the two systems? Why does he believe the system he prefers is superior?

9. Read the article by Houser and Heimer to learn about their "conditions of adequacy" for an individualized learning unit (see the reading list). Then use their checklist to evaluate the adequacy of an individualized unit, either commercially or district produced.

10. Read Thornton's article about math centers (see the reading list). Write a paper in which you illustrate with examples her eight points for effective centers.

11. Pearson's article about the TORQUE tests discusses nonmathematical aspects of tests not commonly considered by teachers as they prepare tests and daily work sheets. Write a paper in which you discuss these aspects. Where possible, illustrate the significance of each one by using contrasting examples of good and bad practices.

## FOR FURTHER READING

*The Arithmetic Teacher*, XIX, No. 1 (January 1972). This issue focuses on individualized instruction in mathematics. Six articles deal with the scope of individualized instruction, ways of providing it, and questions in connection with it.

*The Arithmetic Teacher*, XXI, No. 1 (January 1974). This issue includes four articles dealing with evaluation in mathematics. The articles discuss evaluation of processes as well as products, analyze commercial tests, discuss teacher evaluation of laboratories, and consider the values of teacher-pupil oral interviews.

*The Arithmetic Teacher*, XXIII, No. 7 (April 1976). This issue includes one editorial and eleven articles dealing with classroom uses of the handheld calculator. Information ranges from opinions about their uses, to how to choose a calculator, and classroom games and activities.

Arnold, William R. "Management by Learning Activities: An Alternative to Objectives," *The Arithmetic Teacher*, XXV, No. 1 (October 1977), 50–55. The author rejects the notion of management by objectives which, he says, ". . . features sequential learning of skills at the symbolic level of representation (and employs) pretests and posttests . . . to assess achievement." Rather, he contends that management by learning activities is the way to present material to children. Activities, organized as units, provide children with readiness, developmental, drill, and enrichment experiences that are much broader than the typical objective-based management scheme.

Ashlock, Robert B. *Error Patterns in Computation*, 2nd ed. Columbus, Ohio: Charles E. Merrill Books, Inc., 1976. Children's computational work was studied to determine patterns of errors. This semiprogrammed book gives the reader the opportunity to study these error patterns and establish procedures for helping children overcome them.

Cunningham, Betty. "Individualized Arithmetic Instruction for Fifth and Sixth Graders," *The Arithmetic Teacher*, XXV, No. 8 (May 1978), 44–46. Describes one teacher's plan for individualizing children's work. Includes a sample flow chart and samples of multiplication pre- and posttests.

Danforth, Marion McC. "Aids for Learning Mathematics," *The Arithmetic Teacher*, XXVI, No. 4 (December 1978), 26–27. The "aids" Danforth describes are procedures for helping individuals overcome learning difficulties. Included are such things as giving specific objectives, having a child verbalize thought processes, using error-analysis cards, using analogies, giving directions several ways, using concrete materials, using cues to avoid errors, and removing frustration.

Drake, Paula M. "Calculators in the Elementary Classroom," *The Arithmetic Teacher*, XXV, No. 6 (March 1978), 47–48. A sixth-grade calculator club provided the basis for organized application of children's Christmas gift calculators. The primary use for the calculators was for checking daily work. Additional uses were considered during subsequent meetings of the club.

Gibb, E. Glenadine. "Calculators in the Classroom," *Today's Education*, LXIV, No. 4 (November/December 1975), 42–44. Describes nine classroom uses. Each description can serve as the basis for development of additional calculator activities.

Hopkins, Martha H. "The Diagnosis of Learning Style in Arithmetic," *The Arithmetic Teacher*, XXV, No. 7 (April 1978), 47–50. The author concedes the importance of achievement and diagnostic tests, but she believes that it is just as important for a teacher to obtain information about each child's predominant mode of learning—kinesthetic, auditory, visual, and so on. She describes some examples of how to gather and record information.

Houser, Larry L., and Ralph T. Heimer. "A Model for Evaluating Individualized Mathematics Learning Systems," *The Arithmetic Teacher*, XXVI, No. 4 (December 1978), 54–55. The authors describe a number of "conditions of adequacy" for an individualized learning system. They consider objectives, evaluation procedures, materials, adaptability to individual needs, and other factors. A checklist for evaluating already existing systems or guiding the development of new systems is included.

Immerzeel, George. "It's 1986 and Every Student Has a Calculator," *Instructor*, LXXXV, No. 8 (April 1976), 46–51. A look into the future foretells applications of the calculator in 1986. A variety of activities for all grade levels are described. Some are basic; some are recreational.

Inskeep, James E., Jr. "Diagnosing Computational Difficulty in the Classroom," *Developing Computational Skills*, 1978 Yearbook of the National Council of Teachers of Mathematics. Reston, Va.: The Council, pp. 163–176. Explains a process for diagnosing children's computational skills using teacher-made tests. Gives suggestions for preparing tests, and describes patterned and random errors.

Kranyik, Robert D., and Susan C. Jacoby. "Tape Teaching in the Early Education Classroom: How to Create Your Own Teaching Tapes," *Early Years*, II, No. 3 (November 1971), 67–71. Presents procedures for making teacher-made tapes. A section gives specific help for making mathematics tapes. A nine point tape-making checklist offers an easy-to-follow set of steps for making tapes in any curricular area.

Morris, Janet P. "Problem Solving with Calculators," *The Arithmetic Teacher*, XXV, No. 7 (April 1978), 24–25. Describes several uses of the minicalculator. Included are exploring number patterns, discovering relationships, practicing mental estimation, reinforcing inverse operations, problem applications, developing problem-solving techniques, and individualized exploration and enrichment.

Nichols, Eugene D. "Are Behavioral Objectives the Answer?" *The Arithmetic Teacher*, XIX, No. 6 (October 1972), 419, 474–476. Presents eleven objections to behavioral objectives. Among the objections are that teachers are not responsible for molding behavior, that all objectives cannot be laid out in advance, and that objectives cannot be made to account for all aims of education.

Palmer, Henry B. A. "Minicalculators in the Classroom—What do Teachers Think?" *The Arithmetic Teacher*, XXV, No. 7 (April 1978), 27–28. A survey of teachers and others in the Los Angeles area indicates that teachers view uses of the calculator positively. Some had reservations about the devices' effect on students' computational skills. The need for teacher in-service activities is also cited.

Pearson, Craig. "Teaching Arithmetic is More Than Marking 'Right' and 'Wrong,'" *Learning*, V, No. 6 (April 1977), 30–34. TORQUE (Tests of Reasonable Quantitative Understanding of the Environment) researchers are developing new criterion-referenced tests which they claim will be better for determining children's ability to apply mathematics and help teachers evaluate the ways children think as they perform mathematical operations. Of special interest to teachers is the information about nonmathematical aspects of tests—language, format, formalities of testing, and so on—that impede children's performance.

Rouse, William. "The Mathematics Laboratory: Misnamed, Misjudged, Misunderstood," *School Science and Mathematics*, LXXII, No. 1 (January 1972), 48–56. Rouse relies heavily on the thoughts of Zoltan P. Dienes to indicate his belief about the true purposes and nature of mathematics laboratories. He lists seven principles that should govern materials and activities, and the administration of laboratories.

Schmalz, Rosemary S. P. "Calculators: What Differences Will They Make?" *The Arithmetic Teacher*, XXVI, No. 4 (December 1978), 46–47. We cannot ignore the minicalculator. It is here to stay, so educators should determine its role in the curriculum. According to Schmalz, the calculator is the most efficient means of computing, so children should learn to use it for that purpose. However, they still need to use paper-and-pencil procedures; less efficient but more understandable algorithms than those commonly taught should be used with many children.

Schussheim, Joan Y. "A Mathematics Laboratory, Alive and Well," *The Arithmetic Teacher*, XXV, No. 8 (May 1978), 15–21. Describes the room, its permanent centers, the operating procedures, and sample activities. Includes an extensive list of materials, with annotations and publishers' addresses.

Thornton, Carol A. "Math Centers for Young Learners," *Learning*, VI, No. 1 (August/September 1977), 56–57. Describes the values of math centers and their uses to motivate and reinforce concepts, strategies, and skills. Includes eight pointers for effective use of centers and the distinction between centers and stations.

Walbesser, Henry H. "Behavioral Objectives, a Cause Célèbre," *The Arithmetic Teacher*, XIX, No. 6 (October 1972), 418, 436–440. Walbesser writes with two objectives in mind: to expose the educationally damaging consequences of adopting behaviorism and to attempt to reestablish the credibility of behavioral objectives as a significant educational strategy.

# 4 Problem Solving, Application, and Estimation

Upon completion of Chapter 4, you will be able to:

1. Describe in general terms what a problem is and distinguish between two definitions of problem solving.

2. Distinguish between an "action-sequence" approach and a "wanted-given" approach to problem solving, and tell which is recommended for children's use.

3. Describe at least five materials and procedures that help children become efficient problem solvers.

4. Distinguish between problem solving and applying mathematics in routine ways, and describe how you can foster children's routine use of mathematics.

5. Describe mental arithmetic and its role in a mathematics program.

6. Give examples of mental arithmetic activities for primary and intermediate grades, and explain shortcuts for computation that can be done mentally.

7. Give two reasons why estimation skills are important.

8. Demonstrate how to use a number line to help children round off numbers.

9. Identify two common pitfalls associated with problem solving, application, and/or mental arithmetic, and explain how you can avoid or minimize them.

10. Name five problem-solving skills that children may find to be troublesome.

Key Terms you will encounter in Chapter 4:

problem solving
action-sequence analysis
wanted-given analysis
alternate approach
application
mental arithmetic
shortcut
estimation
rounding off

One of the primary goals of mathematics education has been to help individuals develop the mathematical skills and knowledge they need to perform occupational tasks. In early schools, stress was placed on the computational skills needed to maintain the books for a business, the knowledge of measurement needed to perform the activities of a carpenter, or other highly functional occupations. Over time, occupational skills and knowledge have expanded and become less definite than in the past. A fast-changing world offers few occupations for which mathematical skills and knowledge can be developed entirely within one's school years. Rather, today's student must understand a broad range of concepts and learn skills which have broad application. The specific skills needed for a particular occupation are usually developed in a special training program or on the job itself.

In addition to developing problem-solving skills that can be adapted to many situations, today's student needs frequent opportunities to learn to apply mathematics in everyday situations. Many of the applications are routine, such as maintaining a balanced checkbook and figuring the sales tax to be paid when an item is purchased. The ability to make reasonable estimates in consumer situations and to check the accuracy of minicalculator operations and paper-and-pencil calculations is another skill children should develop. Mental arithmetic, of which estimation is a part, helps children develop skills that are used frequently.

## PROBLEM SOLVING

The term *problem solving* is used in this book in a broader sense than usual. Many writers and teachers consider it to involve only word problems (also called story or verbal problems) in children's mathematics texts. However, a problem should be considered as any situation an individual faces for which no immediate solution is apparent, but for which the possibility of solution exists. Before a problem can be said to exist, the individual must be committed to finding a solution. What is a problem for one person may not be a problem for another.

If this definition is accepted, the word problems in children's books will be recognized for what they are: practice exercises in vicarious problem solving. Word problems are designed to supplement the real problems children encounter during their in-school and out-of-school activities. As far as possible, problem-solving experiences should result from real mathematical problems encountered during lessons in science, social studies, art, and other in-school activities. (Children will also engage in many out-of-school problem-solving activities, but you have little or no control over these.) Alert teachers will see to it that children face many problem-solving experiences in other curricular areas by planning them to include mathematical solutions of problems.

Since children's experiences in other curricular areas do not usually provide enough problem-solving experiences to establish proficiency, textbook word problems and problem cards are a valuable part of the mathematics program.

If children are to become efficient problem-solvers, they must develop a positive attitude toward problem solving. This attitude is not something taught outright. It is developed gradually as children gain confidence in their ability to use mathematics and other skills to solve the problems they face: they will copy this attitude from good teachers who are sure of what they are doing and who display enthusiasm for mathematics as they teach; they gain confidence in their use of mathematics as they learn basic facts and develop skill using the algorithms of arithmetic and instruments of measure; they acquire this attitude as they learn to visualize problem situations, round off numbers, estimate and judge the reasonableness of answers, and perform other mental operations on numbers; they develop positive attitudes as they learn to use the terminology and symbols of mathematics to communicate mathematical ideas.

While no one procedure should be followed to solve all problems, it is possible to indicate in a general way the steps effective problem-solvers follow: they know that before a problem can be solved, its nature must first be recognized; they know that each problem must be analyzed carefully to note similarities and differences between it and other problems to select the correct procedure to solve it; they have developed a habit of doing all work accurately; they know that all work must be carefully checked and the reasonableness of answers judged. Figure 4-1 shows a diagram of these steps.

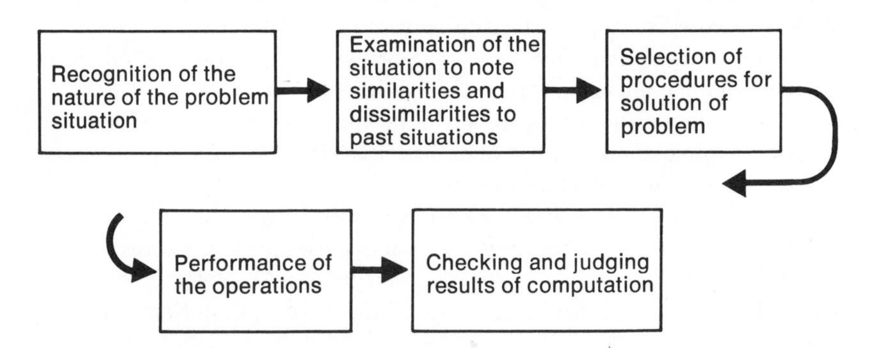


**Figure 4-1** Schematic representation of problem-solving procedure

(G. Bolyn)

## Self-check of Objective 1

Describe orally or in writing what a problem is. Distinguish between the two interpretations of problem solving discussed in this book.

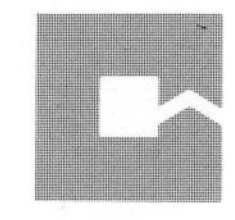

## HELPING CHILDREN BECOME PROBLEM-SOLVERS

No two problem-solvers will approach a problem in exactly the same way. Even so, an awareness of certain procedures will help every child build a repertoire of knowledge and skills useful in problem solving.

### 1. Analysis of the Problem

John Wilson reports a study in which two programs for processing one-step verbal problem information were tested.[1] Both programs use the steps in the schematic representation shown in Figure 4-1. They differ, however, in the action required at the "examination-of-the-situation" stage of the process. One program is based on an "action-sequence" analysis wherein the learner looks for the *action* indicated by the problem. The basic actions are addition, subtraction, multiplication, and division. Once the action has been determined, the *sequence* of the actions and events is noted. Then the sentence(s) for solving the problem is recorded. The second program is based on a "wanted-given" analysis of the problem. The learner examines the problem to find out what is *wanted* and what is *given*. What is wanted will be a sum, difference (unknown addend), product, or quotient (unknown factor), or some other type of answer. In a one-step problem involving one of the basic operations, what is given will be two or more addends, an addend and a sum, two or more factors, or a factor and a quotient. Of course, not all problems are so simple that only one operation is required, so children need to analyze each situation carefully to be sure the proper processes are selected.

Wilson's study was designed to determine if one program is more effective than the other. The results indicate that children instructed to use the wanted-given program were better able to choose correct operations for all types of problems, to obtain correct answers, and to finish the problems more rapidly.[2] The study showed other advantages for the wanted-given program.[3] The findings of Wilson's study suggest that chil-

[1] John W. Wilson, "The Role of Structure in Verbal Problem Solving," *The Arithmetic Teacher*, XIV, No. 6 (October 1967), pp. 486–497.

[2] Wilson, pp. 495–496.

[3] Wilson, p. 496.

dren should be instructed to use the wanted-given process to analyze problems rather than the action-sequence process.

---

## Self-check of Objective 2

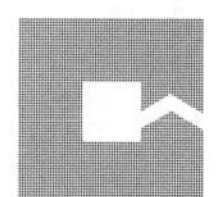

Describe both the action-sequence and wanted-given approaches to problem solving. Include the reasons why Wilson recommends the wanted-given approach for children.

---

### 2. Mathematical Sentences

One of the most important problem-solving skills is the use of a mathematical sentence to express a problem situation. Once children have determined what is wanted, they must use whatever information they have in an appropriate sentence. Then the computational procedures to be used in its solution can be determined. Children will learn to express problems using mathematical sentences through the analysis of problems such as the following one. "For a recent airplane race the route of the flight was from San Francisco to Los Angeles (347 miles), from Los Angeles to El Paso (701 miles), from El Paso to Houston (676 miles), and from Houston to New Orleans (318 miles). What was the total distance of the route?" Once children have expressed this problem in sentence form: $347 + 701 + 676 + 318 = \square$, they can put the numerals in computational form.

Mathematically mature children should learn to write mathematical sentences for word problems involving a combination of operations also. "While on a shopping trip for her mother, Sara bought three bars of soap for 33 cents each, a loaf of bread for 75 cents, and four cans of soup for 26 cents each. How much did the groceries cost?" Before attempting to solve the problem, children should write a mathematical sentence for it: $(3 \times 33) + 75 + (4 \times 26) = \square$. Then they should verify the meaning of each numeral in the sentence: the first 3 represents the number of bars of soap, while the 33 represents the price of each one; the 75 represents the price of a loaf of bread; in the second parentheses, the 4 represents the number of cans of soup, while the 26 represents the price per can. Since the elements in each set are cents, the answer must be written as dollars and cents to represent the meaning correctly.

The following activities provide useful practice with mathematical sentences:

1. Hand out dittoed sheets containing a set of five or six word problems. Ask the children to write a mathematical sentence for each one.

2. Read word problems to children. Ask them to write the mathematical sentence(s) for each one.
3. Write five or six mathematical sentences on the chalkboard. Ask everyone to make up a word problem for each sentence.

Occasionally, you might ask the class to write their own word problems and turn them in for checking. More often, word problems will be presented orally, discussed, and evaluated. In the time it takes children to write one word problem each, they can discuss several orally presented ones.

### 3. Concrete Materials, Pictures, and Diagrams

Many different demonstration materials and learning aids are described in later chapters of this book. As each of these materials is used, children learn their value as problem-solving aids.

Markers and number lines for whole numbers are particularly useful aids to younger children. These aids can illustrate many of the word problems primary-grade children encounter. Number lines that represent fractional numbers either as common fractions or decimal fractions are valuable aids in the intermediate grades. A number line can illustrate a situation involving division with whole numbers and common fractions, as follows: "Jane is helping her father make a boat. They have a piece of wood 12 feet long. If they cut it into pieces each 2½ feet long, how many pieces will they cut from the board?" Once children have read the problem and determined what is wanted and what is given, they can use the number line to illustrate the situation (Figure 4-2). The remainder, 2, can be omitted from the answer, as the number line shows. After children use the number line to see the meaning of the situation, they can write its mathematical sentence: $12 \div 2\frac{1}{2} = \square$, and determine the answer by division.

As children use different learning aids, you should encourage them to rely less on the actual aids and more on mental images of them. For example, children who have used a number line frequently should be encouraged to visualize a number line without actually having one before them. Many children find that they can get a mental image so clear it is almost as if a number line were in the room. When they encounter a problem situation, they can often work it out on their mental number lines. Children should be encouraged to visualize other problem situations using mental images of other learning aids. When you preface re-

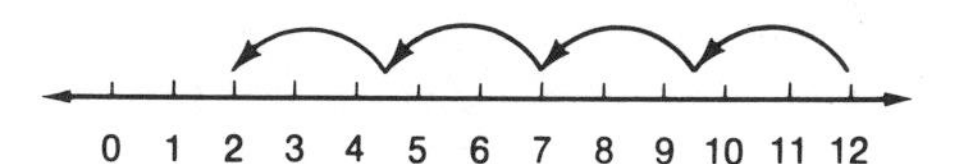


**Figure 4-2** A number line used to illustrate a division word problem

marks about problem solving with comments such as, "Imagine you have a set of markers on the magnetic board," or "Think of the fraction strips we used last week," you will help children construct the proper mental images.

Other activities, to be used occasionally, deal with concrete materials, pictures, and diagrams:

1. Assign five or six word problems. Ask the children to draw a picture or diagram for each one.
2. Show children a picture, map, or diagram. Using the map in Figure 4-3, ask each child to make up a word problem based on it. Those with word problems different from those of others should present them for discussion.
3. Give children department store catalogs and have them make up word problems from them.

### 4. Tables and Graphs

Tables and graphs are problem-solving aids. Once data have been organized in a table and are graphed on a number plane, the problem's solution often becomes obvious. "Sue is knitting a sweater. The directions say that every seven rows of stitches will make 1 inch of sleeve. If each sleeve is to be 16 inches long, how many rows of stitches will one sleeve have?" While the answer can be determined by multiplying 7 and 16, children should be helped to recognize other ways of solving it. A table can organize data related to it. In Figure 4-4, $n$ represents the number of inches knitted and 7 the number of rows of stitches per inch. You should help children prepare their table by asking questions: "When Sue has finished knitting the first inch of a sleeve, how many rows of stitches will there be?" "When she has knitted 2 inches of a sleeve, how many rows of stitches will she have completed?" "How many inches of sleeve will twenty-one rows of stitches make?" As the questions are asked, children should complete enough of the table to graph the data on a number plane.

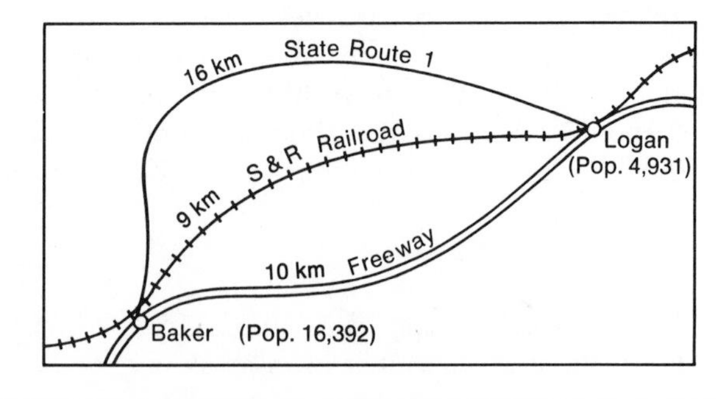


**Figure 4-3** A map serves as the basis for a child-composed word problem

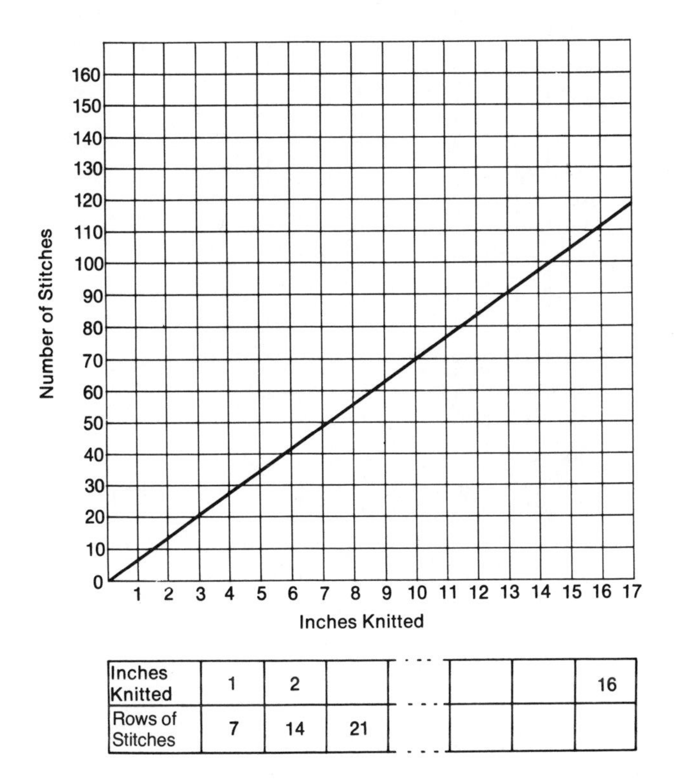


| Inches Knitted | 1 | 2 | | . . . | | | 16 |
|---|---|---|---|---|---|---|---|
| Rows of Stitches | 7 | 14 | 21 | . . . | | | |

**Figure 4-4** Table and graph developed from a word problem

A table and graph are also useful to solve percentage problems. If children are asked to find 23% of 600, the table in Figure 4-5 can be developed. Children should begin by finding 23% of 100, then 23% of 200, and so on. When the table and a graph have been prepared, the answer becomes clear. Children will need help to learn how to read graphs to get answers that are not on their tables. In this example, the approximate value of 23% of any number can be found on the graph. Read across the bottom from the left to the number, go up to the graphed line, and then go back to the percentage figure on the left. The graph shows that 23% of 350 is about 81; 23% of 500 is 115.

## 5. Reading in Mathematics

Many difficulties children encounter in mathematics, particularly in problem solving, stem from their inability to read mathematical material and interpret it properly. Both skillful and weak readers may have this problem. The skills developed during reading lessons are not entirely the same ones required for reading in mathematics, where an analytical approach is necessary.

Since many words used in a mathematical context are already familiar in other contexts, their meanings need careful attention so children are

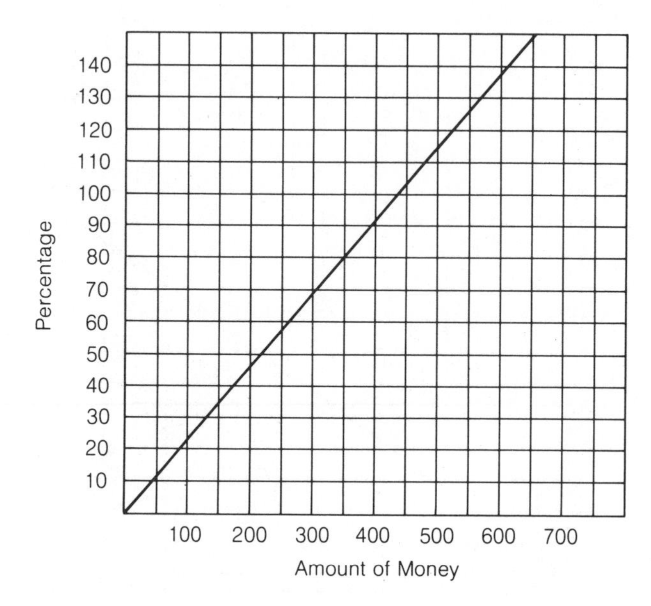


| Amount of Money | 100 | 200 | 300 | | | 600 |
|---|---|---|---|---|---|---|
| Percentage | 23 | 46 | | | | |

**Figure 4-5** Table and graph used to determine 23% of 600

not confused when they see them in a mathematical context. For example, *base, difference,* and *product* have meanings apart from those used in mathematics. Other words, such as *array, quotient,* and *percent,* may be new to children when first encountered in mathematics. Their meanings must be developed carefully. The symbols of mathematics, that is, the signs of operation (+, −, ×, and ÷), and signs of relation (>, <, and =), and others, must be clearly understood. Abbreviations frequently used in mathematics should receive special attention.

Children must learn to read mathematics material slowly enough to determine the meaning of each word and symbol. Activities that enable children to acquire the needed vocabulary and other reading skills are an essential part of mathematics instruction. These activities should supplement any included as part of a basic reading program. Appendix A lists Greek, and Roman prefixes and suffixes commonly used in mathematics and their meanings. Lessons in which their meanings are featured will enhance children's ability to read mathematics material.

Looking for clues in word problems has long been thought to help children determine which procedures to use. Examples of clue words within sentences are: "How many are there in *all* (or *altogether*)?" (Addition). "What is the *difference?*" and "How many are *left?*" (Subtraction).

However, such clues are so limited in value as to be almost useless; many children use them improperly, thinking they suggest one operation when another is appropriate. "Bob had thirty-six marbles. When his father gave him a new bag of marbles. Bob counted all of them and found he had fifty-one altogether. How many marbles were in the bag his father gave him?" The word altogether in this problem incorrectly suggests using addition. Another weakness of the clue-word approach is that many word problems do not contain useful clues. If word clues are the only means children have for selecting a procedure, they often resort to guessing or give up. Teaching children a broad range of reading skills is the best preparation for attacking problems.

### 6. Use Alternate Methods of Solution

The answer to a given problem can frequently be computed in several ways. Children should be encouraged to look for and use these alternate procedures. "Small potted plants are priced three for $1.00. What do twelve plants cost?" After each member of the class has solved this problem, a discussion of the various procedures could include any of the following ones.

1. "I figured that one plant costs 33⅓ cents. Then I multiplied 33⅓ by 12. The product is 400, so the cost of 12 plants is $4.00."
2. "I found the answer by determining that 12 is 4 times 3, so the cost of 12 plants is 4 times as much as the cost of three plants. Four times 100 is 400, so the cost is $4.00."
3. "I used a proportion to get the answer. The proportion I set up is 3 to 100 and 12 to $n$, or $3/100 = 12/n$. I then used cross multiplication to get $3n = 1200$. Finally, I divided 1200 by 3 to get the answer; $n$ equals 400, so the 12 plants cost $4.00."

If your children do not use alternate procedures, encourage them to do so. "You all used the same procedure for determining your answers. Now I want you to think about other ways to get the answer. In earlier lessons, we learned several procedures for solving problems of this kind. See if you can remember any of these to use with this word problem."

### 7. Other Procedures

a. **Simplify Problems.** Some children are perplexed by word problems containing very large or relatively unfamiliar numbers. Simplifying such problems can help them; smaller or more familiar numbers can be substituted as children formulate mathematical sentences. An examination of these should reveal computational procedures, and the original numbers can then be replaced to perform the computation.

**b. Use Word Problems with Too Much or Too Little Information, without Questions, or without Numbers.** Word problems that contain too much or too little information should be assigned from time to time as practice in analyzing problems carefully. Each problem should be carefully analyzed to determine the information that is given. "A car can go 17 miles on 1 gallon of gasoline. How far can it go on a tank of gas?" "Jack had $5.00 in his bank. He then worked for 6 hours and earned $1.50 an hour. How much money did he earn for his work?"

Statements that do not contain questions or numbers must be carefully studied to determine what information is lacking. "An airplane has a top speed of 1100 miles an hour. It flew for 3 hours." "Jack bought a new baseball glove. How much change did he receive?" "Some children went to a movie. How much did it cost them to attend?" After they have decided what information is missing, children should complete the word problems.

---

## Self-check of Objective 3

A variety of materials and procedures for helping children develop their problem-solving skills are discussed. Name at least five different materials and/or procedures, and give an example of how each one contributes to children's ability to solve problems.

---

### APPLICATION

A person routinely uses addition and subtraction to balance his or her checkbook. The balancing of a checkbook cannot be thought of as "problem solving" once it becomes routine. There are many similar situations in which mathematics is applied in routine ways, and children should become familiar with as many of them as possible during their elementary school years.

One of the best ways to help children apply mathematics is to use common situations as you introduce topics. For example, when you introduce the concept of area, consider some of the common situations that involve area—carpeting or tiling a floor, wallpapering or painting a wall, planting grass, and so on. Use these and similar situations to set the stage for working with the concept itself and the formula for finding area. "Before you buy wallpaper, you must know how many square feet of wall you will be covering. If a wall is 8 feet high and 12 feet long, how many square feet of wall do you have?"

A practical and meaningful way to approach this situation is to use a part of a classroom wall to show the 8- and 12-foot dimensions. Then use paper cut in foot-square pieces to tack or tape to the wall along the width and height measures as you discuss the meaning of area. (You should not have to cover all of the 8- by 12-foot part of the wall in order for children to grasp the concept.) You can then talk about how many squares would be needed to cover the entire wall. A similar activity dealing with the area of the classroom floor can also be used. This concrete-manipulative approach will help children understand the meaning of area and, eventually, many applications of the concept.

Later, semiconcrete-pictorial materials can be used as children continue their investigation of area. The eventual use of the area formula—length times width ($l \times w$), or base times height ($b \times h$)—will be a natural outgrowth of the children's activities in real and simulated situations.

Throughout this book you will read examples of problem situations you can use to show children how to apply mathematics effectively and routinely in everyday settings.

---

## Self-check of Objective 4

Explain orally or in writing the distinction between problem solving and the routine application of mathematics. Describe how you can help children to apply mathematics routinely in everyday situations.

---

## MENTAL ARITHMETIC AND ESTIMATION

Mental arithmetic applies to arithmetic done without paper and pencil or other computational aids. Because so many routine applications of mathematics are done mentally, children should learn ways to compute without paper and pencil and to make reasonable estimates. While children cannot be expected to do complex computations mentally, they can experience the satisfaction of learning ways to do some work that way.

### 1. Mental Exercises

Many of the skills identified in later chapters—committing basic facts to memory, multiplying by 10 and powers of 10, estimating quotients in division—can be classified as mental arithmetic skills. Shortcuts sometimes reduce the complexity of computation to the point where all of it can be done mentally. Regular and frequent practice is essential to acquiring these skills as well as others. Spending a few minutes each day with mental arithmetic exercises pays dividends in accurate and quick responses. The following examples might be among your repertoire of exercises:

1. Ask children to give other names for numbers. "Who can tell me another name for 10 besides 5 plus 5?" Limits can be placed on which names will be accepted, such as, "Who can tell me a pair of numbers with a sum of 10 and of which one number is less than 3?"
2. Children can practice the basic facts orally. "What is the sum of 6 and 8?" "If I subtract 9 from 17, what is the remainder?" "What is the difference between 16 and 7?" "What is the product of 9 and 6?" "How much is 5 times 7?" "What is the quotient when 63 is divided by 9?" "How many times larger than 8 is 56?"
3. Children can rename fractions orally after they understand how this is done. "What is the simplest name for $6/8$?" "What are three other names for $1/2$?" "What is the least common denominator for the numbers $1/6$ and $1/8$?"
4. After learning to factor numbers, children should practice factoring orally. "What is a pair of factors for 36?" "What is another pair of factors for 36?" "Another pair?" "What is the prime factorization of 42?"
5. Children should practice naming reciprocals of numbers. "What is the reciprocal of $1/3$?" "What is the reciprocal of 9?" "What is the reciprocal of $7/8$?"
6. Play I'm thinking of a number: "I'm thinking of a number. It is less than 10. It is not even. It's not prime. What is the number?" "I'm thinking of a number. It is greater than 3. It's divisible by 2. It's less than 8. Adding 1 to it gives a prime number greater than 5. What is the number?"

Some children will discover computational shortcuts on their own. Even so, you will need to encourage them and others to learn and use common shortcuts whenever possible. Remember that a shortcut should not be introduced until the operation it applies to is well understood. This means that there are children for whom many (or all) shortcuts are unsuitable. Rather than helping a child whose understanding of an operation is meager, the shortcut may result in further confusion and misunderstanding. The following mental shortcuts are simple enough that many children will understand and use them.

1. Addition

Front-end addition: $56 + 32 = 88$

Add the tens ——→ $50 + 30 = 80$

Add the ones ——→ $6 + 2 = 8$

Add the two sums ——→ $80 + 8 = 88$

Partial addition: $56 + 32 = 88$
Add the first addend to the tens of the second → $56 + 30 = 86$
Add the ones of the second addend to the sum → $86 + 2 = 88$
Compensation: $589 + 233 = 822$
Make the first addend hundreds (by adding 11 in this instance) → $589 + 11 = 600$
Add the second addend to the hundreds → $600 + 233 = 833$
Subtract the number added in step one from the sum → $833 - 11 = 822$

2. Subtraction
Front-end subtraction: $56 - 32 = 24$
Subtract tens of second number from first number → $56 - 30 = 26$
Subtract ones of second number from the remainder → $26 - 2 = 24$

3. Multiplication
Front-end multiplication: $4 \times 36 = 144$
Multiply tens of second factor by first factor → $4 \times 30 = 120$
Multiply ones of second factor by first factor → $4 \times 6 = 24$
Add two partial products → $120 + 24 = 144$
Compensation (This shortcut works for factors that are themselves factors of 100, 1000, and so on. Two examples are given): $25 \times 48 = 1200$
Multiply first factor by 4 → $4 \times 25 = 100$
Multiply second factor by 100 → $100 \times 48 = 4800$
Divide product by 4 → $4800 \div 4 = 1200$
$125 \times 488 = 61{,}000$
Multiply first factor by 8 → $8 \times 125 = 1000$
Multiply second factor by 1000 → $1000 \times 488 = 488{,}000$
Divide product by 8 → $488{,}000 \div 8 = 61{,}000$

As you read the chapters that follow, you will learn about activities with place value devices, tables and charts, and patterns that help children to understand and use computational algorithms. You should continually be alert for ways you can use these and other materials and procedures to further children's abilities to use mental arithmetic.

## Self-check of Objectives 5 and 6

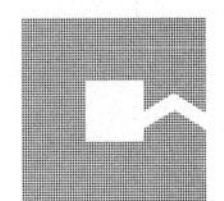

Describe orally or in writing what mental arithmetic is and explain its role in a mathematics program.

Describe some ways you can use mental arithmetic to help children learn basic facts and other important information; explain one mental arithmetic shortcut for each of these operations: addition, subtraction, multiplication.

### 2. Estimation

Skill in estimation is gaining importance as people become more consumer conscious and as they make greater use of minicalculators. A person makes estimates to determine the lower unit price for two boxes of cereal in the supermarket and the price of a milkshake and hamburger at the drive-in. The user of a minicalculator will frequently estimate an answer to judge the reasonableness of a calculation done on the machine. A person cannot be skillful in estimating answers unless he or she has the opportunity to learn and practice the skills needed for processing estimates.

$$\begin{array}{r} 290 \\ 320 \\ 380 \\ +430 \\ \hline \end{array}$$

One important skill is that of rounding off numbers. To process an estimate to the nearest hundred for the addition shown in the margin, a person must first round each number to the nearest hundred. A number line such as the one in Figure 4-6 is useful as a means of helping children see that 290 is nearer to 300 than to 200. By locating 290 on the line and counting the steps (each having a measure of ten) from there to 200 and then to 300, the reason for rounding off 290 to 300 becomes clear. After all four numbers have been rounded off, the answer can be estimated by adding only the hundreds.

When children in elementary school are taught to round off numbers to the nearest ten, they learn that the numbers 31, 32, 33, and 34 are rounded to 30, and the numbers 36, 37, 38, and 39 are rounded to 40. Children can use a number line to see that 31, 32, 33, and 34 are closer to 30 than to 40, and that 36, 37, 38, and 39 are closer to 40 than to 30. When 35 is considered, however, the number line shows that it is as close to 30 as to 40. Children will need to be given a rule for rounding off numbers that end in five. The rule that is most commonly used is to round off the numbers to the next higher ten. By applying this rule to the number 35, the children learn to round it off to 40.

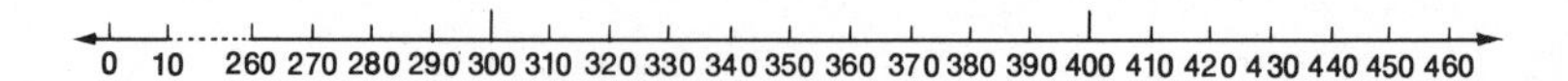


**Figure 4-6** Number line used to help children round off numbers to the nearest hundred

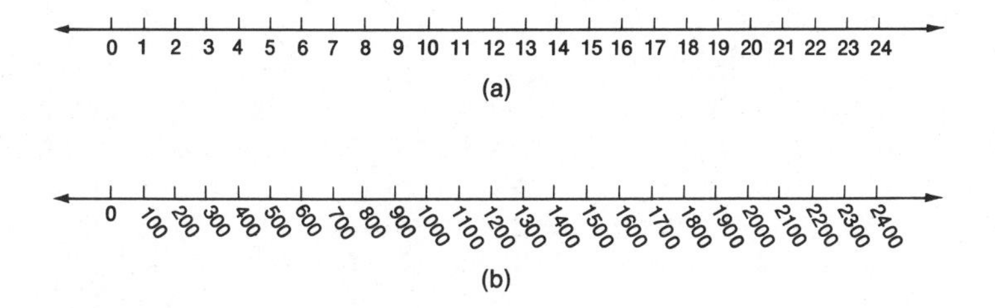


**Figure 4-7** Number lines used to help children learn to round off numbers to (a) the nearest ten and (b) the nearest thousand

For help in rounding off to the nearest ten, a number line similar to the one in Figure 4-7 should show the whole numbers (a). To round off numbers to the nearest thousand, a number line should be marked off by hundreds (b). These lines can be marked off on adding machine tape and fastened to the classroom walls. After they have learned to round off numbers to the nearest ten, hundred, and thousand, help children work with numbers that are to be rounded off first to the nearest ten, then the nearest hundred, and then the nearest thousand. Numbers such as 3291; 4648; and 13,689 are useful. By the time they complete sixth grade, most children will be able to round off numbers of any size through billions to any specified place value position.

Estimating answers should also be developed through exercises similar to these: "What is the approximate cost of a 28-cent can of fruit and a 41-cent quart of milk?" "Each bus carries 48 people. About how many passengers can the six buses carry?" "There are about how many 29s in 612?"

Sometimes estimation makes it unnecessary to compute an exact answer to satisfactorily solve a problem. This is true in the following situations: "Billy wants to buy a baseball glove that costs $6.95 and a ball that costs $2.49. He has $10.00. Is this enough money to purchase the ball and glove?" "A jet plane can carry enough fuel to remain airborne for 5 hours of flight at top speed. If it flies at a top speed of 890 miles per hour, can it make a round trip of 5000 miles without having to refuel?"

---

## Self-check of Objectives 7 and 8

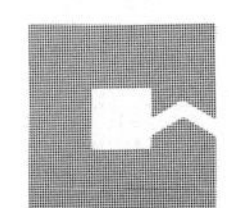

Skill in estimating is important. Describe two situations in which it is commonly used.

Demonstrate how to use a number line to round off each of these numbers: 86 (nearest ten), 460 (nearest hundred), 4450 (nearest thousand).

---

## Common Pitfalls and Trouble Spots

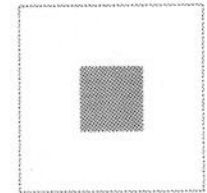

Teachers at all levels continually face pitfalls that get in the way of effective teaching. Unless you avoid the pitfalls mathematics teachers face, you can

lead children to faulty habits, misunderstandings, fear of mathematics, and patterns of failure from which they will recover only slowly, if at all. While you may not succeed with each child as well as you would like, the chances of your being a successful teacher of mathematics are improved if you are conscious of some of the common pitfalls and learn ways to avoid or minimize them. Beginning with this chapter and continuing with each successive chapter, one or more common pitfalls associated with the chapter's subject are discussed.

There are two pitfalls associated with problem solving, application, and mental arithmetic. The first is that a teacher may take a narrow view of problem solving, and rely solely on the story problems in a textbook as the source of all his or her children's activities. While there is a place for story problems, you must remember that all textbook problems are contrived and, even though the situations they portray may be realistic, they do not provide first-hand experiences. In addition to using word problems, your children should be involved in other classroom activities that require them to solve problems. By carefully planning your mathematics lessons, as well as those in science, social studies, art, and other subjects, you can give your children meaningful experiences that go beyond textbook problems.

Second, many programs, both commercial and district-prepared, contain activities that deal in a limited fashion with applications and mental arithmetic. You should examine your program's materials and, if the program is weak in these areas, use the suggestions and examples in this chapter as a basis for developing your own activities.

In addition to the pitfalls for teachers, there are common trouble spots for children. Again, awareness of them is the best safeguard against their affecting your children. Each subsequent chapter ends with a discussion of one or more trouble spots for children (except the last chapter, which discusses a pitfall but no trouble spots).

Five aspects of problem solving are troublesome for many children: (1) reading the problem, (2) recognizing the action involved, (3) visualizing the problem, (4) determining the appropriate operations, and (5) computing the answers. The inclusion of activities that focus on the problem-solving skills discussed in this chapter will help most children avoid these trouble spots. Some children will benefit from participation in special problem-solving groups that meet with you to deal with particular skills.

These guidelines will help you:

1. Determine each child's specific weaknesses.

2. Keep groups flexible, so a child deals with only those skills in which he or she is weak.

3. Consider only one skill at a time.

4. Begin with simple applications of the skill, then move to more complex applications as children become proficient in using it.

5. Provide frequent opportunities for children to apply a skill once it has been learned.

## Self-check of Objectives 9 and 10

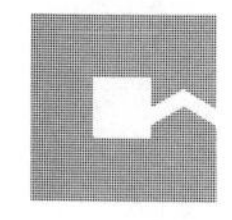

Identify orally or in writing two pitfalls associated with problem solving, applications, and mental arithmetic, and explain how you can avoid or minimize each one.

Name five problem-solving skills that are troublesome to many children, and list some guidelines for activities to overcome them.

## SUMMARY

A problem arises when an individual faces a situation for which an immediate answer is unavailable. A person must also be committed to finding a solution before a problem can exist. To the extent it is possible, the mathematical problems children face should grow out of the real-world situations they face. Since there are some types of problem situations children do not usually encounter, vicarious situations from problem cards, textbooks, and other sources give them exposure to situations they would otherwise miss.

There are strategies children can learn for problem solving as they progress through the grades. Wilson found that a wanted-given strategy is better for most children than an action-sequence strategy, and he recommends that such a strategy be emphasized in the elementary school. The use of mathematical sentences is a part of the recommended strategy. Some other ways children's problem-solving skills can be developed include the use of concrete materials, pictures, diagrams, tables, and graphs, and help with reading word problems carefully and using alternate methods of solution. Other procedures include simplified problems and analysis of problems with too much or too little information.

Many times mathematical operations and concepts are applied in routine ways in everyday situations. You can help children learn these routine uses by frequently using problem situations that show both concrete and semiconcrete examples of them. Mental arithmetic, or arithmetic without paper and pencil or other calculating aids, helps children to develop skills that are useful for meeting many everyday situations. Both exact and approximate answers are found using mental arithmetic skills. There are shortcut procedures for determining some exact answers. Estimation is used for determining approximate answers. Rounding off numbers is necessary for making reasonable estimates.

## STUDY QUESTIONS AND ACTIVITIES

1. It is recommended that children learn to use mathematical sentences to help them interpret the meaning of problem situations,

whether real or contrived. Three activities that involve children with mathematical sentences are described on pages 82–83. Prepare a lesson that deals with one or more of these activities and use it with a group of children for whom it is appropriate. Do their responses indicate that they can use sentences as problem-solving aids? If not, what steps do you believe should be taken to help them?

2. Examine a contemporary mathematics series to determine to what extent word problems are used. Are there other kinds of problem-solving activities, such as puzzles, brain teasers, pattern development, and riddles, that require children to use mathematics in nonroutine ways? Is there any reference to mental arithmetic in either the children's text or the teacher's manual? If so, what is said about it?

3. Prepare a set of word problems that have too much or too little information, or are without questions or numbers. Duplicate copies for children in an intermediate grade. Ask them to study the questions to determine what should be added or deleted to give the word problems meaning. What is your assessment of the children's understanding of work of this type?

4. One way to determine how well children understand a number sentence and the operation it relates to is to have them write a story problem from which a sentence might have arisen. Write a story problem for each of these division sentences: $3 \div \frac{1}{2} = 6$; $\frac{3}{4} \div \frac{1}{4} = 3$; $\frac{1}{2} \div 3 = \frac{1}{6}$. Were you able to do this task without looking in Chapter 12 for help? If not, what does this indicate about your understanding of division involving common fractions? To what do you attribute this lack of understanding?

5. Do you know any shortcuts for mentally computing answers for operations similar to the ones described in this chapter? If so, write an explanation for each one. Be prepared to share your shortcuts with your classmates.

## FOR FURTHER READING

Burgler, Donald R., Jr. "Be A Super Shopper," *The Arithmetic Teacher,* XXV, No. 6 (March 1978), 40–44. Items from the supermarket and corner grocery provide a basis for applying basic skills to a common application of the mathematics children learn. Examples of objectives, problem cards, a work sheet, a record sheet, and flow chart of activities are presented.

Carr, Karen D. "A Common Cents Approach to Mathematics," *The Arithmetic Teacher,* XXVI, No. 2 (October 1978), 14–15. A unit dealing with pay checks, checking accounts, monthly expenses, and emergency expenses is used to give sixth graders a chance to apply some of the skills they learn in mathematics classes.

Cohen, Louis S. "Open Sentences—The Most Useful Tool in Problem Solving," *The Arithmetic Teacher,* XIV, No. 4 (April 1967), 263–267. Emphasizes the importance of number sentences in problem solving. Cohen suggests that before children can write open number sentences, they must learn to use number phrases. Several examples of how children can learn to use phrases are described.

Davidson, James E. "The Language Experience Approach to Story Problems," *The Arithmetic Teacher,* XXV, No. 1 (October 1977), 28. Having children write and solve their own story problems based on personal experiences helps avoid some of the pitfalls of textbook problems—low reading comprehension, lack of experience with the problem situations, unfamiliarity with the language of the situations, and irrelevancy of the problems.

Flournoy, Frances. "Providing Mental Arithmetic Experiences," *The Arithmetic Teacher,* VI, No. 3 (April 1959), 133–139. Flournoy gives evidence of a need for a planned program of mental arithmetic. Such a program should include a variety of experiences, children's use of varied techniques while attacking problems, and opportunities for children to interpret quantitative statements in material they read.

Hannon, Herbert. "Problem Solving—Programming and Processing," *The Arithmetic Teacher,* IX, No. 1 (January 1962), 17–19. Children program problems by developing number sentences for them; they process them by determining correct numbers for open sentences.

Henny, Maribeth. "Improving Mathematics Verbal Problem-solving Ability through Reading Instruction," *The Arithmetic Teacher,* XVIII, No. 4 (April 1971), 223–229. Henny stresses that verbal problem solving requires different skills than other types of reading. Procedures for developing analytical reading skills are discussed and summarized in a two-page table.

Krulik, Stephen. "Problem Solving: Some Considerations," *The Arithmetic Teacher,* XXV, No. 3 (December 1977), 51–52. Presents and elaborates on three criteria for identifying a "problem." Describes seven ways to help children become better problem solvers; some are accompanied by examples.

Lindquist, Mary M., and Marcia E. Dana. "Recycle Your Math with Magazines," *The Arithmetic Teacher,* XXV, No. 3 (December 1977), 4–8. Popular magazines serve as the source of many problem-solving activities. Cut out pictures, ads, and full pages for work dealing with numbers, addition and subtraction, multiplication and division, and measurement.

Maier, Eugene. "Folk Math," *Instructor,* LXXXVI, No. 6 (February 1977), 84, 86–87, 92. "Folk math is the way people handle the math-related problems arising in everyday life." Maier describes two differences between school mathematics and folk math. Then he tells how school math can become more like folk math.

Schall, William E. "Comparing Mental Arithmetic's Modes of Presentation in Elementary School Mathematics," *School Science and Mathematics,* LXXIII, No. 5 (November 1973), 359–366. Children receiving one of three modes of presenting mental arithmetic topics—closed-circuit TV, programmed booklets, and audio-tapes—were compared with children receiving no mental arithmetic and some using programmed booklets in set theory. The results showed that children receiving mental arithmetic, no matter the means of presentation, benefited over those who did not. The attitudes of children receiving mental arithmetic were largely favorable toward the experience.

Schmid, John A. "Experiences with Approximation and Estimation," *The Arithmetic Teacher,* XIV, No. 5 (May 1967), 365–368. Approximation and estimation experiences form a valuable part of the mathematics curriculum. Schmid suggests several experiences for approximation using measures, decimal fractions, common fractions, and estimation with whole numbers and common fractions.

Sims, Jacqueline. "Improving Problem-solving Skills," *The Arithmetic Teacher,* XVI, No. 1 (January 1969), 17–20. The author presents examples of interesting approaches to improving children's problem-solving skills. Includes examples of problems without numbers, problems with irrelevant data, problems with multiple means of solution, uses of number sentences, uses of diagrams, and estimating answers.

Sowder, Larry. "Teaching Problem-solving: Our Lip-Service Objective?" *School Science and Mathematics*, LXXII, No. 2 (February 1972), 113–116. Sowder offers six suggestions for improving children's problem-solving skills and attitudes toward problem solving.

Swenson, Esther J. "How Much Real Problem Solving?" *The Arithmetic Teacher*, XII, No. 6 (October 1965), 426–430. After discussing the nature of problem solving, Swenson suggests sources of problems for children to solve and some procedures for solving them.

Trafton, Paul R. "Estimation and Mental Arithmetic: Important Components of Computation," *Developing Computational Skills*, 1978 Yearbook of the National Council of Teachers of Mathematics. Reston, Va.: The Council, 1978, pp. 196–213. Explains and illustrates the uses of estimation in problem solving and computation, processes for developing estimation skills, and ways to develop mental arithmetic skills.

# 5 Teaching the Foundations of Numeration

Upon completion of Chapter 5, you will be able to:

1. Name some specific ways children's early play experiences help establish foundations for later learning of mathematics.

2. Describe materials and activities dealing with discrete and continuous objects, classification, patterns, and spatial relationships for preschool and kindergarten children.

3. Prepare sample test items for a survey test for preschool, kindergarten, and first-grade children. Explain why directions must be oral and why answers should be given orally or by pointing to test materials.

4. Demonstrate a setting that can be used to introduce children to counting.

5. Use markers and numeral cards to show how children can learn to count the number of objects in a collection.

6. Demonstrate a procedure that introduces the meaning of zero.

7. Illustrate at least five devices or materials that can be used by children who are learning to count.

8. Illustrate two types of materials for providing practice with numeral writing, and explain the differences between the two.

9. Distinguish between the cardinal and ordinal uses of numbers, and describe activities that help children understand ordinal uses.

10. Describe how music and finger-play activities and stories and poems can enhance children's understanding of number concepts.

11. Identify four pitfalls associated with early number work, and describe ways to avoid them.

12. Name two aspects of early number work that are troublesome for some children, and tell how you can help children overcome them.

Key Terms you will encounter in Chapter 5:

rote counting
preoperational stage
discrete object
continuous object
classification
attribute
pattern
spatial relationship
survey test
quantity
comparison
order relation
numeral
ordinal property
manipulate
set
one-to-one matching
markers
one-to-one correspondence
cardinal number
domino pattern
manuscript
cursive
number line
global comparison
lasting equivalence
conservation

Between 1920 and 1950, formal instruction in mathematics was delayed in many schools until at least the third grade. Children in kindergarten and the first and second grades were generally considered too immature to understand number concepts. Teaching of number concepts was often incidental and unplanned. Some teachers made good use of chance opportunities to begin teaching mathematics, some did not, and some taught number concepts only if children exhibited an interest in learning about them.

Today, few believe that children below third grade should be denied a planned program of mathematics instruction. Systematic instruction begins at least as early as grade one, and in many schools, in preschool and kindergarten classes.

## MATHEMATICS IN PRESCHOOL AND KINDERGARTEN

Most children understand some mathematical concepts before they enter school. It is difficult to identify any particular moment in a child's life when he or she acquires his or her first numerical or quantitative concept. Understanding of mathematical concepts begins imperceptibly and grows so gradually it is almost unnoticeable. It has been observed that some children under two years of age have acquired some idea of "more," usually in connection with food, such as taking the larger cookie or piece of candy when given a free choice.[1]

The kinds of understandings children have when they enter school depend upon early experiences. When children participate in activities using numbers, many of them will pick up simple number skills. For example, a skill taught in many homes is reciting number names in sequence without regard to their meanings. Such counting is frequently called "rote counting." It is true that this skill is not particularly valuable in itself, but it is a foundation upon which later work can be built, because children must know the sequence of numbers before they can count objects in a set. If children play games involving spinners, dice, or dominoes, they may be able to recognize the number of objects in small sets, or certain numbers that give small sums.

[1] George E. Hollister and Agnes G. Gunderson, *Teaching Arithmetic in the Primary Grades* (Boston: D. C. Heath and Company, 1964), pp. 12–13.

Recent studies testing children's understanding of number concepts yield these conclusions:

1. Many preschoolers have some skill in counting and numeral recognition and know the uses of common instruments of measure, such as the thermometer, clock, ruler, and calendar.
2. There is a wide variation in skill and understanding among preschoolers.
3. When they start school, boys and girls show no difference in mathematical skills or understanding.
4. Children from high socioeconomic levels generally show more skill in working with numbers and a greater understanding of number concepts than do children from low socioeconomic levels.
5. When they begin school, many children show an interest in mathematics.

## 1. Early Experiences

During the sensorimotor stage, a child's learning develops through movement. The movement is random at first, but later it is directed toward a person, object, sound, or other stimuli. Bodily movements such as kicking, turning, twisting, stretching, crawling, and clapping lead to notions of position, distance, direction, and rhythm, all of which are developed at a later stage.[2]

Between the ages of two and six (the preoperational stage) most children establish foundations for learning mathematics. But some children lack activities that contribute toward building a very firm foundation. A child's environment in general and play experiences and relations with others in particular play a large part in determining mental growth in the preoperational stage.

Children learn through play. Not only do they have fun while they play, but they learn to use words and actions to represent ideas. For example, during play, pieces of wood placed end to end become roadways, while two pieces nailed together crosswise are a supersonic jet. Large blocks become castles, or ships, or bridges. And, the child becomes a king or queen, ship's captain, or bridge tender. All the while, each child is building the vocabulary and background upon which to build later learnings.

[2] Elizabeth Williams and Hilary Shuard, *Elementary Mathematics Today: A Resource for Teachers Grades 1–8* (Menlo Park, Calif.: Addison-Wesley Publishing Company, 1970), p. 8.

## Self-check of Objective 1

Describe how children's play activities help establish foundations for later mathematics learning. What are the implications for a teacher when children have meager play experiences during their preschool years?

### 2. Preschool and Kindergarten Experiences

Children's play experiences should continue in preschool, kindergarten, and first-grade classrooms. A well-equipped room will contain many commercial and teacher-made devices to help children develop a beginning understanding of discrete and continuous objects, classification, patterns, and spatial relationships.

**a. Discrete and Continuous Objects.** Early mathematical activities lead children to recognize that there are both discrete and continuous objects. Discrete objects are ones that can be counted, while continuous objects are those that are measured. Classrooms should offer children opportunities to work with both types. Mary Baratta-Lorton describes simple activities in *Workjobs.*[3] The "outline game"[4] is a discrete-object activity. There are two 12- by 18-inch pieces of tagboard, joined together with tape along a pair of 18-inch sides, and a box of small objects, such as blunt scissors, an eraser, pencil, and different-sized paper clips, rubber bands, bobby pins, and so on. The shape of each object is outlined with colored pencil on the tagboard. Children match each object with its outline, making a one-to-one correspondence between all of the objects and their outlines.

Sand or water play lays a foundation for understanding capacity. You can fill a large plastic tub with water or clean sand. Put several large plastic, tin, or glass containers of various sizes in or near the tub. Allow time for individuals or small groups of children to play at the tub so they can fill smaller containers from a larger one and vice versa. Let them play spontaneously, without directions from you about which container to use, or how to keep track of the number of times one container can be filled from another.

Collections of objects for children to compare for height, length, surface, and volume are easy to assemble. Dowel rods of different lengths can be stood on their ends in pairs and compared to determine which is taller or shorter. Ribbons of different lengths can be laid side by side in

[3] Mary Baratta-Lorton, *Workjobs* (Menlo Park, Calif.: Addison-Wesley Publishing Company, Inc., 1972).

[4] Baratta-Lorton, pp. 50–51.

pairs to determine which is longer or shorter. Collections of cards and boxes of different sizes provide experiences with surfaces and volumes.

These early experiences should be accompanied by questions to develop children's vocabularies. "Which rod is *taller?*" "Which ribbon is *shorter?*" "Which card *covers more* of the red paper?" These and similar questions help children refine their vocabularies by stressing key words. Rather than learn about only *bigger* and *smaller,* children learn to compare things in different ways according to the objects' characteristics.

**b. Classification.** Many different sets of materials have been developed to give children experiences with classification activities. One set of blocks has sixty pieces, with five shapes, three colors, two sizes, and two thicknesses (Figure 5-1). There are many ways children can classify these plastic blocks: by a single attribute—shape, color, size, or thickness; by two attributes—shape and color, size and shape, and so on. Large plastic loops or rope rings form areas in which children can place the blocks as they sort them.

You can make collections of inexpensive materials for classification activities. A box of old buttons offers many attributes—size, color, number of holes, shapes, and the materials of which they are made. Different sizes and colors of plastic autos, airplanes, train cars, and boats invite children to spend long periods of time sorting them. Geometric shapes can be cut from colored railroad board, which can be bought at art supply stores, with all the attributes of the commercial set except thickness. If you put in a middle-sized shape of each color and shape, you will have a forty-five-piece set.

**Figure 5-1** Samples of shapes in a set of commercial attribute materials

A number of books containing classification activities and games have been published. These books contain activities for all grades of the elementary school:

1. Z. P. Dienes and E. W. Golding, *Learning Logic, Logical Games* (New York: Webster/McGraw-Hill Book Company, 1971).
2. Elementary Science Study, *Teacher's Guide for Attribute Games and Problems* (St. Louis: Webster Division, McGraw-Hill Book Company, 1968).
3. Nuffield Mathematics Project, *Logic* [London: John Murray (Publishers), Ltd., 1972]. (This book is sold in the United States by John Wiley & Sons, Inc., New York.)

c. **Patterns.** When children work with patterns, they sharpen their perception and develop awareness of order, sequence, shapes, and aesthetics. They can also be introduced to simple problem cards through pattern activities. For example, each problem card can show a pattern which a child repeats on a pegboard, with beads on a lace, or with other pattern materials. A kindergarten teacher, Ms. Cain, created a pattern center at a large circular table. She used four different pattern materials at seven stations—colored beads and lace, pegboards with colored pegs, pattern blocks, and some pattern materials she made. The beads, pegboard, and pattern blocks are commercial materials, and her pattern shapes were circles, squares, triangles, and rectangles cut from colored railroad board.

She put problem cards at the stations to guide the children's pattern-making activities. Samples of the cards are shown in Figure 5-2. Ms. Cain and her aide looked in on the children from time to time and gave them assistance when it seemed advisable. As each child finished a pattern, the teacher or aide helped him or her compare it with the pattern on the problem card. Later, the children were encouraged to make their own patterns to show their classmates.

d. **Spatial Relationships.** Many play activities lead to preliminary understandings of spatial relationships. A game of follow the leader can lead children *around* the blue-flowered bush, *under* the low monkey bar, and *through* the large cement pipe. In the classroom, children can play "follow the directions": "Go stand *behind* the paint easel." "Stand *inside* the ring formed by the red hula hoop." "Place the yellow block *on top* of the green block."

Later, concepts of *far, near, high, low, left, right,* and so on can be developed by similar games.

These activities for beginners in mathematics do not exhaust the things you can do to lay a firm foundation for later mathematics learning. If you work with children still in the preoperational stage, you are encouraged to use some of these books for additional ideas:

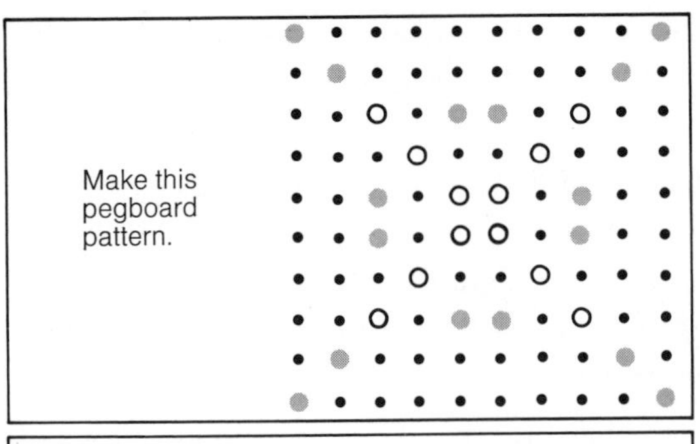


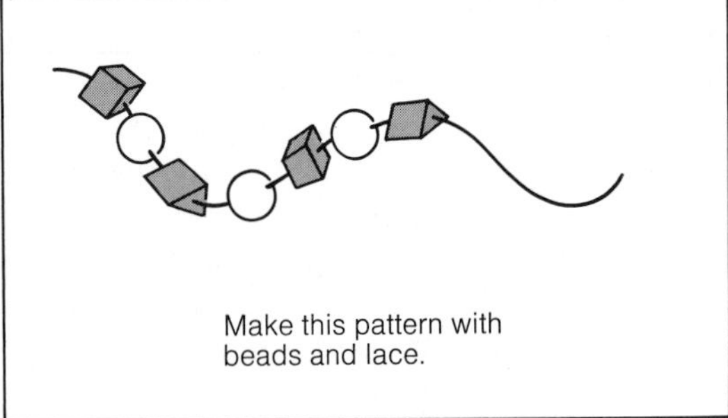

Make this pattern with
beads and lace.


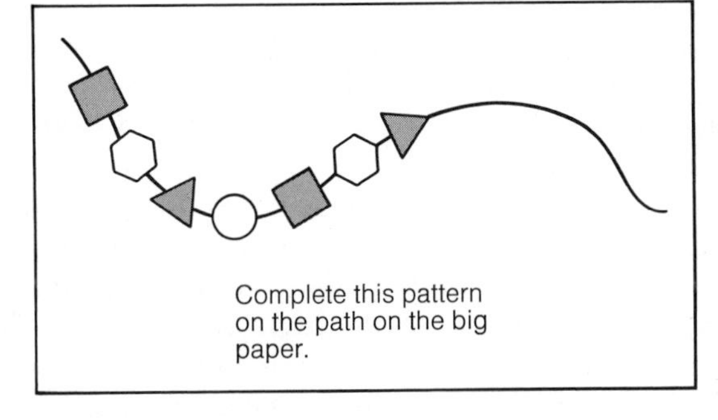

Complete this pattern
on the path on the big
paper.


**Figure 5-2** Samples of pattern cards

1. Baratta-Lorton, Mary, *Mathematics Their Way* (Menlo Park, Calif.: Addison-Wesley Publishing Company, 1976).

2. ———, *Workjobs* (Menlo Park, Calif.: Addison-Wesley Publishing Company, 1972).

3. ———, *Workjobs II* (Menlo Park, Calif.: Addison-Wesley Publishing Company, 1978).

4. Copeland, Richard W., *Math Activities for Children* (Columbus, Ohio: Charles E. Merrill Publishing Company, 1979).

5. Holt, Michael, and Zolton Dienes, *Let's Play Math* (New York: Walker and Company, 1973).

6. Nuffield Mathematics Foundation, *Mathematics, the First Three Years* [London: John Murray (Publishers), Ltd., 1970]. (The

Nuffield Foundation books are published in the United States by John Wiley & Sons, Inc., New York.)

7. Payne, Joseph N., editor, *Mathematics Learning in Early Childhood*, 37th Yearbook (Reston, Va.: National Council of Teachers of Mathematics, 1975).

8. Thyer, Dennis, and John Maggs, *Teaching Mathematics to Young Children* (New York: Holt, Rinehart and Winston, Inc., 1971).

---

## Self-check of Objective 2

Give several examples of both discrete and continuous objects. Describe precounting activities children should have with both types of objects. Also, describe at least one activity for each of these: classification, pattern, and spatial relationships.

---

## SURVEY TEST FOR KINDERGARTEN AND FIRST GRADE

Because kindergartners and first graders differ in their understandings, skills, and interests, you should make a survey of each child's background before you plan their first number experiences. The survey test that follows is one you can use intact or adapt for your class. You should administer the test yourself, since the manner of response, not simply the answer, is revealing. Another person should manage the class so the testing can be done away from the classroom, if possible. If not, a folding screen can be set up in a corner of the room to lessen distractions.

Standardized tests are available,[5] but teacher-made tests can be just as informative. The test should not require any reading or writing; all directions should be given orally; and children should respond orally or by pointing. Responses can be recorded on a 5 × 8 file card dittoed to resemble the one shown in Figure 5-3. The test need not exceed 10 minutes. An effective test must appraise a child's present level of the skills and understandings to be developed during the year.

### 1. Counting

To test a child's ability to count, use a magnetic board or desk top separated into halves (Figure 5-4). Put twenty markers on one side, and instruct the child to count as he moves them one at a time to the other side.

[5] For example, see *KeyMath Diagnostic Arithmetic Test* (Circle Pines, Minn.: American Guidance Associates, 1976).

Figure 5-3 Sample record card for individual mathematics inventory test

Mathematics Inventory Test

Name __________ Grade __________ Teacher __________ Date __________

1. Counting __________
2. Recognition of quantity 4 6 3 7 2 8 5 9 1
3. Comparison of sets $5 > 3$ $2 < 3$ $4 = 4$ $7 < 9$
4. Numeral recognition 4 6 3 7 2 8 5 9 1
5. Order relation of numbers 4 6 Sequence 1 to 9
6. Identifying numerals and sets 4 6 3 7 9
7. Recognition of ordinal property of numbers first last second middle

Demonstrate the process if the child hesitates too long. Note on the record card the highest number of markers the child moves and counts correctly.

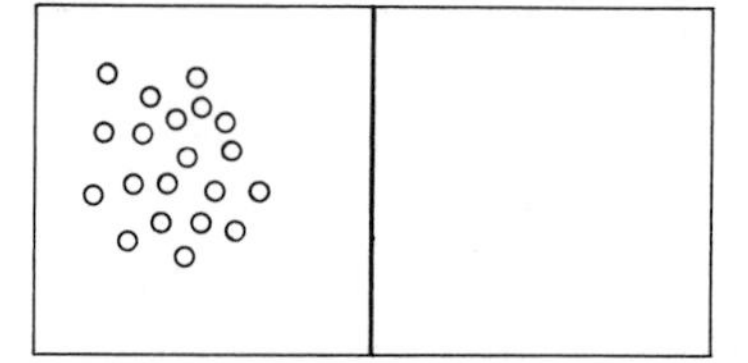

Figure 5-4 Magnetic board set up to test a child's ability to count

## 2. Recognition of Quantity

Arrange a set of cards, like those in Figure 5-5, in the order the numerals appear on the record card. Show each card and ask the child to tell how many dots are on it. If the child answers without hesitation, draw a ring around the numeral on the record card. If the child counts the dots before answering, put a ring around the numeral and mark a "c" beneath it. If the child gives no answer, or an incorrect one, make no mark on the record card.

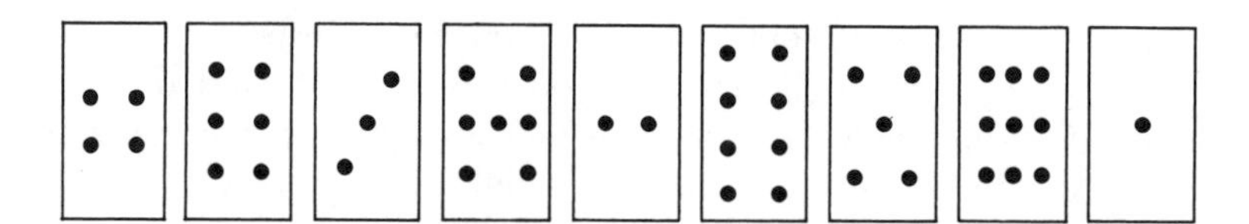

Figure 5-5 Cards for testing child's ability to recognize quantities

## 3. Comparison of Quantity

Use the same cards in this test. Show the child two cards, each with a different number of dots. Ask him or her to point to the card having the greater number of dots, and circle his or her answer on the record card if it is correct. If the child does not respond, or gives an incorrect answer, do not mark the card. To test whether a child recognizes equivalent quantities, show two cards having the same number of dots, such as cards having four dots each. Ask the child to point to the card with more dots on it. If the child says they are the same, circle the "4 = 4" on the record card. If the child points to one or the other set of dots or does not respond, do not mark the record card.

### 4. Numeral Recognition

Arrange numeral cards in the order indicated on the record card. Ask the child to say the numeral as you show each card. Circle each correct answer.

### 5. Order Relation of Numbers

Use the same cards that test recognition of quantity. Place in order the cards showing one, two, and three dots, leaving enough room at each end for one more card. Show the child the remaining cards and ask him or her to put the card that should come next in its place. If the child succeeds without counting, mark a circle around the "4" on the record card. If he or she responds correctly but must count, circle the "4" and put a "c" beneath it. If the answer is wrong or no response is made, do not mark the card. Next, put the cards with seven, eight, and nine dots in order, again leaving room for a card at each end. Ask the child to select the card that should go before the card with seven dots and put it in its proper place. Record the response. Finally, put all the cards on the table in random order and instruct the child to place them in numerical order beginning with the card having one dot on it. Write on the record card the numeral that tells how high the sequence is correctly ordered.

### 6. Identifying Numerals and Quantities

Use the dot cards and numeral cards. Put the dot cards in numerical order on the table. Then put the numeral cards in a pile in random order and instruct the child to look at each numeral card and then point to the equivalent dot card. If the child can answer without counting, mark a circle around the numeral on the record card. If the child counts before responding, circle the numeral and put a "c" beneath it.

### 7. Recognition of Ordinal Property of Numbers

Use picture cards like those in Figure 5-6. Show the picture that has four fish on it and call to the child's attention the direction the fish are facing. Ask the child to point to the first fish. If he or she gives the correct response, mark a circle around "first" on the record card. Do not mark the card if the answer is incorrect. Use the other cards to test the child's recognition of "last," "second," and "middle."

**Figure 5-6** Cards for testing child's understanding of ordinal property of numbers

## Self-check of Objective 3

A sample survey test for preschool, kindergarten, and first-grade children is illustrated and described. Prepare several similar test materials for a survey test. Describe the skills your materials test, and tell how to use them. Why should the children answer orally or by pointing to the test materials?

## FIRST NUMBER EXPERIENCES

A child's early classroom exposure to numbers is a result of those activities planned by you, and others that arise spontaneously. Of course, you plan lessons, storytelling sessions, rhythmic activities, and other mathematics-oriented exercises. In addition, you can arrange to emphasize relationships useful in mathematics during the time primarily devoted to seemingly unrelated activities. For example, children may be asked to count easels during art class, napkins and chairs during snack time, and to compare these counts with the number of children in class. The sizes of objects can be compared during many activities; for example, you might ask which of three things is the smallest or largest, or which jump rope is the longest. Planned mathematics activities can play a much greater role in the child's day than the time officially allotted to them would suggest.

Spontaneous events in the classroom frequently lead to opportunities for activities or discussions that involve numbers. During sharing periods, children often bring cars or dolls, or books that have stories or poems containing numbers or telling about events such as the birth of kittens. You should take full advantage of these situations to engage children in meaningful discussions of numbers.

## 1. Introducing First Counting Experiences

Children's first experiences with counting should be with objects that they can see, manipulate, compare, order in sequence, and finally, count. Collections of suitable objects can be gathered easily and inexpensively.

When there is evidence that several children are ready for activities leading to counting, you can begin work in a familiar setting. Prepare a table with napkins, small paper plates, cups, and plastic utensils. Do not have the same number of each object. Through discussion, guide the children's thinking. For example, ask: "What do you see on the table?" You can direct the children's attention to the different objects and help them describe the characteristics of each one—color, size, shape, texture. You can also introduce the word "set" as one that can be used to identify each collection—"set of dishes," "set of spoons," and so on. (While the use of set concepts and symbolism is not as extensive as formerly, it is helpful for children to consider objects in sets as they learn to count.)

This is an opportune time to have the children match collections of objects and compare the sets' sizes without counting. They can see that by matching one-to-one the objects in one collection with those of another, that there are as many plates as napkins, or more forks than spoons. Because children at this stage do not always reverse their thinking, help them by asking several questions as they compare pairs of sets. "Are there as many plates as napkins?" "Are there as many napkins as there are plates?" "Which set has more, the set of forks or the set of spoons?" "Which set has fewer, the set of forks or the set of spoons?" Do not attempt to have the children use numbers to compare groups of objects at this time.

Children should have other activities that are similar to this introductory lesson. Many kinds of objects, frequently called markers, are available. Commercially prepared disks with magnets for magnetic boards and felt cutouts for flannel boards are attractive and useful for teacher-directed lessons with groups of children. In addition, tongue depressors, popsicle sticks, bottle caps, and small plastic toys are free or inexpensive and readily available. Plastic margarine tubs are suitable containers for children's individual collections of markers.

The research of Piaget and others reveals that children are sometimes confused when the markers in two sets are of disparate size. If markers in one set are noticeably smaller than those of a set with fewer elements, as in Figure 5-7, children tend to be uncertain which is the numerically larger set. Disparate spacing is another stumbling block. Some children tend to think of a set that is spread over a greater area as being larger than a set that is confined to a smaller area, even though a one-to-one correspondence between elements might have been previously established.

Read familiar stories, such as *Snow White and the Seven Dwarfs* and Dr. Seuss' *Cat in the Hat*, and talk about the groups of people and things

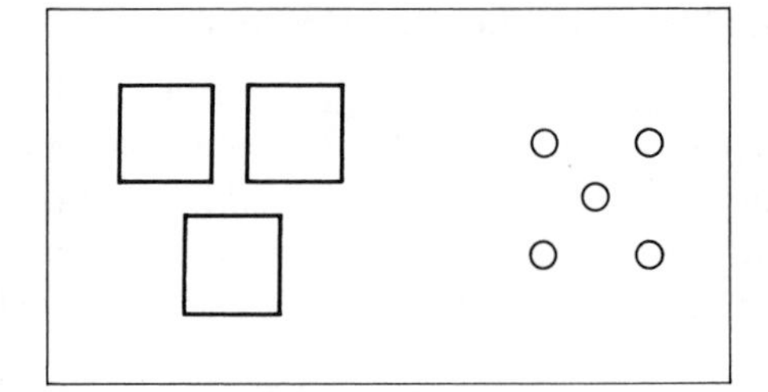

**Figure 5-7** Pairs of sets with markers that vary too much in size are likely to confuse children

described in them. If you use cutout characters from the stories on a magnetic or flannel board, children can manipulate the figures as they group them in various ways and compare the quantities by matching the different groups.

## Self-check of Objective 4

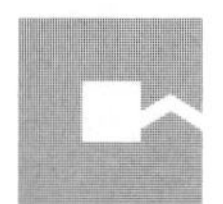

A setting for introducing counting is described. Describe another setting of a similar kind.

### 2. Learning to Count

Counting is based on a one-to-one correspondence between the objects in a set and the set of natural numbers $\{1, 2, 3, \ldots\}$. Children exhibit two levels of maturity when counting. Some children will be able to match each object in a set with cards containing an ordered sequence of number names, but will count the objects before naming their number. More mature children will match the objects to the number names, then will name the number of objects by looking at the last number in the ordered sequence. That number is the number of objects in the set. In each instance, the *cardinal number* of the collection is identified. An understanding of this process is built upon a child's intuitive understanding of the number "one." There is no way to describe the meaning of "one" other than by holding up one finger, by pointing to one marker on a magnetic board, or by isolating a single object in some other way. When an intuitive understanding of "one" is established, two can be thought of as "one-more-than" one, three as "one-more-than" two, and so on, until a child realizes that each number has a successor.

Many children come to school with the ability to count objects in groups up to ten or more. These children do not need to spend time on planned activities to develop this skill. Those who do not count need opportunities to learn to do so in meaningful ways. Their initial counting activities should involve small numbers of objects. Begin with two, such as two markers on a magnetic board. Ask them to name the number of markers. When it is established that there are two, direct the children to look about the room to identify other "groups of two." Those who cannot should be given clues by directing their attention to appropriate sets.

Children can be introduced to "three" by building on a set of "two." Below the two markers, arrange a second set of the same size for the children to identify, as in Figure 5-8. Then place a disk of another color beside it and say, "I have put one more disk with a set of two disks on the board. How many disks are in the new set?" The children should identify it as a set of "three," and other sets in the room should be identified. If your classroom does not already contain enough sets of three objects, you should arrange them beforehand in conspicuous places.

Sets for the rest of the numbers through nine should gradually be developed and studied in a similar manner. Figure 5-9 shows how a magnetic board or chart with geometric shapes might appear after all nine numbers have been studied.

There are many activities for counting centers you can set up once children have begun to count.[6]

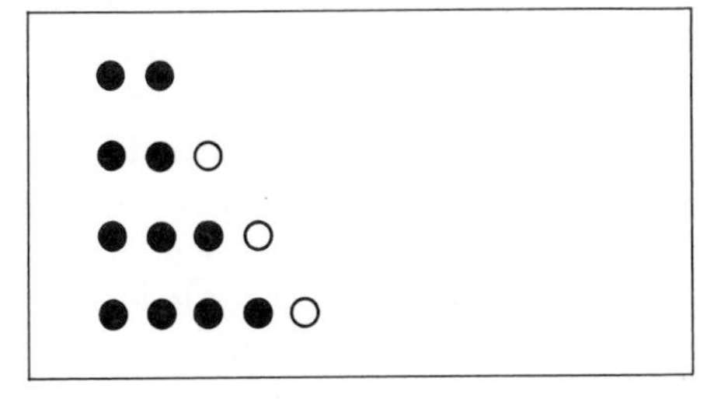

**Figure 5-8** Markers for teaching children to count

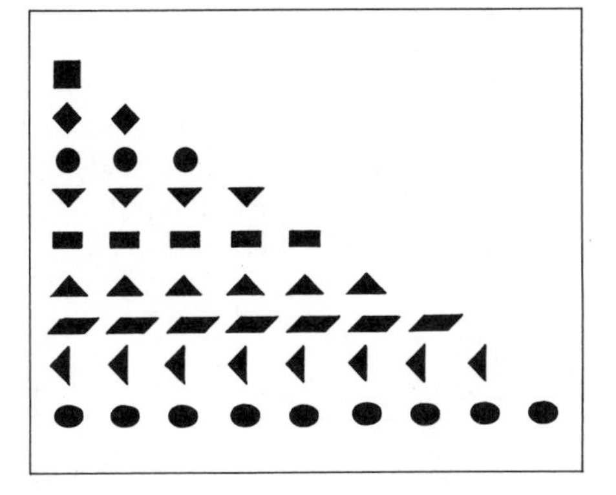

**Figure 5-9** Geometric shapes used on a sequence-of-sets chart to illustrate the "one more than," or successor, pattern

[6] For example, see Section One of Leonard M. Kennedy, *Models for Mathematics in the Elementary School* (Belmont, Calif.: Wadsworth Publishing Company, 1967); Chapter 1 of Mary E. Platts, ed., *Plus*, rev. ed. (Stevensville, Mich.: Educational Services, Inc., 1975); and "Sets" and "Number Sequences" in Baratta-Lorton, *Workjobs*.

Zero can be introduced through a story situation similar to this: Put three felt flowers on a flannel board and say, "These are some flowers that were growing in my garden. How many flowers are there?" (Three.) As you remove one flower, say, "I went to my garden last Saturday and picked one flower. How many were left?" (Two.) Remove the two flowers and say, "I picked two more flowers on Sunday. How many were left?" Children will respond to the question in various ways, such as "There aren't any left," "None," and "They are all gone." Accept all of their responses. Then show them several empty containers, such as bottles and boxes, and ask the children to describe what they see in them. Some will say the containers are empty. At this point, introduce the word zero as the name for the number that tells how many things there are in each of the empty containers.

The initial steps of learning to count take place before children are taught to write the numerals. It is not necessary, however, to teach the meaning of all the numbers 0 through 9 before beginning instruction in reading them. After children have learned the meaning of two, three, and four, the numerals "2," "3," and "4" can be introduced.

Next, children can correlate numerals with their sets through exercises with markers. Magnetic or felt numerals, for use on boards, are particularly useful because children can put a numeral with a set without writing. Placing the numeral with its set, and placing the correct number of markers next to a numeral are the most basic and simplest exercises.

As the "one-more-than" concept is developed, numerical order can be stressed in the correlating activities. A pocket chart such as the one in Figure 5-10 is filled on one side with set cards in numerical sequence, while numeral cards are in random order in front of the chart. As other children serve as "checkers," one child should be called to the chart to place a numeral card in position. Initially, the chart should be arranged so the numeral cards are in sequence beginning with 1. Later, they can be arranged in reverse sequence. After this, the numeral cards can be arranged on the chart, while the set cards are the ones to be placed by the children. Still later, the beginning point can be made at neither end, but at 5, for example, and the children instructed to put in the card that names the number before five, and then after five, and so on.

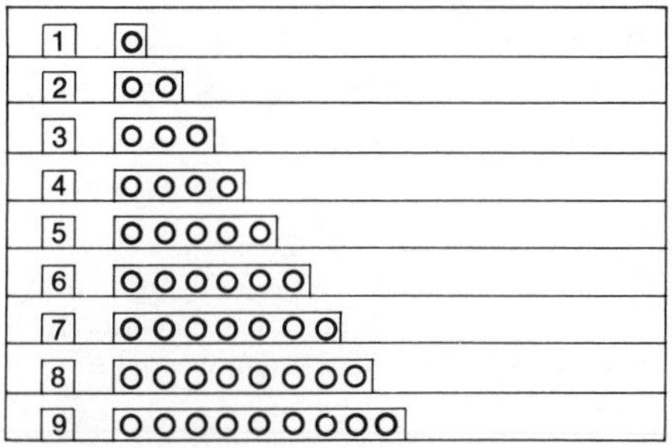

**Figure 5-10** Pocket chart with set strips and numeral cards

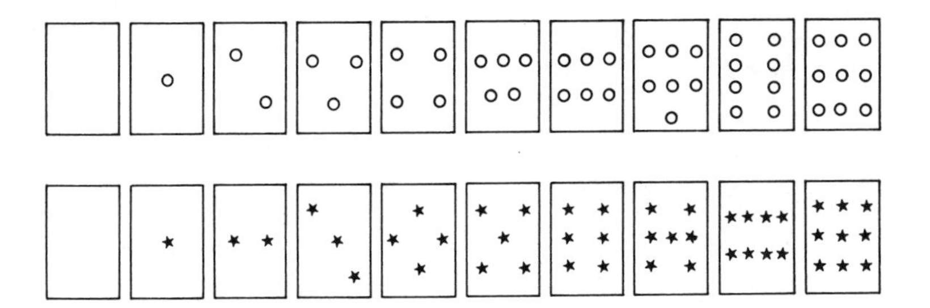

**Figure 5-11** Suggested patterns for domino grouping cards

Following work with numbers shown by the "one-more-than" pattern, children should work with shapes arranged in domino patterns. You can make cards showing shapes arranged in various patterns, as in Figure 5-11. These cards are cut from colored railroad board or another durable material; the shapes are self-adhesive stickers. There is no best way to arrange the shapes; in fact, several arrangements should be made. Make a set of numeral cards to match the domino cards. Use the cards to help children learn the sequence of numbers and to match numerals with domino cards. Encourage children to name the number of shapes on a card by sight, using the pattern as well as by counting.

Another useful set of materials is the set of circle cards shown in Figure 5-12. Each circle is cut in half so children can match the proper halves. Also, have children reproduce sets with markers that match a number you name orally or show by a numeral card. While they are working, have children make their matchings two ways: (1) numeral cards placed with pattern cards or groups of markers, and (2) pattern cards or groups of markers placed with numeral cards. Once you are sure a child can count accurately, he or she can complete practice pages involving counting assigned from a workbook.

Instruction in reading number words should follow instruction in reading numerals. Although these words are usually taught during reading lessons, they should also be taught as part of the mathematics program. Many of the exercises used for associating numerals and pattern cards can be used to learn the words and their meanings. The children should also match cards, numerals, and words by arranging cards showing

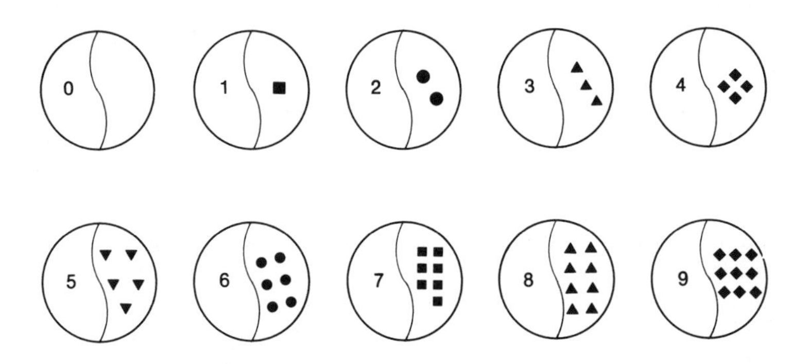


**Figure 5-12** Circular cutouts for matching sets and numerals

each of these in a pocket chart, as shown in Figure 5-10, or on a flannel or magnetic board.

---

## Self-check of Objectives 5, 6, and 7

Prepare a set of materials (include objects and numeral cards) that can be used to introduce children to the process of counting.

The flower-garden situation is only one of many that can be used to introduce zero as a number. Children need more than one activity dealing with the concepts they learn. Make up a similar story and describe the materials you would use with it.

Several devices for providing experiences with counting are illustrated and discussed. Describe at least five such devices, including two besides the ones included here.

---

### 3. Learning to Write Numerals

The manuscript form of writing numerals is used almost exclusively, even though most people change to the cursive form for writing words (Figure 5-13).

1 2 3 4 5 6 7 8 9 0

(a)

1 2 3 4 5 6 7 8 9 0

(b)

**Figure 5-13** (a) The manuscript form for writing numerals and (b) the cursive form

The write-in textbooks used in the first grade in many schools usually include instruction pages on which children can practice writing. A lesson commonly begins with an example of a numeral. Next, the numeral is made of broken lines or rendered in light-colored ink for children to trace. Then the starting point alone appears, showing the child where to begin his or her own numeral. Last, the textbook has spaces in which the child is to practice writing the numeral with no guidance. Figure 5-14 shows a typical exercise.

The authors of the CDA Math Program[7] advise against requiring children to trace ready-made numerals. They have observed that many children constantly check their progress when they are required to trace

[7] Morton Botel, Robert W. Wirtz, and Max Beberman, Memo from the Authors of CDA MATH: *Introducing Children to the Primary CDA Math Program* (Washington, D.C.: Curriculum Development Associates, Inc., 1973), pp. 10–11.

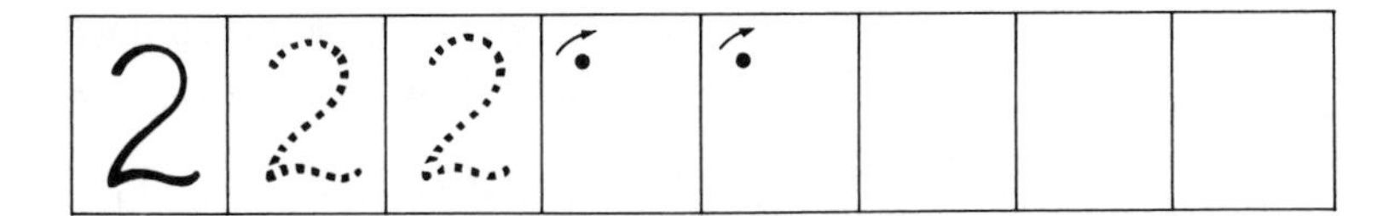

**Figure 5-14** The sequence of steps for instructing children to write a numeral

models and become upset when they stray too far from the models. The authors recommend that children be given sheets of paper that contain four different numeral models. A sample of one practice page is shown in Figure 5-15. A numeral can be introduced as soon as the individual can count the indicated number of objects. A child practices by writing the numerals in the open spaces in each numeral's box. If he or she wishes, a child can trace the models.

As children gain experiences in writing numerals, most of them will make the numerals correctly and neatly during practice sessions. However, some write the numerals carelessly when another phase of mathematics is emphasized. Usually, reminding a child how a particular numeral is formed is all that is needed to prevent a bad habit from developing. If a chart showing the correct way to write each numeral is posted in the room, a child who makes an error can look at it to see how a particular numeral should appear. The child can go to the chart to trace a numeral a few times with his or her finger before writing it on paper.

If a child is unable to write numerals correctly, his or her specific problem should be identified so it can be remedied. Some children may have poor vision that can be corrected with glasses. Some may not have attained the necessary control over the small muscles used for writing. When this is true, there is little that can be done but wait until the child matures. If a child reverses numerals, you should stress the proper configuration of the numerals with which he or she is having difficulty. In severe cases, a child might need the kinesthetic experience of tracing with his or her fingers numerals cut from coarse sandpaper. Some dif-

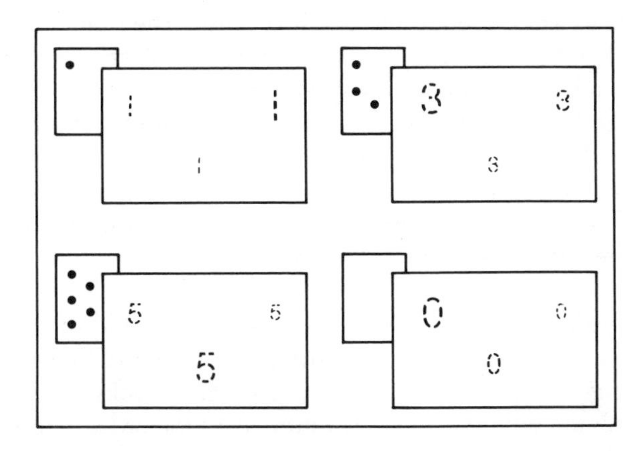

**Figure 5-15** Sample numeral-writing practice page recommended by CDA MATH authors, p. 11 (Used by permission.)

ficulties may be a symptom of a major physical or psychological problem. When you suspect this, you should refer the child to a specialist for help.

The importance of writing carefully should be stressed as children proceed through the grades. Otherwise, their numerals may become habitually poorly formed. You can encourage carefulness by having children proofread their papers before they turn them in, and by refusing to accept carelessly done work.

If children are to write on the chalkboard during activities, they should practice writing numerals there as well as on paper. There is no established standard as to how large the numerals should be, but they should not be extremely small or large. At the chalkboard, a base line and a top line should serve as guides for numeral size. If permanent guides are not marked, and the children are unable to make their own, you should mark them.

---

## Self-check of Objective 8

Describe the two numeral-writing practice pages discussed here, and tell why the authors of CDA MATH prefer theirs.

---

### 4. Learning the Ordinal Uses of Numbers

Number is used in the ordinal sense when the order of objects in a set is established. There are two ways of classifying order which children need to learn. The first is order according to *position*—first, second, or third. The second is to use numbers to identify a year, street address, or ballplayer. When used in the second way, the numerals involved are read differently than when used in the cardinal sense. The numeral in a street address, such as 6764 Elm, is read as "sixty-seven, sixty-four" rather than as "six thousand, seven hundred sixty-four."

Children will usually be instructed in the first way of classifying order before they learn to use the second way. Exceptions occur when they learn to use page numbers in books, or when they are taught their telephone numbers and addresses. However, children often memorize the latter two without the ability to recognize them in written form.

Most textbooks for the primary grades instruct children in the use of ordinal words. However, you should bring up the use of these words and their meanings whenever children line up, when books are ordered according to volume number, and as children learn the order of letters in the alphabet. Children often learn ordinal concepts so well through these incidental experiences that textbook material becomes unnecessary.

---

## Self-check of Objective 9

What are the differences between the cardinal and ordinal uses of numbers? Describe several ways numbers are used in the ordinal sense.

---

### 5. Using the Number Line to Relate Sequence and Counting

The number line is one of the most useful devices for helping children learn the sequence of numbers and counting. Children in kindergarten or first grade can be introduced to the number line by using one that is marked on the floor of the classroom. Such a line can be painted or it can be put down with masking tape. Its points should be about 14 inches apart, and the numerals that name the points should be printed on cards rather than on the floor to increase the number line's versatility.

When children use this number line for the first time, they should walk the length of it, counting their steps from beginning to end. The rest of the children should count silently or in unison with the child on the line. The observers should stand so that movement is from left to right. You must be certain that children count steps as they go from 0 to 1, 1 to 2, and so on, rather than points along the line. Otherwise, children are likely to say "one" as they step on the point marked 0, "two" as they step on 1, and so on. Later, children can order the numeral cards in sequence along the line, fill in unnamed points, and learn to count by twos, threes, and other numbers.

Next, children should use a number line placed on the chalkboard. It can be purchased, or made from a length of adding machine tape which extends as far as the chalkboard allows. The line should be within reach of the children so they can point to the numerals and indicate steps along it with chalk. Each child should also have a small number line attached to his or her desk for individual use.

### 6. Number Rhymes, Songs, and Finger-Play Activities

Number rhymes, songs, and finger-play activities familiarize children with number names and their sequence. Children enjoy a poem like "Five Years Old,"[8] because of its rhythmic sound, the chance for finger play, and because it makes them feel quite grown-up.

[8] From Marie Louise Allen, *A Pocketful of Rhymes* (New York: Harper & Row, Publishers, Incorporated, 1939).

*Please, everybody, look at me!*
*Today I'm five years old, you see!*
*And after this, I won't be four,*
*Not ever, ever, any more!*
*I won't be three—or two—or one,*
*For that was when I'd first begun,*
*Now I'll be five awhile, and then*
*I'll soon be something else again.*

Nursery rhyme books, poetry anthologies, teacher's manuals for some mathematics series, and journals such as *Instructor* are good sources of rhymes and poems.

Songs, such as "One Little Brown Bird,"[9] are enjoyed by children in kindergarten and first grade. The verses of many of these songs give the numbers in sequence. The first three verses of "One Little Brown Bird" are:

*One little brown bird, up and up he flew,*
*Along came another and that made two.*

*Two little brown birds, sitting in a tree,*
*Along came another one and that made three.*

*Three little brown birds, then up came one more,*
*What's all the noise about? That made four.*

Children's records are excellent sources of songs for early number work. Most children listen intently to such songs, eagerly awaiting the chance to join in when the record is replayed.

"Chickadees"[10] is an example of a finger-play activity that can be used with children in the kindergarten and first grade:

*Five little chickadees sitting on the floor;*
(Hold up hand, fingers extended.)

*One flew away and then there were four.*
(Fold down one finger as each bird flies away.)

*Four little chickadees sitting in a tree;*
*One flew away and then there were three.*

*Three little chickadees looking at you;*
*One flew away and then there were two.*

[9] *Sixty Songs for Little Children* (London: Oxford University Press, 1933). (Used by permission of Oxford University Press.)

[10] Marion F. Grayson, *Let's Do Fingerplays* (Washington, D.C.: Robert B. Luce, Inc., 1962), p. 60. (Reprinted by permission of Robert B. Luce, Inc.)

*Two little chickadees sitting in the sun;*
*One flew away and then there was one.*

*One little chickadee sitting alone;*
*He flew away and then there was none.*

Finger-play activities can be found in teacher's manuals, the journal mentioned above, and Scott and Thompson's *Rhymes for Fingers and Flannelboards*,[11] in addition to Grayson's *Let's Do Fingerplays*.

---

## Self-check of Objective 10

Explain why teachers of young children should include number songs, finger-play activities, and stories and poems in their mathematics program.

---

## Common Pitfalls and Trouble Spots

There are four common pitfalls at the level of beginning instruction. First, teachers sometimes assume that children are ready for number work as soon as they enter kindergarten or first grade. Evidence points to the fact that children who are still in the less mature levels of Piaget's preoperational stage, that is, those still making global comparisons and those still uncertain about the lasting equivalence of quantities, are not ready for number work. To avoid having children do work for which they are not ready, administer Piagetian-type tests. Some are described on pages 49 to 51 of Chapter 3 and 399 to 401 of Chapter 14. Give children who cannot yet conserve quantity extensive opportunities to engage in activities like those described in the "Early Experiences" section of this chapter.

Second, children's ability to recite number names in sequence is often misinterpreted as evidence of their understanding of the meaning of numbers. However, children who can count to ten or higher by rote do not always know the meaning of the number names they recite. Administer the mathematics inventory test on pages 107 to 110 to determine the extent of your children's understanding of numbers. Give those who cannot count opportunities to learn this skill through many activities with markers, domino and numeral cards, and other suitable materials.

Third, teachers will present children only one or two teacher-directed lessons with manipulative and/or pictorial materials before assigning paper-and-pencil activities. To avoid having children do paper-and-pencil work for which they are not ready, have each one individually demonstrate his or her understanding of one-to-one matching, more than, less than, counting, and other skills with manipulative materials. Have those who are uncertain continue to work at the concrete-manipulative level until they can do the demonstrations without hesitation or error. Then have them work at the semiconcrete-pictorial level for awhile before beginning their paper-and-pencil work.

[11] Louise B. Scott and Jesse J. Thompson, *Rhymes for Fingers and Flannelboards* (St. Louis: Webster Publishing Company, 1960).

Fourth, a kindergarten-level workbook is used as *the* program. If the book is used exclusively, the children will have had no concrete-manipulative experiences, since workbooks begin children's experiences at the semi-concrete stage. When workbooks are used, care must be taken to precede each of the book's activities with appropriate exploratory activities with manipulative materials. Workbooks are not recommended at all for the many kindergarten children who are still in the preoperational stage of development.

Young children may experience trouble with two aspects of their early number work: The first is that some children have difficulty recognizing on sight the size of a set, even when there are only two or three objects. The inability to recognize the quantity of even small sets hampers later work, such as beginning addition. Children who exhibit this weakness can be helped if you have them work with small groups of objects and with domino cards. Use cards similar to those in Figure 5-11, which you flash for a second or two, then ask the child to name the number of dots. If the child fails to name the number, tell him or her how many there are. Flash the card again and have the child name the number. Repeat this activity from time to time until the child readily names the number of objects in each set. With practice, children can recognize on sight sets of as many as nine objects arranged in regular patterns.

The writing of numerals is often a second trouble spot for some children. Numeral reversals are the most common errors committed as children learn to write them. Sandpaper numerals have already been suggested as an aid for overcoming this problem. Other aids include:

1. Shallow boxes with a half-inch of clean sand covering the bottom in which a child may write numerals with his or her fingers.
2. Finger paints, with which the child may write numerals.
3. Numerals written on the chalkboard for a child to trace with his or her forefinger.

Do not give children directions based on objects in the room, such as, "Begin the two on the side next to the windows." A shift in seating arrangements can confuse a child who has learned according to such cues. Have a child practice only one numeral at a time.

---

## Self-check of Objectives 11 and 12

Four pitfalls associated with early number work are described. Identify at least three of them, and tell how each one can be avoided.

Name two trouble spots associated with early number work, and tell how you can help children overcome them.

---

## SUMMARY

Mathematics begins for most children before they begin school. Their early play experiences provide a basis for later learning. They should continue to have gamelike activities in preschool and kindergarten classes. Many of their activities can concern discrete and continuous objects, and classification, patterns, and spatial relationships. When children are ready for counting, they must have many experiences with a variety of manipulative materials. Carefully planned activities will enable children to understand the meaning of counting and the way numerals are used to record the number of objects in a collection, or set. Once children know the meaning of a number, they are ready to write its numeral. There are two forms of writing numerals, the manuscript and the cursive. Manuscript numerals are used most often. Children learn to write numerals by tracing patterns or making copies alongside already written models. Simple finger-play activities, songs, stories, and poems involving numbers contribute to children's background for learning mathematics. There are four pitfalls to be avoided at this level: (1) having children do number work for which they are not cognitively ready, (2) confusing rote counting with rational counting, (3) having children do paper-and-pencil work too soon, and (4) using a workbook as the only program. Some children have trouble recognizing on sight the sizes of even small groups of objects, and/or make reversals as they write numerals. There are materials and activities you can use to overcome most children's troubles with either of these aspects of early number work.

## STUDY QUESTIONS AND ACTIVITIES

1. Administer a mathematics inventory test similar to the one in this chapter to several five- and six-year-old children. What differences, if any, are revealed in the children's understanding of the number concepts tested? What different materials, if any, are needed in order to provide for the needs of different children?

2. Collect materials suitable for preschool and kindergarten activities. Include at least one for each of these topics: classification and sorting, patterns, spatial relationships, and rational counting. Identify the potential uses of each of the materials.

3. Begin a collection of counting aids to be used as markers. Include objects both for flannel and/or magnetic boards, and for the children's use at their work stations. What are the characteristics of good demonstration aids? What should you keep in mind as you select aids for children?

4. Assist children who have some difficulty learning to write number symbols. What activities prove to be most helpful? Does everyone benefit equally from the same activities?

5. Analyze the written work of children in one of the intermediate grades. What is your reaction to how they write numerals? If you think they could improve, how would you help them?

6. Begin a collection of number rhymes, songs, and finger-play activities.

## FOR FURTHER READING

Brace, Alec, and L. Doyal Nelson, "The Preschool Child's Concept of Number," *The Arithmetic Teacher*, XII, No. 2 (February 1965), 126–133. Understanding of counting, comparison, conservation of number, cardinal and ordinal properties of numbers, and place value was tested in 124 five- and six-year-olds. The authors conclude that the preschool child's ability to count is not a reliable indicator of the development of a true concept of number, that no difference exists between boys and girls in development of the concept of number, and that children from higher-socioeconomic-level homes are superior in number knowledge to children from lower-socioeconomic-level homes.

Bruni, James V., and Helene Silverman. "Making and Using Attribute Materials," *The Arithmetic Teacher*, XXII, No. 2 (February 1975), 88–95. Describes a 54-piece attribute set made from colored oaktag. Explains activities for children of all elementary grades.

Burton, Grace M. "Helping Parents Help Their Preschool Children," *The Arithmetic Teacher*, XXV, No. 8 (May 1978), 12–14. While intended for preschool teachers, the article suggests many experiences for children that are suitable for kindergarten and some first-grade children. Includes activities dealing with sorting and classifying, counting, geometry, and money.

Dunkley, M. E. "Some Number Concepts of Disadvantaged Children," *The Arithmetic Teacher*, XII, No. 5 (May 1965), 359–361. As a part of its work, the School Mathematics Study Group made an inventory of mathematical knowledge of children who were a part of experimental and control groups using SMSG materials. The results show that culturally disadvantaged children had not attained as high a level of understanding of the concepts that were presented as the advantaged children had.

Huey, J. Frances. "Learning Potential of the Young Child," *Educational Leadership*, XXIII (November 1965), 117–120. A brief discussion of how preschool children develop concepts and begin to understand symbols; included are ten suggestions to help young children develop concepts and generalizations.

Kurtz, V. Ray. "Kindergarten Mathematics—A Survey," *The Arithmetic Teacher*, XXV, No. 8 (May 1978), 51–53. Reporting on a survey made in Kansas, the author gives the competencies that are "clearly," "questionable," and "clearly not" for kindergarten children. He also reports that over 75% of the teachers use an adopted series as a basis for their program.

Lettieri, Frances M. "Meet the Zorkies: A New Attribute Material," *The Arithmetic Teacher*, XXVI, No. 1 (September 1978), 36–39. Zorkies, natives of the planet Zorka, differ in color, and number of eyes, legs, and arms; altogether there are 36. Includes illustrated samples of the 34 activity cards, which deal with attribute discrimination, equivalence and difference relations, ordering relations, and transformations.

Liedtke, W. W., and L. D. Nelson. "Activities in Mathematics for Preschool Children," *The Arithmetic Teacher,* XX, No. 7 (November 1973), 536–541. The authors list a variety of activities with manipulative materials. Three different types of activities are discussed: classification, conservation (one-to-one correspondence), and seriation.

Liedtke, W. W. "Rational Counting," *The Arithmetic Teacher,* XXVI, No. 2 (October 1978), 20–26. Presents the bases for developing skill in rational counting through classifying, matching, and ordering and patterns. Describes and illustrates materials and procedures for developing children's background in these skills.

Lindquist, Mary M., and Marcia E. Dana, "Make Counting Really Count—Counting Projects for First and Second Grades," *The Arithmetic Teacher,* XXV, No. 8 (May 1978), 4–11. Counting will count for children who engage in the many counting and recording activities described. The headings include: Listing Counts, Drawing Counts, Hunting Counts, Tallying Counts, Collecting Counts, and others.

Van de Walle, John A. "Track Cards," *The Arithmetic Teacher,* XXV, No. 6 (March 1978), 22–26. Teacher-made track cards provide the basis for many kindergarten and primary grade activities enhancing children's thinking, creativity, verbal skills, understanding of certain topological and symmetry concepts, vocabulary, and so on. Includes ten different types of activities, along with directions for preparing tracking cards.

Williams, Alfred H. "Mathematical Concepts, Skills, and Abilities of Kindergarten Entrants," *The Arithmetic Teacher,* XII, No. 4 (April 1965), 261–268. Eight strands of mathematics were outlined and tested by the Advisory Committee on Mathematics to the California Curriculum Commission. The results of this test, which was administered to 593 kindergarten children, indicate that young children have already developed a number of mathematical concepts, that a program of mathematics should begin in the kindergarten, that individual differences begin to manifest themselves when children are young, and that some forms of grouping might be desirable as early as kindergarten.

# 6 Extending Understanding of Numeration

Upon completion of Chapter 6, you will be able to:

1. Use markers or other objects to demonstrate one-to-many, many-to-one, and many-to-many correspondences, and describe some real-life situations that deal with these correspondences.

2. Describe the major characteristics of the Hindu-Arabic numeration system.

3. Illustrate and describe at least three place value devices, and demonstrate how they are used to help children understand the meanings of numbers greater than 10.

4. Explain why an abacus is a useful device for representing large numbers, and demonstrate how to use one to represent a number such as 647,093.

5. Distinguish between compact and expanded numerals, and write at least three different expanded numeral forms.

6. Tell a story that can be used to introduce children to integers, and illustrate how number lines can be used to clarify the meaning of these numbers.

7. Identify four pitfalls associated with place value and the meaning of numbers, and describe some ways to avoid them.

8. Describe two trouble spots children may encounter as they learn about numbers, and tell how to reduce their impact.

Key Terms you will encounter in Chapter 6:

one-to-many correspondence
many-to-one correspondence
many-to-many correspondence
Hindu-Arabic numeration system
base
place value
place value position
place value device
structured material
Cuisenaire rods
Dienes multibase arithmetic blocks
Unifix cubes
beansticks
abacus
compact notation
expanded notation
regroup
integer
directed (signed) number
vector (directed segment)
absolute value
number opposites

If children are to work effectively with the Hindu-Arabic numeration system, they must understand the system thoroughly and not simply have a rote knowledge of it. Therefore the topics—one-to-many, many-to-one, and many-to-many correspondences; the Hindu-Arabic numeration system; and integers—in this chapter must be presented in a manner that permits children to discover the fundamentals and rationale behind them.

## ONE-TO-MANY, MANY-TO-ONE, AND MANY-TO-MANY CORRESPONDENCES

Counting is a process of making a one-to-one correspondence between a collection of objects and the natural numbers 1, 2, 3, . . . It is important that children understand one-to-one correspondence as the basis for counting. It is important that they recognize that one-to-many, many-to-one, and many-to-many correspondences have useful applications, too.

Examples of one-to-many correspondences are place value (1 ten is equal to 10 ones), money (1 nickel is equivalent to 5 pennies), linear measure (1 meter is equivalent to 100 centimeters), and weight measure (1 gram is equivalent to 1000 milligrams). Many-to-one correspondences are also illustrated by place value (10 tens equal 1 hundred), money (2 nickels are equivalent to 1 dime), linear measure (12 inches are equivalent to 1 foot), and weight measures (16 ounces are equivalent to 1 pound).

Many rate-type problems are based on many-to-one and many-to-many correspondences. An example is a problem involving distance traveled. "A bicyclist can travel at an average rate of 12 miles in 1 hour. What is the time required to travel 36 miles?" The rate of travel is a many-to-one correspondence, while the answer is a many-to-many correspondence. The rate in the problem "If 2 pencils cost 15¢, how many can you buy with 60¢?" is a many-to-many correspondence, as is the answer.

Commercial and teacher-made activities will help children understand these correspondences. *Chip Trading Activities* kits[1] offer a variety of activities. A kit contains five colors of circular plastic chips, chip boards, chip tills, dice, and a teacher's guide. Many games have been devised for the chips. In one game the chips are given values; for example, 5 yellows

[1] Patricia Davidson, Grace Galton, and Arlene Fair, *Chip Trading Activities* (Fort Collins, Colo.: Scott Scientific, Inc.).

are worth 1 blue, 5 blues are worth 1 green, 5 greens are worth 1 red, and 5 reds are worth 1 black. Players take turns rolling a die. Each player gets as many yellow chips as there are spots on his or her roll. When a child has 5 yellow chips, he or she trades them for 1 blue. When a child has 5 blue chips, he or she trades them for 1 green, and so on. Play continues until a player has a black chip, which he or she gets by trading in 5 red chips. This game gives experience with many-to-one correspondence, that is, trading 5 chips of one color for 1 chip of another color. A different game begins with each player having a black chip. Players trade "down," going from black to red, red to green, and so on, until one player trades in all of his or her chips and is declared winner. This game offers experience with one-to-many correspondence. In addition to serving as a means of giving children experiences with many-to-one and one-to-many matchings, these games give them valuable experiences with the basic meaning of place value. In the games described, children work with a base five system. By changing the chips' values, other bases, including base ten, can be used.

---

## Self-check of Objective 1

Can you use poker chips, or other objects, to demonstrate each of these correspondences: one-to-many, many-to-one, many-to-many? Describe a real-life situation that deals with each correspondence.

---

### THE HINDU-ARABIC NUMERATION SYSTEM

In studying our own numeration system, children do not need to be overwhelmed with historical details. Instead, they should acquire an understanding of how the system developed.

Early development took place in India. It is likely that our present system evolved from one having twenty symbols: nine for units, nine for tens, one for hundreds, and one for thousands.[2] We know that numerals were used in a place value scheme by about 600 A.D.[3] Then the forerunners of today's numerals were developed and first appeared in written records in about 700 A.D. These are called Devangari numerals (Figure 6-1).

There are conflicting statements given about the date of the origin of zero. Freebury says zero appears in the Devangari numerals of the eighth

[2] Florian Cajori, *A History of Mathematics* (New York: The Macmillan Company, 1919), p. 89.

[3] H. A. Freebury, *A History of Mathematics* (New York: The Macmillan Company, 1961), p. 72.

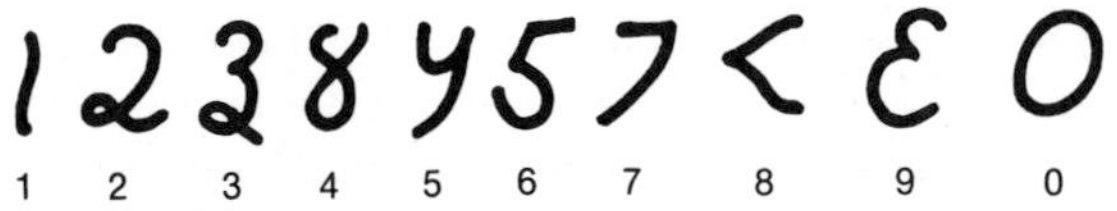

1 2 3 4 5 6 7 8 9 0

**Figure 6-1** The Devangari numerals used in India around the eighth century A.D.

century. Smith says, "The earliest undoubted occurrence of a zero in India is seen in an inscription at Gwalior. In this inscription 50 and 270 are both written with zeros."[4] He goes on to say that the date of origin will probably never be known, but that it most likely was in India in the ninth century.[5]

The Arabs' contribution to the advancement of our numeration system lies more in their transmitting information about it to other parts of the world than from their refinements of it. The Arabs translated into their language much of the knowledge of science and mathematics that had been developed and recorded in Greece, India, and elsewhere. Some of these Arabic translations are our only sources of knowledge about Greek and Indian achievements. One translation, called *The Book of al-Khowarazmi on Hindu Number,* explained the use of Hindu numerals. From the author's name, al-Khowarazmi, came the word *algorithm,*[6] which today means the procedures used to perform number operations. This author also wrote a book called *Al jabr,* which was about reduction and cancellation. *Al jabr* was much used in Europe where the title was eventually corrupted into *algebra,* and came to be used to describe the science of equations.[7] Today the word names one of the important branches of mathematics.

The Crusades, the increased trade among nations of the Mediterranean area, and the Moorish conquest of North Africa and Spain resulted in the spread of the Hindu-Arabic numeration system to many parts of Europe. Roman numerals and the abacus, which had been spread throughout Europe during the Roman conquest, were being used there when the Hindu-Arabic system arrived. Gradually, the Hindu-Arabic system and the use of algorithms were recognized, and, for a time, the two systems coexisted. Eventually, the *algorists* won out over the *abacists,* and by the sixteenth century the Hindu-Arabic numeration system was predominant.

The characteristics of the Hindu-Arabic numeration system are summarized in Chapter 1. The characteristics are repeated and explained in more detail here.

1. There is a base number, which is 10. During their first counting experiences, people undoubtedly used their fingers for keeping

[4] David Eugene Smith, *History of Mathematics, Vol. II* (Boston: Ginn & Company, 1953), p. 69.

[5] Smith, p. 69.

[6] Freebury, *A History of Mathematics,* p. 76.

[7] Freebury, p. 77.

track of the count. After all ten fingers had been used, it was necessary to use a supplementary means for keeping track of the count. It was natural that the grouping was based on ten, the number of fingers available.

2. There is a symbol for zero. It is a place-holder in a numeral like 302, where it indicates that there are no tens. It is also the number that indicates the size, or numerical value, of a set that has no objects in it.

3. There are as many symbols, including zero, as the number indicating the base. The symbols in the Hindu-Arabic system are 0, 1, 2, 3, 4, 5, 6, 7, 8, and 9.

4. The place value scheme has a ones place on the right, a base position to the left of the ones place, a base times base $(b^2)$ position next, a base times base times base $(b^3)$ position next, and so on. Beginning at any place in the system, the next position to the left is ten times greater, while the position to the right is one-tenth as large. This characteristic makes it possible to represent fractional numbers as well as whole numbers with the system.

5. The system makes it possible to make computations using only paper and pencil and a standard algorithm. While the use of the standard algorithms will probably diminish with the advent of the minicalculator, their place in a mathematics program is assured for the foreseeable future because of the role they play in helping children understand the four operations. (Each algorithm is explored fully in later chapters.)

---

## Self-check of Objective 2

State orally or in writing the characteristics of the Hindu-Arabic numeration system.

---

### TEACHING THE HINDU-ARABIC SYSTEM

Before you have children begin work with numbers beyond ten, be certain that they can do rational counting to at least 9 and can read and write the numerals for the numbers 0 through 9. They should also be able to express the numbers 1 through 9 as addition combinations. For example, three can be expressed as "3 + 0," "2 + 1," "1 + 2," and "0 + 3." This ability is not important for the combinations themselves, but as a basis for understanding that a number like fifteen can be expressed as "1 ten and 5 ones" and as "10 + 5."

A classroom should be well equipped with structured materials and place value devices. Structured materials may be either purchased or teacher- or child-made. Cuisenaire rods, flats, and blocks, Dienes multibase arithmetic blocks, and Unifix cubes are commercially prepared materials illustrated in Figure 6-2. Beansticks have gained a wide following among teachers because they are inexpensive, easy to make, and effective (Figure 6-3). Place value pocket charts, frames, and boxes are easily made and are effective, too (Figure 6-4). The place value devices are used similarly, differing only in form. You are not advised to have all of these learning aids, but it is helpful to children if there is more than one. A child who does not grasp the meaning of place value through one physical representation may catch on as he or she works with another.

Children's first work with structured materials should be unstructured; that is, they should play with them for a while. As they play, they will begin to see ways in which the materials are structured. For example, with Cuisenaire materials they will see the stairstep pattern of the rods and that an orange rod is as long as the edge of the orange flat. Once children begin to recognize structure, you should engage them in activities that lead to an understanding of the place value concept. Teachers'

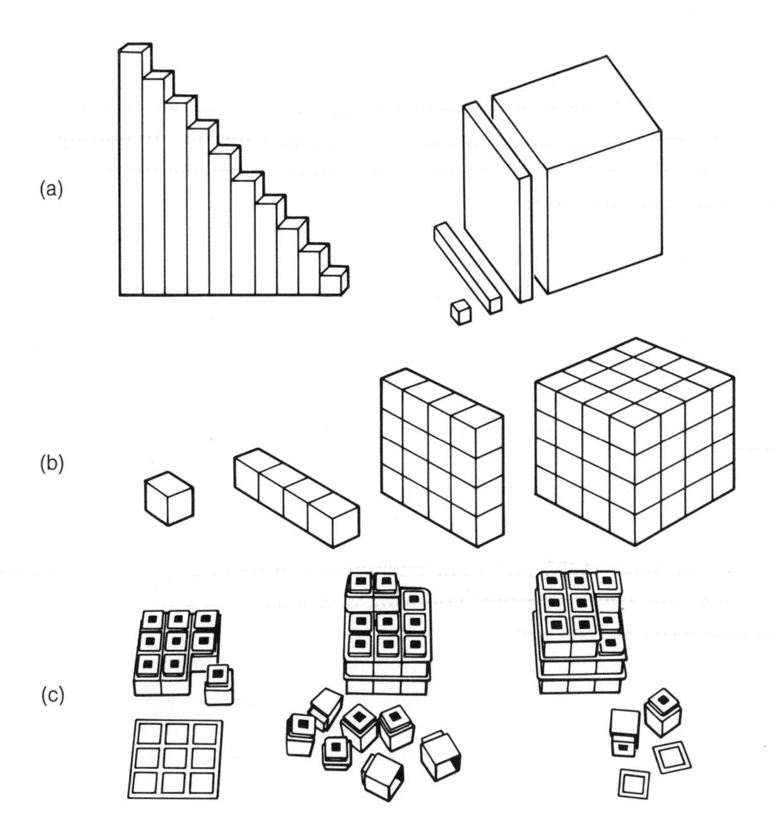


**Figure 6-2** Structured material for place value activities: (a) Cuisenaire rods, squares, and blocks, (b) Dienes multibase arithmetic blocks, and (c) Unifix cubes

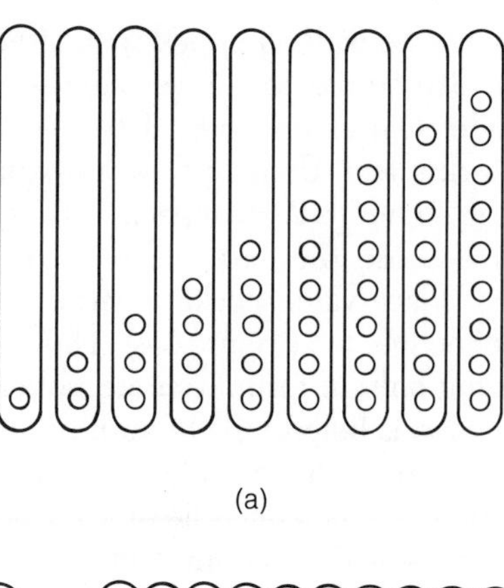


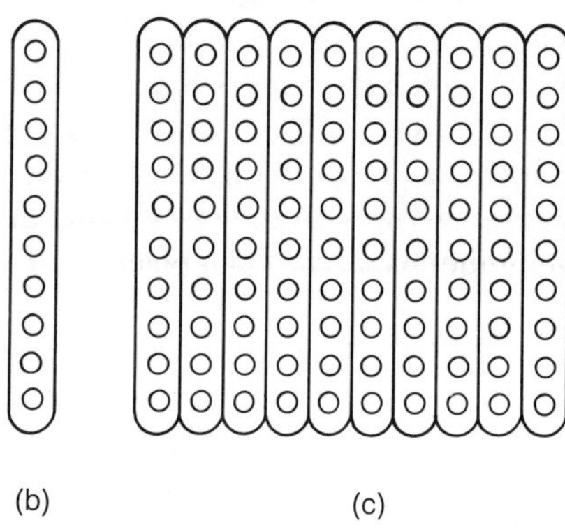


**Figure 6-3** A set of beansticks showing (a) the ones sticks, (b) a tens stick, and (c) a hundreds raft

manuals and problem cards are available for these materials.[8] You can also make your own problem cards for these materials; a sample for each kind is illustrated in Figure 6-5.

Beansticks are teacher- or child-made aids for learning about place value. It is instructive for children to make their own sets. Each child can make a set by gluing beans to popsicle sticks or tongue depressors with white glue. Figure 6-3 shows sticks for the numbers 1 through 9, 10, and "rafts" for 100. The rafts are held together by two sticks glued across the backs of ten sticks. (Some teachers choose not to use sticks for the numbers 1 through 9, preferring to have children use loose beans for these numbers instead.)

Initially, these instructional materials are used as models for (1) the numbers 1 through 9, (2) the number 10 and its multiples, and (3) the numbers 10 through 99.

[8]The Cuisenaire Company of America, 12 Church Street, New Rochelle, N.Y. 10805, has manuals and problem cards for its material. For Dienes material write McGraw-Hill Book Company, 1221 Avenue of the Americas, New York, N.Y. 10020; and for Unifix, write Mind/Matter Corporation, P.O. Box 345, Danbury, Conn. 06810.

You can use a place value device—pocket chart, frame, or box—to introduce children to place value. Put a group of more than ten objects before the children. Select three children: one to count the objects, the second to record the count on the place value device, and the third to record the count with numerals at the chalkboard. As the child who is counting picks up the objects one by one, the second child records the count by placing markers in the ones place of the device. At the same time, the third child records the count by writing numerals in sequence on the chalkboard. When nine objects have been counted, have the children pause to review what they have done so far. When the next object is picked up, a tenth marker will be put on the place value device. Ask the children, "How many markers are on the chart (frame, box)?" "How should we write '10' on the chalkboard?" Explain that ten markers can be bundled together in the tens part of the device. The children should note that the place value device now displays one ten and no ones. Through questions and discussion, help the children see the meaning of the one and zero in "10." As each additional object is counted, the children should talk about the meaning of the markers on the place value device and the ones and tens places of the numerals on the chalkboard.

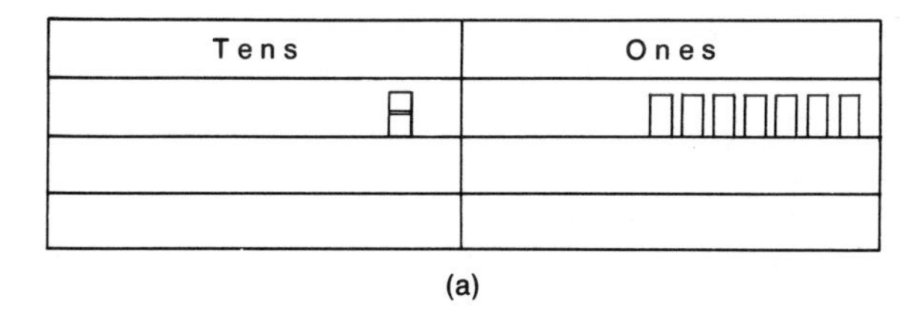


(a)

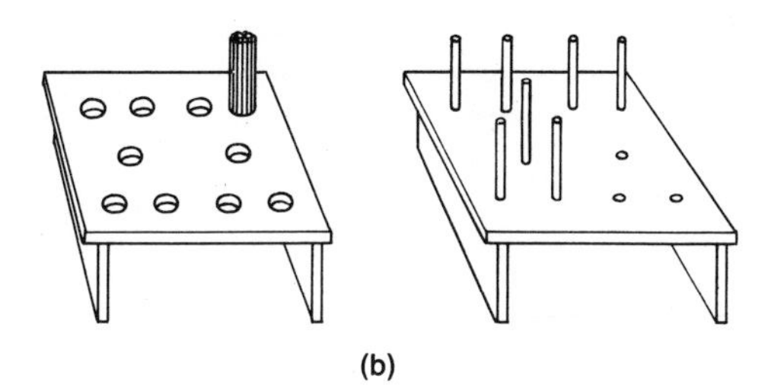

(b)

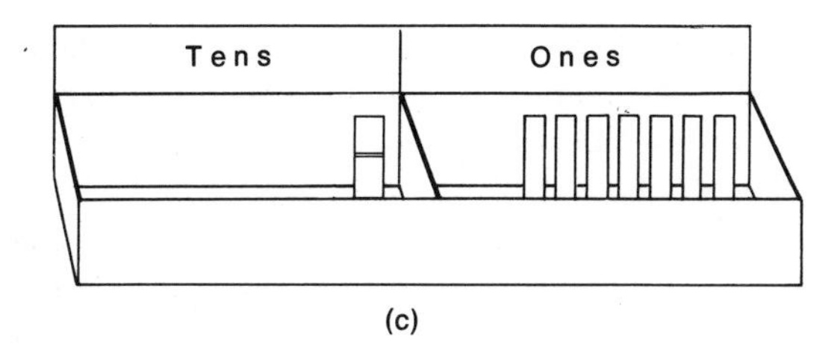


(c)

**Figure 6-4** Place value devices illustrating the meaning of 17. Place value pocket chart (a), place value frames (b), and a place value box (c)

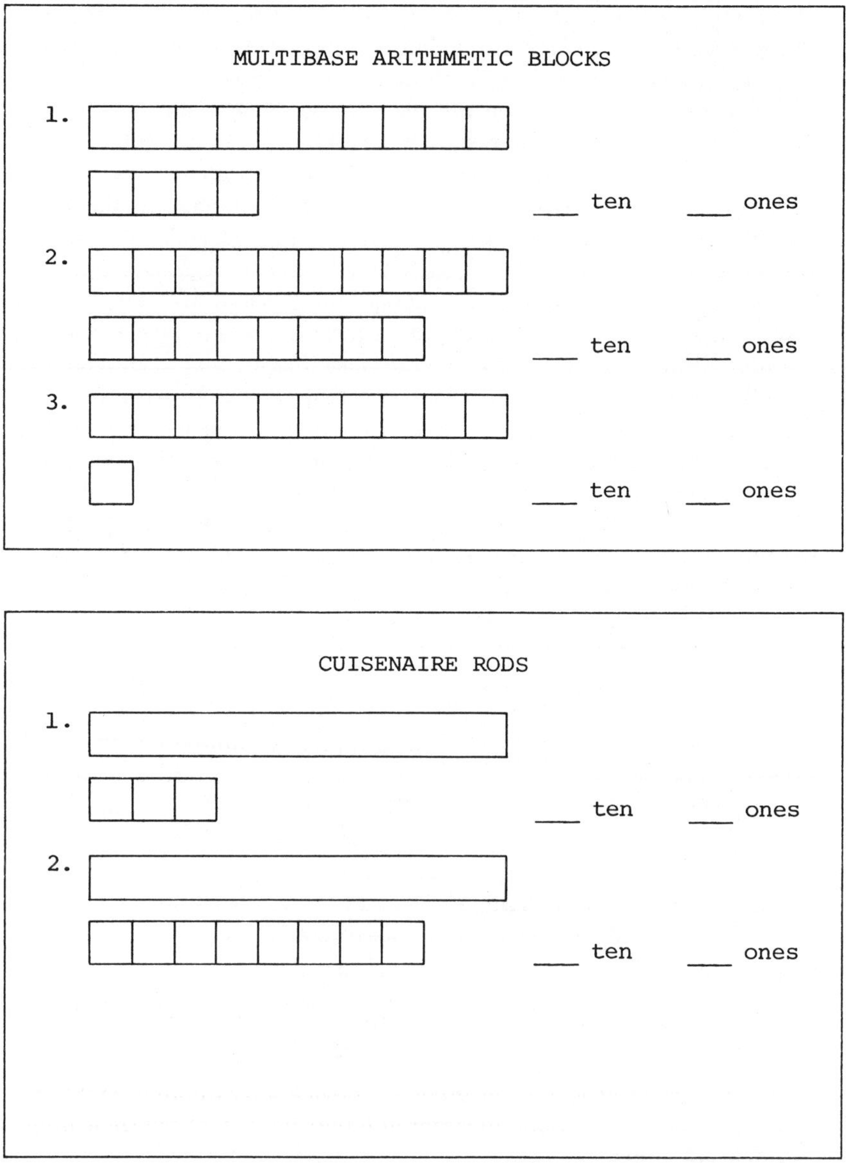

MULTIBASE ARITHMETIC BLOCKS

1. ___ ten ___ ones

2. ___ ten ___ ones

3. ___ ten ___ ones

CUISENAIRE RODS

1. ___ ten ___ ones

2. ___ ten ___ ones

**Figure 6-5** Problem cards for activities with structured materials

After children understand place value through 19, they should continue to work with place value devices and structured materials to learn about place value through 99.

In later grades, children work with larger numbers. Their understanding of the meaning of numbers through 999 should be developed as children engage in activities similar to those used in learning about the numbers through 99. It is not necessary to continue counting objects, but

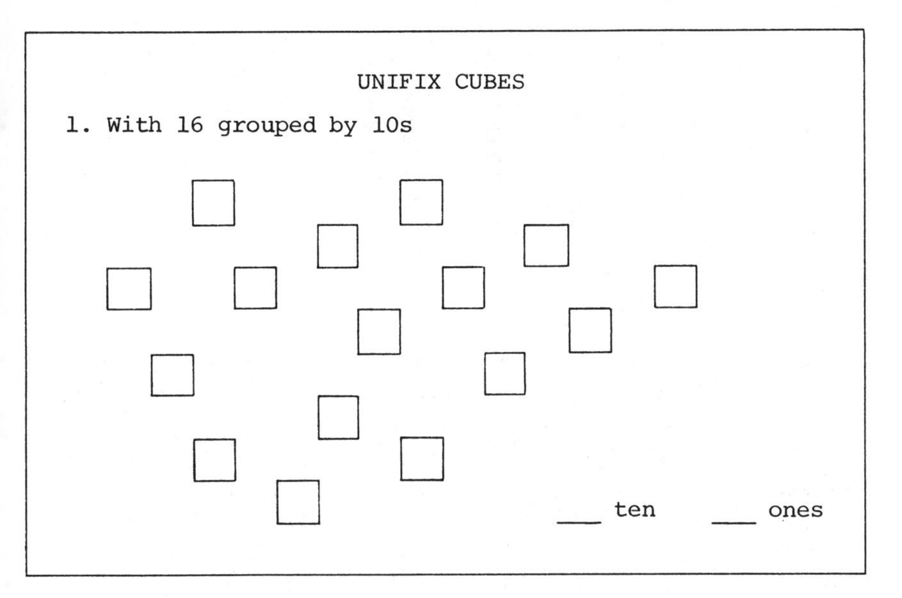

UNIFIX CUBES
1. With 16 grouped by 10s
___ ten ___ ones


**Figure 6-5** (continued)

the place value devices will remain useful in illustrating that 10 tens equal 1 hundred as 10 bundles of 10 markers each are put together to make 1 bundle of 100 in one of the devices.

The flats and rods in Cuisenaire and Dienes sets, and rafts and tens sticks in a set of beansticks, are used to represent numbers between 99 and 999. Cubes are used with flats and rods for numbers between 999 and 9999. Special plastic grids are used to make multibase squares and large cubes of Unifix cubes.

As children work with a device, they also learn how the numeral for 10 tens, or 100, is written. They should note the relationship between the symbol "100" and the markers on the place value device, a single flat in a set of structured material or a beanstick raft. As children continue this study, special attention should be given to the tens place when it contains a zero, as in 101. Some of the larger numbers should be practiced until each student understands the place value concept.

A classroom abacus is an excellent device for extending children's understanding of place value to larger numbers because it can show numbers that are impractical with other devices. To make one for classroom use, insert pieces of stiff wire into holes drilled in a board; wooden beads can be used for markers. Commercially made abacuses are also available. The abacus in Figure 6-6 has enough rods to represent numbers as large as billions.

A child's first abacus should have only three or four rods. You can use this abacus to demonstrate how it represents numbers. Point out that one bead on the tens rod has the same value (that is, represents the same number) as ten beads on the ones rod (10); one bead on the hundreds rod

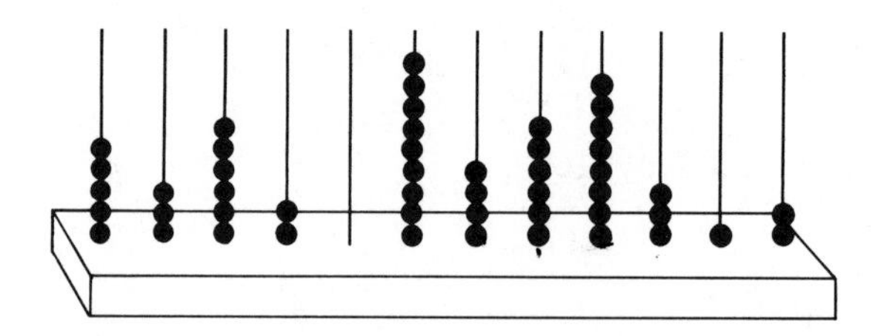

**Figure 6-6** Abacus representing the number 536,209,468,312

represents the same number as ten beads on the tens rod (100). Under your guidance children will conclude that each rod represents a place value position that has a value ten times greater than the position to its immediate right, regardless of where it is on the abacus. By the time they are in the fourth grade, most children will, with your help, be able to reverse their thinking to conclude that each place value position also represents a value that is one-tenth the value of the one to its immediate left. As their understanding of place value increases, children will generalize that the Hindu-Arabic numeration system has a place value scheme based on ten and powers of ten. This same scheme forms the basis of the metric system, a fact which should be capitalized upon as children deal with metric measure.

All the while children are using models for place value, they should have two different, but related types of experiences with them: (1) they should name orally or in writing the numbers represented on the devices, and (2) they should use the devices to represent numbers presented orally or in writing.

"Abacus add on" and "place value machine" are games for reinforcing six- to eight-year-olds' understanding of place value.[9]

A place value chart is a useful means for helping older children understand large numbers (Figure 6-7). Numerals instead of beads are used to represent each place value position. Children should read the numbers represented on the chart and also put numerals on the chart to represent numbers presented orally.

**Figure 6-7** Place value chart showing value of place positions for 536,209,468,312

| Hundred billions | Ten billions | Billions | Hundred millions | Ten millions | Millions | Hundred thousands | Ten thousands | Thousands | Hundreds | Tens | Ones |
|---|---|---|---|---|---|---|---|---|---|---|---|
| 5 | 3 | 6 | 2 | 0 | 9 | 4 | 6 | 8 | 3 | 1 | 2 |

[9]Leonard M. Kennedy and Ruth L. Michon, *Games for Individualizing Mathematics Learning* (Columbus, Ohio: Charles E. Merrill Books, Inc., 1973), pp. 19–22.

## Self-check of Objectives 3 and 4

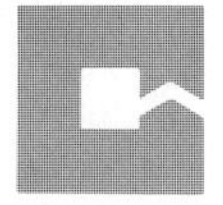

Make models or pictures of at least three place value devices or materials. Demonstrate with your models or pictures how each device is used to represent the meanings of numbers greater than ten.

Give an explanation of how the abacus can be used to represent large numbers. Represent the number 436,488 on a real or pictured abacus.

### EXPRESSING NUMBERS WITH EXPANDED NOTATION

Children use both compact and expanded forms of numerals in elementary school mathematics. The compact form is the one we normally use, as when the number 243 is written as "243." There are times, however, when one of the expanded forms is useful. For example, when children are learning about place value, the expanded forms 2 *hundreds* + 4 *tens* + 3 *ones* and 200 + 40 + 3 emphasize place value in the Hindu-Arabic numeration system. Also, it is best that children know how to express numbers in the simplest expanded form before beginning to add and subtract numbers represented by multidigit numerals, because this form helps children to better understand the algorithms. Another reason to understand expanded notation is that, particularly in its exponential form, it serves as a foundation for the study of extremely large and small numbers and their expression in scientific notation.

The simplest form of expanded notation is the expression of a number using a combination of numerals and words; for example, the number 17 can be expressed as "1 ten and 7 ones." As children count objects and use place value devices, you should repeatedly have them interpret the numbers as so many tens and so many ones and write the numerals in both the standard and simple expanded forms.

From the use of the form "5 tens and 6 ones" to express the number 56, children move to the shorter "50 + 6" form. To develop their readiness for this form, have children count markers grouped by tens, markers on the tens part of a place value device, tens beansticks, or the rods of structured materials. First, have them count "1 ten, 2 tens, . . . , 5 tens," and call the total "5 tens." Then have them use words, "ten, twenty, . . . , fifty," and call the total "fifty." After children can count to 90 by tens, they are ready to express numbers with tens and ones, so that "5 tens and 6 ones" becomes "50 + 6."

Games are good for children who need practice reading numerals in the two expanded forms. Make pairs of cards that contain the compact form on one card and an expanded form on the other. For example, for 96, one card will have "96" printed on it and the other will have "9 tens and 6 ones" or "90 + 6" on it. Pass out half of the deck, giving one card of each pair to the children and keeping the other one. Show your cards one at a time. Have the child who has the matching numeral for your first card hold it up and name it. Ask another child to give a mathematical sentence using the information from the two cards, such as "17 equals 1 ten and 7 ones," or "17 equals ten plus seven." Continue the activity until all of the cards have been matched. "Expanded-notation concentration," "place value probe," "expando land," and "spin a value" are games that reinforce understanding of expanded notation.[10]

In the second through fourth grades, children will apply the expanded form to larger numbers, such as 53,692. As they use an abacus to represent such a number, children should learn to express it in the following ways:

5 ten thousands, 3 thousands, 6 hundreds, 9 tens, and 2 ones
53 thousands, 6 hundreds, 9 tens, and 2 ones
50,000 + 3,000 + 600 + 90 + 2, or

$$\begin{array}{r} 50{,}000 \\ 3{,}000 \\ 600 \\ 90 \\ \underline{2} \\ 53{,}692 \end{array}$$

After children can multiply by ten and powers of ten, they are ready to express numbers in two other forms:

$$(5 \times 10{,}000) + (3 \times 1000) + (6 \times 100) + (9 \times 10) + (2 \times 1)$$

$$[5 \times (10 \times 10 \times 10 \times 10)] + [3 \times (10 \times 10 \times 10)] + [6 \times (10 \times 10)] + [9 \times 10] + [2 \times 1]$$

When children understand the meanings of these forms, they will have little difficulty learning how to express numbers by exponential notation. It is usually in the sixth grade that children learn to use exponents to indicate powers of ten. Using exponential notation, 53,692 is expressed in this way:

$$(5 \times 10^4) + (3 \times 10^3) + (6 \times 10^2) + (9 \times 10^1) + (2 \times 10^0).$$

Children will find it helpful if they can express numbers in many ways when they are learning to add and subtract. For example, 239 can be expressed as:

239 ones
23 tens and 9 ones
2 hundreds and 39 ones

[10] Kennedy and Michon, pp. 22–29.

2 hundreds, 3 tens, and 9 ones
1 hundred, 13 tens, and 9 ones
2 hundreds, 2 tens, and 19 ones

Use structured materials and place value devices to help children understand each of these representations.

Any of the place value devices can be used to regroup numbers to 999, but for numbers larger than 1000, the abacus illustrates the process more conveniently. For this purpose, an abacus must have room for nineteen or twenty beads on each rod. Figure 6-8 shows 23,498 on an abacus. It is in its simplest form in (a). It appears in one of many other forms in (b). The second representation is useful either to show the sum of the addition problem

$$\begin{array}{r} 17{,}349 \\ +\ 6{,}149 \\ \hline \end{array}$$

before it is regrouped or to prepare for subtraction using the decomposition method of the problem:

$$\begin{array}{r} 23{,}498 \\ -\ 9{,}129 \\ \hline \end{array}$$

Particular attention to numbers with zeros builds readiness for regrouping when such numbers are used in subtraction. If children learn that 204 can be expressed as "20 tens and 4 ones," 3006 as "300 tens and 6 ones," and 40,002 as "4000 tens and 2 ones," they will not be confused by this regrouping:

$$\begin{array}{rcrcr} 3006 & \longrightarrow & 300 \text{ tens and } 6 \text{ ones} & \longrightarrow & 299 \text{ tens and } 16 \text{ ones} \\ -2439 & \longrightarrow & -(243 \text{ tens and } 9 \text{ ones}) & \longrightarrow & -(243 \text{ tens and } \ \ 9 \text{ ones}) \\ \hline \end{array}$$

The subtraction can begin after one regrouping. This procedure is easier to do and understand than to regroup step by step from the thousands place to the hundreds place, from the hundreds place to the tens place, and finally from the tens place to the ones place. Such a lengthy process is unnecessary for children who can represent numbers in many expanded forms.

---

## Self-check of Objective 5

Explain the difference between a compact and an expanded numeral. Write at least three different expanded notation numerals for 4863.

---

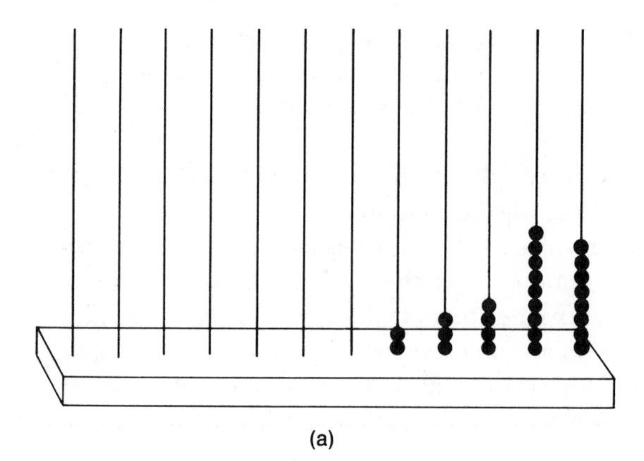


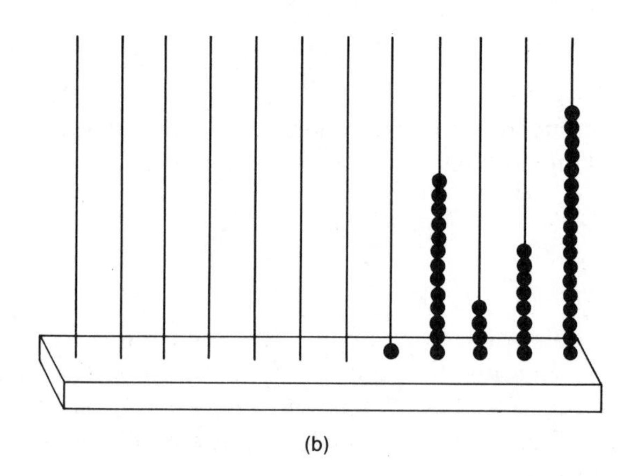


**Figure 6-8** Abacus representing 23,498; in (a) as 20,000 + 3000 + 400 + 90 + 8 and in (b) as 10,000 + 13,000 + 400 + 80 + 18

## INTEGERS

Current elementary school mathematics textbooks introduce integers as early as fourth grade. Study of these numbers by children is based on discovery-type experiences designed to help children develop their understanding of these numbers in an intuitive way.

Even though children have limited encounters with situations that require the use of integers, a few typically do occur in elementary school. When primary-grade children learn to use thermometers, they will note that thermometers can register temperatures below zero. Older children learn about altitudes above and below sea level, read about yardage gained and lost by football teams, and keep scores for games in which gains and losses are recorded.

The study of integers in elementary school should be largely informal, no matter in which grade it begins. Structured, formal, rule-bound teaching is as useless in this area as in the study of other number systems in earlier grades. Children discover the meanings of whole numbers and operations on them by manipulating and counting markers; they should discover the meanings of integers and operations on them through similar meaningful activities.

Stories illustrated with simple drawings are an excellent way to introduce integers. The subjects of the stories can be real or fanciful as long as their sequence of ideas can be readily understood. They should end in a way that leads to the discussion of conclusions that enhance children's understanding of these numbers. The following story, accompanied by simple pictures drawn on the chalkboard or projected with an overhead projector, fits all these requirements.

The Little Hole-Diggers

> There was once a group of little hole-diggers. No one is sure how they looked, but some things are certain: the only time the hole-diggers got any joy out of life was when they were digging holes in the ground. They usually made very special holes that gave them very special pleasure. These special holes were straight down into the ground or straight into the side of a mountain, like this. [Display a drawing similar to Figure 6-9.]
>
> Not only did the hole-diggers enjoy digging these holes, but they got extra special pleasure out of disposing of the dirt in a particular way. When they dug a hole straight down, the dirt from the first foot of the hole was placed right beside the hole in a pile one foot high. The dirt from the second foot of the hole was piled on top of the first foot of dirt, and so on. Also, they always marked the sides of their holes to show how deep they had been dug, and marked the sides of their dirt piles to show how high they had been made. They dug holes into the sides of mountains in about the same way, except that they put the dirt on the ground in front of the hole, as you can see in the picture.

After the story has been told, guide children's thinking by asking questions. "Which numbers tell how deep the hole is?" "Which numbers tell how high the pile is?" "Since we used the same number to answer both the questions, we might find ourselves getting mixed up as we talk about the hole and the pile of dirt. Can you think how we could name each of these numbers so we won't get mixed up as we talk about them?"

Children might suggest "above" and "below" as names for numbers for the vertical number line, and "right" and "left" for numbers on the horizontal line. Some children might know that *positive* and *negative* are mathematical terms used to name numbers in the set of integers. If these names are given by children, they should be used as discussion continues. If no one suggests them, it is not necessary to begin using them at the outset of the study. The names children suggest will have more meaning initially than mathematical terminology and symbolism presented by you. When children begin to use symbols to denote the new numbers, they might begin with $a1$ to indicate the point one foot above ground and $b2$ for

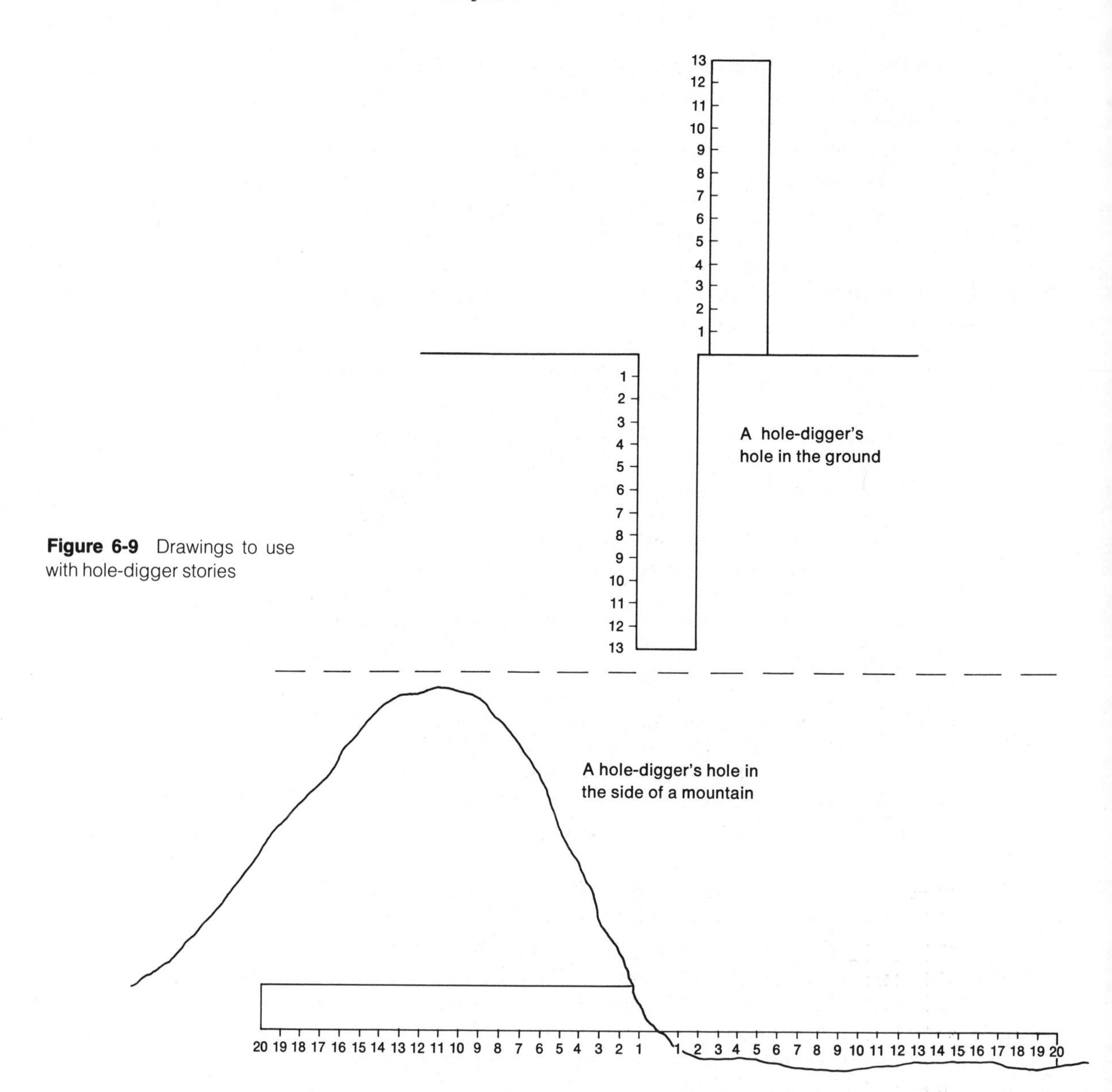


**Figure 6-9** Drawings to use with hole-digger stories

a point that is two feet below the ground. When the hole-in-a-mountain situation is used, the symbolism might be *r*1 for a point one foot outside, or to the right of, the opening and *k*2 for a point 2 feet inside, or to the left of, the opening.

Once children have agreed which terminology and symbolism to use, they should explore the number lines to learn more about directed, or signed, numbers, as integers are also called. They should be told that a vector, or directed segment, is a line segment with an arrow at one end to indicate direction. It represents directed numbers. In Figure 6-10 the vectors represent several directed numbers on number lines. Guide children to note that a vector need not begin at zero, but can begin and end at any point on the line. No matter where a vector appears on a number line, its length and direction indicate a directed number. In Figure 6-10 vectors **a** and **b** indicate the number $a2$ because each is two units long and moves in the $a$ direction. Vectors **c** and **d** each represent $a3$; vectors **e** and **f** represent $b2$. After such an explanation, children should be asked: "Which directed number does each vector on the horizontal line represent?" Later, they should use vectors with number lines to learn about operations on directed numbers. Vectors used for this purpose are discussed in Chapter 8.

Children should also begin to develop their understanding of number opposites and absolute value by examining number lines. Each number found on that part of the line above zero on a vertical line and to the right of zero on a horizontal line corresponds to an opposite number on the line below zero or to the left of it: the opposite of $a2$ is $b2$; the opposite of $k7$ is $r7$. At the same time, children should note that the distance from 0 to $b6$ is the same as the distance from 0 to the opposite of $b6$, or $a6$. The distance from 0 to $r3$ is the same as the distance from 0 to $k3$. An understanding of this concept is the beginning of the children's understanding of absolute value. Children are usually not introduced to its mathematical symbolism until they are in junior or senior high school.

Children should also use their number lines to visualize concepts of "equal to," "greater than," and "less than" as including the directed numbers and relationships among them. On the whole-number line, the

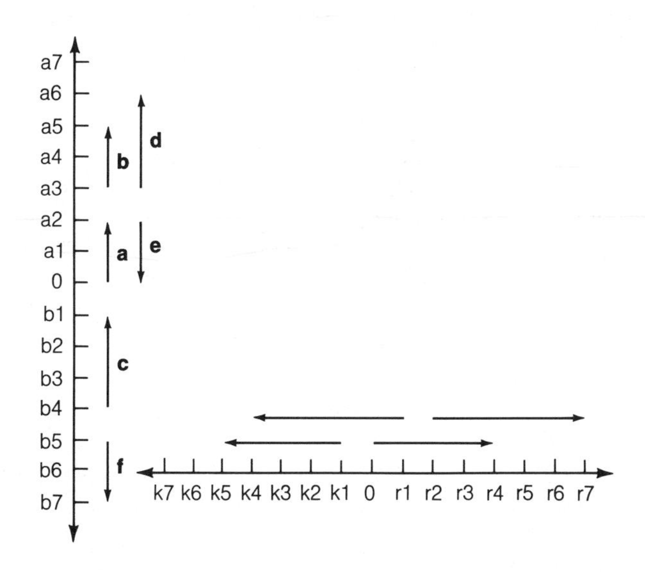


**Figure 6-10** Vectors drawn on vertical and horizontal number lines

number to the right of another number is always the greater one; a number to the left of another number is always the smaller one. The same properties hold for numbers in the set of integers. When the numbers are ordered on a number line, their comparative sizes can be determined by their locations. Children should note that zero and all $a$ or $r$ numbers are greater than all $b$ or $k$ numbers. To compare the sizes of any two $a$ or $r$ numbers, children should recognize that these numbers are treated in the same way as whole numbers. To compare any two $b$ or $k$ numbers, children should note that the number closer to zero on the number line is the greater one.

Various writers have suggested other activities and story themes to use in this context. In his discussion of cricket jumps on the number line, David Page includes the idea that children can discover this property of integers by thinking about jumps that go to the left of zero.[11] Louis Cohen has used postman stories with children in grades four through eight.[12] Charles D'Augustine recommends that children be introduced to integers by using a simple map with houses pictured on it. The map serves as a number line on which children use arrows with numerals to indicate distance and direction to the right and left of a house that serves as zero.[13]

## Self-check of Objective 6

Tell orally or in writing a story you can use to introduce children to integers. Show how a number line can represent positive and negative integers.

## Common Pitfalls and Trouble Spots

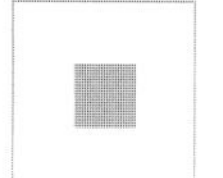

As you help children learn about the Hindu-Arabic numeration system, be aware of the following pitfalls.

First, teachers sometimes fail to recognize that children do not immediately perceive the meaning of a physical model and launch into activities with the model for which the children are not prepared. To avoid this pitfall, allow time for children to have free play with the materials. Then organize work so the structure of the model is revealed in stages; do not overwhelm children by presenting too much information about a model too rapidly.

[11] David A. Page, *Number Lines, Functions, and Fundamental Topics* (New York: The Macmillan Company, 1964), chap. 6.

[12] Louis S. Cohen, "A Rationale in Working with Signed Numbers," *The Arithmetic Teacher*, XII, No. 7 (November 1965), pp. 563–567.

[13] Charles D'Augustine, *Multiple Methods of Teaching Mathematics in the Elementary School*, 2nd ed. (New York: Harper & Row, 1973), pp. 288–289.

Second, teachers may forget that the reason for using a place value model is to help children understand the Hindu-Arabic numeration system, and not to learn about the model itself. You should continue using a model until children can use it with confidence to represent numbers presented orally or in writing and can name and write numbers represented on it.

Third, children's ability to recite or write the multiples of ten up to ninety is often interpreted as an indication that they understand the numbers themselves. As children deal with ten and its multiples, be sure they have opportunities to count the individual beans on tens beansticks and individual sticks in bundles of ten, as well as count the beansticks and bundles of ten.

Fourth, many teachers assume that children will always reverse their thinking. The natural progression for developing children's understanding of place value is to begin with ones, go to tens, then hundreds, and so on. Children see that each position has a value that is ten times that of the one to its right. However, they will not automatically reverse their thinking to consider that each position has a value that is one-tenth of the one to its left. Be sure that your children have opportunities to work in the left-to-right direction so that they will see this relationship as well as the right-to-left ("ten times") relationship.

The following two aspects of work with place value and the meaning of numbers cause trouble for some children.

First, the writing of the numerals for numbers between ten and twenty is troublesome for many children. These numbers have names that are unlike those for numbers larger than twenty, and the difference can be confusing. For the larger numbers, the numerals are named in order as each number is spoken—46 is "forty-six" and 28 is "twenty-eight." (Even though "twenty" does not sound like "two," the association is clear to most children.) However, the order of the numerals is reversed when the numbers thirteen through nineteen are spoken. Some children write "31" when they mean "13" because they first hear a sound like "three" when thirteen is presented orally. The other teen numbers may be reversed in the same way. (Eleven and twelve have no identification of their numerals at all when they are given orally. They must be learned as special cases.) Close observation of children as they work with place value devices representing these numbers and the immediate and careful checking of written work will help you spot potential problems and identify children who need special help.

Second, the meaning of large numbers is not always well understood. Other than during the infrequent times when numbers like 513,694,069 appear in their textbooks, children seldom encounter large numbers. The articles by Makurat and by West and Hass in this chapter's reading list offer suggestions for experiences with large numbers. One way to provide more frequent experiences with these numbers is to include activities with them as part of daily five- and ten-minute review and mental arithmetic sessions. From time to time, display a large number on an abacus or place value chart, then have children identify the number, its place value positions, and the number in each position, and write the number. Also, have children represent on a device and in writing the numerals for numbers you dictate.

## Self-check of Objectives 7 and 8

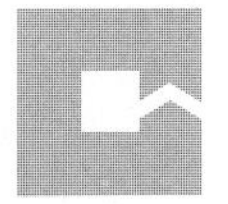

Identify orally or in writing four pitfalls associated with place value, and describe at least one way to avoid each one.

Describe two trouble spots associated with numbers and their meanings, and tell how their impact may be reduced.

## SUMMARY

Children must have a well-developed understanding of the Hindu-Arabic numeration system if they are to be mathematically literate. In addition to understanding one-to-one correspondence, which is the basis for counting, they must know and understand one-to-many, many-to-one, and many-to-many correspondences, for which numerous applications are found in mathematics. Place value pocket charts, bean-sticks, abacuses, and other place value devices, along with structured materials such as Cuisenaire rods, Dienes blocks, and Unifix cubes are useful for clarifying the meaning of the Hindu-Arabic and other place value numeration systems. Numerals can be written in both compact and expanded forms. Expanded forms are useful for explaining place value and the meaning of algorithms for such operations as addition and subtraction.

Children who are introduced to integers should learn about them through stories and activities that are meaningful to them. There are four pitfalls to be avoided while working with numbers and their meanings: (1) not taking sufficient time for children to understand the structure of a place value model, (2) using models as ends in themselves rather than to teach the meaning of the Hindu-Arabic numeration system, (3) assuming that rote counting by ten signifies understanding of the multiples of ten, and (4) assuming that children will automatically consider place value positions from left to right as well as from right to left. The writing of numerals for numbers between ten and twenty is a trouble spot for some younger children, while the meaning of large numbers is not well understood by many older ones. Special care and continual attention to these numbers must be given to lessen the impact of misunderstandings on children's progress.

## STUDY QUESTIONS AND ACTIVITIES

1. A number of different devices for helping children understand the Hindu-Arabic numeration system are described. Use at least three different devices to represent each of these numbers. (If the devices

are unavailable, use drawings to show the representation of each number.)
(a) 136 (b) 95 (c) 908

2. Use four different expanded notation forms to represent these numbers. List each of your answers in the order that is followed as expanded forms are presented to children.
(a) 2346 (b) 62,894 (c) 40,203

3. What is the value of the "3" in each of these numerals?
(a) 3468 (b) 4238 (c) 36,956 (d) 3,698,465,257

4. Write in simple expanded form the following numbers, using two different regrouped forms that might arise in subtraction. For example, 6384 would be regrouped as 6000 + 200 + 180 + 4 to subtract 6384 – 4291 and as 5000 + 1300 + 70 + 14 to subtract 6384 – 4726. Show a subtraction combination for which each of your regroupings might be used.
(a) 2345 (b) 60,349 (c) 267

5. Give the values of the vectors shown on the vertical number line in Figure 6-10.

6. Draw vectors for the following numbers on a number line.
(a) ($^{+}3$) (b) ($^{+}5$) (c) ($^{-}12$) (d) ($^{-}3$) (e) ($^{+}6$)

## FOR FURTHER READING

Anderson, Alfred L. "Why the Continuing Resistance to the Use of Counting Sticks?" *The Arithmetic Teacher,* XXV, No. 6 (March 1978), 18. Children who use counting sticks (popsicle sticks) in conjunction with abstract numbers and pictorial representations in textbooks and work sheets performed with more understanding than those restricted to workbooks, work sheets, finger counting, and other methods of tallying.

Ashlock, Robert B., and Tommie A. West. "Physical Representations for Signed-Number Operations," *The Arithmetic Teacher,* XIV, No. 7 (November 1967), 549–554. Presents the number line, several simple story situations, and their uses for helping children learn about integers. An extensive bibliography, with brief annotations, is included.

Cajori, Florian. *A History of Mathematical Notations,* Vol. I. LaSalle, Ill.: Open Court Publishing Company, 1928. The history of symbols and notation for many ancient systems is presented in Chapter 2. Of special interest is a section on fanciful hypotheses about the origins of the symbols for the Hindu-Arabic system.

———. *A History of Mathematics.* New York: The Macmillan Company, 1919. In the early chapters, the history of many numeration systems and mathematics in the countries where the systems had their origin is presented; included are Babylonian, Egyptian, Greek, Roman, Mayan, and Hindu-Arabic numeration systems.

Cohen, Louis S. "A Rationale in Working with Signed Numbers," *The Arithmetic Teacher,* XII, No. 7 (November 1965), 563–567. Cohen suggests using stories involving a postman and his delivery and picking up of checks and bills as means of involving children in situations that require the addition and subtraction of directed numbers.

———. "A Rationale in Working with Signed Numbers—Revisited," *The Arithmetic Teacher,* XIII, No. 7 (November 1966), 564–567. The postman stories are extended to include the operations of multiplication and division.

Freebury, H. A. *A History of Mathematics.* New York: The Macmillan Company, 1961. In a rather brief, but very interestingly written book, Freebury describes the origins of the Hindu-Arabic and other numeration systems. This book is especially recommended for the elementary school teacher's library.

Fremont, Herbert. "Pipe Cleaners and Loops—Discovering How to Add and Subtract Directed Numbers," *The Arithmetic Teacher,* XIII, No. 7 (November 1966), 568–572. Fremont presents a full discussion of pipe cleaner loops and their uses to help children visualize the meaning of addition and subtraction with directed numbers.

Makurat, Phillip A. "A Look at a Million," *The Arithmetic Teacher,* XXV, No. 3 (December 1977), 23. A computer printout of a million "$" signs is relatively easy to acquire—much easier than a million bottle caps. With such a printout children can determine how long it would take each one to print a million dots. Other possible activities are suggested.

March, Rosemary. "Georges Cuisenaire and His Rainbow Rods," *Learning,* VI, No. 3 (November 1977), 81–82, 84, 86. Gives a brief account of the man and how he became inspired to develop the Cuisenaire rods. Describes the struggle to gain acceptance of the rods, and includes testimonials to their value.

Page, David A. *Number Lines, Functions, and Fundamental Topics.* New York: The Macmillan Company, 1964. Part II discusses in detail various ways children can learn about negative numbers and operations on them.

Ronshausen, Nina L. "Introducing Place Value," *The Arithmetic Teacher,* XXV, No. 4 (January 1978), 38–40. Explains carefully sequenced steps for teaching place value. Although they were developed in one-to-one tutorial situations, the author claims the activities are suitable for small group work, too.

Skokoohi, Gholam-Hossein. "Manipulative Devices for Teaching Place Value," *The Arithmetic Teacher,* XXV, No. 6 (March 1978), 48–51. Acorns and glasses provide the means for developing the place value concept as children record the count of passing cars by one-to-one correspondence. What to do when the first glass is full becomes the basis for determining that a second glass is needed, and the place value concept is introduced.

Smith, David Eugene. *History of Mathematics,* Vol. II. Boston: Ginn and Company, 1953. The second chapter presents excellent material about the history of numeration systems.

———, and Jekuthial Ginsburg. *Numbers and Numerals.* Washington, D.C.: The National Council of Teachers of Mathematics, 1937. This is one of the early publications of the National Council; it is a highly readable account of numeration, including its history, suitable for older children as well as teachers.

Van Arsdel, Jean, and Joanne Lasky, "A Two-dimensional Abacus—the Papy Minicomputer," *The Arithmetic Teacher,* XIX, No. 6 (October 1972), 445–451. The Papy Minicomputer is a simple, easily made device. It is useful for teaching place value in base ten and other bases, and for performing addition, subtraction, multiplication, and division.

West, Mike, and Ken Hass. "It's Neat Being Surrounded by Peanuts," *The Arithmetic Teacher,* XXVI, No. 1 (September 1978), 22. Mike, a ten-year-old student, bought a 100-pound bag of peanuts and held a contest; fellow students guessed the number of peanuts. To determine the number, Mike and a few friends counted the peanuts, grouping by tens and powers of tens—there were 16,870. Guesses ranged from as low as 800 to well over a billion, prompting Ken, the teacher, to observe that even some high school students who made estimates lack good number sense.

# 7 Introducing Addition and Subtraction

Upon completion of Chapter 7, you will be able to:

1. Give either orally or in writing a definition of addition and subtraction, and name the parts of addition and subtraction sentences.

2. Describe at least two types of activities that can be used with young children to develop readiness for addition and subtraction.

3. Describe procedures that can be used to introduce addition and subtraction and number sentences for these operations.

4. Describe some manipulative materials that should be available for children to use during the beginning stages of work with addition and subtraction.

5. Distinguish between the sentence and vertical forms of writing addition and subtraction combinations, and illustrate at least one procedure for making the transition from the sentence to vertical form.

6. Name four real-life situations that give rise to subtraction, and illustrate each with a simple problem story.

7. Demonstrate with poker chips, or similar markers, at least two ways to introduce basic addition facts with sums greater than 10.

8. Demonstrate with poker chips, or similar markers, at least two ways to introduce basic subtraction facts with sums from 11 through 18.

9. Define a basic addition fact and a basic subtraction fact.

10. Illustrate with examples two ways addition and subtraction combinations are grouped as "families."

11. Illustrate the meaning of addition and subtraction on a number line.

12. Reproduce the addition table and explain how it can be used as an aid for learning basic addition and subtraction.

13. Describe and demonstrate materials and activities that can be used to provide practice with the basic addition and subtraction facts for children who understand their meaning.

14. Demonstrate with markers and other devices how the commutative, associative, and closure properties and the identity for addition can be introduced and developed so their meanings are clear to children.

15. Identify two pitfalls associated with addition and subtraction with whole numbers, and explain how each one can be avoided.

16. Identify two trouble spots children encounter with addition and subtraction, and describe how each can be minimized.

Key Terms you will encounter in Chapter 7:

| | | |
|---|---|---|
| addition | addend | addition fact |
| subtraction | sum | subtraction fact |
| inverse operation | missing addend | number line |
| minuend | place-holder box | addition table |
| subtrahend | minus | commutative property for addition |
| difference | open sentence | associative property for addition |
| remainder | set removal | identity element for addition |
| sign of operation | compare | closure for addition |

A considerable amount of the time devoted to mathematics in the elementary school is spent on the two operations of addition and multiplication and their inverses, subtraction and division. By the time children complete the elementary school, they have learned to use algorithms for these operations with whole numbers and the common and decimal forms of rational numbers. Some children also learn to do the operations with small negative numbers.

In Chapter 1, addition is described in terms of numbers and defined as the operation used to assign a sum to an ordered pair of numbers, which are called addends. Subtraction is described as the inverse of addition and is defined as the operation used to find a missing addend when a sum and the other addend are known. These definitions give us useful ways of summarizing addition and subtraction at the abstract level. However, they are not useful definitions for children, who need concrete ways of visualizing the two operations. This chapter presents methods and materials for introducing these operations to children.

Numbers that are added are called addends, and their answer is a sum. When subtraction is defined in terms of addition, the numbers are a sum, given addend, and missing addend. In the example $236 - 162 = \square$, 236 is a sum, 162 a given addend, and the answer a missing addend. The numbers in subtraction also have special names. In the example, 236 is the minuend, 162 is the subtrahend, and the answer, or missing addend, is the difference, or remainder. During beginning work, children learn the relationship between addition and subtraction by using the words *sum, addend,* and *missing addend.* Later, as their understanding of subtraction matures, children can learn its terminology.

## Self-check of Objective 1

Define addition and name the parts of an addition sentence. Do the same for subtraction. Why are these definitions unsuitable for young children?

## DEVELOPING READINESS FOR ADDITION AND SUBTRACTION

Activities that develop children's intuitive understanding of addition and subtraction are an integral part of the work they do to learn to count to ten. These initial addition and subtraction activities are done without definitions, explanations of the operations, or written numerals and other symbols.

Even while they are learning to count, children develop readiness for later addition and subtraction work. Many will recognize that within larger groupings there are subgroupings of various sizes. For example, some children will observe that a grouping of four objects is made up of two sets of two objects each and that a grouping of five objects is made up of a set of two objects and a set of three objects. You should encourage children to search larger groupings for different subgroupings within them.

As children work with manipulative materials, ask them questions to stimulate their thinking about addition and subtraction situations. For example, "You have six clothespins on that card. If I give you two more, how many do you think you will have? Put these two on and check to see if there are eight." "You have five acorns in that basket. If I take three of them from the basket, how many do you think you will have left? You take out three and see if two are left." "You have six golf tees in the holes in that board, and four in the holes in the other board. Which board has fewer tees in it? How can you tell? How many fewer tees does it have?"

Children can use a walk-on number line for readiness activities, too. When the sequence of numbers 0 through 9 has been learned, children can play "train" on the walk-on line. Each numeral card is a "station." Direct a child to begin at 0 and walk to station number 3. Now tell the child to move four more stations along the "track." Have the child tell you the number of the station where he or she is now standing. Another child can begin at station 8. Have that child go back three stations and tell you where he or she is standing.

After the children have satisfactorily completed these and other readiness activities, they are ready for more formal work with addition and subtraction. However, do not stop using manipulative materials too quickly after formally introducing addition and subtraction. Many teachers are inclined to move children away from concrete materials and into written activities too rapidly. An especially bad practice is the one of having children engage in drill activities to memorize addition and subtraction combinations before they know the combinations' meanings.

## Self-check of Objective 2

Two different types of activities for developing readiness for addition and subtraction are described. Explain how each can be used with kindergartners and first graders. Can you think of other activities for developing children's readiness for these operations?

## ADDITION AND SUBTRACTION SENTENCES

### 1. Introducing Sentences

Children who are ready for addition and subtraction should not be delayed because of less mature classmates. When you have a group of children who are ready for this new work, introduce addition to them by using a concrete situation. You might begin with a set of two dolls and a set of three dolls. First, have the children tell the number of dolls in each set. Then join the two sets and have the children determine the total number. Use other things, such as toy cars, dishes, blocks, and books, to provide additional experience in joining other collections. These early experiences are designed to teach the meaning of addition in terms of joining sets of objects. They also offer terminology, or oral descriptions, to represent addition situations: "A set of two dolls joined with a set of three dolls make a set of five dolls." "Two dolls and three dolls make five dolls." "Two plus three is five." "Two and three equal five."

After using realistic objects in sets, you should have children use materials they can manipulate on a magnetic or flannel board and at their desks. As they begin using markers, children should be introduced to the mathematical sentence as a means of describing addition situations. Magnetic or felt numerals, signs of operation, and place-holder frames can be used to show sentences on the magnetic or flannel board, as in Figure 7-1.

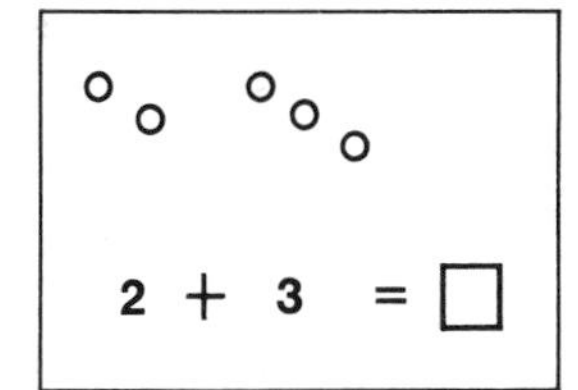


**Figure 7-1** Markers and symbols used to illustrate an addition sentence

You can help children understand the meaning of the "+" and "=" signs in an addition sentence by having them relate each sentence to markers on a magnetic or flannel board. In this way, children can build up mental images of situations involving real objects and the related number sentences. A child who has a good store of mental images will have little difficulty in writing mathematical sentences when confronted by similar but not identical situations at a later time.

Children need not be working with all of the possible combinations of number pairs having sums through 9 before they are introduced to subtraction. As long as they work with manipulative materials, and it is clear to them that a situation is either a joining or separating one, there is no reason to delay introduction of subtraction.

To begin work with subtraction, you should use materials and procedures similar to those you use to introduce addition. For example, say, "Here are six books. I am going to give two of them to Billy. How many books will I have left?" The children should count the books in the original set to confirm that there are six of them. Then you should give two books to Billy. Many of the children will immediately recognize that four books remain. Even so, either one child or all of them in unison should count the books. This kind of lesson can introduce other examples using different objects in sets of varying sizes smaller than ten. The objects in each set should be counted by the children before any of them are taken away. As objects are removed, their number should be carefully noted by each child, either by recognition on sight or by counting. Finally, the number of objects still remaining in the set should be determined by each child, again either through recognition of the quantity or by counting. Unless the remaining number is quite small, the objects should be counted so that each child is sure of the quantity.

During introductory lessons, children need not use the sentence form for subtraction. Neither is it necessary to use the terms "sum," "addend," and "missing addend." It is better to first establish an understanding of the meaning of the "take-away" subtraction situation.

Once children recognize the meaning of "take-away" subtraction and can distinguish it from addition, the subtraction sentence can be introduced. Ask the children to recall the subtraction situations of earlier lessons. Ask, "If I take away four markers from this set of seven, how many will be left?" Have the children tell the answer. After they do, write the sentence $7 - 4 = \square$, and ask the children to tell what they think it means. As the children discuss the meaning of the sentence, they should note that the "7" indicates the size of the set before any markers are removed, the "4" indicates the number of markers to be removed, and the place-holder box indicates how many markers will be left. Tell them that "−" is read as *minus*. Children can now read the sentence as "7 minus 4 equals place-holder." After the answer has been written in the place-holder box, have them read the sentence as "7 minus 4 equals 3." The other sets of markers can be separated and their subtraction sentences written.

## 2. Using Manipulative Materials to Reinforce the Meaning of Sentences

In her book *Workjobs*,[1] Mary Baratta-Lorton offers many suggestions for activities dealing with addition and subtraction. Key features of these

[1] Mary Baratta-Lorton, *Workjobs* (Menlo Park, Calif.: Addison-Wesley Publishing Company, 1972), pp. 194–211.

activities are that each uses manipulative materials, and a child writes his or her own number sentences most of the time. For example, for one activity a child has a set of 9-inch by 12-inch cards with beads strung across the face of each one, as shown in Figure 7-2(a). A child separates the beads on each card into two groups and then writes an addition sentence that describes the arrangement of beads, as illustrated in (b). Another set of materials includes colored beads and blocks of wood with two finishing nails in each one. A child puts beads on the two nails of each block and then writes an addition sentence for each finished block.

For a subtraction situation, small aluminum pie tins and small blocks of wood are used. A child chooses a number, say 6, and puts that many blocks into each tin. Next, the child takes some blocks from each tin, turns the tin upside down over them, and puts the rest of the blocks on top of the upturned tin. If a child puts the same number of blocks under each of the tins, he or she should be encouraged to put a different number under each one. The child then writes two number sentences for each tin. If there are two blocks beneath a tin and four on top, the sentences are $4 + ? = 6$ and $4 + 2 = 6$.[2] Or, the child might be taught to write the sentences $6 - 2 = 4$ or $6 - 4 = 2$.

It is not difficult to collect and organize materials for activities like these. Pencil, shoe, and similar-sized boxes make good containers for individual activities. Classroom bookshelves and closets provide convenient storage space. Storage shelves can even be made of large cardboard boxes if more conventional space is not available. Be sure each child knows where to get the box containing an activity and how to return it when finished.

---

## Self-check of Objectives 3 and 4

Demonstrate with manipulative materials or by drawing a sequence of pictures how to introduce addition and its sentence. Do the same for subtraction.

Explain why a wide variety of manipulative materials should be available for children who are learning addition and subtraction. Describe some of these materials.

---

### 3. Using Different Types of Sentences

When children are ready for practice pages in their texts or workbooks, you must be certain that they understand each of the sentence forms they will encounter. These forms are found in most books: Addition sentences will be written as $6 + 2 = \square$, $\square + 2 = 8$, and $6 + \square = 8$; subtraction sentences will be written as $8 - 2 = \square$, $8 - \square = 6$, and $\square - 2 = 6$. You

[2] Baratta-Lorton, p. 199.

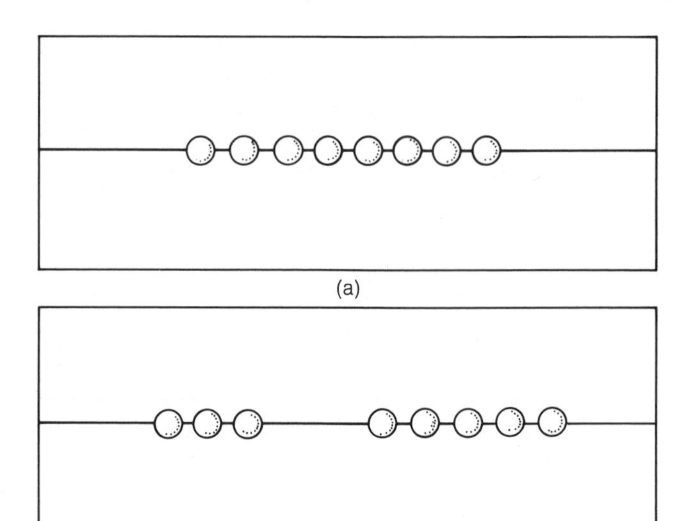


**Figure 7-2** Bead card for working with addition combinations for 8

will need to help children understand the uses of the place-holder box in different positions. Some children have a tendency to add whenever they see + and to subtract whenever they see −, regardless of the position of the place-holder box. This leads to errors in sentences such as $\square + 2 = 8$ and $\square - 2 = 6$. Dramatized situations, markers arranged on a flannel or magnetic board, and teacher-prepared illustrations should help children see the meaning of each of the open addition and subtraction sentences.

### 4. Introducing the Vertical Notation Form

The mathematical sentence, rather than the vertical form of notation, is most often used when children are introduced to addition. There are at least two reasons for this:

1. It can be considered a shorthand form of the longer word sentence; hence its meaning can easily be recognized.
2. It is a tool that helps solve problems involving numbers.

However, children must also learn to use the vertical form of notation because it is used in the algorithms for addition, subtraction, and multiplication. One way to introduce it is by using clothespins on a card. A number of clothespins should be placed in two groups along one edge of a large card. The card should then be placed against the chalkboard. The

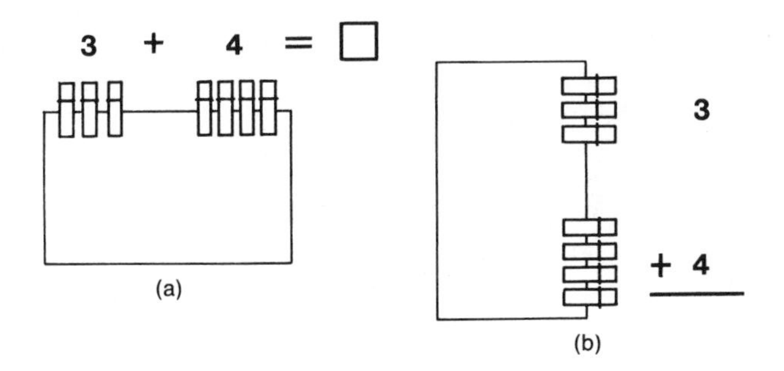


**Figure 7-3** Clothespins on a card used to show addition with (a) mathematical sentence notation and (b) vertical notation

number of clothespins in each set should be determined by the children and the numeral for each set written above it as in Figure 7-3(a). The symbols for completing an addition sentence should then be written, with a place-holder frame used in place of the sum. The card should be turned, as in (b), so the clothespins are to the right. Again, the numeral for each set is written on the board. Then its vertical notation form is completed and its similarity to the mathematical sentence is discussed. Once children are acquainted with the vertical notation for addition, you can show them the vertical form for subtraction and explain it.

## Self-check of Objective 5

Write examples of the sentence and vertical forms for addition involving the addends 6 and 3 and their sum. Demonstrate with materials or illustrate with a sequence of pictures how you can help children make the transition from the sentence to vertical form.

## DIFFERENT SUBTRACTION SITUATIONS

There is only one type of addition situation—the joining of two or more sets. Subtraction, however, arises from several kinds of situations. Children should have experience with these types:

1. Subtraction is used when part of a set is removed. This is illustrated by the following situation. "A merchant began a sale with 396 boxes of greeting cards. During the first day, he sold 218 boxes. How many boxes did he have left?" Situations such as this are of the take-away type, which are most frequently encountered by children.

2. Subtraction is used to compare the sizes of two sets. In this situation, subtraction is used to find how much larger or smaller one set is than another. "There are two schools in a town. One school has

684 students in it; the other has 478 students. The second school is how much smaller than the first?" (Or, "The first school is how much larger than the second?")

3. Subtraction is used to determine the size of a set to be united with another set of known size to make a third set of known size. "Mrs. Jones is filling her trading stamp book. It holds 1200 stamps when filled. After putting in all her stamps, Mrs. Jones finds that she has 700 stamps in her book. How many more stamps does she need to fill the book?"

4. Subtraction is used to determine the size of a group within a set. "A school of 537 pupils has 249 boys in it. How many girls attend the school?" This is not a take-away situation.

Naturally, you will not use examples with such large numbers for children's first experiences with subtraction. However, they should have frequent experiences with all four types. Examples of the take-away situation have already been described.

This example might help children use subtraction to find how much larger or smaller one set is than another: "In a classroom, the children put their lunches on a shelf. One day Julie counted them and found there were ten in lunch boxes and seven in paper sacks. How many more children brought lunches in lunch boxes than in paper sacks?" Children should have time to work out their answers. Illustrations of different kinds should be encouraged. Alternate ways to determine the answer should be discussed. If none of the children shows that the answer can be determined by using two sets of markers (one with ten members and the other with seven) and matching them one-to-one, then you should help them see this possibility. Markers representing the two types of lunches should be matched one-to-one on a flannel or magnetic board by putting them close together or connecting them with yarn as in Figure 7-4. Point out that the unmatched markers in the larger set show how much larger that set is than the second. Or, these same markers can be used to indicate how much smaller the second set is than the first. When children compare sets, the point of view of the questions should vary; they should compare the smaller set with the larger one as well as the larger with the smaller.

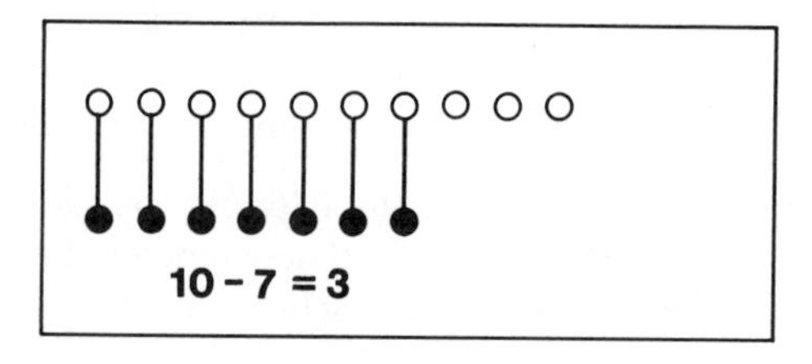


**Figure 7-4** Markers can be used to compare two sets

Examples similar to those that follow should help children see subtraction used in "how many more are needed" situations. "Sally is saving the inner seals from the lids of her favorite soft drink for a free kite. She must have ten seals for the kite she wants. She has six seals now. How many more does she need to get her kite?" Time should be allowed for use of markers and other devices to illustrate the situation and, later, so that some of the children can explain their method to the rest of the children.

An example of the final type of subtraction situation follows: "Billy was watching ten airplanes fly overhead. Before they flew out of sight he saw that four of the planes were red. How many planes were not red?" The children again use markers to demonstrate the situation. The difference between this situation and a take-away situation is that while one group within the set is identified, perhaps encircled with yarn as in Figure 7-5, it is not removed from the set as is done in a take-away situation.

## Self-check of Objective 6

Identify orally or in writing each of the real-life situations that give rise to subtraction. Make up a simple problem situation for each. Describe materials that can be used to illustrate each situation.

## ADDITION AND SUBTRACTION FACTS WITH SUMS GREATER THAN 10

### 1. Addition

Understanding base ten numeration is a foundation for understanding addition facts having sums greater than 9. First, children should learn the basic facts that involve pairs of addends with 10 as their sum by using real objects, markers, and charts showing patterns that make them easy to learn.

Markers on a flannel or magnetic board are effective because they can be arranged to show patterns of combinations that make 10. After

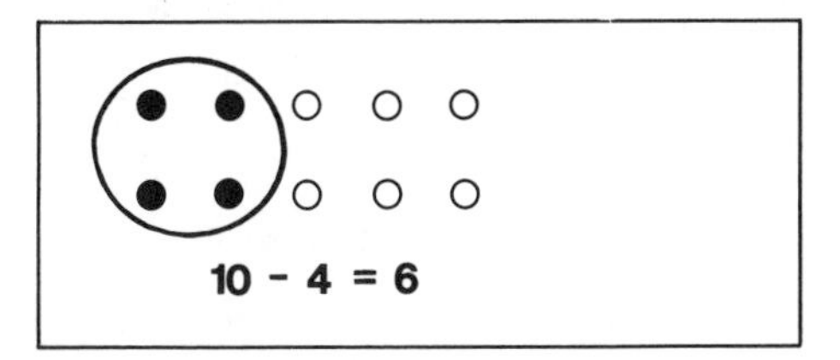


**Figure 7-5** Markers can be used to determine the size of a group within a set

work with markers on a board, a chart might be prepared for future reference as facts with sums from 11 through 18 are learned. You can begin a chart similar to the one in Figure 7-6 by marking the lines and putting in the combinations of addends on the left, the tens on the right, and the circles for the sets. Then as children arrange markers to show the set combinations on a flannel or magnetic board, they can also mark the sets on the chart by coloring in the correct number of circles for each pair of addends. (You can prepare a chart such as this one much faster than the children can. However, the value of the chart lies as much in the experience of making it as in its use during future lessons. Children at all grade levels should make most of the charts they will use. There are times when teacher-made or commercial charts are useful, but such charts should not be used to the exclusion of children-made charts.)

Children who have learned the meaning of the addition facts with sums of 10 and less, and who have some understanding of the concept of place value, are ready to learn those basic facts with sums of 11 through 18.

An understanding of addition of pairs of numbers having these sums is built on the idea that the sum will be a ten plus some number of ones. This can be illustrated with sets of disks on a magnetic board. For example, to complete the addition sentence 8 + 4 = □, shown in Figure 7-7, put a set of eight white disks and a set of four red disks on a magnetic board. Tell the children to count the number of disks in each set and to

| | | |
|---|---|---|
| 10+0 | ● ● ● ● ● ● ● ● ● ● | 10 |
| 9+1 | ● ● ● ● ● ● ● ● ● ○ | 10 |
| 8+2 | ● ● ● ● ● ● ● ● ○ ○ | 10 |
| 7+3 | ● ● ● ● ● ● ● ○ ○ ○ | 10 |
| 6+4 | ● ● ● ● ● ● ○ ○ ○ ○ | 10 |
| 5+5 | ● ● ● ● ● ○ ○ ○ ○ ○ | 10 |
| 4+6 | ● ● ● ● ○ ○ ○ ○ ○ ○ | 10 |
| 3+7 | ● ● ● ○ ○ ○ ○ ○ ○ ○ | 10 |
| 2+8 | ● ● ○ ○ ○ ○ ○ ○ ○ ○ | 10 |
| 1+9 | ● ○ ○ ○ ○ ○ ○ ○ ○ ○ | 10 |
| 0+10 | ○ ○ ○ ○ ○ ○ ○ ○ ○ ○ | 10 |

**Figure 7-6** Chart showing pairs of addends and sets for combinations making ten

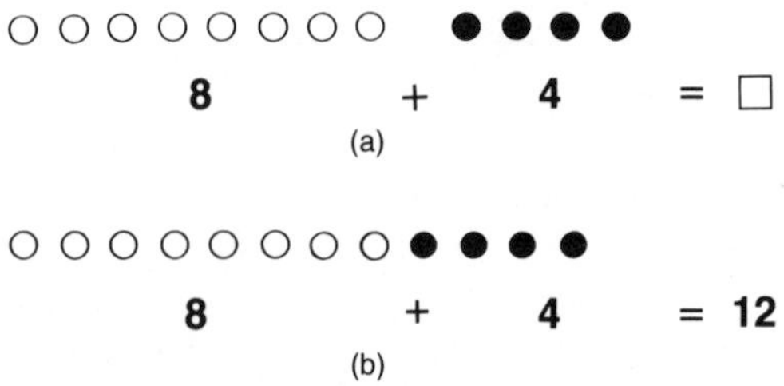

○ ○ ○ ○ ○ ○ ○ ○   ● ● ● ●

**8 + 4 = □**

(a)

○ ○ ○ ○ ○ ○ ○ ○ ● ● ● ●

**8 + 4 = 12**

(b)

**Figure 7-7** Sets and sentences illustrating the steps in addition for 8 + 4 = □

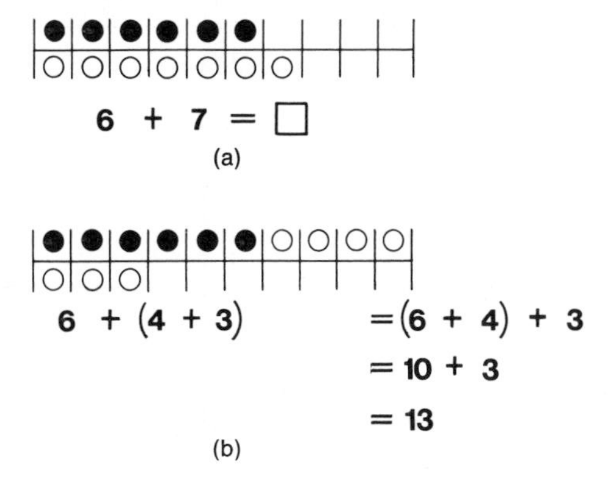


**Figure 7-8** Markers in a lattice to show steps in addition for 6 + 7 = □

determine the addition sentence that describes the situation. Next, have the children put the two sets together and determine the total number by counting.

A lattice can also show the meaning of the steps in this process. The sets for the sentence 6 + 7 = □ are illustrated with a lattice in Figure 7-8 and show that the answer consists of a ten plus 3 ones, as shown in (b).

---

## Self-check of Objective 7

Demonstrate with poker chips, or similar materials, two ways to introduce combinations having sums greater than 10.

---

### 2. Subtraction

The most straightforward way to present the facts with minuends 11 through 18 is to relate the subtraction directly to the addition facts with sums greater than 10. After children use markers to show sentences such as 8 + 4 = 12 and 6 + 9 = 15, have them use the same markers to show that 12 – 4 = 8 and 15 – 9 = 6.

Breaking the subtraction down into steps may prove to be helpful to some children, particularly older ones who have had trouble mastering the more difficult combinations. To subtract 15 – 7 = □, the thinking might be: "What do I subtract from 15 to make 10?" (5) "Since I am subtracting 7, I must take away 2 more; 10 – 2 = 8, so 15 – 7 = 8." Figure 7-9 illustrates how markers can be used to show this thought process. If this process is to be useful, children must eventually be able to do a complete sequence mentally; that is, without markers or paper and pencil: "16 – 9 = □, 16 – 6 = 10, 10 – 3 = 7, so 16 – 9 = 7."

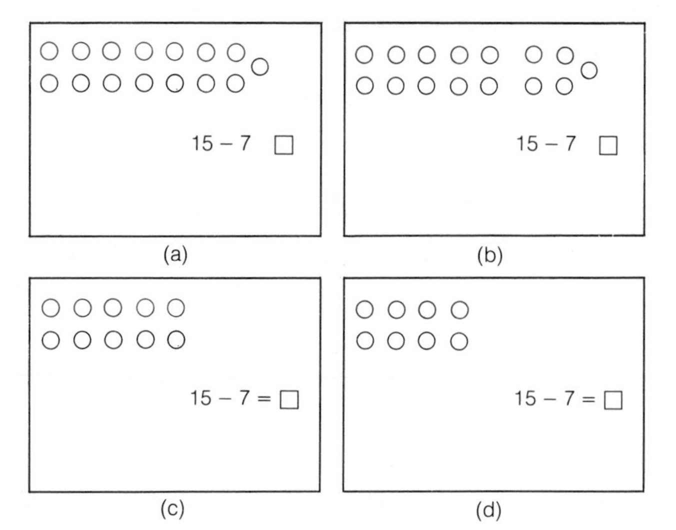


**Figure 7-9** Markers used to show meaning of 15 − 7 = 8

## Self-check of Objective 8

Demonstrate with markers at least two ways to introduce the subtraction facts with sums from 11 through 18.

## REINFORCING LEARNING OF THE BASIC FACTS

A basic *fact* is an ordered pair of whole number addends each smaller than 10 and their sum. Altogether, there are 100 addition facts which include all of the possible combinations using pairs of numbers represented by one-digit numerals and their sums. Every addition fact has a subtraction fact which is its inverse, so there are also 100 subtraction facts.

Some writers have suggested that a given procedure will reduce the number of facts to be learned by children. For example, Irwin suggests that the process of changing the minuend to a ten and ones will reduce the number of facts children must learn; once they learn the facts that have 10 as their minuend, they have the knowledge they need to determine the answers for the facts that have larger minuends.[3] Other writers have suggested that children who know the addition facts well and who understand that subtraction is the inverse operation of addition can use their knowledge of addition to determine the answers to subtraction facts and will therefore not need to memorize them. The claim is also made that

[3] Evelyn S. Irwin, "An Approach to Subtraction Using Easy Facts," *The Arithmetic Teacher,* XI, No. 4 (April 1964), p. 260.

knowledge of the commutative property of addition reduces the number of addition facts to be mastered by nearly one half. While it is true that all of this information will help children understand the meaning of addition and subtraction better, no part of it can be considered as a substitute for mastery of the facts. Ultimately, each child who is capable of doing so must commit each of the facts to memory for immediate recall. Otherwise, the child will be handicapped when working with algorithms for the various other operations.

---

## Self-check of Objective 9

Define a basic addition fact, and give an example. Do the same for subtraction.

---

Once your children are ready for the work, you must provide practice activities that help children focus their attention on patterns and relationships associated with the addition and subtraction facts.

### 1. Using "Families" to Learn the Facts

Addition and subtraction facts are frequently organized by "families" in children's textbooks and workbooks. Families help children relate addition and subtraction in two ways:

1. All sentences for a given sum are grouped together: $3 + 0 = 3$, $2 + 1 = 3$, $1 + 2 = 3$, $0 + 3 = 3$, $3 - 1 = 2$, $3 - 2 = 1$, and $3 - 3 = 0$.
2. Or, the four combinations involving a pair of addends and their sum are grouped together: $6 + 2 = 8$, $8 - 2 = 6$, $2 + 6 = 8$, and $8 - 6 = 2$.

Either way, the grouping of combinations helps children to see the relationships between addition and subtraction and to commit each combination to memory.

Cuisenaire rods are useful for organizing facts by families. When two rods are put end to end, they are called a "train." Trains can show different combinations of addends that equal 6 (Figure 7-10).

---

## Self-check of Objective 10

One way to group number combinations as a "family" is to group combinations having a given sum. Another way is to group a pair of addends and their sum in their four possible arrangements. Write the family for numbers having a sum of 4; write the family for the addends 7 and 2 and their sum.

---

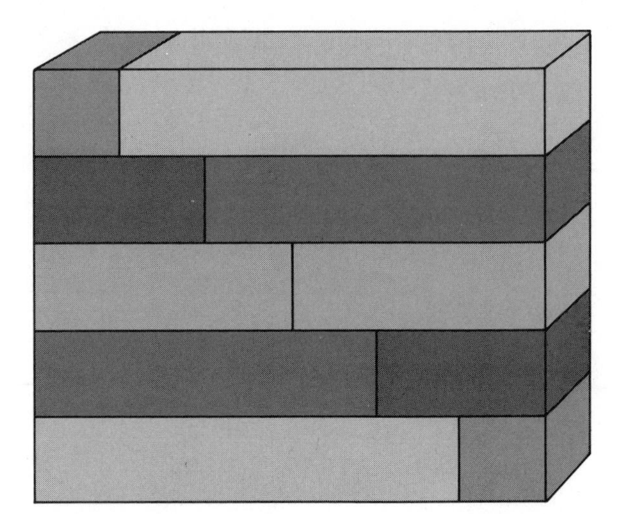

**Figure 7-10** Five different trains for 6 can be made with pairs of Cuisenaire rods

## 2. Using the Number Line to Learn Facts

A chalkboard number line is useful to reinforce children's understanding of addition and subtraction and to build mental images of the operations. Some teachers use the idea of a cricket or frog jumping along the line to introduce a chalkboard number line. To illustrate the sentence $3 + 4 = \square$, children can see the cricket or frog start at 0 and jump to 3, and then jump four more units to 7, as shown in Figure 7-11.

The number line is also useful for showing the relationship between addition and subtraction. First, use an addition sentence. The sentence $6 + 3 = 9$ is illustrated in Figure 7-12(a). Children should think of ways the number line can show subtraction. Those who have suggestions should have an opportunity to present their ideas to the class. The procedure illustrated on the number line in (b) follows logically from the use of the "jumps" in addition. Children should note that the jump showing the number to be subtracted begins at the point that indicates the number from which it is to be subtracted and moves to the left. The fact that subtraction jumps move to the left and addition jumps move to the right should be stressed.

The number line can help children understand the meaning of basic addition facts with larger sums. A sentence, such as $7 + 8 = \square$, should be

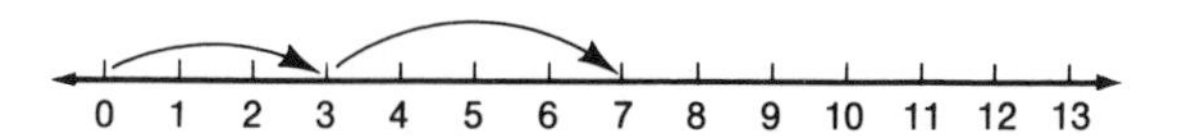


**Figure 7-11** Number line showing how to illustrate the sentence $3 + 4 = 7$

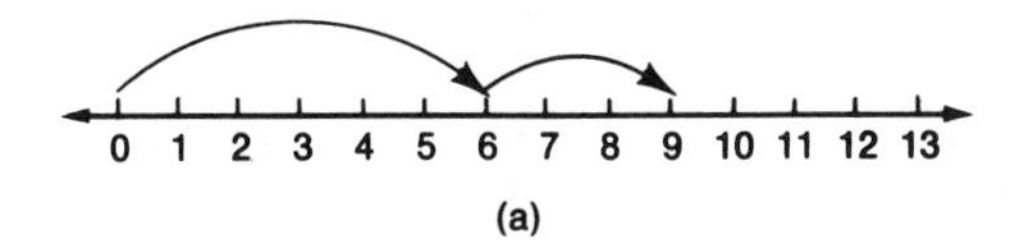


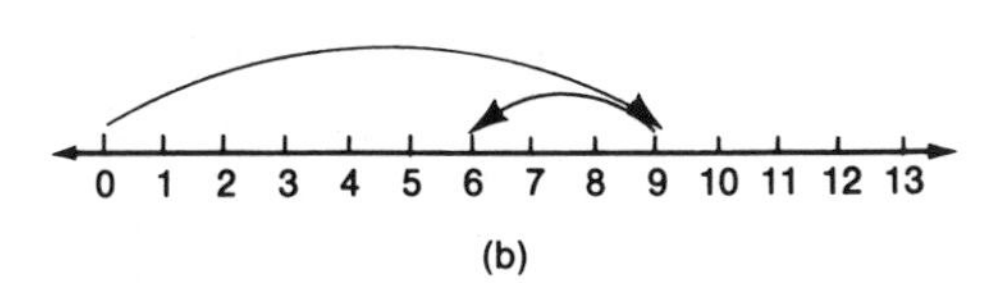


**Figure 7-12** Number lines used to illustrate inverse relationship between addition and subtraction. In (a), the addition sentence $6 + 3 = 9$ is shown; the subtraction sentence $9 - 3 = 6$ is shown in (b)

displayed. Call a child to the number line to show the first addend using a cricket jump. Then the second addend should be shown with a second jump of eight units (Figure 7-13). The addition sentences should be written on the chalkboard as the steps are shown on the number line.

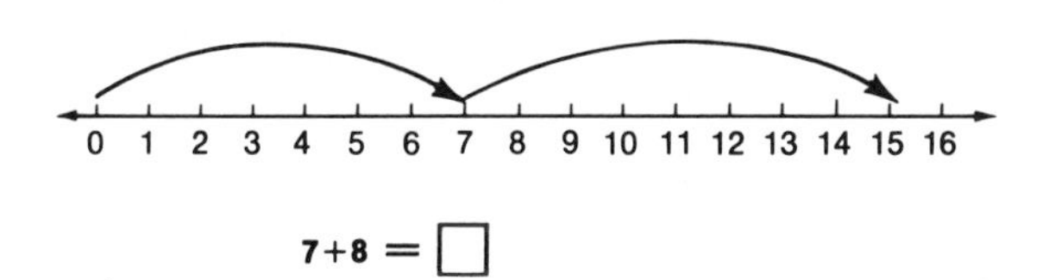


**Figure 7-13** The addition $7 + 8 = \square$ shown on the number line

---

## Self-check of Objective 11

Draw two number lines from 0 to 10. Illustrate the addition sentence $4 + 5 = \square$ with cricket jumps on one. Illustrate the subtraction sentence $8 - 3 = \square$ with cricket jumps on the other.

---

### 3. Using the Addition Table to Learn Facts

Sometime during their work with addition and subtraction, children should see the facts organized in a table (Figure 7-14). The table should not be presented already filled in; neither should the children simply fill it in as a seat-work activity. Rather, children should complete it as part of a group lesson during which you discuss with them features that will help them master the facts. Among the ideas to be stressed are these:

1. The first row at the top and the first column at the left contain facts involving zero—the identity for addition. (Ways of teaching the

| + | 0 | 1 | 2 | 3 | 4 | 5 | 6 | 7 | 8 | 9 |
|---|---|---|---|---|---|---|---|---|---|---|
| 0 | 0 | 1 | 2 | 3 | 4 | 5 | 6 | 7 | 8 | 9 |
| 1 | 1 | 2 | 3 | 4 | 5 | 6 | 7 | 8 | 9 | 10 |
| 2 | 2 | 3 | 4 | 5 | 6 | 7 | 8 | 9 | 10 | 11 |
| 3 | 3 | 4 | 5 | 6 | 7 | 8 | 9 | 10 | 11 | 12 |
| 4 | 4 | 5 | 6 | 7 | 8 | 9 | 10 | 11 | 12 | 13 |
| 5 | 5 | 6 | 7 | 8 | 9 | 10 | 11 | 12 | 13 | 14 |
| 6 | 6 | 7 | 8 | 9 | 10 | 11 | 12 | 13 | 14 | 15 |
| 7 | 7 | 8 | 9 | 10 | 11 | 12 | 13 | 14 | 15 | 16 |
| 8 | 8 | 9 | 10 | 11 | 12 | 13 | 14 | 15 | 16 | 17 |
| 9 | 9 | 10 | 11 | 12 | 13 | 14 | 15 | 16 | 17 | 18 |

**Figure 7-14** An addition table helps children master the addition and subtraction facts

addition identity are discussed later in this chapter.) These facts are easily mastered.

2. The second row and second column contain sums for facts involving 1 as an addend. The sums in each row and column result from adding 1 to the other addend.

3. Sums for facts involving 2 and 3 as addends are in the third and fourth rows and columns, respectively.

4. The diagonal from the upper-left to lower-right corners contains the sums of the doubles, that is, pairs of like addends.

5. The diagonals immediately above and below the one containing doubles contain the "near" doubles. Sentences like $5 + 6 = 11$ and $7 + 8 = 15$ are near doubles. Once children have mastered the doubles, they can use them to learn the near doubles.

6. By now most of the table has been filled in and there remain only a few facts to consider. These are the combinations that give children the most difficulty. Most children will be impressed with how few difficult combinations there are. These should be identified and put on flashcards for future practice.

7. Since subtraction is the inverse of addition, the table also contains the subtraction facts. Instead of using the numbers in the top row and left column as addends to determine sums, use a sum and addend to find a missing addend. To find the answer to $9 - 6 = \square$,

locate 9 opposite 6 in the left column and go to the top of the column containing 9; the answer—3—is located at the top of the column. For $9 - 3 = \square$, locate 9 opposite 3 in the left column and go to the top of the column to locate the answer—6.

Point out the roles of 0 and 1 in subtraction; discuss the doubles and near doubles in a manner similar to the discussion of them in addition; this will help children master them for subtraction.

---

## Self-check of Objective 12

Make a copy of the addition table. Discuss some ways it can be used to help children master the addition and subtraction facts.

---

## PROVIDING PRACTICE WITH THE BASIC FACTS

Children's first experiences with the basic facts of addition and subtraction should be through activities that develop understanding of the operations. Once you are sure that children understand the facts' meanings, give them activities that encourage them to commit the facts to memory. A child should begin practice activities as soon as he or she is ready for them, and continue until he or she can give all of the answers without hesitation.

Five generalizations regarding practice in mathematics are given in Chapter 2; they are repeated here:

1. Practice and understanding go hand in hand.
2. The reasons for practice must be clear to children.
3. The kind and amount of practice are not likely to be the same for all children at the same time.
4. The practice sessions should be brief and occur often.
5. A variety of materials and procedures should be used.

In addition to these generalizations, which govern the way practice sessions are organized, there are principles which are helpful for guiding specific strategies during practice sessions.[4] During drill sessions, stress memorization. If a child hesitates when shown a flashcard or given a

[4]These principles are adapted from Edward J. Davis, "Teaching the Basic Facts," *Developing Computational Skills*, 1978 Yearbook, National Council of Teachers of Mathematics, 1978, pp. 52–58.

combination orally, do not give an explanation or allow time for the child to count on his or her fingers. Instead, show or name the answer. Then show the flashcard or repeat the combination orally and have the child respond immediately. Provide practice with only a few combinations at a time; give frequent review of previously learned combinations. Provide constant encouragement and recognition of progress through verbal praise, badges (see page 65, 67), and individual record charts.

### 1. Practice Materials

There are many sources of practice materials. The most commonly used are children's textbooks or workbooks. These sources are good ones, and should not be overlooked. At the same time, other materials are required if the above principles are to be observed.

Cassette and reel-to-reel tape programs are an effective way to provide for individual children's needs. Among recently developed programs are the following:

*Arithmetricks*—a set of twelve tapes for the primary grades. Coronet Media, 65 E. South Water Street, Chicago, Ill. 60601.

*Computapes*—a set of tapes organized into six modules for grades one through six. Science Research Associates, Chicago, Ill. 60611.

*Drill tapes*—a set of eighty tapes for grades one through six. Educational Progress Corporation, Tulsa, Okla. 74145.

*Mathematics Teaching Tape Program*—a set of thirty-two primary- and forty intermediate-grade tapes. Houghton Mifflin Company, Boston, Mass. 02107.

*Skillseekers*—a set of three separate kits, each with twelve tapes and practice cards. Addison-Wesley Publishing Company, Reading, Mass. 01867.

*Wollensak Teaching Tapes—Mathematics*—a set of more than 140 tapes for grades one through twelve. Mincom Division, 3M Company, St. Paul, Minn. 55101.

In addition to tapes, these programs have work sheets, and some have diagnostic-placement tests.

### 2. Games

Games also give variety to a practice program. The following games are popular among primary-grade children:

**a. Deliver the Mail.**[5] The outlines of houses are cut from tagboard or colored construction paper. Each house has a one-digit numeral printed

[5] See *GCMP—Greater Cleveland Mathematics Program*, Teacher's Guide for First Grade (Chicago: Science Research Associates, 1962), p. 146.

on it. The houses are pinned to a bulletin board. Cards printed with open addition sentences, such as 5 + 4 = □, are distributed to the children. Some children are "letter carriers" who deliver the mail by pinning the sentence cards beneath the proper houses, while the other children check to see if the mail is delivered correctly.

**b. Railroad Board Flowers.** Colorful sheets of railroad board provide the material for flowers. Cut a center for each flower from one color of material. Mark a numeral to represent a sum on each center. Make flower petals from another color of railroad board, and mark combinations that equal that sum, such as 4 + 1 and 3 + 2, on separate petals. Make a stem and two leaves from green railroad board. Mark a combination on each leaf. Attach a small magnet to the back of each flower part. Put the stem and center of several flowers on a large magnetic board. Children make the flowers "grow" by putting the petals and leaves on them (Figure 7-15).

**c. I Am Thinking of a Sentence.** A child is chosen to be "It." She is instructed to think of a number sentence. When she has thought of one, for example, 6 + 3 = 9, she says "I am thinking of an addition sentence for 9." The rest of the children try to guess which sentence "It" has in mind. When called on by "It," a child will ask "Are you thinking of __ + __ = 9?" (filling in two numbers for the blanks). The child who gives the correct sentence replaces "It" and the game continues.

*Deliver the Mail* and *I Am Thinking of a Sentence* can be used for subtraction practice, too. The subtraction games that follow are also good for primary children.

**d. Cardboard Ring.** Prepare a cardboard ring, such as the one in Figure 7-16, and fasten it to the chalkboard; write a numeral, such as "9," in its center. Children practice the facts by giving the answer for each combination as you point to different numerals on the ring. Answers can be given in unison or by individual children. Or, a child can write the answers on the chalkboard around the outside of the ring. Other rings

**Figure 7-15** Railroad board flowers for addition combinations for 8, 9, and 10

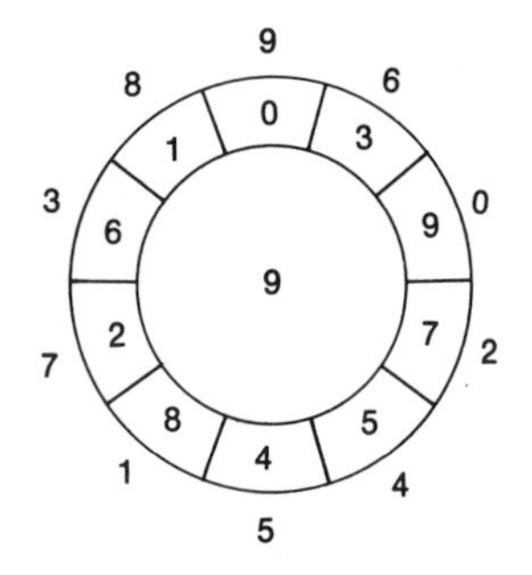


**Figure 7-16** Ring for practicing subtraction and addition facts

should be prepared to provide practice with different minuends. For example, to show subtraction with minuends of 11, the ring would not have the numerals "0" and "1" on it since the numbers represented by these numerals are not a part of the basic facts that have 11 as a minuend. Similarly, the facts that involve 8 as a minuend do not have 9 as a number in them. This ring can also be used to practice with addition facts. One of the numbers 0 through 9 can be indicated by a numeral in the center. Children add the number pairs named by the numeral in the center and the numerals on the ring.

b. **Stranger in the Family.**[6] Write on the chalkboard sets of addition and subtraction facts for families.

| | |
|---|---|
| $4 + 3 = 7$ | $4 + 5 = 9$ |
| $7 - 4 = 3$ | $9 - 4 = 5$ |
| $3 + 4 = 7$ | $9 - 6 = 3$ |
| $4 - 3 = 1$ | $9 - 5 = 4$ |
| $7 - 3 = 4$ | $5 + 4 = 9$ |

Have the children check the sentences in each family to determine the one sentence that does not belong.

c. **Set Cards.** Folding set cards can be used to practice both subtraction and addition facts. Prepare cards, such as those shown in Figure 7-17, large enough so the sets of dots are easily seen from all parts of the room. To practice the subtraction facts, show an unfolded card. After the children have determined the number of dots on it, fold one side of the card back. Then ask them to count the dots that still show. Write the sentence for the card on the chalkboard after one side has been folded back. The pairs of sentences for the three cards in Figure 7-17 are: $7 - \square = 3$ or $7 - \square = 4$, $8 - \square = 1$ or $8 - \square = 7$, and $16 - \square = 7$ or $16 - \square = 9$. The place-holder is in the middle position in each of the sentences because the children see how many dots there are altogether and how many are left after the dots on one half have been folded back; they determine the size of the group that was folded back.

The teachers' manuals for primary-grade texts and workbooks contain suggestions for many games, as do mathematics games books.[7] Commercial games are also available. Most school supply companies stock mathematics games along with those for other subjects.

[6] Caroline H. Clark, et al., *Elementary Mathematics: Two by Two*, Teacher's Edition (New York: Harcourt, Brace & World, Inc., 1965), p. 15.

[7] See, for example, Mary E. Platts, ed., *Plus*, rev. ed. (Stevensville, Mich.: Educational Services, Inc., 1975), pp. 69–122; and Leonard M. Kennedy and Ruth L. Michon, *Games for Individualizing Mathematics Learning* (Columbus, Ohio: Charles E. Merrill Books, Inc., 1973), pp. 31–54.

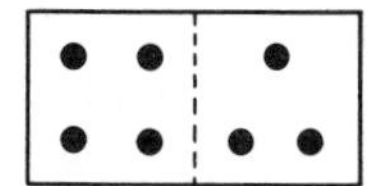
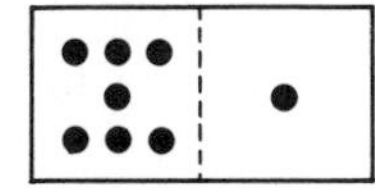
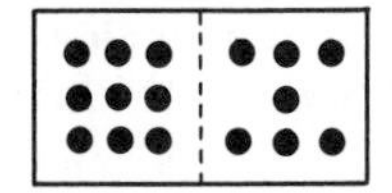

**Figure 7-17** Folding set cards for studying addition and subtraction facts

As children progress through the grades, the activities are on a more mature level. The following activities can be used with older children.

a. **'Round the World.** A child is chosen "It." She stands beside a seated classmate. Show the two children an addition flashcard, such as

$$\begin{array}{r} 3 \\ +7 \\ \hline \end{array}$$

or give a combination orally, "3 plus 7 equals ?" The child who gives the correct answer first is the winner. She takes her place standing beside another seated child. The game continues, with the winner of each pair going to the next child. The object is to get "'round the world" by moving from seat to seat without being beaten by one's classmates.

| 0 | 6 | 18 | 9 | 15 |
|---|---|---|---|---|
| 4 | 3 | 11 | 8 | 7 |
| 5 | 7 | Free | 17 | 16 |
| 6 | 15 | 13 | 2 | 4 |
| 10 | 1 | 12 | 14 | 6 |

**Figure 7-18** Sample card for addition bingo

b. **Addition Bingo.** Each child is given one or two bingo cards on which numerals from 0 to 18 are printed (Figure 7-18). A numeral may appear more than once on a card. One at a time, show addition flashcards. Children cover the numerals that name sums for combinations on the flashcards. A winner is declared when a child has covered a row, column, or diagonal of five squares. Sets of cards for addition bingo can be purchased or made by you.

c. **Speed Drills.** Different procedures for giving speed drills should be used. Duplicate the 100 addition combinations in the vertical form and in random order. Instruct the children to write the answers to the combinations as rapidly as possible without sacrificing accuracy. After 2 minutes, or longer if the children are working slowly, record the time on the chalkboard by 10-second intervals until everyone has completed the work. Have each child record the time on his or her paper. Each child should make a study list of the facts missed. Children who do not know the combinations well should have special help and additional practice. Repeat the process periodically, suggesting that children try either to lessen the time it takes to do the work or to increase their number of correct answers.

Another type of speed drill requires each child to have a sheet of ruled paper that has been lined vertically to make five or six columns. The instructions are to record the answers to addition combinations you display. The first answer should be placed on the first line of the first column, the second beneath the first, and so on. A line is to be left blank if the sum for a combination is not known. Show addition flashcards arranged in random order one at a time, regulating the speed according to the capabilities of the class. Do not allow children to count or use other aids to determine an answer. The purpose of the practice is to help increase speed, so the drill may go a little faster than the class may wish. Children who become confused, ask for more time, or miss too many answers are probably not ready to concentrate on developing speed; this procedure should not be used for them.

Minicalculators can be used to provide practice to increase speed. See page 68 of Chapter 3 for a description of one way to do this. These activities can be adapted to provide practice with the subtraction facts as well.

---

## Self-check of Objective 13

A variety of practice materials and activities have been described. Name at least five different types of practice activities. Demonstrate some games that can be used for practice with the facts.

---

## TEACHING THE PROPERTIES OF ADDITION AND SUBTRACTION

Less emphasis is placed presently on the properties of addition, subtraction, multiplication, and division than during the "modern math" era. The commutative, associative, and closure properties and identity element need not be stressed in highly formal ways to be useful to children; however, their meanings and applications should be developed informally.

### 1. The Commutative Property

The following procedure is one of the many effective ways to informally introduce the commutative property. Prepare some paper plates by marking a line through the middle to separate each into halves. Give each child a plate and a handful of markers. Have each child put some markers on one side of the plate and some more on the other side. Then have the children write the addition sentence for the markers on their plates, as shown in Figure 7-19. Have each child show his or her plate and read the sentence.

Next, have the children turn their plates halfway around to reverse their sets. Have them write the new sentences. When the children com-

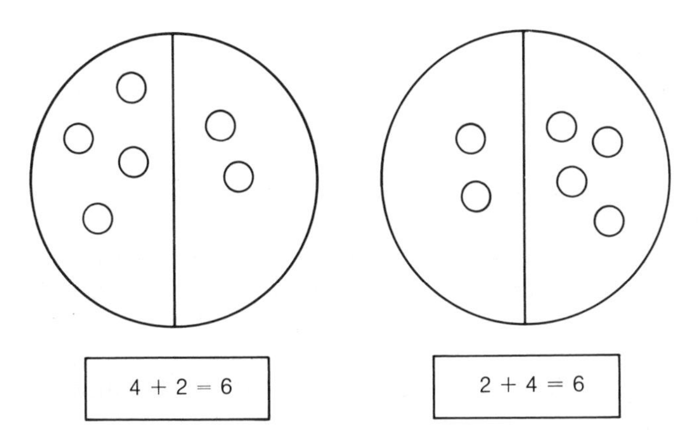


**Figure 7-19** Paper plates and markers illustrating the commutative property

pare their sentences, the fact that the sums are identical, even though the order of addends is reversed, is readily seen. If a child has two sets the same size, the addends will be equal, so both sentences will be the same after the sets are reversed, and the commutative property will not apply. The children should discuss this property in terms that are meaningful to them.

Later, children should learn the correct name for the property and a more precise definition of it. By the time they are fourth or fifth graders, they should use the property and its name correctly as they learn to add new kinds of numbers, such as fractional numbers. You should regularly raise questions about commutativity to help children see its use in a variety of situations. To enlarge their understanding of this property, children should look for additional applications of it as they work with new kinds of numbers.

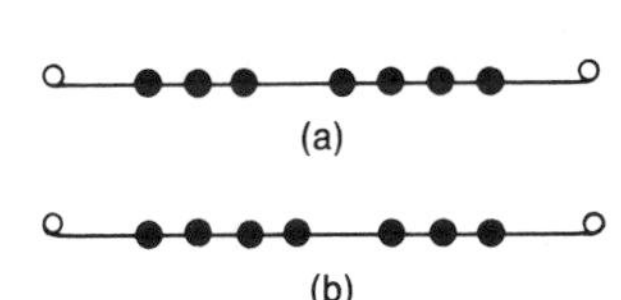


**Figure 7-20** Beads on a wire to show sentences (a) 3 + 4 = □ and (b) 4 + 3 = □

Other examples of how to help young children learn the commutative property of addition are:

1. Put two sets of clothespins on the edge of a large card. Hold the card against the chalkboard and write the sentence, including the sum, on the board above the card. Turn the card so the order of the sets is reversed and write the new sentence.
2. Use beads on pieces of stiff wire like those shown in Figure 7-20. Show the two sets (a), and then reverse their order (b).
3. Use Cuisenaire rod trains to show commutativity. Put a train of a 3 rod and a 4 rod next to one made of a 4 rod and a 3 rod (Figure 7-21).
4. Use a number line such as the one in Figure 7-22.

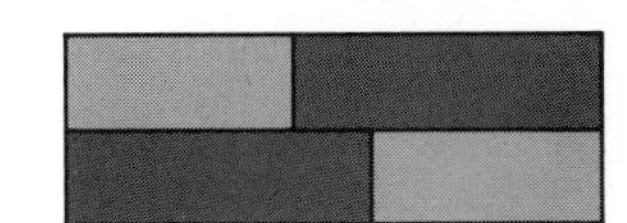

**Figure 7-21** A Cuisenaire rod train illustrates the commutative property

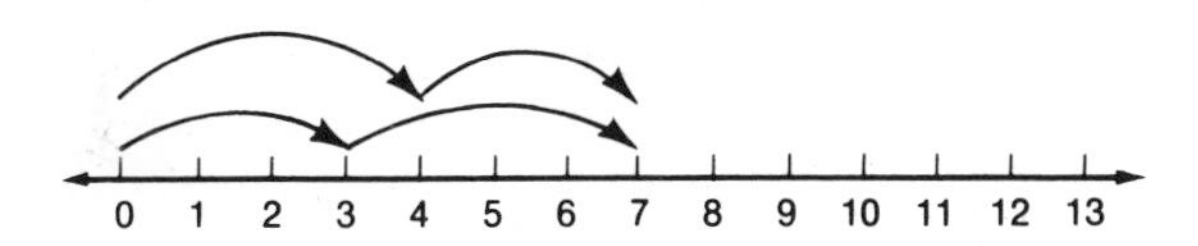


**Figure 7-22** The number line used to show the commutative nature of addition

## 2. The Associative Property

The associative property of addition is used when three or more addends are added. Because addition is a binary operation, it is possible to add only two numbers at a time. Thus when there are three addends, two of the three must be added first; then the third number is added to the sum of the first two.

Addition using three addends is usually introduced before the 100 basic addition facts have all been presented. No specific number of basic facts is a prerequisite, but children must previously have worked with all the basic facts they will use as they first work with three addends. As with other new topics, this one should be introduced by illustration with sets of markers. Three small sets can be arranged horizontally across a magnetic or flannel board and the mathematical sentence describing this situation shown (Figure 7-23). As their attention is directed to the first and second sets of markers, these two sets are joined, and children should think of the resulting number sentence:

$$3 + 2 = 5.$$

This sentence should be put on the board beneath the original sentence. Then the children should note that there is a set of 5 to be joined with a set of 4. The sentence that now describes the situation,

$$5 + 4 = \square,$$

should be shown. When all the sets have been joined, the steps in the process will be shown by these sentences:

$$\begin{aligned} 3 + 2 + 4 &= \square \\ 3 + 2 &= 5 \\ 5 + 4 &= \square \\ 5 + 4 &= 9. \end{aligned}$$

The meaning of each sentence should be reviewed before the process is repeated with other sets of markers.

During their first experiences with three addends, children should add them in order, reading from left to right. After children have learned to add in this manner, the associative property of addition can be introduced through such situations as these:

1. Group children in three small sets at the front of the room. Have the first and second sets unite; then have the third set unite with the combined set. The children should then form in the original groups. The second and third sets unite; then the first set unites with the combined set. Discuss the action as it takes place, and write the addition sentences on the chalkboard.

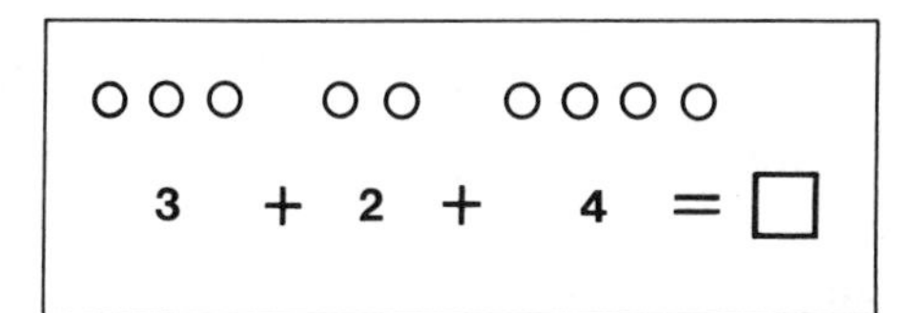


**Figure 7-23** Markers illustrating the sentence 3 + 2 + 4 = □

2. Use clothespins on a card. Put three sets of pins on one edge of the card. Place the card against the chalkboard and write the addition sentence above the pins. Combine the pins in the two sets to the children's left; then unite the third set with the combined set. Put the pins in the three original groupings, and unite the second and third sets. Then unite the first set with the combined sets. Complete the addition sentences as the sets are united.
3. Use sets of beads on a wire, following the procedure suggested for clothespins on a card.
4. Use Cuisenaire rod trains. Make two three-car trains, each composed of the same number and types of rods arranged in the same order. Have children consider the two rods on the left, then the third rod. Next, have them consider the two rods on the right, then the first rod. In both instances, the addition sentences have the same sum, regardless of the order in which the addends are considered.

After the use of this property is clear in simple situations, it should be used in increasingly sophisticated ways. The associative property and its applications for number expressions such as fractional numbers and the simpler signed numbers should be discussed in the intermediate grades of elementary school.

Once children are acquainted with the commutative and associative properties, give them situations in which both properties are applied in useful ways. Children in the third grade and beyond can refine their thinking about the properties through work with examples like this. Instruct the children to determine the sum of

$$246 + 398 + 2 = \square$$

without using paper and pencil. It may be necessary to assist the children by directing their attention to the addends 398 and 2. Once the sum for these addends has been determined, the children should easily determine the sum for the entire sentence. Other sentences, such as those that follow, should be displayed for children to complete without paper and pencil:

$$396 + 241 + 4 = \square$$
$$10 + 367 + 390 = \square$$

$$16 + 83 + 7 + 4 = \square$$

After children have solved each of the sentences, they should discuss how they paired addends and the sequence of steps they used as they added. The discussion should bring out the ideas that the commutative property permits them to think of numbers in any order, and the associative property permits them to pair addends in any way they wish, as long as they use each addend only once and remember to use every addend. Thus in the first example a child might say he or she used the associative property by thinking

$$(398 + 2) = 400, \qquad 246 + 400 = 646.$$

For the second example, the child might say he or she commuted the 241 + 4, and then associated (396 + 4) to get a sum of 400, to which 241 was added for a sum of 641. Of course, the child would have been just as correct to say he or she had commuted the 396 + 241, then associated (396 + 4) and added them to get 400, and then 241 + 400 = 641. Similar steps would be outlined for the third and fourth sentences.

### 3. The Identity Element of Addition

Children who understand the meaning of zero will have little difficulty learning its role in addition. As they deal with markers to learn about addition, they will realize that when zero is an addend, there will be no markers representing that number. To make this more obvious, some primary-grade teachers prefer clearly identifying each group by putting a ring around the markers on a display board. The two addition sentences illustrated in Figure 7-24 are represented by ringed markers. The ring above the 0 in the second sentence indicates the absence of markers when zero is an addend.

Eventually, children will see that when zero is one of a pair of addends, the sum is always the number of the other addend. Zero is the identity element for addition.

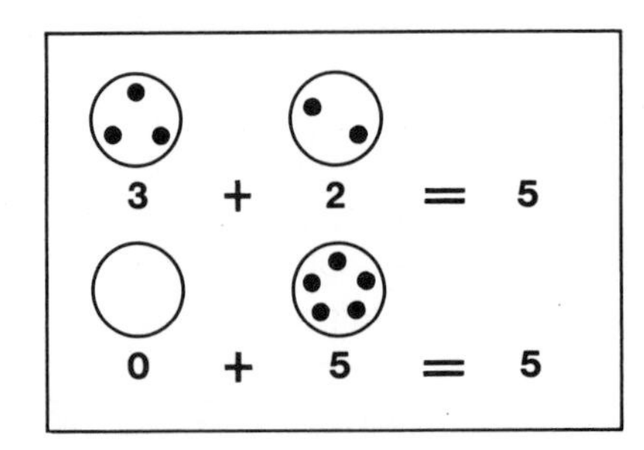


**Figure 7-24** Sets illustrating addition sentences 3 + 2 = 5 and 0 + 5 = 5

During work with subtraction children should be given opportunities for the role of zero to become clear to them. They should see that when zero is subtracted from another number, the answer is always the other number. And when a number is subtracted from itself, the answer is always zero.

### 4. Closure in Addition and Subtraction

The addition table (Figure 7-14) is useful for helping children develop an intuitive understanding of closure for addition. During a discussion of the completed table, you can help children realize that the series of addends across the top and the ones down the left side can be extended indefinitely. A question like "What can we say about the sums that would appear in the table if we had room to extend the addends across the top and down the side without end?" will lead to these conclusions: (1) There would be a sum for every ordered pair of addends, regardless of their sizes, and (2) every sum would be a whole number. These conclusions summarize the closure property for addition. (The table can be used to summarize other properties, too. See pages 165 and 166 for ways the table can be used to review commutativity and the identity element.) The commutative property for addition is emphasized by examining the halves above and below the upper-left to lower-right diagonal. For every sum above the diagonal, there is a corresponding sum involving the same pair of addends below the diagonal.

The closure property does not apply to subtraction of whole numbers. Examples like $6 - 7 = \square$ and $3 - 9 = \square$ show children that for every ordered pair of whole numbers there will not always be a whole-number answer.

---

## Self-check of Objective 14

Demonstrate with manipulative materials or another suitable device how each of these number properties for addition can be demonstrated: (a) commutative, (b) associative, (c) identity, and (d) closure. Show that subtraction is not commutative or associative, and why closure does not apply.

---

## Common Pitfalls and Trouble Spots

Two common pitfalls to avoid during initial work with addition and subtraction include the following.

The first, and most common mistake teachers make as children learn to add and subtract is to introduce the sentences prematurely and/or too rapidly.

Since sentences are abstract, or symbolic, ways of representing actions with real objects, children who are still looking at collections of objects in global ways and do not understand that quantity remains the same regardless of actions on the parts will not fully understand addition and subtraction sentences. Be sure each child is a conserver of quantity before abstract sentences are introduced.

Children who are conservers are often given sentences at too rapid a pace. Even conservers should work with markers until they are familiar with all of the concrete representations of the larger combinations before they deal with them in the semiconcrete and the abstract modes. Then concrete-manipulative, semiconcrete-representational, and paper-and-pencil activities should continue simultaneously until children are ready for activities that lead to mastery of the facts.

Secondly, some teachers misinterpret a child's ability to repeat statements explaining a number property as evidence of his or her understanding of the property. Children may state that "Addition is a commutative operation" and still not know either what the statement means or how to apply the property when adding. Once a number property is introduced, children need frequent and appropriate opportunities to use it and to advance their understanding of it in new situations.

Two aspects of early work with addition and subtraction are troublesome to many children.

The first trouble spot is the difficulty many children have in mastering each operation's basic facts for immediate recall. Difficulties arise from different causes. Many children become "finger counters" because they have insufficient opportunities to work with markers during the early stages of addition and subtraction. Even though a child may become very skillful and quick at adding or subtracting on his or her fingers, if the process is habitual, it will interfere with memorization of the facts. Ample experiences with manipulative materials and the delayed use of pictures and paper-and-pencil work will reduce the incidence of finger counting.

Another cause of failure in memorizing the basic facts is the lack of opportunity to practice them in systematic ways. Procedures for doing this have been discussed.

A second trouble spot for children is working with sentences with the place-holder in other than the terminal position; that is, sentences such as $\square - 5 = 3$ and $5 + \square = 7$. It is not uncommon for children to put a "2" in the place-holder box in the subtraction sentence and a "12" in the box for the addition sentence. Children who are introduced to such sentences prematurely rely on their incomplete knowledge of addition and subtraction to obtain their answers. Wait until the two operations are well understood before introducing such sentences, then use manipulative materials that illustrate each type of sentence, along with meaningful story-problem situations. Then if you find yourself resorting to telling children *rules* for completing them, hold off until a later time.

## Self-check of Objectives 15 and 16

Two pitfalls associated with early work with addition and subtraction are discussed. Identify each one; then describe a way each can be avoided.

Describe two trouble spots for children during early work with addition and subtraction; then tell how each can be minimized.

## SUMMARY

Addition and subtraction are usually introduced during the first grade, and once each is introduced, they are usually studied simultaneously. Children develop readiness for addition and subtraction as they learn to count using manipulative materials and the number line. Working in the concrete-manipulative mode with real objects and markers gives children experiences joining and separating groups of objects, which is the basis for understanding the import of addition and subtraction and the use of abstract sentences.

The real-life situation that gives rise to addition is the joining of two sets having no members in common; there are four situations that give rise to subtraction. There are 100 basic addition facts, which are all the possible combinations of pairs of addends smaller than 10 and their sums. Each addition combination has a corresponding basic subtraction fact.

Once children know the meaning of addition and subtraction, they need to practice the basic facts until they can give the answer for each combination without hesitation. Practice pages in workbooks, taped practice materials, large- and small-group games, and other materials provide a variety of practice opportunities. The basic properties of addition—commutative, associative, identity, and closure—can be introduced and developed through work with manipulative materials, number lines, and an addition table.

The premature and too rapid introduction of addition and subtraction sentences, and failure to recognize evidence of understanding of basic addition and subtraction properties are pitfalls for a teacher to avoid. Children frequently find that mastering the basic addition facts—because of interference from finger counting—and understanding sentences with the place-holder box in other than the terminal position are trouble spots.

## STUDY QUESTIONS AND ACTIVITIES

1. Many nonmathematical operations undo what has been done by a preceding operation, for example, when a child dresses and then undresses a doll. Make a list of other such situations children might

consider as they discuss the meaning of inverse operations. Besides addition and subtraction, what other examples of inverse operations in mathematics can you identify?

2. The operation of addition is commutative. One way to help children remember this property's meaning is to encourage them to think of commutative and noncommutative nonmathematical operations. An example of a commutative nonmathematical operation is to apply first salt, then pepper, to a fried egg. The result will be the same if the procedure is reversed. An example of a noncommutative operation is mixing a cake and then baking it. List other operations of both types.

3. In an intermediate grade, give a speed test of the basic addition facts. As children take the test, observe for signs of finger counting or other immature procedures for determining sums. Check the papers to determine how accurately the children responded. Do they respond as readily and accurately as you believe they should? If not, what procedures do you recommend to help them?

4. Begin a collection of games and activities suitable for developing children's mastery of the basic addition and subtraction facts. Include group/individual, competitive/noncompetitive, and self-correcting/teacher-corrected types of games and activities. If possible, find at least one game and/or activity that serves each of these purposes: (a) helps develop a child's understanding of addition and/or subtraction, (b) provides practice, and (c) helps evaluate a child's understanding of addition and/or subtraction.

5. Some strategies for helping children learn the basic addition and subtraction facts are presented in this chapter. Read appropriate articles from this chapter's reading list to identify other strategies. Write a brief description of each strategy.

## FOR FURTHER READING

Ashlock, Robert B. "Teaching the Basic Facts: Three Classes of Activities," *The Arithmetic Teacher*, XVIII, No. 6 (October 1971), 359–364. According to Ashlock, three classifications of activities are involved in learning the basic facts: (1) understanding the facts, (2) relating the facts, and (3) mastering the facts. Illustrated examples of activities at each level are given.

———, and Carolynn A. Washbon. "Games: Practice Activities for the Basic Facts," *Developing Computational Skills*, 1978 Yearbook of the National Council of Teachers of Mathematics. Reston, Va.: The Council, pp. 39–50. Describes the role of practice activities and guidelines for selecting and preparing them, along with examples of games. A short list of steps for creating games helps the teacher design his or her own games.

Bernard, John E. "Constructing Magic Square Number Games," *The Arithmetic Teacher*, XXVI, No. 2 (October 1978), 36–37. The magic square games capitalize on children's interest in tic-tac-toe games. Squares involving whole numbers, common fractions, and factors of numbers are illustrated. Those for whole numbers provide practice with the basic facts.

Davidson, Patricia S. "Rods Can Help Children Learn at All Grade Levels," *Learning*, VI, No. 3 (November 1977), 86–88. Cuisenaire rods provide a manipulative model for developing a wide variety of mathematical concepts. Not only are they useful as models for addition and subtraction, but also for multiplication and division, place value, fractions, and metric units of measure.

Herald, Persis Jean. "Helping the Child Who 'Can't Do Math,' " *Teacher*, LIX, No. 7 (March 1974), 46–47, 89. Several ways of helping children who find mathematics difficult are discussed: pinpointing problems, providing "hands-on" learning, developing understanding, overcoming messiness, and developing strengths.

Hollister, George E., and Agnes G. Gunderson. *Teaching Arithmetic in the Primary Grades*. Boston: D. C. Heath and Company, 1964. The authors present ideas about teaching addition to children in the primary grades in Chapter 6. Many useful teaching aids are illustrated.

Hutchinson, James W., and Carol E. Hutchinson. "Homemade Device for Quick Recall of Basic Facts," *The Arithmetic Teacher*, XXV, No. 4 (January 1978), 54–55. An overhead transparency using a grid of random digits and paper shields that reveal rows and columns of digits offers opportunity for several skill-development activities; for example, recall of basic facts (sums and products), comparing numbers, reading two- and three-place numbers.

Macy, Joan M. "Drill?—Deadly?—Never," *School Science and Mathematics*, LXXIII, No. 7 (October 1973), 595–596. Ways to spice up and create enthusiasm for drill are suggested. All are simply prepared and used by a teacher with whole or part of a class.

Myers, Ann C., and Carol A. Thornton. "The Learning Disabled Child—Learning the Basic Facts, *The Arithmetic Teacher*, XXV, No. 3 (December 1977), 46–50. Even though the title of the article refers to learning disabled children, the strategies for helping children master the basic addition and multiplication facts and applying them to work with larger numbers are applicable to all children. Explains examples of five different strategies.

Rathmell, Edward C. "Using Strategies to Teach the Basic Facts," *Developing Computational Skills*, 1978 Yearbook of the National Council of Teachers of Mathematics. Reston, Va.: The Council, pp. 13–38. Presents strategies for organizing basic facts for aiding children's understanding and retention. Includes many illustrations.

Thompson, Charles S., and William P. Dunlop. "Basic Facts: Do Your Children Understand or Do They Memorize?" *The Arithmetic Teacher*, XXV, No. 3 (December 1977), 14–16. Discusses two simple procedures for diagnosing and analyzing children's knowledge of basic facts, level of understanding, and rate of learning facts. Both require materials easily and inexpensively made by a teacher.

Weill, Bernice F. "Mrs. Weill's Hill: A Successful Subtraction Method for Use with the Learning Disabled Child," *The Arithmetic Teacher*, XXVI, No. 2 (October 1978), 34–35. Mrs. Weill's hill is a method of helping children visualize subtraction with minuends from 10 to 19. It has been used successfully with learning disabled, slow learning, and average children.

# 8 Extending the Operations of Addition and Subtraction

Upon completion of Chapter 8, you will be able to:

1. Explain orally or in writing the meaning of higher decade addition, and describe a minimum of two materials and procedures for helping children understand and perform it accurately.

2. Demonstrate with place value devices, such as a pocket chart, beansticks, and an abacus, how to represent the addition of two numbers greater than 10 when regrouping ("carrying") is not required and when it is.

3. Explain two different processes for completing the algorithm for subtracting when regrouping ("borrowing") is necessary, and demonstrate with materials and procedures how to make each process meaningful to children.

4. Demonstrate a one-step process for subtraction involving minuends that contain zeros, such as in the example

$$\begin{array}{r} 6002 \\ -4365. \\ \hline \end{array}$$

5. Explain the low-stress algorithms for addition and subtraction, and tell for whom they are useful.

6. Explain at least three ways of checking addition involving three or more addends, and two ways of checking subtraction.

7. Describe briefly five addition and subtraction activities for the minicalculator.

8. Use a number line to represent the meaning of addition and subtraction of integers.

9. Identify four pitfalls common to work with addition and subtraction, and describe a way to avoid each one.

10. Explain some of the common errors children make as they add and subtract whole numbers, and describe general guidelines for overcoming such errors.

Key Terms you will encounter in Chapter 8:

higher decade addition
adding by endings
algorithm
sum
addend
regrouping ("carrying")
regrouping ("borrowing")
hundreds chart
decomposition subtraction
"crutch"
equal additions subtraction
minuend
subtrahend
difference
low-stress algorithm
palindrome
palindromic number
magic square
Nim
integer
negative number
directed number
vector

Materials and procedures for introducing addition and subtraction and their properties are discussed in Chapter 7. The present chapter extends the discussion to include addition and subtraction of larger whole numbers and small negative integers.

## TEACHING HIGHER DECADE ADDITION

Higher decade addition is adding a number represented by a one-digit numeral to a number represented by a two-digit numeral. The importance of this skill becomes evident when we examine situations in which it is commonly used:

1. Social situations, such as adding to determine the cost of two items that are priced at 23¢ and 9¢.
2. Addition involving three or more addends, such as $8 + 9 + 6 = \square$, where 6 is added to 17 as the addition proceeds from left to right.
3. Multiplication that involves pairs of factors such as 6 and 98, where 4 is added to 54 as the multiplication is completed.

You should realize that unless children are readily able to add numbers such as these in one step—that is, without any regrouping procedures—they will be handicapped in much of their future work with addition and multiplication. (This type of addition is also called *adding by endings* because an answer can be determined by using the one-digit numeral and the ones place, or "ending," of the two-digit numeral.)

Before children participate in activities that help them learn higher decade addition, they must know the basic addition facts well. When children are ready for higher decade addition, you can introduce the following activities.

### 1. Use a Number Line

A number line is useful to help children grasp the meaning of higher decade addition; one that extends through 100 should be available. To show the meaning of the sentence, $29 + 5 = \square$, a child should begin by locating the 29 on the line. He or she next moves along the line by taking

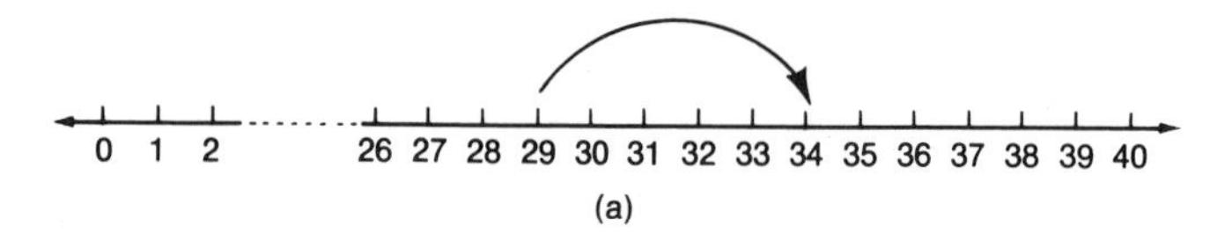


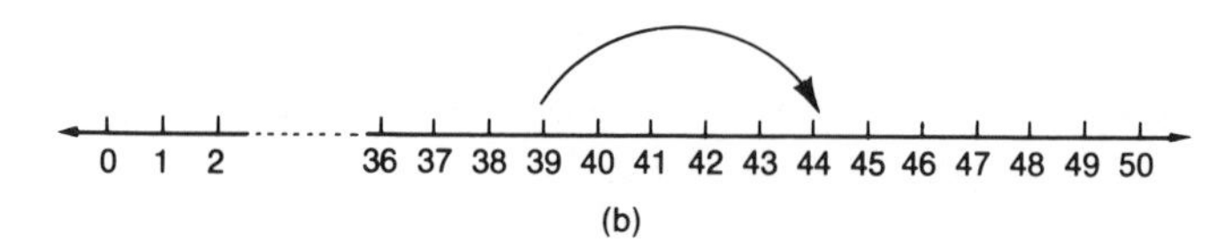


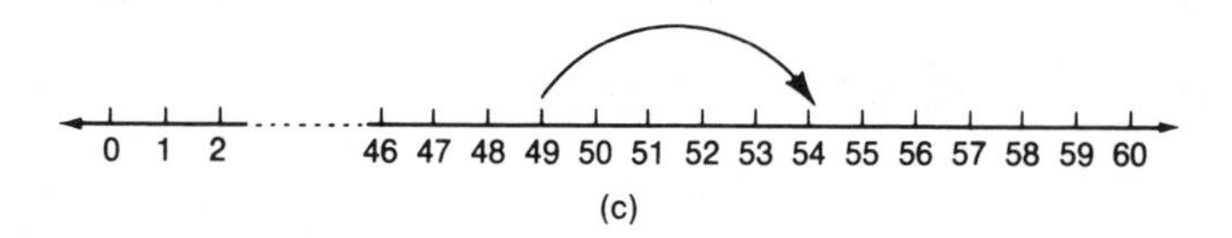


**Figure 8-1** Number line used to illustrate higher decade addition sentences (a) $29 + 5 = \square$, (b) $39 + 5 = \square$, and (c) $49 + 5 = \square$

a jump that is five units long, ending at 34 (Figure 8-1). The number line will help determine the answers to other higher decade addition combinations that involve 9 and 5 in the ones place, as (b) and (c) demonstrate for $39 + 5 = \square$ and $49 + 5 = \square$.

### 2. Use a Hundreds Chart

Children can use a hundreds chart similar to the one in Figure 8-2 to determine answers for higher decade addition. For the sentence $45 + 6 = \square$, a child would locate 45 on the chart and then count six more places, stopping at 51. Then the sums for other sentences involving 5 and 6, such as $55 + 6 = \square$ or $65 + 6 = \square$, can be determined.

### 3. Use Patterns

After the number line and hundreds chart are used, encourage a one-step thought process in doing higher decade addition. Addition sentences organized according to a pattern will help:

$$5 + 3 = \square,\ 15 + 3 = \square,\ \ldots,\ 95 + 3 = \square.$$
$$7 + 8 = \square,\ 17 + 8 = \square,\ \ldots,\ 97 + 8 = \square.$$

After children have seen a number of sentences involving higher decade addition represented on the number line and a hundreds chart, they should discover the following:

1. When the sum of the numbers represented by numerals in the ones place is less than 10, the number in the tens place of the sum is the

| 1 | 2 | 3 | 4 | 5 | 6 | 7 | 8 | 9 | 10 |
|---|---|---|---|---|---|---|---|---|---|
| 11 | 12 | 13 | 14 | 15 | 16 | 17 | 18 | 19 | 20 |
| 21 | 22 | 23 | 24 | 25 | 26 | 27 | 28 | 29 | 30 |
| 31 | 32 | 33 | 34 | 35 | 36 | 37 | 38 | 39 | 40 |
| 41 | 42 | 43 | 44 | 45 | 46 | 47 | 48 | 49 | 50 |
| 51 | 52 | 53 | 54 | 55 | 56 | 57 | 58 | 59 | 60 |
| 61 | 62 | 63 | 64 | 65 | 66 | 67 | 68 | 69 | 70 |
| 71 | 72 | 73 | 74 | 75 | 76 | 77 | 78 | 79 | 80 |
| 81 | 82 | 83 | 84 | 85 | 86 | 87 | 88 | 89 | 90 |
| 91 | 92 | 93 | 94 | 95 | 96 | 97 | 98 | 99 | 100 |

**Figure 8-2** A hundreds chart can be used for determining answers for higher decade addition

same as the number in the tens place of the larger addend, for example,

$$33 + 4 = 37.$$

2. When the sum of the numbers represented by numerals in the ones place is ten or more, the number in the tens place of the sum is one greater than the number in the tens place of the larger addend, for example,

$$38 + 5 = 43.$$

From time to time, children should practice higher decade addition to maintain proficiency. The following exercises are useful.

1. Sentences similar to the above examples but arranged in random order can be written on the chalkboard or duplicated for children so they can write the sums.
2. Children can give oral responses to such dictated sentences as "Add 6 to 45." Because much higher decade addition is encountered during multiplication, many of the numbers dictated should be the product of two numbers smaller than 10, like 35, 42, 49, or 56. The number to be added to these numbers need not be greater than 8 since this is the largest number that is "carried" in multiplication.
3. Have children count by 5s, 6s, 7s, 8s, and 9s beginning at any number represented by a two-digit numeral; for example, to count by 5 beginning at 18, one would say "18, 23, 28, 33, . . ." The counting can be done in unison or by individual children.
4. Have children respond orally to dictated instructions, such as, "Add 6 to the product of 6 and 8."

## Self-check of Objective 1

Explain the meaning of higher decade addition, and describe at least one situation giving rise to it. What materials and procedures can be used to help children understand this type of addition and perform it rapidly?

## TEACHING ADDITION OF LARGER WHOLE NUMBERS

Children are usually ready to learn to add whole numbers represented by two- and three-digit numerals when they are in second grade; however, they must have a good understanding of subordinate concepts, such as the meaning of addition, the basic addition facts, and place value.

An introductory lesson should begin with a familiar situation. You might start with a story: "Yesterday John had thirty baseball trading cards. Today he has twenty more. If he has all his cards at school today, how many does he have altogether?" Each child should have time to determine the answer in his or her own way. Many children will be able to solve this problem without supplementary aids or pencil and paper. Others may be able to determine it only after they count, use markers or their fingers, or make marks on paper. After the children have had time to determine the answer, one of them should be asked to use it in the addition sentence

$$30 + 20 = 50.$$

Write the sentence on the chalkboard, and then write the problem in vertical, or algorithm, form, as in Figure 8-3. Next, have a child represent the meaning of the first addend with markers in a pocket chart. Write the 30 as "3 tens." After the second addend is illustrated in the pocket chart and by the numeral "2 tens," the problem will appear as in Figure 8-3. One of the children should be called to the pocket chart to join the sets of markers and count them. Then she should write the answer beneath the algorithm. Additional similar problems should be assigned to be worked out with the pocket chart and the algorithm in simplified form.

| Tens | Ones | |
|---|---|---|
| ▯ ▯ ▯ | | 30 → 3 tens |
| ▯ ▯ | | +20 → +2 tens |
| | | |

**Figure 8-3** Place value pocket chart and simplified algorithm for showing meaning of 30 + 20 =50

In lessons that follow, children can use work sheets that contain pictures of the place value pocket chart or another place value device. Instead of using markers (concrete mode), they use pictures of them (semiconcrete mode). Figure 8-4 illustrates an example of the type of work children might complete on a work sheet. To illustrate $40 + 30 = \square$, a child draws markers to show the 4 tens in one addend and the 3 tens in the other. Then the child writes numerals at the bottom to indicate the total number of tens—7—and ones—0. Six or eight examples can be put on one work sheet.

Addition of only multiples of ten will not present any special problems to most children. This should be soon followed by addition of numbers without zero in the ones place. Most textbooks introduce the addition of these numbers with examples that do not require regrouping, or "carrying," as the numbers are added. Some authorities believe children should begin regrouping right away. Their argument states that when children add two numbers, such as 23 and 45, using the conventional algorithm form

$$\begin{array}{r} 23 \\ +45, \\ \hline \end{array}$$

they can begin their addition in either the tens or the ones place and get the correct sum either way. Thus they do not learn the value of beginning addition in the ones place. If an addition requires regrouping, as

$$\begin{array}{r} 28 \\ +47 \\ \hline \end{array}$$

does, they will learn from the beginning that it is convenient to start the addition in the ones place and then go to the tens place. You will probably decide which kind of example to use on the basis of what the textbooks used in the classroom include. This is reasonable in most cases because the textbook will provide most of the practice materials the children will use once they have been introduced to the new work. Of course, it is not necessary to introduce a topic according to the text's order of presentation. If your children have a good understanding of numbers and the simpler

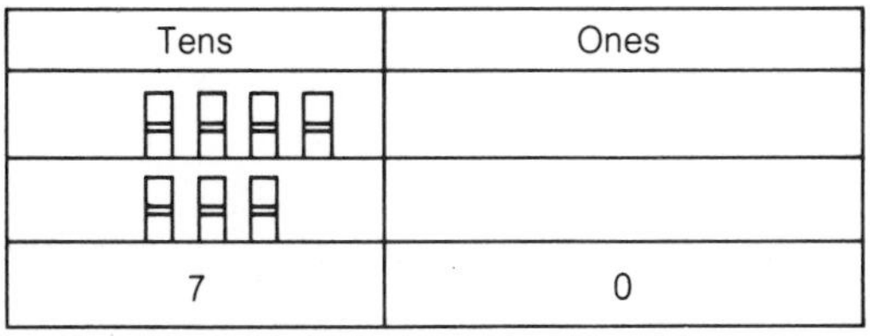

| Tens | Ones |
| --- | --- |
| ▯▯▯▯ | |
| ▯▯▯ | |
| 7 | 0 |

**Figure 8-4** Illustration from a child's work sheet showing that 40 + 30 = 70

addition processes, you may choose to use examples that require regrouping regardless of the text's order of progression. Then addition practice with and without regrouping can proceed as you select material from the textbook as needed.

When addition with regrouping is introduced, a place value pocket chart or abacus is useful to make the regrouping process meaningful. The following situation might begin a lesson: "Sally has twenty-six paper doll dresses for one doll and seventeen dresses for another. How many dresses does she have for both of them?" Children should first be asked to think of the addition sentence for the story, 26 + 17 = □. Then they should determine how to place the markers in a pocket chart, such as the one in Figure 8-5, or an abacus to indicate the two addends and how to write the

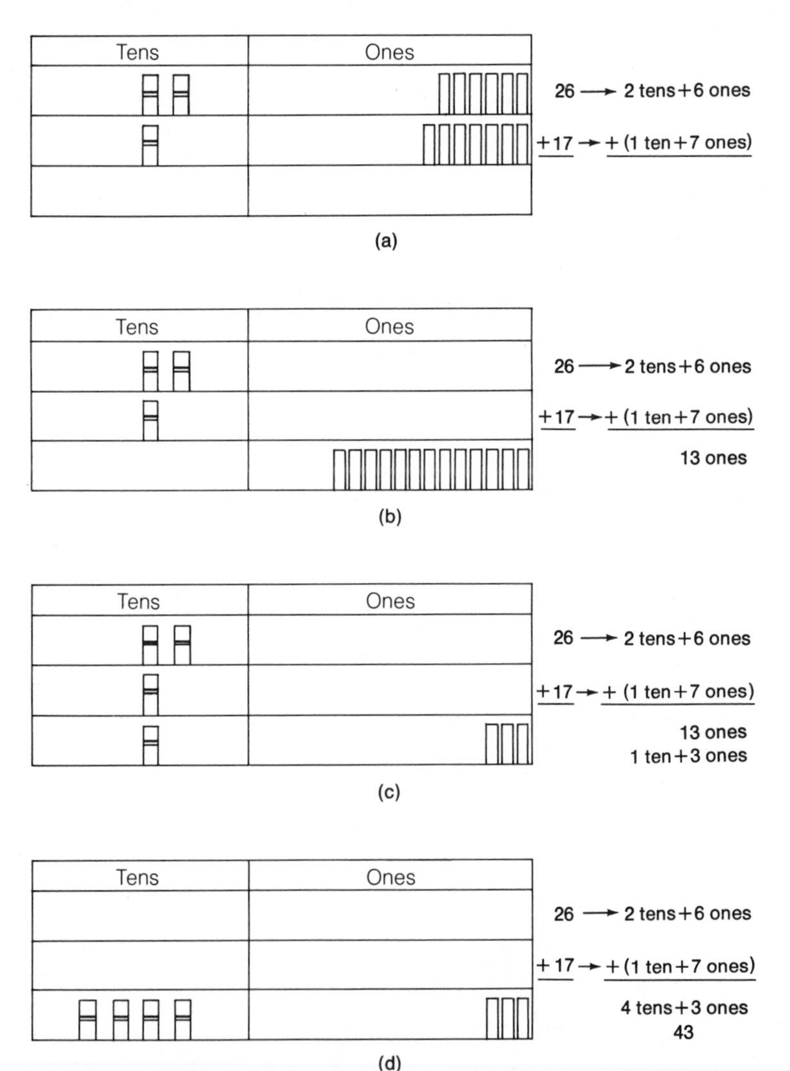


**Figure 8-5** Place value pocket chart illustrating (a) markers and expanded notation for 26 + 17 = □, (b) joining of markers and sum for ones, (c) regrouping of 13 ones as 1 ten and 3 ones, and (d) markers and sum at end of process

expanded form of the addends (a). A child should join the markers in the ones place of the device and write the sum for the ones place in the expanded algorithm (b). Children will note that the 13 markers can be regrouped to give 1 bundle of ten and 3 ones, as in (c), and the sum rewritten as "1 ten + 3 ones." Finally, the bundles of ten should be put together, as in (d), giving 4 bundles of ten and 3 ones. The sum is written as "4 tens + 3 ones" and then as "43."

Children should see further examples of addition with regrouping illustrated on the place value devices and in the expanded form. As children acquire an understanding of the process, the expanded form of writing the algorithm should be shortened to the conventional form. The progression will proceed in this manner.

$$\begin{array}{rcrcr}
2\text{ tens} + \ \ 9\text{ ones} & \rightarrow & 20 + \ \ 9 & \rightarrow & 29 \\
+\,3\text{ tens} + \ \ 5\text{ ones} & \rightarrow & +30 + \ \ 5 & \rightarrow & +35 \\
\hline
5\text{ tens} + 14\text{ ones} & \rightarrow & 50 + 14 & \rightarrow & 64 \\
(5\text{ tens} + 1\text{ ten} + 4\text{ ones}) & & (50 + 10) + 4 & &
\end{array}$$

You must be certain that children understand that the carried 1 stands for 1 ten, which results from renaming 14 after 9 and 5 are added. One good way to check their understanding of the process is to occasionally have children reverse it: They should begin with the algorithm, determine the answer, and then show the meaning of the steps on a place value device.

As children proceed through the grades, you should review with them the meaning of regrouping with larger numbers. A good way tc do this is with an abacus, because it can illustrate numbers through thousands. However, before it is to be used for addition, the meaning of the abacus itself must be clear; therefore its uses should be reviewed following procedures like those in Chapter 6.

To use the abacus to illustrate addition, put an example such as

$$\begin{array}{r}
4358 \\
+2926 \\
\hline
\end{array}$$

on the chalkboard and go through the steps shown in Figure 8-6. Ask a child to set beads on the abacus to represent the addend in the lower portion of the algorithm, 2926. Then the other addend, 4358, can be represented with beads on the abacus. In the figure, small clips, such as tiny plastic clothespins, separate the beads on the wires, illustrating the two addends. The abacus would appear as shown in (a). The meaning of the beads and their relationship to the addends in the algorithm should be discussed. Next, a child should unite the beads on the ones wire by removing the clip between the two sets (b), while a second child works with the algorithm on the board. The regrouping of the sum in the ones place is demonstrated as ten beads from the ones wire are exchanged for one bead on the tens wire. (The one bead on the tens wire is separated from the others by a clip.) The beads on the tens wire are then combined while the numbers in the tens place of the algorithm are added. The process continues as the beads on the hundreds wire are united. Again the regrouping is shown as ten beads from the hundreds wire are exchanged for one bead on the thousands wire (c). The process is completed as the beads on the thousands wire are combined and the algorithm is finished (d). The addition of pairs of numbers having sums to 9999 can be shown on a four-wire abacus. For larger numbers, an abacus with more wires is needed.

---

## Self-check of Objective 2

Select any two place value devices described in this chapter and demonstrate how to use them to represent the addition examples in the margin. (Use a sequence of pictures to represent the steps if the two devices are unavailable for your demonstrations.) Write the steps as they appear in the algorithm for the sequence of actions with the devices.

| 64 | 74 |
|---|---|
| +33 | +87 |

---

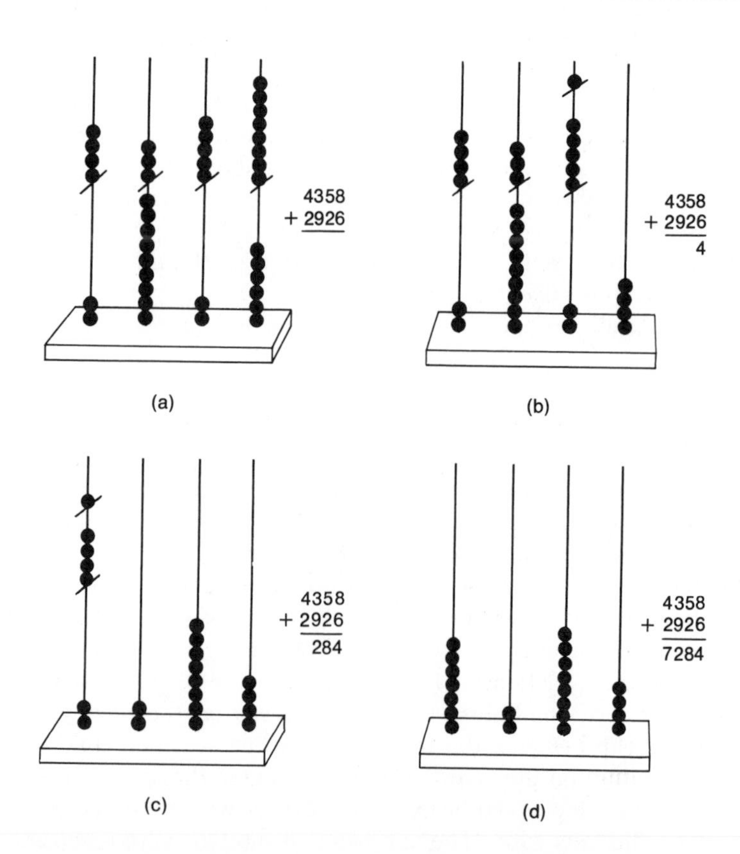


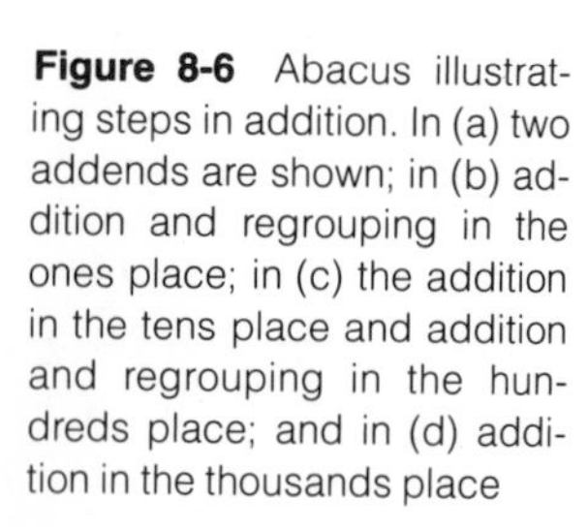

**Figure 8-6** Abacus illustrating steps in addition. In (a) two addends are shown; in (b) addition and regrouping in the ones place; in (c) the addition in the tens place and addition and regrouping in the hundreds place; and in (d) addition in the thousands place

## INTRODUCING SUBTRACTION OF LARGER WHOLE NUMBERS

Introductory work with subtraction of larger whole numbers should parallel introductory work for addition. That is, first work should be with realistic situations and adequate time for children to explore the operation's meaning. "A rope 70 feet long is being cut into smaller jump ropes. If 40 feet of rope has been cut from it, how many feet remain to be cut?" After children have used their own methods to determine the answer, have different ones share their procedures. Place value devices should be brought into the discussion either by a child or by you. Beansticks are illustrated here, but any of the devices can be used. In Figure 8-7(a) the 70 feet of rope is represented by 7 ten sticks. The removal of 40 feet of rope is indicated by removing 4 of the ten sticks. The answer—30 feet of rope—is represented by the 3 ten sticks that remain (b).

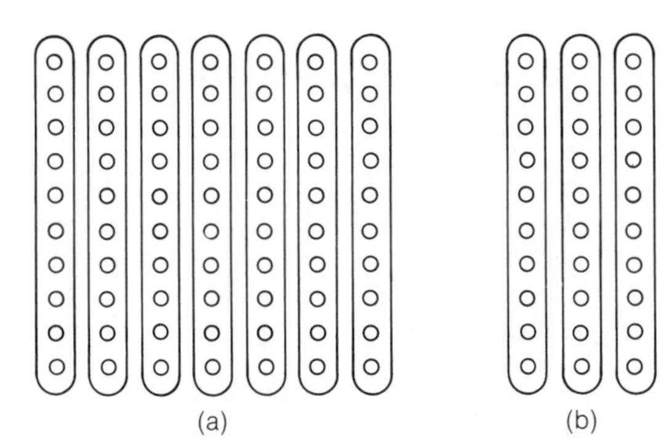


**Figure 8-7** Beansticks show the meaning of 70 − 40 = 30

Other similar problem situations should be illustrated with devices until the meaning of the subtraction is clear. The children should also represent situations by drawing markers on place value devices illustrated on work sheets. Figure 8-8 illustrates an example of how a child can show the subtraction 80 − 50 = □ on a picture of a place value pocket chart. In (a), the 80 is represented by 8 bundles of ten markers drawn on the chart. In (b), 5 bundles have been marked out to indicate the subtraction of 50, leaving the 3 bundles that represent the answer. The algorithm form

$$\begin{array}{r} 60 \\ \underline{-20} \end{array}$$

should be introduced only after children indicate that they are ready for it. Check the children's understanding from time to time by having them illustrate further examples with place value devices and drawings of them.

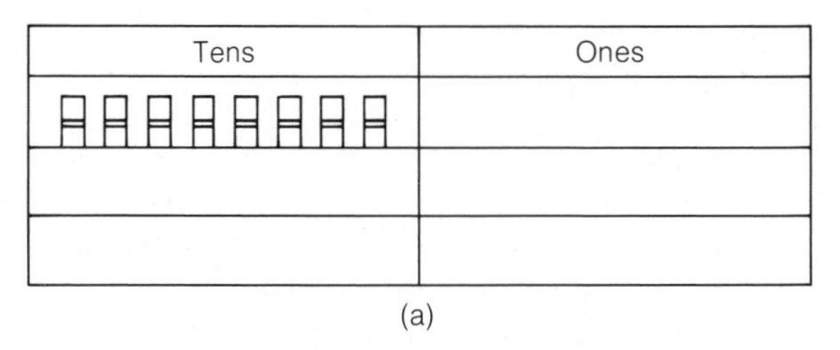


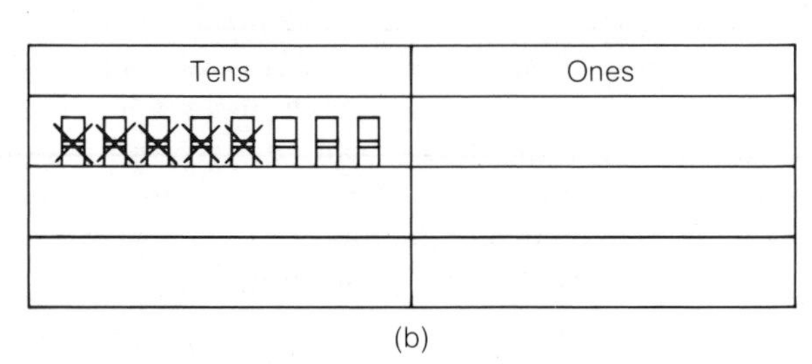


**Figure 8-8** Illustration from a child's work sheet showing that 80 − 50 = 30

## TEACHING SUBTRACTION WITH REGROUPING

Subtraction with regrouping has been the subject of much debate: which of two methods of computation should children be taught?[1] The decomposition method is most commonly used, but there are many advocates for the equal additions method.

### 1. The Decomposition Method

The decomposition method of subtraction is generally favored because children easily rationalize its process of regrouping as they use markers in a place value device or abacus. The relationship between subtraction and addition is emphasized by this method because the steps of subtraction by decomposition are the reverse of those for addition by regrouping.

A good understanding of place value and addition of larger whole numbers, with and without regrouping, is a prerequisite of the decomposition method of subtraction. An understanding of expanded notation is also important. For example, children should know that the number 36 can be expressed as "36," "3 tens and 6 ones," "30 + 6," and "20 + 16."

The following example can introduce children to subtraction using decomposition. "There were eighty-two boxes of apples at a fruit stand when it opened. During the day, fifty-three boxes were sold. How many boxes were left at the end of the day?" Children should determine the

[1] Actually, there are more than two processes. Not commonly used in school mathematics programs and subject to less debate are those described by Hitoshi Ikeda and Masui Ando, "A New Algorithm for Subtraction?" *The Arithmetic Teacher*, XXI, No. 8 (December 1974), pp. 716–719, and by Paul R. Neureiter, "The 'Ultimate' Form of the Subtraction Algorithm," *The Arithmetic Teacher*, XII, No. 4 (April 1965), pp. 277–281.

mathematical sentence, 82 – 53 = □. Different learning aids should be available during this exploratory period so children can work out their own procedures for determining the answer. Some children may use a hundreds board or a place value pocket chart. Each child with a different means of finding the answer should explain that procedure to the other children.

While the hundreds board can be used to find the answers for sentences such as 82 – 53 = □ and 75 – 48 = □, its use involves a process of counting backward with no regrouping. So that children will understand the decomposition method, encourage them to use a place value device. Use of a place value pocket chart is explained as follows:

The algorithm in the margin should appear as in Figure 8-9. The number 82 is represented in the tens and ones pockets (a) and in the

$$\begin{array}{r} 82 \\ -53 \\ \hline \end{array}$$

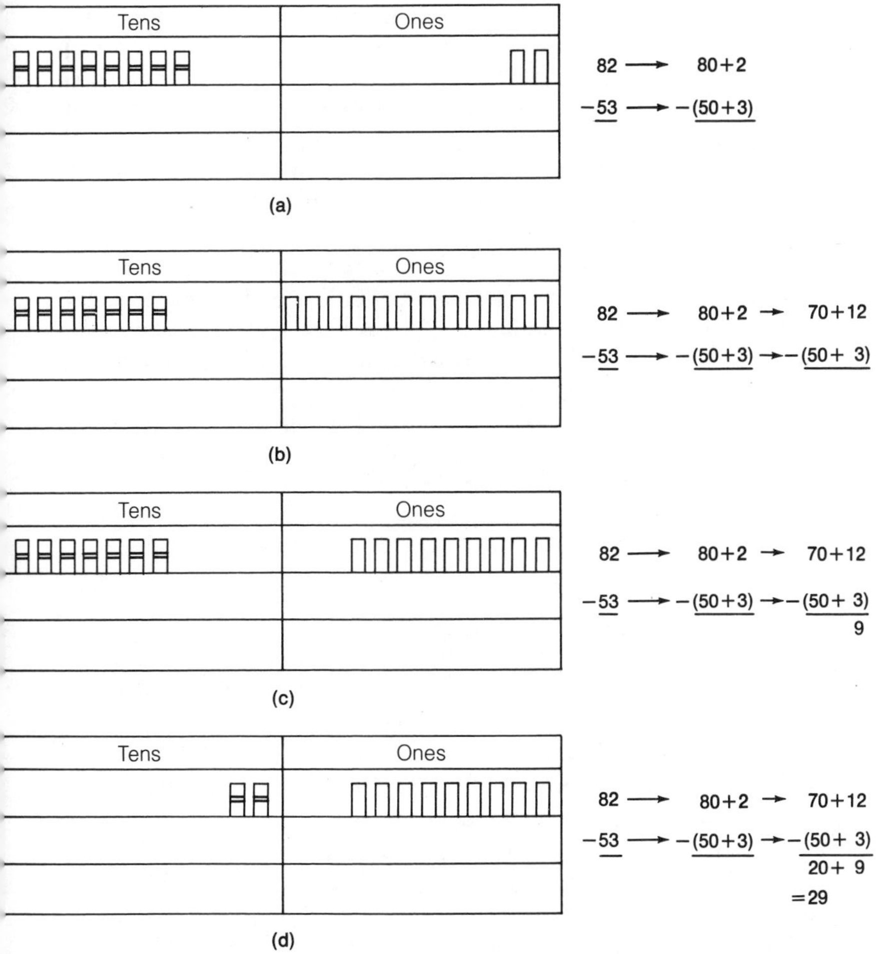

| Tens | Ones |
|---|---|

$$\begin{array}{rl} 82 \longrightarrow & 80+2 \\ -53 \longrightarrow & -(50+3) \end{array}$$

(a)

| Tens | Ones |
|---|---|

$$\begin{array}{rll} 82 \longrightarrow & 80+2 \rightarrow & 70+12 \\ -53 \longrightarrow & -(50+3) \rightarrow & -(50+\ 3) \end{array}$$

(b)

| Tens | Ones |
|---|---|

$$\begin{array}{rll} 82 \longrightarrow & 80+2 \rightarrow & 70+12 \\ -53 \longrightarrow & -(50+3) \rightarrow & -(50+\ 3) \\ & & 9 \end{array}$$

(c)

| Tens | Ones |
|---|---|

$$\begin{array}{rll} 82 \longrightarrow & 80+2 \rightarrow & 70+12 \\ -53 \longrightarrow & -(50+3) \rightarrow & -(50+\ 3) \\ & & 20+\ 9 \\ & & =29 \end{array}$$

(d)

**Figure 8-9** Pocket charts and algorithm in expanded notation form for showing meaning of regrouping for the subtraction 82 – 53 = 29

subtraction algorithm, which is shown in both the regular vertical and the expanded forms. The first step is to subtract three ones; and children will see that this is not possible as the numbers in the ones place of the algorithm are now expressed. By questioning, you can show the children that they can use one of the tens and the 2 ones in the pocket chart to indicate 12 ones; then 82 can be renamed as 70 + 12, as in (b). Three markers can then be removed from the ones place on the pocket chart and 3 subtracted from 12 in the algorithm (c). Finally, five of the tens bundles should be removed from the pocket chart and 50 subtracted from 70 in the algorithm (d).

Give children other numbers that involve regrouping so they can use place value devices and expanded notation forms of writing the algorithms until they have a clear understanding of the process. Eventually, the number of recorded steps used to complete a subtraction can be reduced to a minimum. The most mature way to subtract the example in the margin is to compute the algorithm mentally, without writing anything other than the answer. A procedure that can be used as a step between the expanded form and the mature form is to write in "crutches," or to indicate the regrouping as the work is done, as shown in the margin.

$$\begin{array}{r} 362 \\ -136 \\ \hline \end{array}$$

$$\begin{array}{r} {}^{5\,1} \\ 3\not{6}2 \\ -136 \\ \hline 226 \end{array}$$

Once children understand the decomposition process with the algorithm for numbers with tens and hundreds, they should be able to subtract with still larger numbers. An abacus with twenty beads on each rod can be used to review the process with larger numbers for children in the middle grades. The process is illustrated in Figure 8-10 for the example in the margin. Only the minuend, 6342, is shown on the abacus in (a). The steps taken to show decomposition on an abacus are comparable to those used with a place value pocket chart.

$$\begin{array}{r} 6342 \\ -3528 \\ \hline \end{array}$$

In the middle grades, subtraction involving 0 in the minuend should receive special attention since problems such as

$$\begin{array}{r} 306 \\ -148 \\ \hline \end{array} \qquad \text{and} \qquad \begin{array}{r} 6003 \\ -4298 \\ \hline \end{array}$$

are difficult for some children. There are two ways decomposition can be done with examples of this type. To subtract 148 from 306, the thinking might be: "There are no tens, so go to the hundreds. Regroup one of the hundreds as 10 tens, leaving 2 hundreds. Then regroup one of the tens as 10 ones, giving a total of 16 ones and 9 tens." The subtraction can then be done. The same procedure would be used for the number 6003, except that more regroupings would be required.

Children who have a good understanding of numbers can use another method: 306 can be considered as 30 tens and 6 ones. When the regrouping is done, 1 ten is taken from the 30 tens, leaving 29 tens. The 1 ten is renamed as 10 ones and is added to the 6 ones in the ones place. In

$$\begin{array}{r} {}^{2\,9\,1} \\ \not{3}\not{0}6 \\ -148 \\ \hline 158 \end{array}$$

one step, all of the required regrouping is completed. If the regrouping is indicated in the algorithm, it appears as shown in the margin. To subtract 4298 from 6003, the regrouping is also done in one step. The number 6003 is thought of as 600 tens and 3 ones. When 1 ten is renamed as 10 ones, there are 599 tens left. The algorithm in the margin shows how the regrouping might be indicated. By the time children are in the fifth or sixth grade, they should be able to subtract using zeros in this more mature way.

$$\begin{array}{r} {\scriptstyle 5\,9\,9\,1} \\ \not{6}\not{0}\not{0}3 \\ -4298 \\ \hline 1705 \end{array}$$

## 2. The Equal Additions Method

The equal additions method of subtraction is based on the mathematical concept that there is an infinite number of equivalent subtraction problems for a given remainder. The sentences 68 − 14 = □ and 70 − 16 = □ are equivalent because the answer, or remainder, is the same for both. The strength of the equal additions method of subtraction is the ease and speed with which computation can be done once the process has been mastered. The weakness of this method is that many children find it difficult to understand.

If the method is to be taught, it should be preceded by activities showing that the same number can be added to both subtrahend and minuend without changing the remainder. Figure 8-11 demonstrates this mathematical principle; two sets are put before the children (a), and they are asked to compare the sizes of the two sets to determine how much larger one is than the other. The sentence that describes the situation should be written for the children to see: 13 − 7 = 6. Then put some

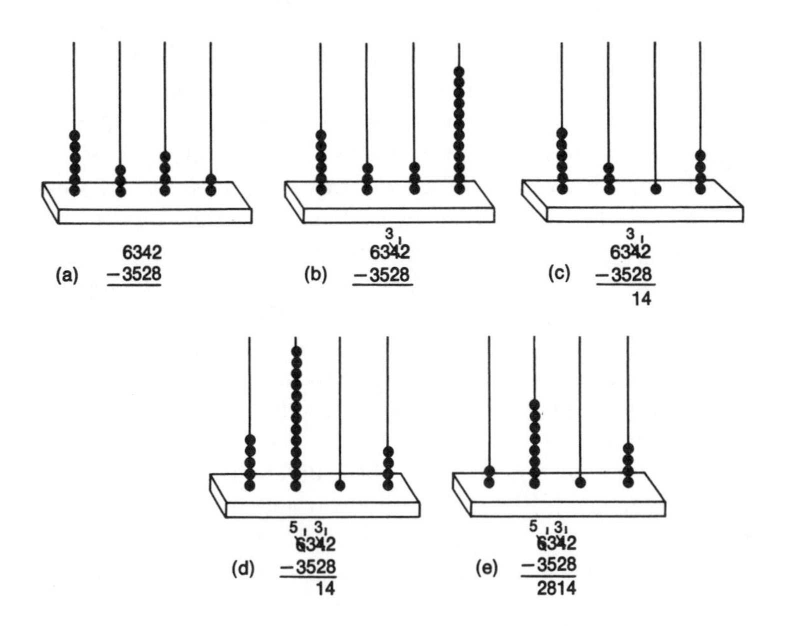


**Figure 8-10** Abacus and algorithm showing decomposition for the subtraction sentence 6342 − 3528 = 2814

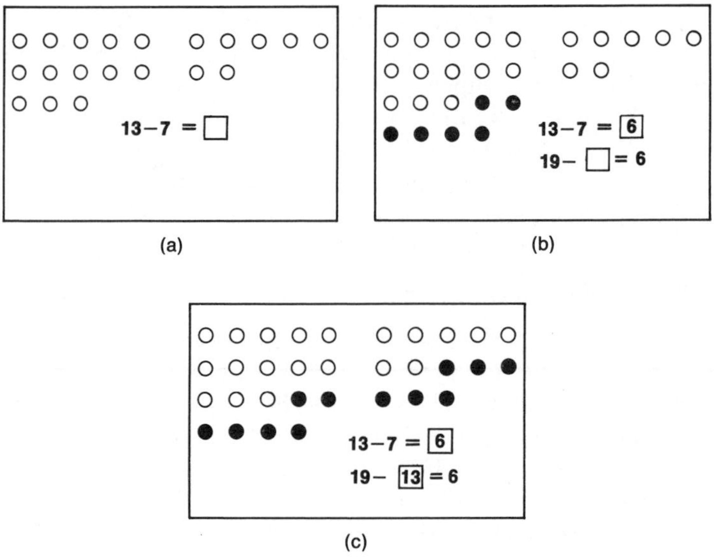

**Figure 8-11** Markers and sentences showing that the same number must be added to both minuend and subtrahend to keep the difference unchanged

additional markers with the larger set. Use questions to guide the children to observe that the difference between the sizes of the two sets has changed (b). "What must we do to the smaller set so there will once again be a difference of 6 in the sizes of the two sets?" The children should see that if they increase the second set by the same number of markers as were put with the first set, the difference will again be 6. Changes in sets should always be reflected in changes in the sentences (c). Other examples, using different-sized sets and alternating which set is changed first, will help children discover this principle.

The number line is another useful aid for discovering this principle, as Figure 8-12 shows. Write a subtraction sentence, such as $19 - 13 = \square$, and represent the minuend on the number line with an arrow extending from 0 to 19, and the subtrahend by an arrow extending from 19 backward thirteen steps to the 6 (a). Then increase the minuend by a given amount, say 6, so it becomes 25. Show the new minuend on the number line (b). The children can then discuss what must be done to the subtrahend so the difference, 6, will remain unchanged. An arrow drawn from the new minuend and extending backward to 6 on the line will show the new subtrahend. A look at this arrow reveals that the new subtrahend is the number of the old subtrahend increased by the same amount as was the original minuend (c).

In addition to understanding that the difference remains the same if a given number is added to both minuend and subtrahend, an understanding of place value is also essential before the equal additions method of subtracting should be introduced. Once an adequate background for

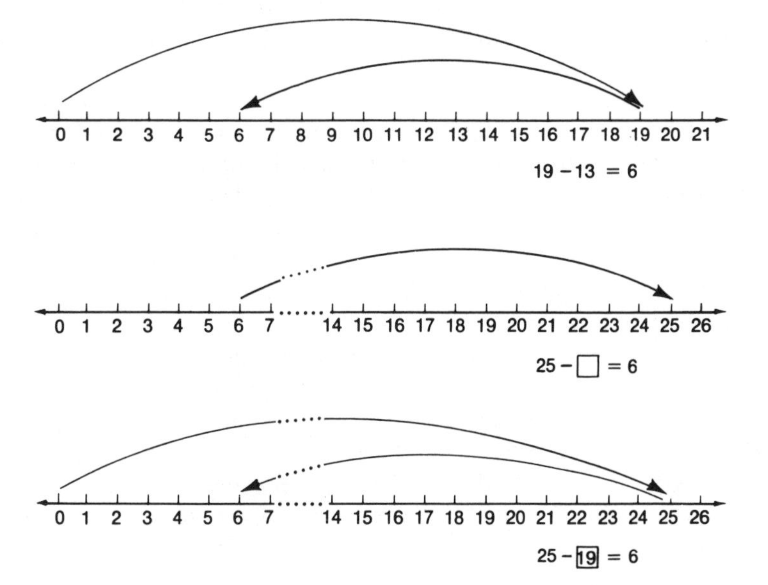


**Figure 8-12** Number line and sentences showing that the same number must be added to both minuend and subtrahend if the difference is to remain unchanged

the process has been established, proceed as follows. Write a subtraction example on the chalkboard:

$$\begin{array}{r} 632 \\ -216 \\ \hline \end{array}$$

Children see that it is not possible to subtract 6 ones from 2 ones. The subtraction is made possible when 10 ones are added to the 2 ones in the minuend to make 12 ones. The children then subtract 6 from 12 and record a "6" in the ones place of the algorithm. Call the children's attention to the fact that since 10 was added to the minuend, 10 must also be added to the subtrahend. This is done by adding 1 ten to the number in the tens place of the subtrahend. Then 2 tens are subtracted from 3 tens and the answer is recorded in the tens place. The subtraction is completed by subtracting 2 hundreds from the 6 hundreds. The completed algorithm would appear as shown above in the margin. The sentence 3246 − 1937 = □ is shown below in the algorithm in the margin. Once children learn the process, the thinking is 16 − 7 = 9, 4 − 4 = 0, 12 − 9 = 3, and 3 − 2 = 1. When the number in a place value position in the minuend is smaller than the number in the same place value position of the subtrahend, it is automatically increased by 10, and the subtraction for that position is completed. Each time this is done, the number in the subtrahend in the place value position to its immediate left is increased by 1. The process becomes automatic when it is well understood and sufficiently practiced. This process presents no particular difficulties with zeros in the minuend.

$$\begin{array}{rrr} 6 & 3 & {}^{1}2 \\ -2 & {}^{2}1 & 6 \\ \hline 4 & 1 & 6 \end{array}$$

$$\begin{array}{rrrr} 3 & {}^{1}2 & 4 & {}^{1}6 \\ -{}^{2}1 & 9 & {}^{4}3 & 7 \\ \hline 1 & 3 & 0 & 9 \end{array}$$

In addition to the controversy over which method of computation should be taught, there has also been debate over which types of examples should be used to introduce subtraction involving numbers greater than 10. Many programs of instruction begin with a sentence that consists of multiples of 10, such as $80 - 60 = \square$. These examples are usually followed by subtraction involving a sentence such as $86 - 42 = \square$, in which 2 ones are subtracted from 6 ones and 4 tens from 8 tens, to get the answer, 44. Some writers contend that once subtraction involving multiples of ten has been introduced, the next work should begin with a pair of numbers which need regrouping (or decomposition).[2] The argument in favor of this holds that children will be more aware of the steps involved in using the algorithm and doing the computation if regrouping is required than if simpler examples are used. This leads children to work meaningfully rather than in a rote way. Actually, children's understanding of subtraction of whole numbers greater than 10 depends as much on you and the materials and procedures you have them use as on the type of examples used during the introductory lessons. Thus the most important consideration remains the need to have children learn the steps in the new algorithm forms meaningfully.

---

## Self-check of Objectives 3 and 4

Explain how to perform decomposition and equal additions processes for the example in the margin. Demonstrate decomposition with a place value chart, beansticks, or abacus. Describe materials and procedures that can be used to make the equal additions process meaningful to children.

Show how renaming the number 5004 can be done to permit a one-step regrouping process for the subtraction example.

$$\begin{array}{r} 432 \\ -196 \\ \hline \end{array}$$

$$\begin{array}{r} 5004 \\ -1368 \\ \hline \end{array}$$

---

### LOW-STRESS ALGORITHMS

In spite of a teacher's careful instruction, there are some children who may not master the conventional algorithms for addition and/or subtraction. One possible solution for this problem is the introduction of "low-stress" algorithms. These algorithms ". . . appear to permit easy mastery after brief training, to provide greater computational power than conventional algorithms, to operate with much less stress on the user than conventional algorithms, and to enjoy other advantages. . . . The role of these procedures in the skills component of the mathematics curriculum may eventu-

[2] See, for example, Howard F. Fehr and Jo M. Phillips, *Teaching Modern Mathematics in the Elementary School*, 2nd ed. (Reading, Mass.: Addison-Wesley Publishing Company, 1972), pp. 90–93.

ally be major and fundamental. It appears that immediate applications of the procedures should be directed toward students having severe remedial needs in the upper elementary and junior high school, as well as high school and adults."[3] The algorithms for addition and subtraction will be discussed briefly in this chapter; the one for multiplication is discussed in Chapter 10. For additional information refer to Chapter 16 of the 1976 National Council of Teachers of Mathematics Yearbook referred to in footnote 3.

## 1. Addition Algorithm

The addition algorithm is similar to the conventional algorithm, but it has a slightly different form of notation, called half-space notation. The difference is shown in the margin, where the conventional notation for a basic fact is shown on the left and half-space notation for the same fact is on the right. For a combination having a sum less than 10, the single numeral is written slightly below and to the right of the two addends. In half-space notation the line separating the addends from their sum is omitted in basic facts. The reason for this notation will become clear as the description continues.

$$\begin{array}{r} 5 \\ +7 \\ \hline 12 \end{array} \qquad \begin{array}{l} 5 \\ +7_{1\;2} \end{array}$$

$$\begin{array}{r} 6 \\ +2 \\ \hline 8 \end{array} \qquad \begin{array}{l} 6 \\ +2_{8} \end{array}$$

The low-stress algorithm is particularly useful for adding when there are three or more addends. The example in the margin shows that all addition is recorded as the work is completed. Each fact in the set of addends has its sum recorded; no mental accumulation of sums is necessary. In the example, the process is this: add 8 + 4, record 12; add 2 + 6, record 8; add 8 + 9, record 17; add 7 + 8, record 15. Now all addends have been added, so record 5 below the line. Count the number of times a ten was recorded at the left of the column; record 3 below the line. The sum for this addition is 35.

$$\begin{array}{r} 8 \\ {}_1 4_2 \\ 6_8 \\ {}_1 9_7 \\ +{}_1 8_5 \\ \hline 35 \end{array}$$

Addition with addends greater than 10 is illustrated in the margin. Add 6 + 4, 10; 0 + 8, 8; 8 + 9, 17; record 7 below the line. Count the number of times ten was recorded, then "carry" 2 to the tens column. Add 2 + 2, 4; 4 + 8, 12; 2 + 7, 9; 9 + 4, 13. Record 3 below the line. Count the tens (these are actually hundreds since the addition is in the tens column), then record 2 below the line. The answer is 237.

$$\begin{array}{rr} \text{②} & \\ 2_4 & 6 \\ {}_1 8_2 & {}_1 4_0 \\ 7_9 & 8_8 \\ {}_1 4_3 & {}_1 9_7 \\ \hline 2\;3 & 7 \end{array}$$

Once a student has mastered the process with numbers smaller than 10, and then smaller than 100, he or she can quickly move to addition with quite large numbers and several addends. For an older student who has experienced failure with the conventional algorithm, success with the low-stress computation will perhaps bolster the child's self-esteem and help change a negative attitude toward mathematics to a positive one.

[3] Barton Hutchings, "Low-Stress Algorithms," *Measurement in School Mathematics* (Reston, Va.: National Council of Teachers of Mathematics, 1976), p. 219.

## 2. Subtraction Algorithm

When doing subtraction, the student completes all regrouping first. The example below illustrates the process. The original combination is shown in (a). The regrouping is placed between the minuend and subtrahend, as shown in the series of steps in (b), (c), and (d).

| (a) | (b) | (c) | (d) | (e) |
|---|---|---|---|---|
| 6241 | 6 2 4 1 | 6 2 4 1 | 6 2 4 1 | 6 2 4 1 |
| −3675 | $3\ {}^{1}1$ | $1\ {}^{1}3\ {}^{1}1$ | $5\ {}^{1}1\ {}^{1}3\ {}^{1}1$ | $5\ {}^{1}1\ {}^{1}3\ {}^{1}1$ |
| | −3 6 7 5 | −3 6 7 5 | −3 6 7 5 | −3 6 7 5 |
| | | | | 2 5 6 6 |

Finally, the subtraction is completed and the answer recorded.

Subtraction with zeros in the minuend is done this way:

| (a) | (b) | (c) | (d) |
|---|---|---|---|
| 4002 | 4 0 0 2 | 4 0 0 2 | 4 0 0 2 |
| −2146 | $3\quad\ {}^{1}2$ | $3\ 9\ 9\ {}^{1}2$ | $3\ 9\ 9\ {}^{1}2$ |
| | −2 1 4 6 | −2 1 4 6 | −2 1 4 6 |
| | | | 1 8 5 6 |

The original example is shown in (a). In (b) the ones are regrouped and the 4 in the thousands place is changed to 3, with the zeros being ignored at this point. After this regrouping is completed, each 0 is replaced by a 9 (c). Now the subtraction can be completed (d). The process with a larger minuend having zeros is illustrated below, with the original problem shown in (a). In (b) the regrouping of numbers greater than 0 is completed. In (c) each 0 is replaced by a 9; the answer is shown in (d).

| (a) | (b) | (c) | (d) |
|---|---|---|---|
| 430002 | 4 3 0 0 0 2 | 4 3 0 0 0 2 | 4 3 0 0 0 2 |
| −246483 | $3\ {}^{1}2\qquad {}^{1}2$ | $3\ {}^{1}2\ 9\ 9\ 9\ {}^{1}2$ | $3\ {}^{1}2\ 9\ 9\ 9\ {}^{1}2$ |
| | −2 4 6 4 8 3 | −2 4 6 4 8 3 | −2 4 6 4 8 3 |
| | | | 1 8 3 5 1 9 |

The primary advantage of this algorithm for the student is that all of one part of the work—the regrouping—is completed before the next part—the subtraction—is done. This reduces the cognitive load and the possibility of error because the child does not shift from thinking about regrouping to subtracting, back to regrouping, then subtraction again, and so on. An advantage of both algorithms for the teacher is that it is easy to detect errors. By examining a child's work, it is possible to spot errors with basic addition and/or subtraction facts, failure to "carry" in addition, and regrouping errors in subtraction.

## Self-check of Objective 5

Write an addition example having at least six addends larger than 10. Determine the sum using the low-stress method of computing. Write a subtraction with both minuend and subtrahend larger than 1000 and for which regrouping is necessary. Complete it using the low-stress method of subtracting. Indicate for whom these algorithms are especially useful.

## CHECKING ADDITION AND SUBTRACTION

The value of accuracy needs to be stressed as soon as children begin to perform operations with numbers. At first, they should be permitted and encouraged to check their addition by using markers, counting, and any other procedures meaningful to them. Later, children should be encouraged to use more mature ways of checking their work. You can demonstrate procedures they might use, and encourage them to devise their own.

1. Point out that grouping by tens is sometimes useful when adding a column of numbers. In the example in the margin, the 7 and 3 and the 8 and 2 can each be grouped to make 10. When this is done, the sum of the ones column is easily determined to be 24. In the tens column, the 4 and 6 can be added, then the "carried" 2 can be added to the 2 and 3, and their sum to 9. Finally, 16 and 10 are added, giving 26, which is recorded.

$$\begin{array}{r} 63 \\ 48 \\ 97 \\ 24 \\ +32 \\ \hline 264 \end{array}$$

2. Another scheme is to add tens and ones alternately. For the addition shown above, this might go $63 + 40 = 103$, $103 + 8 = 111$, $111 + 90 = 201$, $201 + 7 = 208$, $208 + 20 = 228$, $228 + 4 = 232$, $232 + 30 = 262$, $262 + 2 = 264$.

3. Children should be encouraged to add the subtrahend and remainder (difference) to check subtraction, as shown in the margin. If the sum of the addition is the same as the original minuend, the subtraction is probably correct.

$$\begin{array}{r} 634 \\ -269 \\ \hline 365 \end{array} \qquad \begin{array}{r} 365 \\ +269 \\ \hline 634 \end{array}$$

4. Subtraction can be checked by subtracting the remainder from the original minuend, as shown in the margin. If the new remainder is the same as the original subtrahend, the subtraction is probably correct. (There are times when a subtraction can be wrong, even though a check indicates that it is right. For example, if a child says that $14 - 9 = 6$, he or she might say that $9 + 6 = 14$. If this is done in the example in 3, the answer will check out to be correct, when it is

$$\begin{array}{r} 426 \\ -138 \\ \hline 288 \end{array} \qquad \begin{array}{r} 426 \\ -288 \\ \hline 138 \end{array}$$

not. Children will sometimes "force" a check. When they do this, they don't actually add or subtract to check their work, they merely write the numerals from the original algorithm in the algorithm for the check without thinking about them as they do.)

5. The low-stress algorithms for addition and subtraction can be used as checks for work done using the conventional algorithms.
6. The minicalculator makes it easy to check addition and subtraction.

---

## Self-check of Objective 6

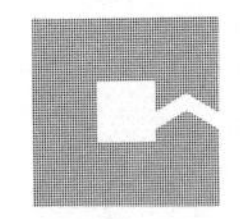

Explain at least three different ways of checking the accuracy of the addition example in the margin. Explain two ways of checking the accuracy of the subtraction example in the margin.

---

```
        62
        47
        59
 642    66
-397   +82
 ----  ---
 245   316
```

## THE MINICALCULATOR AND ADDITION AND SUBTRACTION

There are many interesting activities dealing with addition and subtraction that can be carried out using a minicalculator. In addition to using it for checking their work, many children will want to try activities like the ones described below. It is recommended that you refer to the calculator books and other materials listed in Appendix C for further activities.

### 1. Estimating Answers

```
 46    50
 53    50
 91    90
+37   +40
      ---
      230

 9342
-6962
 ----
```

The ability to make reasonable estimates of the expected answers in a problem-solving situation and when using a minicalculator is becoming increasingly important. The calculator itself can provide help in developing children's skill in estimating. First, children must be skillful in rounding off numbers. Ways to help them develop this skill are given on pages 92 to 93. Give them an example to estimate, as in the margin. The estimated answer is 230. Addition on the calculator gives the sum 227, indicating that the estimated answer is reasonable. In the subtraction example in the margin, the estimate can be made to the nearest hundred or the nearest thousand. An estimate made to hundreds is 2300; to the nearest thousand it is 2000. The calculated answer—2380—indicates that each estimate is reasonable. Give your children practice with work sheets containing addition and subtraction examples to first estimate, then check with a minicalculator.

## 2. Palindromic Numbers

A palindrome is a number that reads the same backwards as forwards—232, 46,564, 1,234,321. An interesting investigation involves addition to generate palindromes. Select any whole number. Add that number to the one you get when you reverse its digits. If the sum is a palindrome, the addition is complete, as in the examples in the margin. If the first addition does not give a palindrome, use the sum and its reversed digits as new addends. Add these numbers. If this sum is not a palindrome, repeat the process until a palindrome appears. Note that palindromes can be generated from numbers smaller than 10.

| | |
|---|---|
| 42 | 421 |
| +24 | +124 |
| 66 | 545 |

| | | |
|---|---|---|
| 69 | 596 | 3 |
| +96 | +695 | +3 |
| 165 | 1291 | 6 |
| 561 | 1921 | 6 |
| 726 | 3212 | 12 |
| 627 | 2123 | 21 |
| 1353 | 5335 | 33 |
| 3531 | | |
| 4884 | | |

Interested children will test many numbers. Often the numbers they use will exceed the machine's capacity, so they may revert back to paper-and-pencil computation or attempt to devise ways to do the work on the calculator.

## 3. Magic Squares

Magic squares have a long history.[4] Their mystery has interested professional and amateur mathematicians for centuries. Until now, children who have been interested in magic squares have found the computation required in connection with them so tedious they do not extend their investigations beyond the more simple squares. The minicalculator provides the means for taking the burden out of the computing, so interested students can investigate much more sophisticated squares. Magic squares with 3 × 3, 4 × 4, and 5 × 5 cells are illustrated in Figure 8-13. Add the numbers in each column, row, and diagonal. If the sum is the same for each group of numbers, you have a magic square. Children should be encouraged to work with these squares and then to generate their own, if

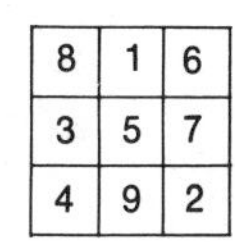

| | | |
|---|---|---|
| 8 | 1 | 6 |
| 3 | 5 | 7 |
| 4 | 9 | 2 |

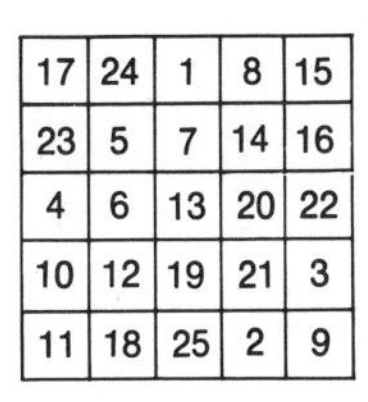

| | | | | |
|---|---|---|---|---|
| 17 | 24 | 1 | 8 | 15 |
| 23 | 5 | 7 | 14 | 16 |
| 4 | 6 | 13 | 20 | 22 |
| 10 | 12 | 19 | 21 | 3 |
| 11 | 18 | 25 | 2 | 9 |

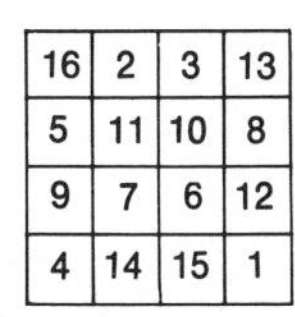

| | | | |
|---|---|---|---|
| 16 | 2 | 3 | 13 |
| 5 | 11 | 10 | 8 |
| 9 | 7 | 6 | 12 |
| 4 | 14 | 15 | 1 |

**Figure 8-13** Magic squares

[4] See William Heck and James Frey, "Magic Squares," *Historical Topics for the Mathematics Classroom* (Reston, Va.: National Council of Teachers of Mathematics, 1969), pp. 80–82.

possible. The book *Adventures With Your Hand Calculator*[5] contains examples of more complex magic squares and a commentary about them.

### 4. Nim

Nim is an old game that has been played in many ways with a variety of materials. The minicalculator offers a new means of playing the game. To begin, select a target number, say 50. One player enters a one-digit number to the display. The second player adds a second number to the first. Players alternate adding one-digit numbers until one reaches 50 and is the winner. Different target numbers should be selected. Players should be encouraged to work out a strategy for winning. A variation is to start at a number, say 50, and subtract one-digit numbers to reach 0.

### 5. Subtracting to 495

This activity is based on the fact that a series of subtractions involving the largest and smallest numbers created from three digits, with at least one digit different from the other two, will eventually yield the number 495. Try the digits 8, 6, and 4. The largest three-digit number is 864, while the smallest is 468. Subtract 468 from 864: $864 - 468 = 396$; subtract 369 from 963: $963 - 369 = 594$; subtract 459 from 954: $954 - 459 = 495$. Try the digits 9, 8, and 2: $982 - 289 = 693$; $963 - 369 = 594$; $954 - 459 = 495$. Try 8, 9, and 9: $998 - 899 = 99$; $990 - 099 = 891$; $981 - 189 = 792$; $972 - 279 = 693$; $963 - 369 = 594$; $954 - 459 = 495$. Note that zeros are used to make 99 a three-digit number. This practice is followed any time the difference is less than 100. Encourage your children to test two- and four-digit numbers to see if a similar result occurs.

---

## Self-check of Objective 7

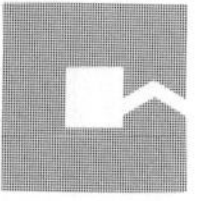

Describe briefly each of the five addition and subtraction activities involving the minicalculator.

---

## ADDITION AND SUBTRACTION OF INTEGERS

### 1. Addition of Integers

When children are introduced to addition and subtraction of integers, they should use a number line on which vectors represent the directed

[5]Lennart Rade and Burt A. Kaufman, *Adventures With Your Hand Calculator* (St. Louis: CEMREL, Inc., 1977), pp. 9–12, 57.

numbers to be added or subtracted. They should recognize that to add a pair of positive numbers is the same as adding a pair of whole numbers. The addition sentence $(^+2) + (^+4) = \square$ is illustrated in Figure 8-14. The two addends are represented by the abutting vectors with the solid shafts, while the sum is represented by the single vector with the broken shaft.

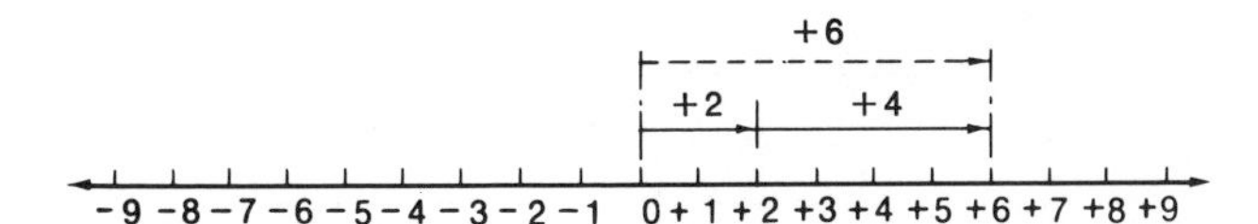


**Figure 8-14** Number line used to illustrate the addition sentence $(^+2) + (^+4) = \square$

To add a pair of negative numbers, children should note that the vector representing the first addend begins at 0 and goes to the left. The second vector abuts the first and also goes to the left. The sum is represented by the single vector with the broken shaft that begins at 0 and goes to the left for the number of units indicated by the two abutting vectors. The addition sentence $(^-4) + (^-3) = \square$ is illustrated in Figure 8-15.

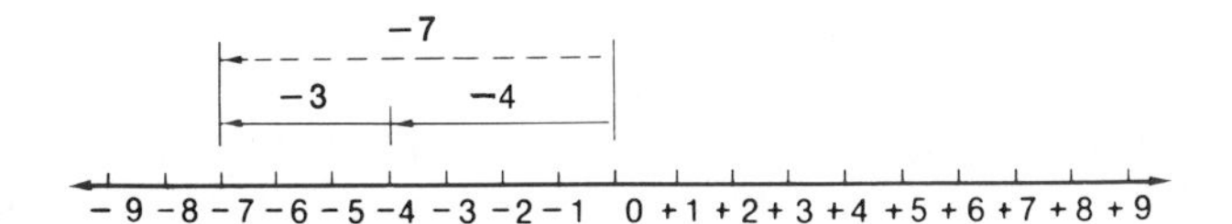


**Figure 8-15** Number line used to illustrate the addition sentence $(^-4) + (^-3) = \square$

Children should add a sufficient number of pairs of like directed numbers to understand the meaning of the operation with these numbers before they begin to add pairs of unlike directed numbers. Then ample opportunities to use vectors on their number lines will permit a clear understanding of the process of addition of numbers with unlike signs. The addition sentence $(^+6) + (^-3) = \square$ is illustrated in Figure 8-16. The

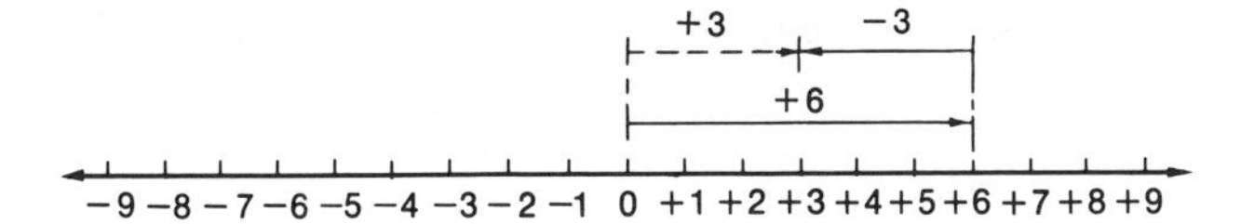


**Figure 8-16** Number line used to illustrate the addition sentence $(^+6) + (^-3) = \square$

two addends are again represented by the vectors with solid shafts, while the sum is represented by the vector with the broken shaft. Three other addition sentences with pairs of directed numbers with unlike signs are illustrated in Figure 8-17. These three sentences and the one in Figure 8-16 represent different kinds of sentences that occur when pairs of directed numbers are added.

A number of other ways to introduce the addition of directed numbers have been suggested. Cohen's postman stories create situations that

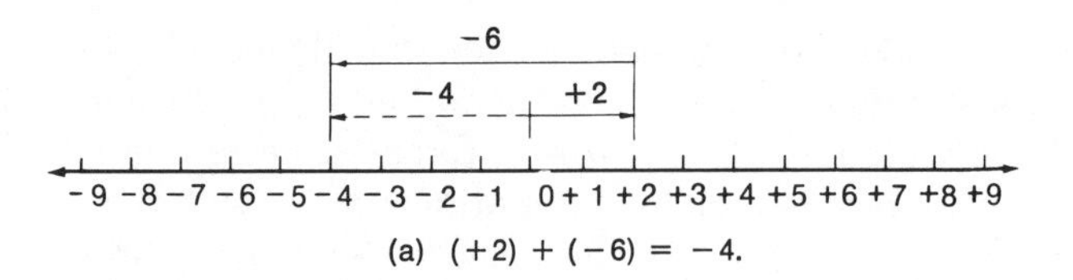


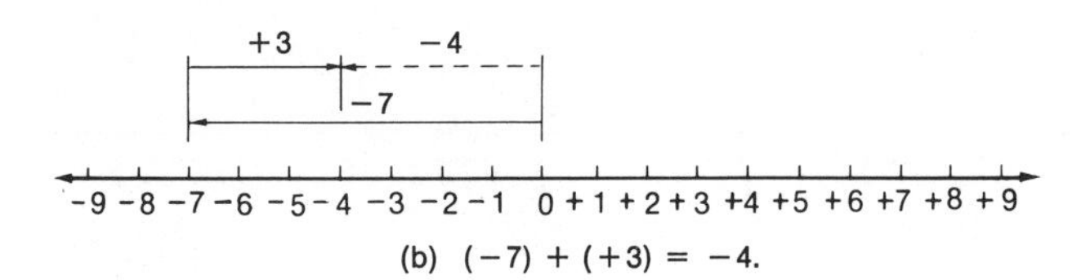


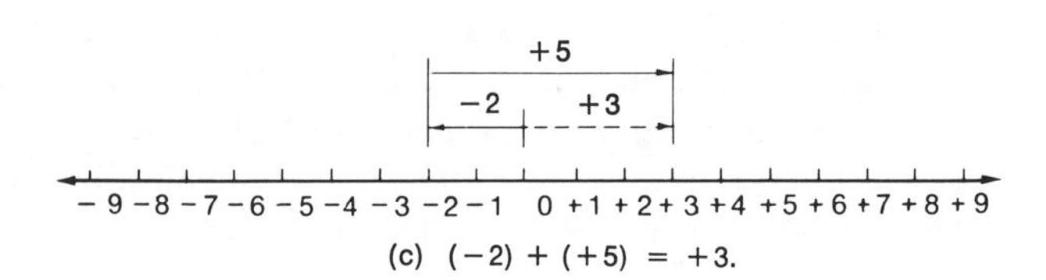


**Figure 8-17** Number lines used to illustrate three addition sentences

provide a basis for discussing this operation.[6] These stories concern a postman who haphazardly delivers letters regardless of the addresses that appear on them. Later he comes back and picks up some or all of the incorrectly delivered mail. Among the mail he delivers are checks and bills. When a housewife receives a check, she is happy (apparently even if it does not have her name on it), and when she receives a bill, she is sad (again apparently even if it is not hers). The receipt of a check for $6.00 is represented by the symbolism "$+^{+}6$." The receipt of a bill for the same amount is represented by "$+^{-}6$." Varied mixtures of bills and checks are described in simple stories so children can determine whether the housewife is richer or poorer, happier or sadder.

Herbert Fremont suggests another interesting approach.[7] He describes how short pieces of pipe cleaner can be bent and used to represent

+1 −1
(a) +2 (b) −2
+3 −3

**Figure 8-18** Positive and negative numbers represented by pipe cleaner loops.

[6] Louis S. Cohen, "A Rationale in Working with Signed Numbers," *The Arithmetic Teacher*, XII, No. 7 (November 1965), pp. 563–567.

[7] Herbert Fremont, "Pipe Cleaners and Loops—Discovering How to Add and Subtract Directed Numbers," *The Arithmetic Teacher*, XIII, No. 7 (November 1966), pp. 568–572.

directed numbers (Figure 8-18). Positive numbers are represented by pieces like those in (a). Negative numbers are represented by pipe cleaners placed like those in (b). Procedures for using the loops for addition are described by Fremont.

Still another way to introduce addition with directed numbers is suggested by David Page.[8] *Positive money* and *negative money* are used: Positive money is an asset, so its possession is valued; negative money is a liability, so its possession is to be avoided. A debtor can pay debts by giving positive money to or accepting negative money from a creditor. One can collect debts by receiving positive money from or giving negative money to a debtor. Stories built around these two kinds of money provide a context for learning about addition of directed numbers.

Children should discover that the commutative and associative properties and the identity element apply to addition of integers as well as to the set of whole numbers. They should use vectors to test these properties with integers.

## 2. Subtraction of Integers

When whole numbers are subtracted, the minuend must be equal to or larger than the subtrahend. Children discover this when they try to solve sentences like $3 - 5 = \square$. As they use number lines, pipe cleaner loops, and story situations to learn the meaning of subtraction with integers, children discover that the operation is possible with all pairs of these numbers.

To use vectors on number lines to represent subtraction of directed numbers, children must keep in mind that subtraction is the inverse operation of addition. They should use vectors to represent a sum and a known addend and then determine the direction and length of the vector representing the missing addend. A subtraction sentence is shown with vectors on the number line in Figure 8-19. The original sentence, $(^-6) - (^+4) = \square$, is rewritten as an addition sentence, $(^+4) + \square = {^-6}$, and is read as, "What number is added to $^+4$ to obtain a sum of $^-6$?" The vectors indicate the answer, $^-10$. In Figure 8-20, other subtraction sentences are represented by vectors on number lines. In (a), sentence $(^+4) - (^-6) = \square$ is interpreted after it has been rewritten as $(^-6) + \square = {^+4}$. In (b) and (c), the sentences $(^-5) - (^-2) = \square$ and $(^+4) - (^+7) = \square$, respectively, are represented with vectors on number lines.

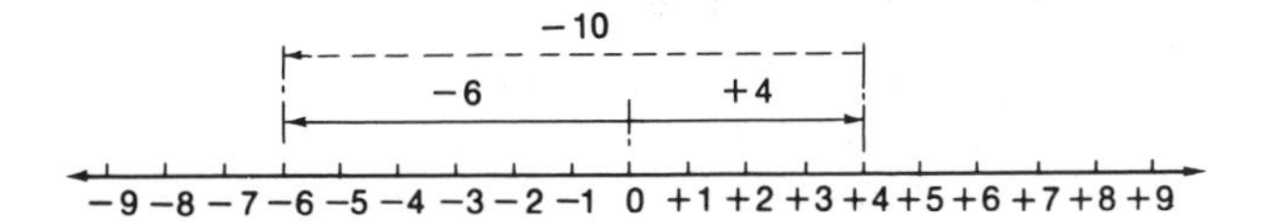


**Figure 8-19** The subtraction sentence $(^-6) - (^+4) = {^-10}$ represented with vectors on a number line.

[8] David Page, *Number Lines, Functions, and Fundamental Topics* (New York: Macmillan Company), chap. 10.

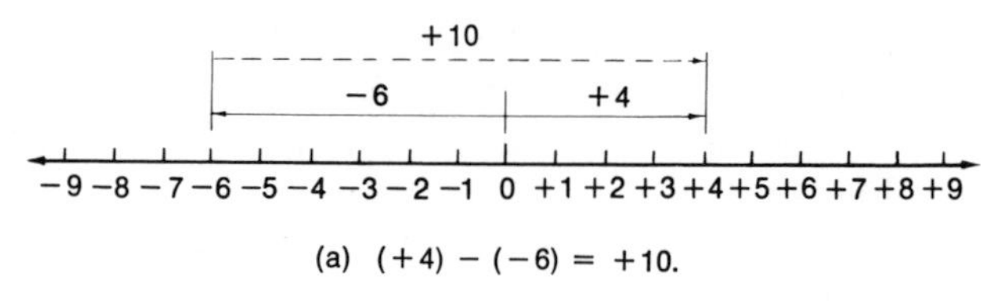


(a) $(+4) - (-6) = +10$.

−5
−3 −2
−9 −8 −7 −6 −5 −4 −3 −2 −1 0 +1 +2 +3 +4 +5 +6 +7 +8 +9

(b) $(-5) - (-2) = -3$.

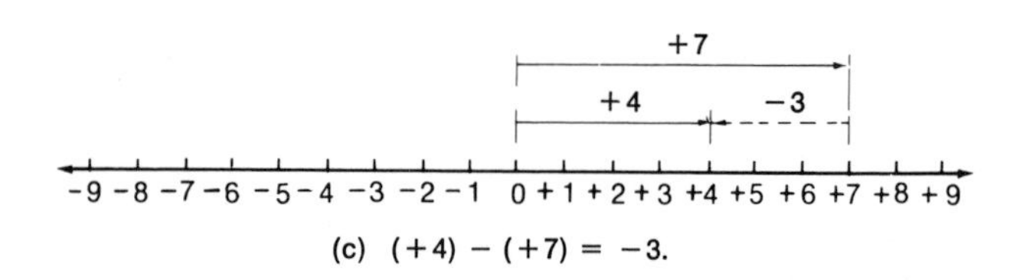


(c) $(+4) - (+7) = -3$.

**Figure 8-20** Number lines used to illustrate three subtraction sentences

Children should also explore the meaning of subtraction with integers by using postman stories, pipe cleaner loops, and positive and negative money. Subtraction occurs in the postman stories when the postman picks up mail. It also occurs when he delivers one check and picks up another. The subtraction sentence $(^{+}4) - (^{+}5) = \square$ describes the delivery of a check for \$4.00 and the reclaiming of a check for \$5.00. The delivery of a bill for \$3.00 and the picking up of a check for \$6.00 is described by the sentence $(^{-}3) - (^{+}6) = \square$. Processes using the loops are described by Fremont.[9]

---

## Self-check of Objective 8

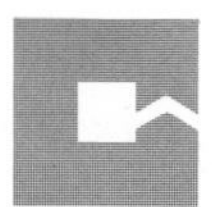

Make eight number lines for the integers $^{-}10$ through $^{+}10$. Use vectors on one line at a time to represent these addition and subtraction sentences: $(^{+}4) + (^{+}3) = {^{+}7}$, $(^{-}3) + (^{-}3) = {^{-}6}$, $(^{+}5) + (^{-}3) = {^{+}2}$, $(^{+}3) + (^{-}6) = {^{-}3}$, $(^{+}6) - (^{+}3) = {^{+}3}$, $(^{-}7) - (^{-}3) = {^{-}4}$, $(^{-}3) - (^{-}4) = {^{+}1}$, $(^{+}3) - (^{-}6) = {^{+}9}$.

---

## Common Pitfalls and Trouble Spots

Below are the four common pitfalls you should avoid as you help children learn addition and subtraction with whole numbers and integers.

First, teachers will sometimes use an unfamiliar place value device to illustrate regrouping processes. An unfamiliar device is more likely to confuse

[9] Fremont, pp. 571–572.

children than to help them. Be certain your children can read numbers represented on a device and can use it to represent numbers before it is used to explain processes for addition and subtraction.

Second, place value devices are sometimes used incorrectly. Two errors are frequently made: (1) The algorithm will be completed first and then be illustrated on a device. During their first work with an algorithm, children should see steps completed on the place value device first, then see them in the algorithm. (2) The device is used improperly. Steps in a process may be done out of sequence, or too rapidly for children to follow their meaning, or they may be omitted entirely. Be sure you know how to use and demonstrate a device properly before you use it with children.

Third, the regrouping process is taught by rule only. While rules may be sufficient for some children, there are many for whom they are not. Do not hurry children into work with algorithms; use both the concrete and semiconcrete modes to gradually develop procedures, ensuring that each step in the algorithm is clear.

And fourth, addition and subtraction of integers are taught by rote. It is easy to present rules for this work and have children practice them, but their usefulness is limited because they are easily forgotten. Use procedures similar to those suggested in this chapter to help children formulate their own rules for adding and subtracting integers.

The troubles children experience with addition and subtraction usually stem from instruction that progresses to the abstract mode too rapidly. Carefully sequenced and properly paced activities with appropriate devices will reduce the amount of trouble children have in any area of mathematics. However, even with good instruction, there may be some children who experience difficulty with addition and subtraction of whole numbers and who need special help. These are some of the common errors they make:

1. Addition

   a. Adding the numbers' digits as though each digit is a separate number:

$$\begin{array}{r} 32 \\ +65 \\ \hline 16 \end{array} \qquad \begin{array}{r} 94 \\ +27 \\ \hline 22 \end{array} \qquad \begin{array}{r} 87 \\ +49 \\ \hline 28 \end{array}$$

   b. Beginning at the left:

$$\begin{array}{r} 3^{2}4^{3}2 \\ +9\ 7\ 2 \\ \hline 1\ 1\ 7 \end{array} \qquad \begin{array}{r} 4\ 6^{3}9 \\ +2\ 7\ 5 \\ \hline 6\ 117 \end{array} \qquad \begin{array}{r} 3^{1}4\ 2\ 9 \\ +8\ 1\ 5\ 6 \\ \hline 1\ 6,\ 715 \end{array}$$

   c. Failure to carry:

$$\begin{array}{r} 342 \\ +972 \\ \hline 12{,}114 \end{array} \qquad \begin{array}{r} 439 \\ +275 \\ \hline 61{,}014 \end{array} \qquad \begin{array}{r} 5798 \\ +6204 \\ \hline 119{,}912 \end{array}$$

2. Subtraction

   a. Beginning at left:

$$\begin{array}{r} \not{4}\,{}^{1}7 \\ -2\;\;9 \\ \hline 2\;\;6 \end{array} \qquad\qquad \begin{array}{r} \not{7}\,{}^{1}\not{3}\,{}^{1}6 \\ -5\;\;8\;\;7 \\ \hline 2\;\;5\;\;9 \end{array}$$

   b. Subtracting the smaller number in a column from the larger, regardless of where it is in the algorithm:

$$\begin{array}{r} 47 \\ -29 \\ \hline 22 \end{array} \qquad\qquad \begin{array}{r} 736 \\ -587 \\ \hline 251 \end{array}$$

Each of these errors is usually committed by a child who has only a partial knowledge of the process and has used his or her incomplete knowledge to develop a way of processing the algorithm that seems to make sense. Because each error is systematic, it must be detected early and corrected immediately before it becomes habitual. These are general guidelines for correcting such errors:

1. Detect the error pattern.
2. Analyze the child's knowledge and understanding of subordinate concepts and skills.
3. Reteach the processes, beginning with the concrete mode at the level of the hierarchy where the child's lack of knowledge inhibits further growth.
4. Select materials and procedures that provide a slightly different approach from the ones that led to the faulty thinking in the first place.
5. Select practice materials carefully. Choose games, puzzles, work sheets, and other materials that emphasize the particular step a child is learning.
6. Provide frequent feedback and recognition of growth.

---

## Self-check of Objectives 9 and 10

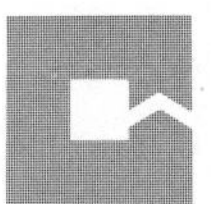

Four pitfalls related to addition and subtraction with whole numbers and integers are discussed. Identify each one, and explain orally or in writing how it can be avoided.

Identify at least two errors children commonly make as they do addition and subtraction. State orally or in writing a list of guidelines for helping children overcome systematic errors.

---

## SUMMARY

When children have a good understanding of place value for numbers through 99 and of the processes of addition and subtraction, they are ready for addition and subtraction of numbers with sums up to 99. Later, addition and subtraction are extended to whole numbers of any size. Skill in adding by endings is important so children can do certain addition without paper and pencil, as when making purchases in stores, adding columns of numbers, and doing some types of multiplication. Addition of larger numbers will involve situations where no regrouping (carrying) is required and other situations where it is. Place value pocket charts, bean-sticks, an abacus, and other devices will make this addition meaningful to children.

Subtraction of larger numbers can be performed by decomposition, equal additions, and other processes. Decomposition is generally favored by teachers because its meaning can be illustrated with place value devices. Equal addition is performed rapidly by persons who learn to do it. Eventually, children learn to use the addition and subtraction algorithms without benefit of manipulative aids. There are processes for checking the accuracy of addition and subtraction; children should learn to habitually use one or more for each operation.

The minicalculator is useful for engaging children in interesting and instructive learning activities and investigations dealing with addition and subtraction. Activities include making estimates of sums and differences, investigating palindromes and magic squares, a subtraction activity, and the game of Nim.

Not all elementary school children will have sufficient background to learn to add and subtract integers. Children who are ready to learn how to perform these operations on integers should use number lines and other devices as they work, rather than learn rote procedures that lack meaning.

Four pitfalls to be avoided are: (1) using unfamiliar place value devices for showing regrouping, (2) improper use of place value devices, (3) teaching regrouping by rules, and (4) teaching the addition and subtraction of integers by rules and in a rote manner. Some children fail to learn to add and/or subtract properly and may develop faulty procedures. Careful reteaching, beginning at the point where the faulty procedures emerged, is necessary to overcome such errors.

## STUDY QUESTIONS AND ACTIVITIES

1. Demonstrate with a number line and a hundreds chart how children might learn higher decade addition, or adding by endings, involving the following sentences:
   (a) $43 + 2 = \square$, $53 + 2 = \square$, $63 + 2 = \square$

(b) $27 + 5 = \square$, $37 + 5 = \square$, $47 + 5 = \square$

2. Practice with a place value pocket chart, beansticks, and a classroom abacus until you are proficient in using each aid to illustrate the following examples.

(a) $\begin{array}{r} 23 \\ +41 \\ \hline \end{array}$ (b) $\begin{array}{r} 68 \\ +26 \\ \hline \end{array}$ (c) $\begin{array}{r} 203 \\ +429 \\ \hline \end{array}$ (d) $\begin{array}{r} 688 \\ +247 \\ \hline \end{array}$

3. Use the decomposition and equal additions methods of subtraction to solve the following problems. Practice each procedure until you become proficient. With which procedure can you subtract most rapidly?

(a) $\begin{array}{r} 62 \\ -48 \\ \hline \end{array}$ (b) $\begin{array}{r} 436 \\ -209 \\ \hline \end{array}$ (c) $\begin{array}{r} 943 \\ -387 \\ \hline \end{array}$ (d) $\begin{array}{r} 4003 \\ -3829 \\ \hline \end{array}$

4. Use a place value pocket chart, beansticks, and abacus to represent the following examples. Practice until you can represent the decomposition process of subtraction meaningfully.

(a) $\begin{array}{r} 43 \\ -19 \\ \hline \end{array}$ (b) $\begin{array}{r} 441 \\ -236 \\ \hline \end{array}$ (c) $\begin{array}{r} 836 \\ -447 \\ \hline \end{array}$ (d) $\begin{array}{r} 3006 \\ -1948 \\ \hline \end{array}$

5. Practice the low-stress algorithms for addition and subtraction until you are proficient with each one. Make up your own examples, including four or more addends larger than 100 for addition and numbers greater than 1000 and with and without zeros in the minuend for subtraction.

6. Begin a collection of activities dealing with addition and subtraction of whole numbers for the minicalculator. Use books such as those in Appendix C and articles from *The Arithmetic Teacher* as sources.

7. The articles and books in footnotes 6, 7, and 8 and Bennett's article reviewed in the reading list for this chapter describe other ways than the number line for helping children understand addition and subtraction of integers. Read one of the articles or in Page's book to learn one alternate method, then use that method to illustrate the meaning of each sentence in this chapter's self-check 8.

## FOR FURTHER READING

Backman, Carl A. "Analyzing Children's Work Procedures," *Developing Computational Skills*, 1978 Yearbook, National Council of Teachers of Mathematics. Reston, Va.: The Council, pp. 177–195. Diagnosis requires more than analysis of tests. Children's daily work—both results and processes—needs to be analyzed, too. Suggestions for doing this and for correcting errors in computation are explained and illustrated.

Beardslee, Edward C. "Teaching Computational Skills with a Calculator," *Developing Computational Skills*, 1978 Yearbook, National Council of Teachers of Mathematics. Reston, Va.: The Council, pp. 226–241. Explains a variety of activities with a calculator dealing with counting, addition and subtraction, multiplication and division, problem solving, pattern investigations, and decimals.

Bennett, Albert B., Jr., and Gary L. Musser. "A Concrete Approach to Integer Addition and Subtraction," *The Arithmetic Teacher*, XXIII, No. 5 (May 1976), 332–336. Black chips serve as whole numbers while red chips represent negative integers for activities that develop the meaning of addition and subtraction with integers.

Bradford, John W. "Methods and Materials for Learning Subtraction," *The Arithmetic Teacher*, XXV, No. 5 (February 1978), 18–20. A process for teaching subtraction using white Cuisenaire rods, ten strips, and hundreds and thousands grids shows subtraction's close relationship to addition. The author claims success in helping third and fourth graders overcome systematic errors in subtraction.

Cox, L. S. "Diagnosing and Remediating Systematic Errors in Addition and Subtraction Computations," *The Arithmetic Teacher*, XXII, No. 2 (February 1975), 151–157. Identifies and explains systematic errors in addition and subtraction. Describes ways to diagnose and remediate such errors.

Hutchings, Barton. "Low-Stress Algorithms," *Measurement in the Classroom*, 1976 Yearbook, National Council of Teachers of Mathematics. Reston, Va.: The Council, pp. 218–239. Explains low-stress algorithms for all four operations. The processing of multiplication and division facts using the low-stress addition process are also possible.

———. "Low-Stress Subtraction," *The Arithmetic Teacher*, XXII, No. 3 (March 1975), 226–322. Discusses the low-stress algorithm for subtraction. Includes several examples and an extensive bibliography dealing with research into the addition and subtraction algorithms.

Lichtenberg, Donovan R. "The Use and Misuse of Mathematical Symbolism," *The Arithmetic Teacher*, XXV, No. 5 (January 1978), 12–17. Makes the distinction between expressions used to name numbers—$6 + 3$ and $(4 \times 6) + 2$—and sentences—$6 + 3 = 9$ and $(4 \times 6) + 2 = 26$. These expressions should be distinguished from vertical notation used for the purpose of computing answers, or finding the simplest expression for a number name.

Logan, Henrietta L. "Renaming with a Money Model," *The Arithmetic Teacher*, XXVI, No. 1 (September 1978), 23–24. Children who have difficulty grasping the meaning of subtraction with regrouping may be aided by using dollars, dimes, and pennies as models for place value. First, different ways of representing a given amount of money are practiced, then the money is used as a model for the subtraction algorithm.

Merseth, Katherine K. "Using Materials and Activities in Teaching Addition and Subtraction Algorithms," *Developing Computational Skills*, 1978 Yearbook, National Council of Teachers of Mathematics. Reston, Va.: The Council, pp. 61–77. Explains readiness activities with base ten blocks, record-keeping activities with mats and blocks, and addition and subtraction activities with the mats and blocks. Activities leading to abstractions of the processes are part of the discussion.

Nichol, Margaret. "Addition Through Palindromes," *The Arithmetic Teacher*, XXVI, No. 4 (December 1978), 20–21. There are both words and numbers that are palindromes. Of particular interest are the suggestions for using palindromes to provide addition practice. Also of interest is the fact that it takes 24 steps to change 89 to a palindrome.

Quast, W. G. "Method or Justification," *The Arithmetic Teacher*, XIX, No. 8 (December 1972), 617–622. Quast considers expanded notation and other forms of performing the operations of addition, subtraction, multiplication, and division. He then discusses whether each is a method of computing or a process for justifying the basic algorithm for each operation. He concludes that teachers should recognize that these forms are justifications and need to be considered in proper perspective. Teachers who recognize the place of alternate procedures will use only the ones required to make the basic algorithm meaningful for particular children.

Rheins, Gladys B., and Joel J. Rheins. "A Comparison of Two Methods of Compound Subtraction," *The Arithmetic Teacher,* II, No. 3 (October 1955), 63–69. Reports the results of a study comparing groups of children taught by two subtraction methods, the decomposition and equal additions methods. The conclusion is that the decomposition method is a better way to introduce compound subtraction.

Schwartsman, Steven. "A Method of Subtraction," *The Arithmetic Teacher,* XXII, No. 8 (December 1975), 628–630. The algorithm employs both "regular" subtraction and subtraction using complements of numbers. It has been used in a process-oriented way (that is, without structural meaning) with remedial students in upper grades and as enrichment for those who understand the conventional algorithm well.

Trafton, Paul R., and Marilyn N. Suydam. "Computational Skills: A Point of View," *The Arithmetic Teacher,* XXII, No. 7 (November 1975), 528–537. Reporting for the Editorial Panel of *The Arithmetic Teacher,* the authors identify ten tenets of the teaching of computation. Each is explained and its implications for curriculum makers and teachers are described.

# 9 Beginning Work with Multiplication and Division

Upon completion of Chapter 9, you will be able to:

1. Describe three types of multiplication situations.

2. Distinguish between measurement and partitive division situations.

3. Demonstrate with materials some procedures that can be used to build children's readiness for the introduction of multiplication and division.

4. Explain orally or in writing how children's knowledge of addition can be used to develop their understanding of multiplication, and demonstrate how the symbolism of multiplication can be explained.

5. Use the terms *factor, product, multiplier,* and *multiplicand* properly in connection with a multiplication sentence.

6. Demonstrate at least three types of materials and procedures to reinforce children's understanding of the repeated addition concept of multiplication.

7. Arrange objects or draw pictures to illustrate arrays for sentences such as $6 \times 3 = 18$ and $4 \times 9 = 36$.

8. Describe and/or demonstrate materials children can use to learn about the array concept of multiplication.

9. Describe at least three examples of Cartesian, or cross, product situations suitable for elementary school children.

10. Tell orally or in writing a story for both a measurement and a partitive situation in division, and demonstrate how manipulative materials can be used to represent each of your stories.

11. Demonstrate at least three activities that can be used to provide practice with the multiplication and division facts.

12. Demonstrate how these properties of multiplication can be made meaningful to children: commutative, associative, distribution of multiplication over addition, the roles of one and zero in multiplication.

13. Identify a common pitfall associated with early work in multiplication and division.

14. Describe two ways to help children who have difficulty learning the basic multiplication and division facts.

Key Terms you will encounter in Chapter 9:

repeated addition
array
Cartesian (cross) product
measurement division
partitive division
dividend
divisor
quotient
product
known factor
missing factor
multiplier
multiplicand
multiplication table
commutative property for multiplication
associative property for multiplication
distributive property
identity element for multiplication
closure for multiplication

The definition of multiplication in Chapter 1 states that it is the operation that assigns to an ordered pair of numbers, called factors, a single number, called their product. Division is defined as the operation used to find a missing factor when one factor and a product are known. Once again we have examples of abstract definitions that are useful at a mature level but that cannot be used with children just beginning work with operations that are new to them.

## MULTIPLICATION SITUATIONS

Children's initial work must be at the concrete-manipulative level so they can develop an intuitive understanding of multiplication. Later, more mature definitions can be introduced. Activities with concrete materials will introduce children to three situations involving multiplication.

### 1. Repeated Addition

Multiplication is used to determine the answer for a repeated addition situation: "Sue has a book containing her coin collection. She has nine pages filled with fifteen coins on each page. How many coins are on the nine pages?" The problem can be solved by addition using 15 as an addend nine times, or by multiplying 9 times 15.

### 2. Arrays

Multiplication is used in array situations: "In a store, boxes of shoes are stored on shelves. Each shelf has four rows of boxes with twelve boxes in each row. How many boxes does each shelf hold?" The boxes may be visualized in a 4 by 12 array, as shown in Figure 9-1. The problem can be solved by multiplying 4 times 12. (The answer can also be found by using 4 as an addend twelve times, or 12 as an addend four times. However, children should be encouraged to think of an array in this situation and to use multiplication.)

### 3. Cartesian, or Cross, Product

Multiplication is used in Cartesian, or cross, product situations: "Automobiles come from an assembly line painted red, gold, white, or brown. Then each car is given a white, black, or brown vinyl roof. How many possible combinations of paint and vinyl colors are there for these cars?" The answer can be determined by matching red, gold, white, and brown cards that represent the cars one at a time with white, black, and brown cards that represent the vinyl roof colors to show the twelve combinations. The multiplication 4 times 3 can also be used.

---

## Self-check of Objective 1

Describe repeated addition, array, and Cartesian product situations for multiplication.

---

## DIVISION SITUATIONS

Children's first activities with division also require that they use manipulative materials to grasp the meaning of the situations that give rise to division and to develop an understanding of the operation. Activities should be used to introduce children to these two situations:

### 1. Measurement

Division is used to determine the answer for a measurement situation: "There are 8 eggs. How many servings of scrambled eggs, each using 2 eggs, can be made from these eggs?" The answer can be determined by

| | | | | | | | | | | | |
|---|---|---|---|---|---|---|---|---|---|---|---|
| Shoes | Shoes | Shoes | Shoes | Shoes | Shoes | Shoes | Shoes | Shoes | Shoes | Shoes | Shoes |
| Shoes | Shoes | Shoes | Shoes | Shoes | Shoes | Shoes | Shoes | Shoes | Shoes | Shoes | Shoes |
| Shoes | Shoes | Shoes | Shoes | Shoes | Shoes | Shoes | Shoes | Shoes | Shoes | Shoes | Shoes |
| Shoes | Shoes | Shoes | Shoes | Shoes | Shoes | Shoes | Shoes | Shoes | Shoes | Shoes | Shoes |

**Figure 9-1** Four rows of shoe boxes with twelve in each row illustrate a 4 by 12 array

successively using 2 eggs and counting the number of servings. This can be shown by repeated subtractions:

$$\begin{array}{r} 8 \\ -2 \\ \hline 6 \end{array} \quad \begin{array}{r} 6 \\ -2 \\ \hline 4 \end{array} \quad \begin{array}{r} 4 \\ -2 \\ \hline 2 \end{array} \quad \begin{array}{r} 2 \\ -2 \\ \hline 0 \end{array}$$

The number of times 2 is subtracted is four. Using a division sentence, the same situation can be expressed as $8 \div 2 = 4$. In a measurement situation, the divisor indicates the size of a subunit that is measured against an original unit to determine how many times it is contained in the original unit.

### 2. Partitive

Division is used in a partitive situation: "There are twenty-four candy bars in a box. If the candy bars are to be distributed among three children so that each gets the same number of bars, how many will each child get?" The answer can be determined by distributing the twenty-four candy bars one by one into three groups so that there are the same number of bars in each group. In a partitive situation, the divisor indicates the number of equal-sized groups to be derived from a given set, while the quotient indicates the size of each group.

During early work, measurement situations help children see the relationship between multiplication and division. Also, during early work, do not use the special names given to the numbers in a division sentence. Later, you can introduce these special names: For the sentence $36 \div 9 = 4$, the first number is the dividend, the second number is the divisor, and the answer is the quotient. When the operation is defined in terms of multiplication, the dividend is a product, the divisor is a known factor, and the quotient is a missing factor.

---

## Self-check of Objective 2

Define a measurement situation and a partitive situation in division.

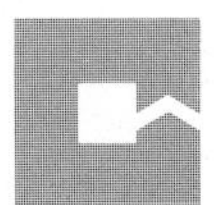

---

## DEVELOPING READINESS FOR MULTIPLICATION AND DIVISION

Children will not have been introduced to or have mastered addition and subtraction at all levels of difficulty before they begin activities with multiplication and division. In fact, activities that build background for multiplication and division can be begun with first- and second-grade children. As they use markers in sets to learn to count, you should show them groupings of equal-sized sets to facilitate counting by 2s, 3s, 4s, and so on. When children use set cards, such as those in Figure 9-2, and a number line, as in Figure 9-3, the meaning of a sequence like 4, 8, 12, . . . , 36 becomes clear to them. Children should also count markers arranged on magnetic or flannel boards and pictures of familiar objects—birds, flowers, or books—arranged in patterns that show sets of equal size. Groups of objects in the classroom—panes of glass in windows, desks in rows, or sets of books—can provide further counting practice.

Set cards like those used for counting by 4s can also be used to build background for division. After a set of cards has been displayed, as shown in Figure 9-2, children can count backward from 36 as the cards are removed one by one. Sets of cards with two, three, five, or six dots on each card give practice in backward skip-counting for each of these numbers.

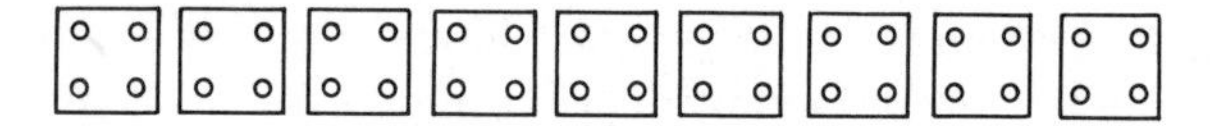

**Figure 9-2** Set cards used to give children experience with counting by 4s. These can be made by sticking adhesive labels to pieces of tagboard or colored railroad board

Backward counting on the number line can begin while children are still using a walk-on line. A child can begin at 10 on the line and take jumps of two units each until he has reached 0. The rest of the children should count the number of jumps he has taken. The children should also begin at other points on the line and use jumps that make it impossible to land on 0. A child might begin at 8 and take three-unit steps. When he reaches 2 on the line, he and the rest of the children should stop to think. They will conclude that three-unit steps cannot begin at 8 and end on 0. The two units between 2 and 0 will have to be a "remainder." A walk-on line can also help to work out situations involving units too large for a child to actually step off. For example, after the above exercise, you can ask: "How many steps from 10 to 0 would Billy take if he could take steps that are five units long?" Later, a chalkboard or other type of longer number line should be used for work with larger numbers. A number line that can be marked with chalk or an erasable felt pen is best so children can actually see and count the jumps they make.

Pictures of familiar objects and manipulative aids—tongue depressors, felt shapes, or magnetic disks—grouped in equal-sized sets will also help build background for learning the measurement and partitive concepts of division, if used frequently.

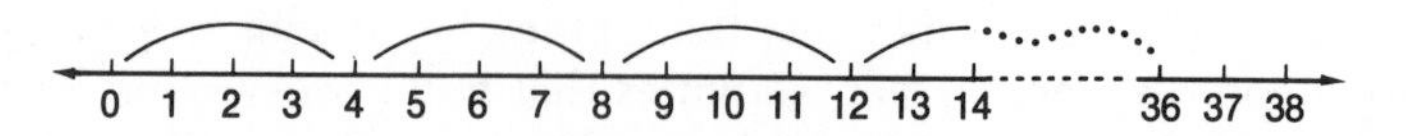


**Figure 9-3** Counting by 4s from 0 to 36 illustrated on the number line

## Self-check of Objective 3

Demonstrate with materials or drawings at least one activity that will build children's readiness for multiplication and division. Which types of multiplication and division situations are involved in your activities?

## INTRODUCING THE REPEATED ADDITION CONCEPT

### 1. Introduction

Multiplication may be introduced in late second grade, although it is frequently delayed until grade three. Children use their knowledge of addition as the foundation for building understanding of the new operation. For example, when confronted with four sets, each of which contains two elements, children can first determine the answer by adding $2 + 2 + 2 + 2$. They can then be introduced to multiplication by learning that the expression "$4 \times 2$" can also be used to determine the answer. They are guided to note that $2 + 2 + 2 + 2 = 4 \times 2$; that is, that the meaning of the multiplication sentence $4 \times 2 = 8$ is equivalent to that of the addition sentence $2 + 2 + 2 + 2 = 8$.

You should plan lessons to encourage children's use of familiar materials and devices in exploratory activities. A number line, hundreds board, magnetic or flannel board with markers, place value device, and markers for use at individual desks should all be available during this introduction. The following problem might lead children into exploratory activities. "At the store, Sally bought stick candy in bundles of six sticks each. If she bought four bundles of the candy, how many sticks did she get?" Each child should determine the answer with whatever aids or procedures are meaningful. As the children work, you should move among them to observe their work and offer encouragement and praise for their efforts. Later, some of the children should discuss their procedures with others.

Children who used the most concrete materials—bundles of sticks, markers on a magnetic board, or marks drawn on a piece of paper—should explain their procedures first. Users of a hundreds board or number line should report next. If some children used a mathematical operation, such as addition or multiplication, they should discuss their methods last. As children tell what they did, they should show which materials were used and how they were used. Give the children several examples to work out before you introduce the symbolism for multiplication.

Once you are satisfied children have the background they need to understand multiplication, you can introduce symbolism for the operation by using a problem already discussed by the children. For the situation with the bundles of candy, the answer might have been determined by using four sets of six objects, the addition sentence $6 + 6 + 6 + 6 = 24$, and the column addition shown in the margin. Through a questioning process, you can guide children to note the common elements in each of the examples: there are always four sets or addends and each always has six members or is the number 6.

$$\begin{array}{r} 6 \\ 6 \\ 6 \\ +6 \\ \hline 24 \end{array}$$

Finally, introduce the multiplication sentence $4 \times 6 = 24$. As the children compare the multiplication sentence with the sets of markers, the number line, and the addition sentence and column addition, they will see that the number 4 indicates the number of sets or the number of addends. The number 6 indicates the size of each set, or the name of each addend. The number 24 indicates the size of the combined sets and the sum for the addition. The children should also learn to read a multiplication sentence. The sentence $4 \times 6 = 24$ is read "four times six equals 24," or "four sixes are 24." The displays for other problem situations should also be reviewed and extended to include multiplication as a means of determining their answers.

Following introduction of multiplication sentences, children should learn the terms that name the parts of a sentence. The two numbers multiplied are called factors, while the answer is a product. When they use a sentence to express multiplication arising from a repeated addition situation, children can refer to the "first factor" as the number that indicates the number of sets and the "second factor" as the number that indicates the size of each set. Later, they should learn that the first factor is also called the multiplier and the second factor is called the multiplicand.

To show children the value of multiplication, you can ask them which process they would use to find the answer for a problem involving nine sets, each of which has seven members. If you start to write 7s in a column on the chalkboard to indicate addends, the children will realize it is easier to use multiplication than addition even before you have finished writing nine of them.

When an introductory procedure such as the foregoing is used, each child can participate in the activities and discussions, following progressively from the use of concrete materials to the abstract operations. Another feature of this procedure is that the number and sizes of sets used during the introductory activites are larger than are often recommended in introductions to multiplication. Children with many experiences that provide readiness for learning multiplication need not be restricted to the use of only two or three sets with few members.

## 2. Reinforcement Activities

The following items used as sets provide various ways to use the repeated addition concept of multiplication:

1. *Set cards.* Have children organize the facts with a given number as one factor. For example, with the cards illustrated in Figure 9-2, children can organize the following combinations containing 4 as they begin with one card and use additional cards one at a time.

| | |
|---|---|
| $1 \times 4 = 4$ | $6 \times 4 = 24$ |
| $2 \times 4 = 8$ | $7 \times 4 = 28$ |
| $3 \times 4 = 12$ | $8 \times 4 = 32$ |
| $4 \times 4 = 16$ | $9 \times 4 = 36$ |
| $5 \times 4 = 20$ | |

   When only one card is exposed and the sentence $1 \times 4 = 4$ is written, children should note that the first factor indicates the number of exposed sets while the second factor indicates the size of the set. The product tells how many objects there are altogether. As two, three, and more cards are exposed, children should relate them to the sentences. You should also help children note the progression of factors and products so they will see that each successive product is four larger than its predecessor.

2. *Sets on a magnetic or flannel board.* Equal-sized sets of markers can be placed on a magnetic or flannel board to display the mathematical sentences for each group (Figure 9-4).

3. *Children.* Equal-sized groups of children can be called to the front of the room. As different numbers of groups are brought forward, the multiplication sentences should be displayed.

4. *Sentences illustrated on the number line.* Write a multiplication sentence on the chalkboard. Have a child mark a number line with arrows to show the meaning of the sentence. Or, you or a child can

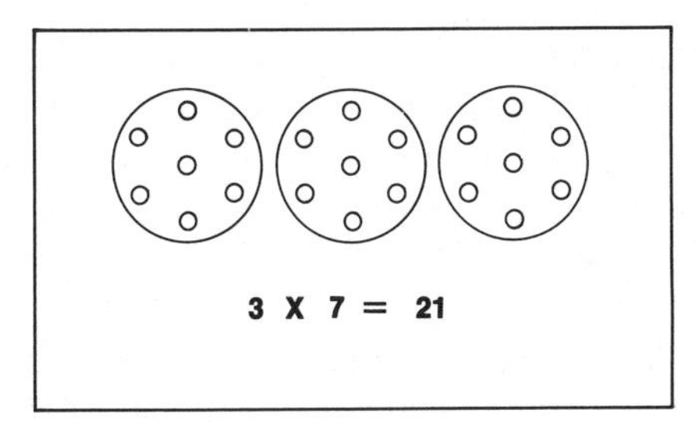


**Figure 9-4** Markers and symbols used to illustrate the multiplication sentence 3 × 7 = 21

mark jumps or arrows on a number line and children can determine the sentence they illustrate. (Be sure the first factor of the sentence indicates the number of jumps or arrows and the second factor the size of each; this is consistent with the way repeated addition is interpreted.)

## Self-check of Objectives 4, 5, and 6

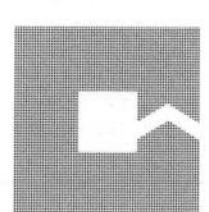

Explain how children's knowledge of addition can be used to introduce the repeated addition concept of multiplication. Use markers or pictures to illustrate how the meaning of a multiplication sentence can be explained.

Define each of these words: *factor, product, multiplier,* and *multiplicand.*

Demonstrate materials and procedures that can be used to reinforce children's understanding of the repeated addition concept of multiplication.

## INTRODUCING THE ARRAY CONCEPT

An array is an arrangement of items in a number of rows, each row containing the same number of items. A 3 by 5 array is illustrated in Figure 9-5, where there are three rows of five dots each. The generally accepted sentence for this array is $3 \times 5 = 15$, and is read "3 times 5 equals 15." The first factor in the sentence indicates the number of rows in the array, while the second factor indicates the number of elements in each row. The product indicates the number of elements in the entire array.

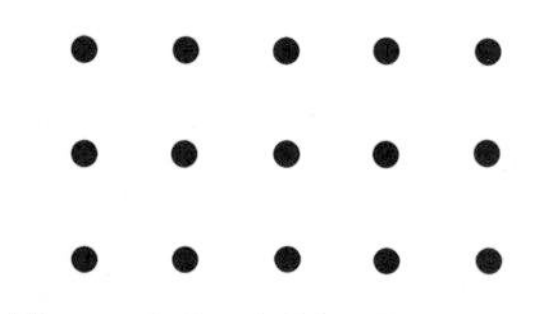

**Figure 9-5** A 3 by 5 array

### 1. Array Patterns

Six children in Ms. Torres' third grade had developed a good understanding of the repeated addition concept of multiplication. Now she wanted them to work with arrays. She reasoned that work with arrays not only would introduce the children to a different way of viewing multiplication, but also would give them additional experiences with basic multiplication combinations they should master.

Ms. Torres set up a learning center for these six children. (She knew that the materials for the center would be used later by other children.) The center contained six masonite pegboard squares, each with 100 holes arranged in ten rows of ten holes each, a box of colored golf tees, a set of problem cards, and paper and colored pencils. The problem cards are shown in Figure 9-6.

Her plans called for the children to write and complete the underlined sentences on each problem card, work the multiplication sentences, and draw pictures of the array patterns as they worked the card's activities.

MULTIPLICATION ARRAYS (1)

1. Use 6 red golf tees to make this pattern on your pegboard.

   O O O
   O O O

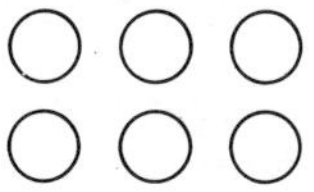

   This pattern has 2 rows of tees, with 3 tees in each row. Altogether there are ___ tees.

2. Make this pattern on your pegboard with green tees.

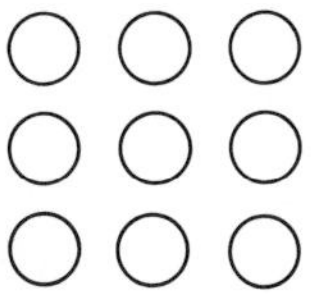

   O O O
   O O O
   O O O

   This pattern has ___ rows of tees, with ___ tees in each row. Altogether there are ___ tees.

3. Make a pattern that has 3 rows, each with 4 yellow tees. Draw the pattern on your paper with a yellow pencil. There are ___ tees in this pattern.

**Figure 9-6(a)** First of a series of three problem cards dealing with arrays

When the activities were completed, Ms. Torres planned to discuss them with the six children. You can set up a similar center for children in your classroom, or lead the children's investigation of similar activities.

Nearly every classroom contains array patterns, such as panes of window glass, rows of ceiling and/or floor tiles, and sets of lights. If your classroom has no apparent patterns, you can create some by arranging books on a table, geometric shapes on a wall board, or pictures on a bulletin board. Once children know about arrays, they should search their room and other parts of the school environment for examples. Children's interest in this activity can be heightened by challenging them to find unusual arrays in and around their school.

Array patterns made from adhesive labels arranged in rows and columns on colored railroad board provide simple aids for showing different arrays. The array for $7 \times 9$ is shown in Figure 9-7. In (a), sixty-three labels are arranged in seven rows with nine labels in each row. An array for seven rows of one each appears in (b) to show a 7 by 1 array.

(2)

4. Make a pattern that has 5 rows, with 3 blue tees in each row. Draw the pattern on your paper with a blue pencil. There are ___ tees in this pattern.

5. Each pattern is an <u>array</u> of tees. Another way to write the name of the pattern of blue tees is to call it a 5 by 3 array. <u>The pattern of blue tees is a ___ by ___ array</u>.

6. Use the tees to make these arrays. Draw pictures of arrays on your paper, using any colors you wish.
   a. 2 by 4   d. 5 by 4
   b. 3 by 6   e. 4 by 4
   c. 4 by 5   f. 6 by 2

7. Tell how many tees are in each of the arrays in problem 6.

8. Use your arrays to answer these multiplication sentences:
   a. 2 x 4 = ___   d. 5 x 4 = ___
   b. 3 x 6 = ___   e. 4 x 4 = ___
   c. 4 x 5 = ___   f. 6 x 2 = ___

**Figure 9-6(b)** Second of the series of problem cards dealing with arrays

As children use an array card, they should uncover successive rows of disks and write the multiplication sentences for the arrays: $7 \times 1 = 7$, $7 \times 2 = 14$, and so on as they move the cover card from left to right, as in Figure 9-7(b); and $1 \times 7 = 7$, $2 \times 7 = 14$, and so on as they move the card from top to bottom after giving the array card a quarter turn, as in (c). Direct the children's attention to the two lists of sentences so they will notice the commutative property; for example, they will see that $7 \times 1 = 7$ and $1 \times 7 = 7$.

## 2. Squared Paper Activities

One application of the array concept is to find the area of a plane surface enclosed by a rectangular figure. The area of this surface can be found by counting the number of square units along one edge of the surface and the number of rows of square units along the other edge. If there are eight square units along one edge and nine rows of units, there is a 9 by 8

(3)

9. Use colored pencils to draw some arrays of your own on your paper. Write a multiplication sentence for each of your arrays.

10. Look about the room for patterns that form arrays. Make a picture of the arrays you see, then describe each of your pictures. Write a multiplication sentence for each of your pictures.

**Figure 9-6(c)** The third of three problem cards dealing with arrays

array, and the number of square units (area) can be found by multiplying 9 times 8.

A meaningful way to introduce this use of the array is through activities with squared paper and colored pens or pencils. Use paper with either half-inch or centimeter squares, with directions written on problem

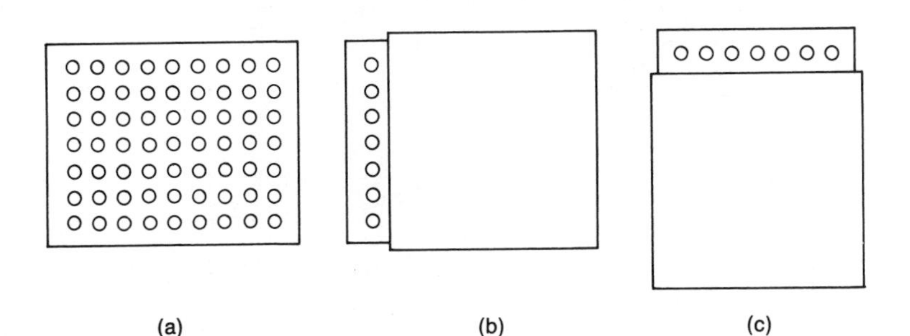


**Figure 9-7** A card that shows a 7 by 9 array (a). In (b) the card is used to show a 7 by 1 array; and in (c) it shows a 1 by 7 array

cards. (Sample arrays are shown in Figure 9-8.) Children should identify each array with the proper array symbolism and a multiplication sentence, as shown in the illustration.

---

## Self-check of Objectives 7 and 8

Use dot patterns to make arrays for these sentences:

$$6 \times 4 = 24, 3 \times 9 = 27, 1 \times 7 = 7, \text{ and } 7 \times 1 = 7.$$

Describe materials children can use to learn about the array concept of multiplication. Can you think of materials other than the ones described in this book?

---

## INTRODUCING THE CARTESIAN PRODUCT CONCEPT

A Cartesian, or cross, product is found by matching each element from one set with each element from another set to make a set of *ordered pairs*. The example of auto colors and vinyl roofs mentioned earlier can be used to introduce this concept.

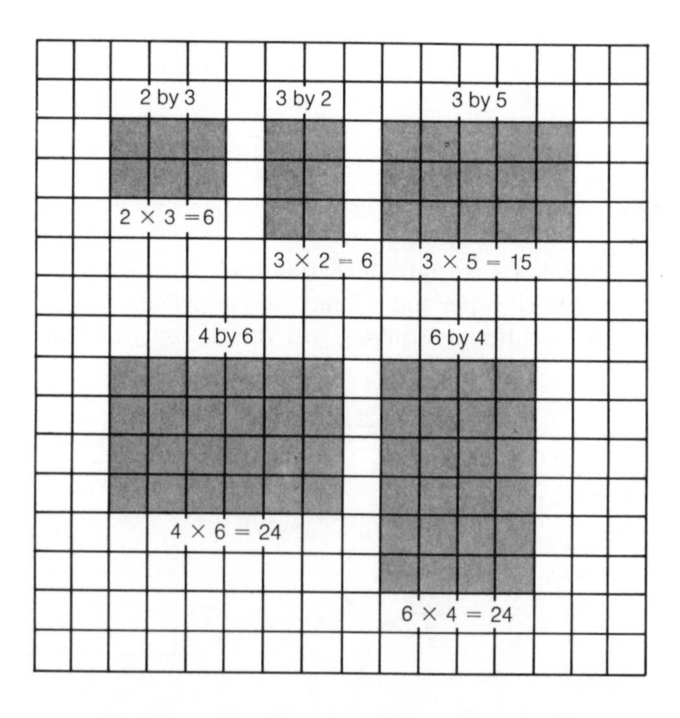


**Figure 9-8** Samples of arrays drawn on squared paper

### 1. Problem Situations

Another situation that can be used concerns ice cream and toppings. Have children name their favorite ice cream flavors and some toppings to go on them. When the ice creams are listed across the top of a chart and the toppings down the left side, the combinations are easy to record (see Figure 9-9). In the figure, when the ordered pairs are made by matching toppings to ice cream, the sentence $6 \times 4 = 24$ is a meaningful representation of the situation. Of course, the ice cream flavors can be matched with the toppings, so the sentence $4 \times 6 = 24$ can also be used.

The Cartesian product interpretation is useful to explain the roles of 1 and 0 in multiplication. Where one of the sets involved in forming a Cartesian product contains only one element, there can be only as many ordered pairs as there are elements in the other set (Figure 9-10). When there is one kind of ice cream and six toppings, there will be six possible combinations of ice cream with topping. The multiplication sentence is $1 \times 6 = 6$. If all of the ice cream is eaten, however, it is obvious that there can be no matchings of ice cream with toppings. Such a situation clearly shows that 0 times 6 and 6 times 0 equal 0.

The study of Cartesian products in the elementary school is limited to simple situations that are within the children's realm of interest and understanding. Nevertheless, there is ample opportunity for children to understand the Cartesian product concept. This understanding, if developed in the elementary school, will provide a foundation for later study of the concept in junior and senior high school.

### 2. Developing Other Ordered Pairs

The use of the following additional situations extends the understanding of the Cartesian product concepts.

| Ice Cream → <br> Toppings ↓ | Vanilla | Coffee | Banana | Walnut |
|---|---|---|---|---|
| Marshmallow | M-V | M-C | M-B | M-W |
| Cherry | Ch-V | Ch-C | Ch-B | Ch-W |
| Chocolate | Choc-V | Choc-C | Choc-B | Choc-W |
| Strawberry | S-V | S-C | S-B | S-W |
| Caramel | Car-V | Car-C | Car-B | Car-W |
| Raspberry | R-V | R-C | R-B | R-W |

**Figure 9-9** The ordered pairs that result from matching six flavors of topping with four flavors of ice cream

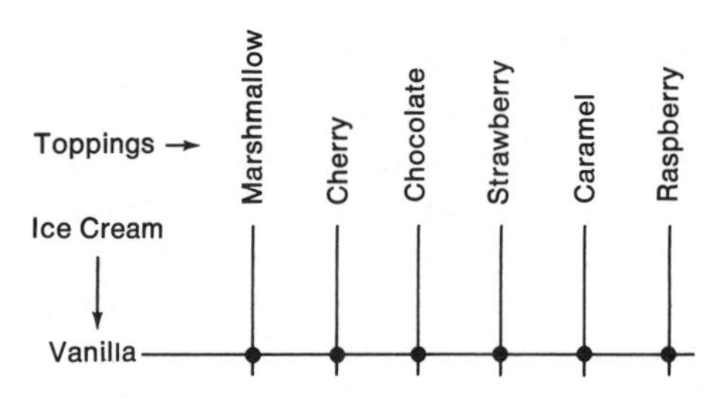


**Figure 9-10** The ordered pairs resulting from matching one flavor of ice cream with six flavors of topping

1. Children can determine how many boy-girl pairs can be matched from a certain set of boys and a certain set of girls. The two sets can be called to the front of the room and the ordered pairs of names written on the chalkboard as the children are paired.
2. Children can determine the number of outfits that can be assembled by matching shirts and pants from a person's wardrobe.
3. Children can determine the number of different cakes made possible by matching cakes and frostings.

---

## Self-check of Objective 9

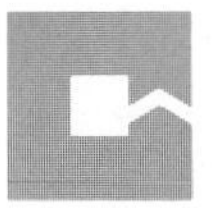

Several Cartesian product situations have been described. Describe at least two, and develop another that you think is suitable for children. Explain why Cartesian product situations are useful for explaining the role of zero in multiplication.

---

## INTRODUCING DIVISION

### 1. Measurement

The measurement concept is most often used to introduce children to division. After studying how second-grade children respond to different types of division situations, Zweng concluded that the measurement-rate type is the easiest one for children to understand.[1] The following example is a measurement-rate situation. "Susan has fifteen marbles. If she puts the marbles into bags, so there are five marbles in each bag, how many bags will she need?" The difference between measurement-rate and measurement-basic problems, as Zweng defines them, is that in rate problems the container—bag, box, or basket—is named, while in basic prob-

[1]Marilyn J. Zweng, "Division Problems and the Concept of Rate," *The Arithmetic Teacher,* XI, No. 8 (December 1964),pp. 547–556.

lems it is not. Zweng found that it is easier for children to determine an answer when the container is named.

A lesson to introduce division should offer sets of objects and containers for children to use as they work. With each computation, the repeated subtractions can be shown along with markers and other objects (Figure 9-11). Before children leave a problem, they should see that the number of sets resulting from the action with the objects and the number of repeated subtractions are the same.

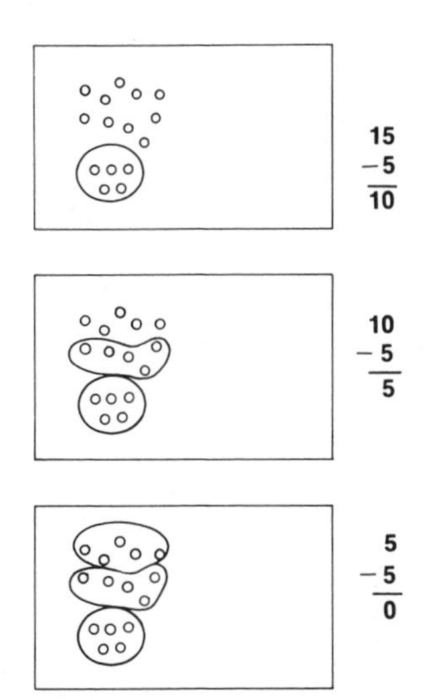


**Figure 9-11** A repeated subtraction situation can be illustrated with markers and the subtraction algorithm

After several problems have been solved by subtraction, it will be obvious to the children that repeated subtraction is a rather long process for obtaining answers. You might say, "We have used repeated subtraction to show with numerals what we have been doing with the marbles, flowers, and pencils. It has taken quite a bit of writing just to show what we have done in these rather simple situations. Suppose we want to find out how many boxes it will take to hold 100 pencils if we put four pencils in each box. How can we find the answer to this problem?" Children will probably suggest repeated subtraction. As the repeated subtraction is done, children should conclude that a shorter way of determining answers to problems of this type is desirable.

The children should then be introduced to the division sentences for the problems they have already worked. The sentence $15 \div 5 = 3$ would be used for the situation illustrated in Figure 9-11. As the sentence is presented, the meanings of its different numbers should be discussed: The first numeral in a sentence indicates the size of the original set of objects; the second numeral indicates the size of each of the equal-sized groups that are removed from the original set; the last numeral, or quotient, indicates the number of groups. Later, when division is used in partitive situations, the meanings of the divisor and quotient will be reversed. The divisor will then indicate the number of equal-sized groups, while the quotient will indicate the size of each of the groups. It is not necessary to use the words *dividend, divisor,* and *quotient* at the time children are introduced to division. It is better to wait until they have developed a good understanding of the process before the words are introduced.

The study of division should continue simultaneously with the study of multiplication after introductory lessons for both processes have been completed. The problem situations used to learn about the repeated addition concept of multiplication can be reversed to learn about division's repeated subtraction concept. Objects on magnetic and flannel boards, arrays, the number line, set cards, markers at desks, and other materials can be used to study both operations.

## 2. Partitive Situations

An introduction to the partitive concept should use familiar materials and situations. "Sally has twenty-four small dolls that she wants to put into a display case that has three shelves. If she puts the same number of dolls on each shelf, how many dolls will each shelf hold?" You should provide dolls and a case—the dolls can be paper with sandpaper backs and the "case" can be a flannel board with three lines to represent shelves—so the children can act out the situation. One possible procedure is to distribute the twenty-four dolls one by one to the shelves until all have been distributed and each shelf holds eight dolls. However, Zweng found that children rarely use the one-to-one procedure.[2] In most instances, they begin by placing a larger number of objects in each of the groups. Once they have done this, they then distribute the remaining objects one by one, two by two, or in whatever manner seems best to them. They will sometimes end up with groups that are not equivalent and that need to be made equivalent; if so, you should help them. After children have acted out several other situations, you can help them learn the meaning of the dividend, divisor, and quotient for division from partitive situations. You should stress the advantage of dividing by helping them see how time consuming the process of actually distributing the objects is. The sets involved need not be very large for children to recognize that the process of physically distributing the objects of a set is a slow one.

Once children are familiar with the two situations, give them further activities with instructions written on problem cards. They should use learning aids as they complete these activities and keep records with pictures, word descriptions, and division sentences that tell what they have done. In this way they reinforce their understanding of each type of situation and learn to use division sentences in meaningful ways.

---

## Self-check of Objective 10

Make up a story for both a measurement and a partitive situation. Demonstrate with manipulative materials, or a series of drawings, how your problems can be illustrated.

---

## PRACTICING THE MULTIPLICATION AND DIVISION FACTS

The study of multiplication and division must include time for practice on the basic facts for both operations. There are 100 multiplication facts, which are the ordered pairs of whole-number factors smaller than 10 and their products. There are only 90 division facts because the ten inverses of the multiplication facts that give 0 as a divisor are excluded. During introductory and exploratory lessons, some children will learn

[2]Zweng, pp. 547–556.

many of the facts without special practice. However, it is not likely that everyone will remember all of the facts during these lessons. Therefore children in the third grade and beyond need activities that help them commit the facts to memory for immediate recall. The generalizations in Chapter 7 in connection with practice with the addition and subtraction facts apply here.

## 1. The Multiplication Table

Use the multiplication table to help children master the 100 multiplication facts and the 90 division facts. Begin with an unfinished table as shown in Figure 9-12(a) drawn on a large sheet of butcher paper or tagboard so the completed table (b) can be saved for later use. As children complete the table under your direction, stress these points:

1. The factors for each combination are in the top row and the left-hand column.
2. In the second row and second column all products are 0 because at least one factor in each combination is 0. There are 19 of these facts.
3. The third row and third column contain products that are the same as the factor that is not 1 (except for 1 itself). These combinations involve the multiplication identity, and account for 17 more facts.
4. Altogether, combinations involving either 0 or 1 account for more than one-third of all of the multiplication facts.
5. Combinations yielding the products in the fourth row and fourth column have 2 as one factor. Each product is twice the factor that is not 2. They account for 15 more facts.
6. Products involving 5 as a factor are in the seventh row and seventh column.

| X | 0 | 1 | 2 | 3 | 4 | 5 | 6 | 7 | 8 | 9 |
|---|---|---|---|---|---|---|---|---|---|---|
| 0 | | | | | | | | | | |
| 1 | | | | | | | | | | |
| 2 | | | | | | | | | | |
| 3 | | | | | | | | | | |
| 4 | | | | | | | | | | |
| 5 | | | | | | | | | | |
| 6 | | | | | | | | | | |
| 7 | | | | | | | | | | |
| 8 | | | | | | | | | | |
| 9 | | | | | | | | | | |

(a)

| X | 0 | 1 | 2 | 3 | 4 | 5 | 6 | 7 | 8 | 9 |
|---|---|---|---|---|---|---|---|---|---|---|
| 0 | 0 | 0 | 0 | 0 | 0 | 0 | 0 | 0 | 0 | 0 |
| 1 | 0 | 1 | 2 | 3 | 4 | 5 | 6 | 7 | 8 | 9 |
| 2 | 0 | 2 | 4 | 6 | 8 | 10 | 12 | 14 | 16 | 18 |
| 3 | 0 | 3 | 6 | 9 | 12 | 15 | 18 | 21 | 24 | 27 |
| 4 | 0 | 4 | 8 | 12 | 16 | 20 | 24 | 28 | 32 | 36 |
| 5 | 0 | 5 | 10 | 15 | 20 | 25 | 30 | 35 | 40 | 45 |
| 6 | 0 | 6 | 12 | 18 | 24 | 30 | 36 | 42 | 48 | 54 |
| 7 | 0 | 7 | 14 | 21 | 28 | 35 | 42 | 49 | 56 | 63 |
| 8 | 0 | 8 | 16 | 24 | 32 | 40 | 48 | 56 | 64 | 72 |
| 9 | 0 | 9 | 18 | 27 | 36 | 45 | 54 | 63 | 72 | 81 |

(b)

**Figure 9-12** Multiplication table (a) ready to be completed by children and (b) ready for study

7. Products in the diagonal from the upper-left to lower-right corners are the result of multiplying a number by itself.

By the time you are finished filling in these rows and columns, most children will be impressed with the fact that there are so few of the harder combinations and that they know more of the facts than they realized. The harder combinations can be put on flashcards (see Figure 9-20) for future practice.

The completed chart should be displayed for future practice by those who have not memorized all the facts. For quick reference at their desks, each child can make a copy by completing a dittoed page of the chart.

The chart is also useful for learning the division facts. Use it this way: To determine the quotient for $56 \div 7 = \square$, locate the 56 that is in the column beneath the 7 in the row of factors across the top. Then move across the chart to the left column. The numeral that represents the quotient, 8, is found to the left of 56. To find the quotient for $56 \div 8 = \square$, locate the 56 in the column beneath 8 in the row across the top. Next, move across the chart to the 7 in the left column. The use of the chart in this manner is consistent with the way it is used to learn the multiplication facts.

## 2. Multiplication Games and Activities

The following activities provide additional practice:

a. **Multiplication and Division Baseball.** Divide the children into two teams. Place three chairs to serve as bases at the front of the room. Select a pitcher and catcher for the team that is in the field. All the players on the other team are batters. The pitcher assumes a position at the front of the room. The catcher stands beside the first batter. The pitcher makes a pitch by showing a sentence card,

$$\boxed{6 \times 3 = \square},$$

or by giving a sentence verbally, "6 times 3 equals ______ ." The batter is safe with a hit if he or she can respond with the correct answer before the catcher does. The batter is out if the catcher responds first. You, an aide, or a child should serve as umpire. If the batter is safe, he or she moves to the first chair. The catcher stands beside the second batter, and the game continues until a side has made three outs. A run is scored each time a batter rounds the bases and reaches home. A batter moves up one base each time a teammate gets a hit. No runs are counted for runners who are left on the bases. The game continues with the teams alternating at bat until nine innings have been completed or until a predetermined amount of time has elapsed. New pitchers and catchers are selected after each inning to allow as many children as possible to play these positions. Change the combinations to division facts to give practice with that operation. The pitcher then shows flashcards or gives oral statements such as "63 divided by 7 equals ______ ."

**b. Multiplication or Division Relay.** Divide the children into two teams. Then write a number of open sentences, such as the following multiplication sentences, on the chalkboard, making two identical lists.

| | |
|---|---|
| $6 \times 9 = \_\_$ | $6 \times 9 = \_\_$ |
| $3 \times \_\_ = 21$ | $3 \times \_\_ = 21$ |
| $2 \times 6 = \_\_$ | $2 \times 6 = \_\_$ |
| $9 \times \_\_ = 81$ | $9 \times \_\_ = 81$ |
| $\_\_ \times 7 = 35$ | $\_\_ \times 7 = 35$ |
| $9 \times 7 = \_\_$ | $9 \times 7 = \_\_$ |

A child from each team is selected to go to the board. At a signal, each child writes the numerals for the missing factors or products. The child who is first to finish is the winner for that round, unless he or she has fewer correct answers than the other child. A point is scored for the winner's team. The game continues with other pairs coming to the board. The lists of sentences should be changed frequently.

**c. Multiplication or Division Wheels.** One child or a pair of children can use multiplication wheels like those in Figure 9-13 to practice the facts. When one child uses a wheel, he or she says the products or lists them on a piece of paper. Answers are checked by turning the wheel over to see the answers in the frame on the tab. When two children work together, one turns the tab as the other responds. The child who is answering should say the entire sentence, or "4 times 6 equals 24," rather than just the answer. When the wheels are used for division, use the side that shows division sentences and read answers on the multiplication side.

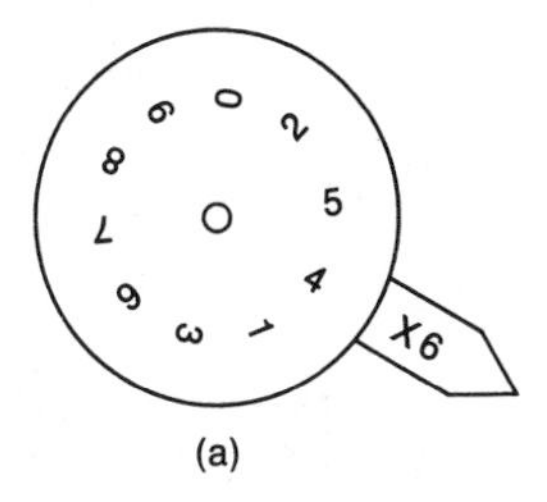


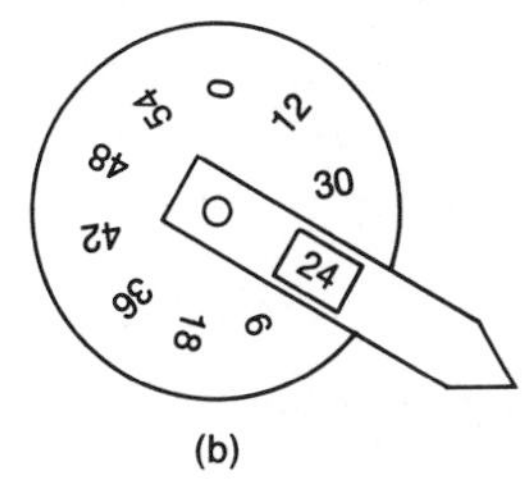


**Figure 9-13** Multiplication and division wheel with 6 as one factor showing (a) front view and (b) reverse side with answer framed in tab

**d. 'Round the World.** This game, described in Chapter 7 as a drill activity for addition and subtraction, can also be used for multiplication and division. Show multiplication or division flashcards or orally present open sentences for these operations to pairs of children. A variation is to use cards that show the products of two numbers smaller than 10. Children respond by giving pairs of factors for each product.

**e. Bingo.** Bingo for addition is also described in Chapter 7. For multiplication, use multiplication flashcards while the children cover numerals that represent products. Five numbers down, across, or in a diagonal give a winning card, or play for "black-out" and have the first child to cover an entire card be the winner. For division, use division flashcards and answer cards with numerals for quotients for children's responses.

**f. Speed Drills.** The speed drills described in Chapter 7 can be used for multiplication and division.

**g. Verbal Exercises.** Short verbal exercises at odd moments during the day are also useful. For example, you might say, "I'm thinking of a pair of numbers. One is a product and the other is a factor. The product is 56 and the factor is 7. What is the other factor?" "How many 9s in 36?" "Six times what number is 42?" Use oral exercises like these, and others dealing with addition and subtraction, expanded and compact numerals, rounding off, and so on, frequently, to review and extend previously learned concepts and facts and to sharpen children's responses.

**h. Beat the Calculator.** This game, described on page 68, can be used for practice with both the multiplication and division facts.

The cassette and reel tape programs mentioned in Chapter 7 contain tapes dealing with multiplication and division as well as addition and subtraction. Game books, such as *Plus* and *Games for Individualizing Mathematics Learning*,[3] contain games for practicing multiplication and division.

---

## Self-check of Objective 11

Demonstrate at least three activities that can be used to provide practice with the multiplication and division facts. If materials for the activities are unavailable, give oral or written explanations of the activities.

---

[3] Mary E. Platts, ed., *Plus*, rev. ed. (Stevensville, Mich.: Educational Services, Inc., 1975), pp. 102–122; and Leonard M. Kennedy and Ruth L. Michon, *Games for Individualizing Mathematics Learning* (Columbus, Ohio: Charles E. Merrill Books, Inc., 1973), pp. 59–73.

## TEACHING THE PROPERTIES OF MULTIPLICATION AND DIVISION

### 1. The Commutative Property

The commutative property of multiplication is readily observed by children when they use arrays because many opportunities are available for them to see that changing the order of a pair of factors does not affect their product. For example, the 2 by 3 and 3 by 2 arrays on squared paper in Figure 9-9 show that $2 \times 3 = 6$ and $3 \times 2 = 6$. You can include exercises on problem cards to call children's attention to the commutative property as they work with the pegboards and golf tees and the squared paper and colored pens and pencils.

The number line is a useful aid for understanding the commutative property. When an open sentence with a pair of small factors is displayed, a child can put arrows above the number line to illustrate the sentence and show the product (Figure 9-14). The factors in the sentence should then be reversed and the new sentence's meaning demonstrated.

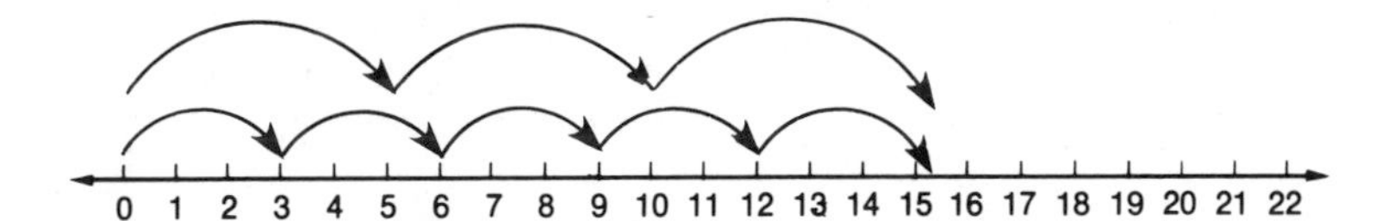


**Figure 9-14** Number line used to show that $5 \times 3 = 3 \times 5$

Division is not a commutative operation. Children who understand that multiplication is commutative can see that division is not by examining a few examples such as $72 \div 9 = \square$ and $9 \div 72 = \square$. These examples make it clear that changing the numbers' order changes the answer.

### 2. The Associative Property

An introduction to the associative property of multiplication should help children recognize its significance. This property frees them to deal with pairs of factors in any way they choose in operations involving three or more factors.

Button cards are useful for an activity that emphasizes the associative property of multiplication. In Figure 9-15(a) the card contains six four-hole buttons, arranged in two rows of three buttons each. Two views of the card should be considered. For the first view, consider one row of three buttons and ask, "How many holes are there in these three buttons?" The answer—12—is determined by multiplying 3 times 4. "How many holes are there in the buttons in both rows?" The answer—24—is determined by multiplying 2 times 12. The sentence for this view is $2 \times (3 \times 4) = \square$. For the second view, consider the two rows of three buttons and ask, "How many buttons are there on this card?" The answer—6—is determined by multiplying 2 times 3. "How many holes are there in these six buttons?" The answer—24—is determined by multiplying 6 times 4. The sentence for this view is $(2 \times 3) \times 4 = \square$. Now reconsider the two sentences to see how parentheses are used to group different pairs of factors and show which pair to multiply first. A similar pair of views should be taken of the card in Figure 9-15(b).

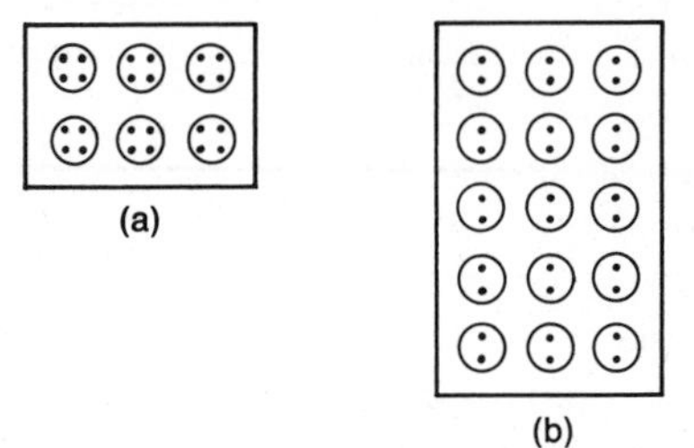


**Figure 9-15** Button cards illustrate the associative property of multiplication: (a) shows the sentences $2 \times (3 \times 4) = \square$ and $(2 \times 3) \times 4 = \square$; (b) displays the sentences $5 \times (3 \times 2) = \square$ and $(5 \times 3) \times 2 = \square$

Sets of array cards, such as those in Figure 9-16, show how to combine the commutative and associative properties of multiplication. The sentence $4 \times (4 \times 2) = \square$ is illustrated in (a); (b) illustrates $2 \times (4 \times 4) = \square$.

## 3. The Distributive Property

The distributive property of multiplication over addition can be taught before all the multiplication facts have been introduced. This knowledge will be useful as children study multiplication facts with larger products.

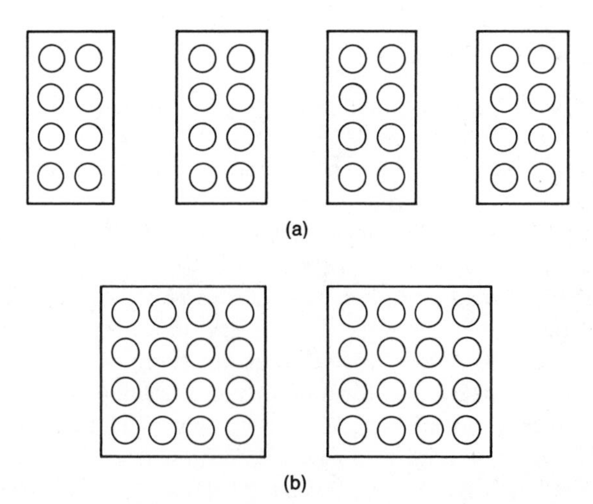


**Figure 9-16** Array cards used to illustrate the associative and commutative properties of multiplication used together. In (a) the multiplication sentence $4 \times (4 \times 2) = \square$ is illustrated; in (b) the multiplication $2 \times (4 \times 4) = \square$ is illustrated

It can help find the answer to $4 \times 8 = \square$ if the multiplication fact $4 \times 4 = 16$ is known. The 8 in $4 \times 8 = \square$ can be renamed and the sentence expressed as $4 \times (4 + 4) = \square$. The answer is then determined by completing the sentence

$$4 \times (4 + 4) = \square \rightarrow (4 \times 4) + (4 \times 4) = \square \rightarrow 16 + 16 = 32.$$

Before children can be expected to understand how to apply the property in this way, they should use aids that will make the property meaningful to them. For example, an array can be arranged on a magnetic board for the sentence $4 \times 8 = \square$ (Figure 9-17). Children first verify that

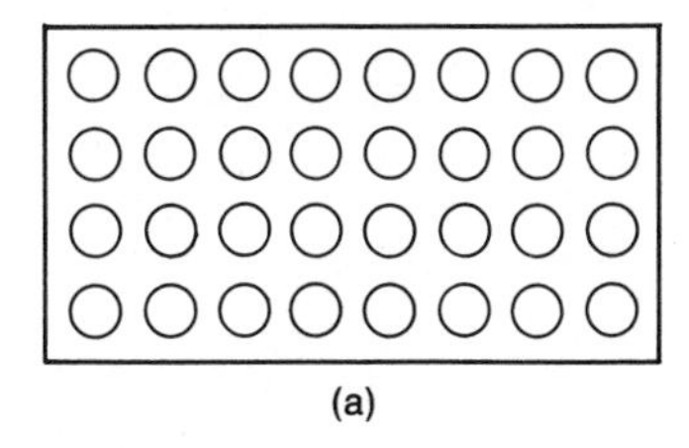
(a)

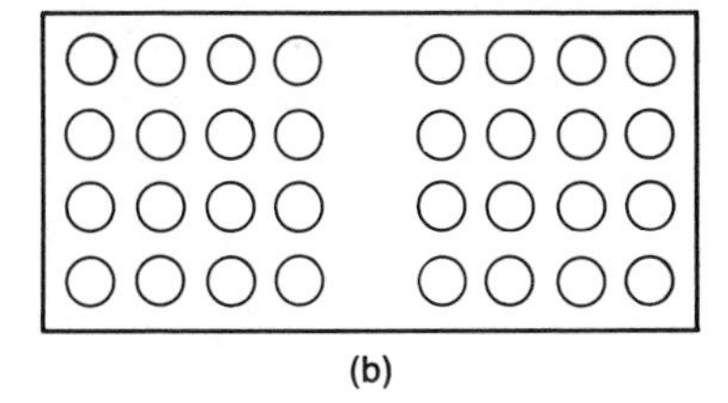
(b)

**Figure 9-17** Arrays used to illustrate (a) $4 \times 8 = 32$ and (b) $(4 \times 4) + (4 \times 4) = 16 + 16 = 32$

it is a 4 by 8 array (a). Then rearrange the array as in (b), and display it to show

$$(4 \times 4) + (4 \times 4) = \square.$$

The children determine the products for the two expressions of $4 \times 4$, and the sum of the products is determined. The sequence of steps in the sentence is

$$4 \times 8 = 4 \times (4 + 4) = (4 \times 4) + (4 \times 4) = 16 + 16 = 32,$$

because the second factor, 8, has been renamed $4 + 4$. This sentence now corresponds to the array shown in (b). Finally, they should see the original array (a) again and its sentence:

$$4 \times 8 = \square \rightarrow 4 \times 8 = 32.$$

By seeing different arrangements of the 4 by 8 array, children will recognize that 8 can be renamed in other ways, such as $6 + 2$ and $5 + 3$, and that the sentence can be expressed as

$$4 \times (6 + 2) = (4 \times 6) + (4 \times 2) = 24 + 8 = 32$$

and

$$4 \times (5 + 3) = (4 \times 5) + (4 \times 3) = 20 + 12 = 32.$$

Another way to show the distributive property is with an open-end abacus. Two colors of beads on an abacus used as a bead frame easily separate an array into two arrays to show distribution of multiplication over addition. In Figure 9-18(a) a 6 by 7 array is shown as a 6 by 2 array

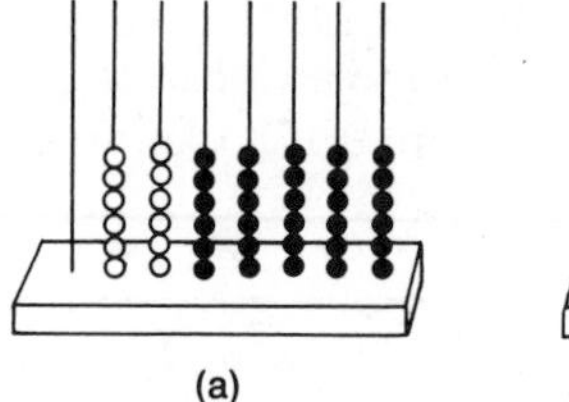


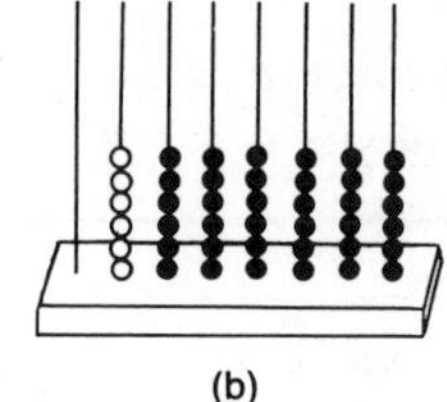


**Figure 9-18** The open-end abacus illustrates the distribution of multiplication over addition. In (a), the beads show $6 \times 7 = 6 \times (2 + 5) = (6 \times 2) + (6 \times 5)$; and in (b), $6 \times 7 = 6 \times (1 + 6) = (6 + 1) + (6 \times 6)$

and a 6 by 5 array to show that

$$6 \times 7 = 6 \times (2 + 5) = (6 \times 2) + (6 \times 5) = 12 + 30 = 42.$$

In (b) the arrays 6 by 1 and 6 by 6 show that

$$6 \times 7 = 6 \times (1 + 6) = (6 \times 1) + (6 \times 6) = 6 + 36 = 42.$$

In these examples the left factor has been distributed over the right factor. To show distribution of the right factor over the left, an array must be separated into two sets of rows, as in Figure 9-19, rather than into two sets of columns. First, the illustration shows a 6 by 7 array. In (b), the array is changed to show two 3 by 7 arrays. The beads can also show any other way of renaming 6: 5 + 1, 4 + 2, 2 + 4, or 1 + 5.

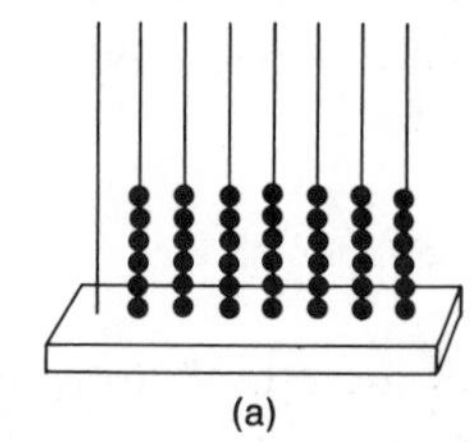


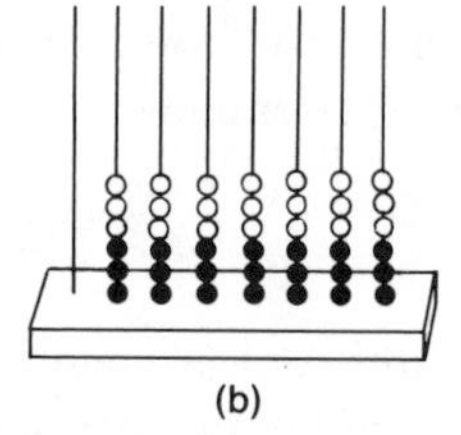


**Figure 9-19** Open-end abacus used as bead holder to illustrate the distribution of multiplication over addition. In (a), it shows $6 \times 7 = 42$; and in (b), $6 \times 7 = (3 + 3) \times 7 = (3 \times 7) + (3 \times 7) = 21 + 21 = 42$

Division is also distributive over addition, as indicated in Chapter 1. Applications of this property are included in Chapter 10 during discussion of the algorithm for division.

## 4. The Identity Element

An identity element is a number which when operated on with another number results in an answer that is the same as the second number. *One* is the identity element for multiplication. Ways of introducing children to the role of 1 in multiplication have already been discussed in the section dealing with arrays and Cartesian products.

The role of 1 in division is also important. You should provide children with activities that help them generalize that whenever a number is divided by 1, the quotient is always the number that is divided. They should also learn that any number divided by itself results in the quotient of 1, with 0 divided by 0 excepted.

## 5. The Role of 0 in Multiplication and Division

Zero has a special role in multiplication and division. In multiplication, the product is always 0 when 0 is one of the factors. The best way to illustrate this is with examples of Cartesian product situations like the one dealing with ice cream and toppings discussed earlier in this chapter. It is easy to show that no matter how many flavors of ice cream remain, there can be no combinations if there are no toppings. Another example can involve cakes and frostings, where children can see that no frosted cakes can be made if there are cakes but no frostings.

The role of 0 in division is a special one. When the dividend is 0, the quotient is also 0 if the divisor is a number other than 0. This is illustrated by the example $0 \div 9 = 0$. This sentence is true because $0 \times 9 = 0$. However, 0 is never used as a divisor. The reasons for 0 not being a divisor can be explained this way: When division is defined in terms of multiplication, $c \div b = a$, if and only if $a \times b = c$. If $b$ is 0 and $c$ is not 0, the sentence $c \div 0 = a$ implies that $a \times 0 = c$. But the product of $a$ and 0 is 0, which means that $c$ could not be a nonzero dividend. This contradiction indicates that the divisor cannot be 0 when the dividend is a whole number other than 0. For the sentence $0 \div 0 = a$, it is implied that $a \times 0 = 0$. Since this latter sentence is true for any whole number $a$, there is an ambiguity that indicates that 0 cannot be a divisor when the dividend is 0.

Since children are not mature enough to understand the reasons why 0 is never used as a divisor when division is introduced, it is best to exclude discussion of the reasons until later grades. There are mathematically mature children in grades five and six who can follow an explanation similar to the one given here, but most children in these grades cannot. It is not uncommon, then, for discussion to be delayed beyond the elementary school years.

### 6. Closure in Multiplication and Division

The procedure described in Chapter 7 for helping children learn about closure for addition by whole numbers can be used to help them learn about closure for multiplication. A multiplication table, such as the one in Figure 9-12, can be displayed. Have children imagine an indefinite extension of factors across the top and down the left side. Ask, "Will there be a product for every pair of numbers on our imagined chart?" Children should recognize that there will be a product for each pair, even though they do not actually determine each one of them. This will help them realize that for every pair of whole-number factors, there is a whole-number product.

When children work with division in the set of whole numbers, they find that there are pairs of numbers for which there are no whole-number answers. For example, there are no whole-number answers for either of these sentences: $1 \div 3 = \square$ or $14 \div 5 = \square$, even though in both instances the dividends and divisors are whole numbers. Since many pairs of whole numbers cannot yield a whole-number quotient, there is no closure for the operation of division in this set of numbers.

---

## Self-check of Objective 12

Demonstrate at least one procedure, along with materials, that can be used to help children understand these properties of multiplication: commutativity, associativity, distribution of multiplication over addition, and the roles of 1 and 0 in multiplication.

---

## Common Pitfalls and Trouble Spots

The premature introduction of mathematical sentences and acceptance of children's verbalizations of basic properties as evidence of their understanding of the properties, discussed earlier as pitfalls associated with addition and subtraction, apply to early work with multiplication and division, too.

In addition to these two, there is a third pitfall that is particularly critical in connection with multiplication and division. Many teachers are unable to generate realistic situations and to use concrete and semiconcrete materials properly to illustrate the three interpretations of multiplication—repeated addition, array, and Cartesian product—and the two kinds of division—measurement and partitive. As you plan activities dealing with these operations, carefully review the situations and materials described in this chapter, and/or in the teacher's manual of your textbook series, so that you have each interpretation clearly in mind and are aware of how to use materials to illustrate each one.

Memorization of the multiplication and division facts is a major trouble spot for many children. The problem stems partly from the fact that representation of products greater than 25 with markers and pictures is not conveniently done, so children see models of the larger products infrequently, if at all. The array model in Figure 9-7 is one way for children to see both factors and products as they deal with larger products. Models should be used frequently by children who have difficulty with the facts.

Another reason children have trouble is the lack of both assistance from their teacher and systematic practice. These are ideas you might try as children practice the facts:

1. Continue using the multiplication table, (a) just as it is in Figure 9-12; (b) in parts, with only those factors needing special attention; or (c) with the factors mixed up rather than in sequence.

2. Point out these facts about certain products: (a) the sum of the digits is nine when 9 is a factor; (b) an even or odd number times an even number results in an even product, while an odd number times an odd number results in an odd product; and (c) the product ends in 5 or 0 when 5 is a factor.

3. Use a variety of flashcards. Typical flashcards are illustrated in Figure 9-20(a). The cards in (b) and (c) are used for both multiplication and division. In (b) the product—63—is marked in one color, while the factors—7 and 9—are marked in another. For multiplication, cover the "63" and have the child name the product. For division, cover one factor, which the child is to name when shown the product and other factor. The card in (c) has all the division and multiplication facts in which 8 is one factor. It is used by two children at a time, one being "student" and the other "teacher." The "teacher" puts a pencil in a notch to indicate the

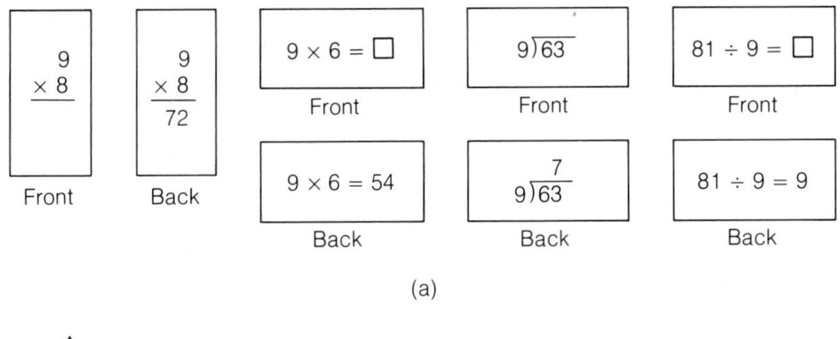


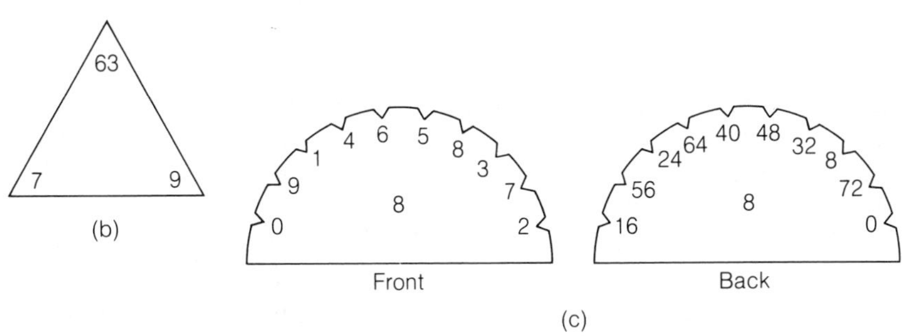


**Figure 9-20** Three types of flashcards for practicing multiplication and division facts

combination of numbers to be used. If the pencil is put in the "6" notch, the combination is 6 times 8, to which the "student" responds by giving the product—48. The answer is on the card's back, so the "teacher" can check the accuracy of the response. For division, the card is reversed, so the "student" sees a product and the "8." Again, the "teacher" puts a pencil in a notch to indicate which pair of numbers is being considered. The "student" names a factor when shown a product and "8."

## Self-check of Objectives 13 and 14

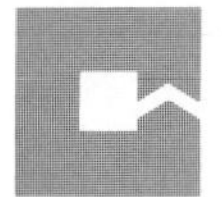

Identify a pitfall teachers face during the time they introduce the meanings of multiplication and division.

Explain two ways you can help children who have difficulty learning the multiplication and division facts.

## SUMMARY

There are three interpretations of multiplication with which children should become acquainted. Each one can be demonstrated with manipulative materials and story situations to make them meaningful. These interpretations are repeated addition, the array, and Cartesian, or cross, product. Repeated addition is used most often to introduce multiplication because it can be related to the already understood addition operation. Arrays are easily represented by objects such as poker chips and dot patterns and are useful for showing basic multiplication combinations that involve a given factor, such as all combinations with a factor of 7. Cartesian product situations are useful for showing the role of 0 in multiplication.

There are two types of situations that give rise to division—measurement and partitive. Children should be introduced to both types of situations through meaningful real-life problems and manipulative-material activities. The formal introduction of multiplication and division should be preceded by informal activities with number lines, dot-pattern cards, and other materials.

A wide variety of interesting practice materials and activities should be available so children can commit the basic multiplication and division combinations to memory for immediate recall. The basic properties of multiplication—commutativity, associativity, distribution of multiplication over addition, closure, and the roles of 1 and 0—can be presented through activities with markers, arrays, multiplication tables, number lines, blocks, and other manipulative devices.

The three interpretations of multiplication and the two division situations are not always illustrated effectively by teachers. Practice with both concrete and semiconcrete representations of these two operations should precede their use with children. Many children have difficulty learning the multiplication and division facts for immediate recall. The facts should be illustrated frequently; array cards are convenient to use. Practice should also be systematized and allocated the necessary time if children are to learn the facts well.

## STUDY QUESTIONS AND ACTIVITIES

1. Examine a modern mathematics textbook series to note the different situations involving multiplication that are included. Which one is used during introductory lessons? Are all the situations discussed in this chapter included in the series?
2. Examine a modern textbook series to see if both partitive and measurement situations are used as examples of division. Give an example of a word problem for each type of situation, if they are both included. Check the teacher's manual to see if it includes a discussion that distinguishes between the two situations. Does the manual suggest procedures for introducing both situations?
3. Examine a modern textbook series to see if the commutative, associative, and distributive properties of multiplication are included. How is the role of 0 handled? What materials and procedures for introducing these properties are recommended in the teachers' manuals?
4. Several of the articles in this chapter's reading list deal with mastery of the multiplication and division facts. Use them as sources of activities—games, puzzles, and so on—to include in your collection of mathematics learning aids.
5. The importance of understanding the various interpretations of multiplication and division has been stressed. Demonstrate your understanding of ways to represent the operations by writing answers or drawing pictures to illustrate the self-checks of objectives 1, 2, 4, 6, 7, and 9.

## FOR FURTHER READING

Ando, Masue, and Hitoshi Ikeda. "Learning Multiplication Facts—More than Drill," *The Arithmetic Teacher*, XVIII, No. 6 (October 1971), 359–364. Presents activities that develop understanding of the basic multiplication facts, ways of organizing the facts in tables, and procedures for memorizing them.

Bruni, James V., and Helene J. Silverman. "The Multiplication Facts: Once More, With Understanding," *The Arithmetic Teacher*, XXIII, No. 6 (October 1976), 402–409. Several activities and games using arrays provide the means for developing understanding and memorization of the basic multiplication facts.

Cacha, Frances B. "Exploring the Multiplication Table and Beyond," *The Arithmetic Teacher*, XXVI, No. 3 (November 1978), 46–48. Close study of the multiplication table reveals the properties of multiplication, information about odd and even factors, patterns on diagonals, and digit sums, all of which help children understand and master the facts.

Fishback, Sylvia. "Times Without Tears," *The Arithmetic Teacher*, XXI, No. 3 (March 1974), 200–201. Describes the experiences with multiplication of one teacher and her class. It is an excellent example of how fourth graders can master many multiplication skills with understanding when a teacher is aware of the operation's properties.

McDougall, Ronald V. "Don't Sell Short the Distributive Property," *The Arithmetic Teacher*, XIV, No. 7 (November 1967), 570–572. Explains applications of the distributive property of multiplication over addition to the processes of multiplying a pair of whole numbers, adding fractional numbers, and multiplying two binomials.

Smith, C. Winston, Jr. "Tiger-bite Cards and Blank Arrays," *The Arithmetic Teacher*, XXI, No. 8 (December 1974), 679–682. Arrays with "bites" taken from them serve as a basis for children's searches for missing factors when the total (product) and number of rows (given factor) are known. The cards lead to a useful way of investigating the meaning of the division algorithm.

Souviney, Randall J. "Giving Division Some Meaning," *Learning*, V, No. 6 (February 1977), 68–69. The two types of division situations are described and illustrated. Representational materials, in the form of tens strips and small squares, are used to help children understand the meaning of the algorithm for each type of division situation.

Spitzer, Herbert F. "Measurement or Partition Division for Introducing Study of the Division Operation," *The Arithmetic Teacher*, XIV, No. 5 (May 1967), 369–372. Spitzer argues that partitive situations should be used when division is introduced, because they are more closely related to multiplication than are measurement situations; the relationship between multiplication and division should be emphasized as division is introduced.

Zweng, Marilyn J. "Division Problems and the Concept of Rate," *The Arithmetic Teacher*, XI, No. 8 (December 1964), 547–556. This is a report on research done with second graders to determine which types of division situations are easiest for them to understand. Zweng found that measurement-rate situations were the most easily understood.

# 10 Extending the Operations of Multiplication and Division

Upon completion of Chapter 10, you will be able to:

1. Demonstrate with manipulative materials or a series of drawings at least two sets of materials children can use as they learn the multiplication algorithm with no regrouping.

2. Make up a real-life story problem for a sentence such as $4 \times 16 = 64$; then demonstrate with markers, a pocket chart, or a series of drawings of one of the devices the meaning of the algorithm for this example.

3. Demonstrate materials and procedures that can be used to introduce multiplication where both factors are greater than 10.

4. Demonstrate how the low-stress addition algorithm is used to generate multiplication and division facts.

5. Demonstrate how to use the low-stress algorithm for multiplication, from either the left or the right.

6. Explain processes that help children determine quotients when the division algorithm is introduced.

7. Explain orally or in writing why the ability to multiply by ten and its powers and by multiples of ten is an important prerequisite skill for learning to use the division algorithm, and describe ways these skills can be developed.

8. Demonstrate with manipulative materials or a series of drawings how division with quotients greater than 10 can be introduced meaningfully.

9. Make up a real-life story problem involving division with regrouping. Then write a sample dialogue that might take place between you and several children to help them understand how the algorithm is used to solve the problem.

10. State orally or in writing several problem situations involving division with a remainder, and explain how the remainder should be handled in each situation.

11. Demonstrate materials and procedures that can be used to help children understand division with large numbers.

12. Explain orally or in writing how multiplication and division can be checked.

13. Describe orally or in writing five minicalculator activities with multiplication and division.

14. Identify a common pitfall associated with division and its effect upon children, and describe how it can be avoided.

15. Identify a common trouble spot for multiplication and one for division, and describe a procedure for overcoming each one.

Key Terms you will encounter in Chapter 10:

expanded notation
partial product
multiplication with regrouping
distributive property
adding by endings
low-stress multiplication algorithm
Russian peasant multiplication
dividend
divisor
quotient
division with regrouping
multiplication patterns
division patterns
remainders

Before children learn to work with the algorithms for multiplication and division, they must know the various meanings of the operations and ways to represent them with manipulative materials and pictures. They should also know the basic facts for both multiplication and division. Chapter 9 presents activities for introducing the two operations and ways of providing the practice children need to achieve mastery of the basic facts. The present chapter extends the work with multiplication and division to include the common algorithms for larger numbers, a low-stress algorithm for multiplication, and activities for the minicalculator.

In school, instruction about the algorithms for multiplication and division is spread over several years, beginning in the third grade and continuing to the sixth grade, or beyond. Thus work with the two operations alternates between one and the other as increasingly more difficult aspects of each one are considered. Even though this is true, the operations are treated separately in this chapter. The discussion of multiplication is completed before division is considered. This treatment offers a sequential development to help you to recognize and understand the hierarchy of steps for each operation.

## INTRODUCING THE MULTIPLICATION ALGORITHM

Multiplication using two-or-more-digit numerals can begin before introduction of the 100 basic multiplication facts has been completed. For example, children can multiply

$$\begin{array}{r} 32 \\ \times 2 \\ \hline \end{array}$$

before they have learned the facts that involve larger pairs of numbers. However, before they are introduced to multiplication that involves larger numbers, children should have a good understanding of expanded notation and regrouping.

An introduction to this algorithm should be made by using a problem for children to solve. For example, "Billy was arranging some books in a new bookcase. He found that twelve books fit on each of the three shelves. How many books did the case hold?" Children should have time to determine the answer with structured materials, place value pocket charts, and other devices, if they need them. A discussion of the various procedures used enables them to share the ways the devices help determine the answer. It is good practice to begin with children who used the concrete devices and move to the more abstract procedures as the discussion progresses. The following devices and procedures help children understand how to multiply larger numbers.

### 1. Markers on Magnetic or Flannel Board

Markers can illustrate multiplication via the array concept, as shown in Figure 10-1. A 3 by 12 array (a) is also described by a sentence and by an algorithm form,

$$3 \times 12 = \square \quad \text{and} \quad \begin{array}{r} 12 \\ \times 3 \\ \hline \end{array}.$$

The array should then be changed to reflect its separation into 3 by 10 and 3 by 2 arrays; the sentence and algorithm are rewritten with expanded notation (b). As children determine the number of markers in each array, the sentence and the algorithm should be completed.

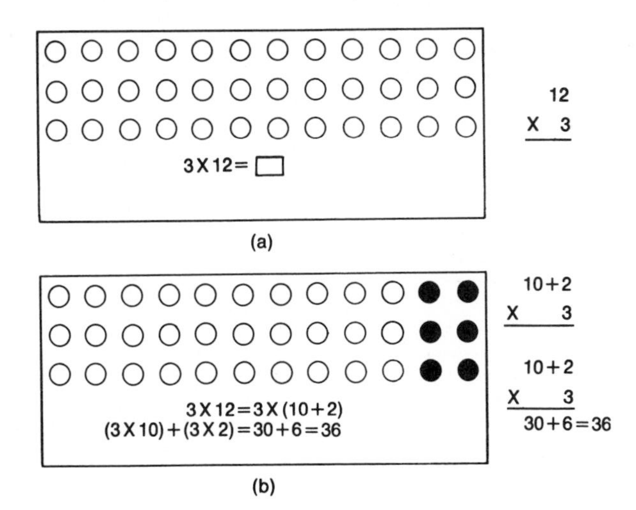


**Figure 10-1** Markers used to form an array to illustrate (a) $3 \times 12 = \square$ and (b) the expanded notation form of the sentence for showing distribution of multiplication over addition

## 2. Structured Materials

Cuisenaire, Dienes, and Unifix materials offer a different manipulative-materials approach. Figure 10-2(a) shows Cuisenaire rods set up to illustrate the sentence $3 \times 12 = \square$, where each 12 is represented by 1 tens rod and 2 ones rods. The sentence and algorithm forms can be used to reflect actions performed on the rods, so that from

$$3 \times 12 = \square \quad \text{and} \quad \begin{array}{r} 12 \\ \times 3 \\ \hline \end{array}$$

in (a), the sentence and algorithm are expanded to become

$$3 \times (10 + 2) = \square \quad \text{and} \quad \begin{array}{r} 10 + 2 \\ \times 3 \\ \hline \end{array}$$

to represent the pattern of rods in (b). When children complete the action with the rods, they see that altogether there are 3 tens rods and 6 ones rods representing the product of 3 and 12, which is 36. Dienes and Unifix materials are used in a similar way.

## 3. Place Value Pocket Chart

In Figure 10-3 markers in the pocket chart represent three sets of twelve books (a). Both the sentence $3 \times 12 = \square$ and the algorithm form

$$\begin{array}{r} 12 \\ \times 3 \\ \hline \end{array}$$

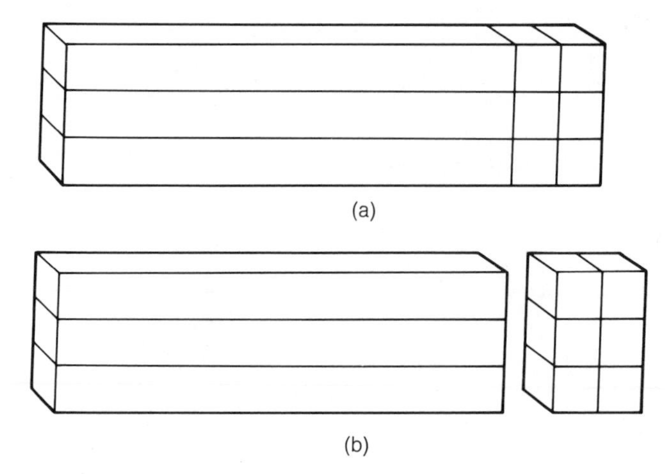


**Figure 10-2** Cuisenaire rods used to illustrate the multiplication of 3 × 12 = □ with compact numerals (a) and expanded notation (b)

should be displayed. The markers in the pocket chart will help children see how to rename the 12 as (10 + 2) so they can rewrite the sentence and algorithm form as

$$3 \times (10 + 2) = \square \quad \text{and} \quad \begin{array}{r} 10 + 2 \\ \times 3 \\ \hline \end{array}$$

The simpler form of expanded notation, "1 ten and 2 ones," might also be used, if necessary, so everyone can see the meaning of what is being done.

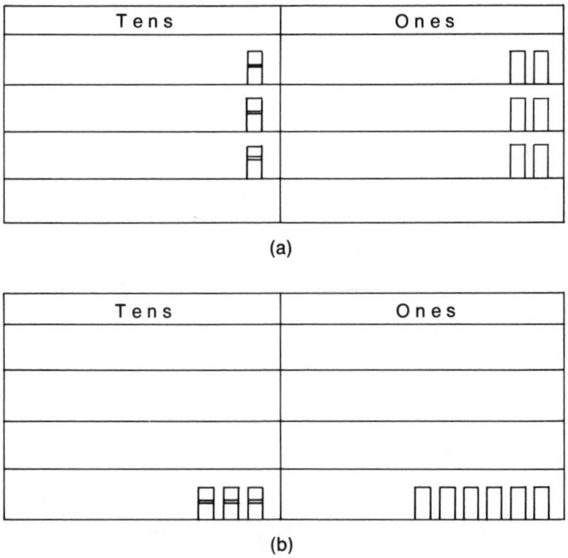

| Tens | Ones |
|---|---|

(a)

| Tens | Ones |
|---|---|

(b)

**Figure 10-3** Place value pocket charts used to show the arrangement of markers that illustrate the sentence 3 × 12 = □ (a) and the markers after the product has been determined (b)

Also, the addition shown in the margin can serve as a reminder that it is possible to find the answer by adding. As the markers in the pocket chart are grouped to show 3 tens and 6 ones (b), children should see that, in the multiplication sentence and the algorithm, the distributive property is employed as the sentence is completed:

$$\begin{array}{r} 12 \\ 12 \\ +12 \\ \hline \end{array}$$

$$3 \times 12 = 3 \times (10 + 2) = (3 \times 10) + (3 \times 2) = 30 + 6 = 36$$

and

$$\begin{array}{r} 10 + 2 \\ \times 3 \\ \hline 30 + 6 = 36. \end{array}$$

## 4. The Abacus

The abacus is used in the same way as a place value pocket chart, with 12 represented as a factor three times (Figure 10-4).

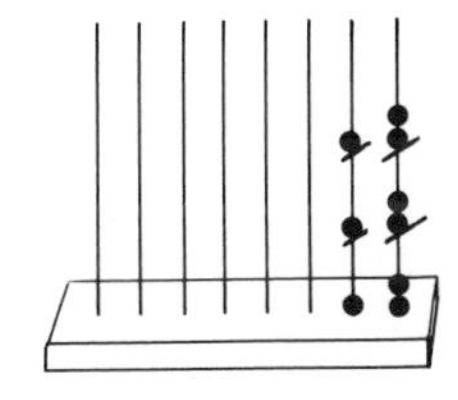

**Figure 10-4** The multiplication sentence 3 × 12 = ☐ illustrated on an abacus

As children use the various devices during exploratory activities, they see the array, structured materials, place value pocket chart, and other devices illustrating the meaning of the multiplication algorithm in a variety of ways. Use of different materials in this way is recommended because a device that is meaningful to some children may not have the same meaning for others. Under your careful guidance children will see how the devices clarify the meaning of expanded notation and the way the distributive property makes multiplication possible. Children must have an understanding of the distributive property before they can understand multiplication of numbers represented by two-or-more-digit numerals. Those who have not been introduced to the distributive property prior to the time they begin multiplying these larger numbers will need special help with it as they use the devices and expanded forms of expressing numbers in sentences and algorithms.

Working a sufficient number of examples with larger numbers using these devices helps to make the meaning of this operation clear. After multiplication with numbers larger than 10 as multiplicand and numbers smaller than 10 as multiplier, examples with numbers larger than 100 as the multiplicand and requiring no regrouping should be tried. No particular difficulties are involved in this multiplication. A few examples with numbers such as 2, 3, and 4 used as the multiplier should be demonstrated on the place value devices and with structured materials.

Writing numerals in the algorithm in the expanded form should be discontinued once the use of the algorithm for multiplication is understood. Children will realize that the short, or common, form of the algorithm is an efficient way to do multiplication.

## Self-check of Objective 1

Demonstrate with one of the place value devices and either markers in an array or structured materials how the multiplication algorithm can be introduced, using $3 \times 12 = 36$ as your example. Show the steps in the algorithm that accompany the sequence of steps with the manipulative materials. (Draw a series of pictures for the devices if the materials themselves are not available.)

## MULTIPLICATION WITH REGROUPING

Children who understand the meaning of expanded notation and regrouping with addition and subtraction, and who also understand simpler multiplication should have no particular difficulties learning to use the algorithm for multiplication with regrouping.

As in all cases when new steps in a process are presented, you should employ meaningful devices to introduce multiplication with regrouping. Children should use these devices as they explore and discuss the meanings of the steps involved in regrouping. The use of the place value pocket chart and array will be described.

### 1. Place Value Pocket Charts

The sentence and algorithm for a problem involving 3 and 24 are written on the chalkboard.

$$3 \times 24 = \square \qquad \begin{array}{r} 24 \\ \times 3 \\ \hline \end{array}$$

Ask children to make up a problem situation such as, "There are three boxes of candy bars, each with 24 bars. How many candy bars are in the three boxes?" Then have a child represent the situation with a place value pocket chart, as shown in Figure 10-5. The algorithms showing the multiplication as repeated addition, in both the usual and expanded forms, is also shown, as in the margin. As a child combines the markers in the ones place of the pocket chart, another can do the computation in the ones place of the addition algorithm. The regrouping of the 12 as 1 ten and 2 ones is shown in the pocket chart and the algorithm (b). Then the addition of the 20s should be shown as the bundles of ten markers are combined. After they have been collected, the bundle of ten from the 10 ones should be put with them to show seven bundles of ten, and the algorithm should then be completed (c).

$$\begin{array}{rr} 24 \rightarrow & (20+4) \\ 24 \rightarrow & (20+4) \\ +24 \rightarrow & +(20+4) \\ \hline \end{array}$$

The markers should be replaced in the pocket chart and the process repeated as the multiplication sentence and algorithm are completed. The steps for each follow.

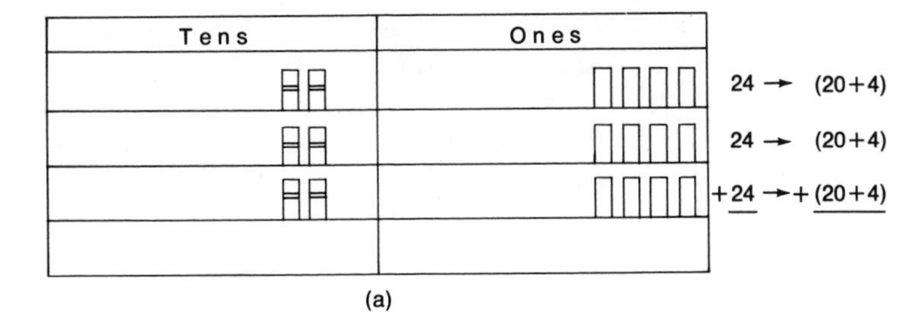


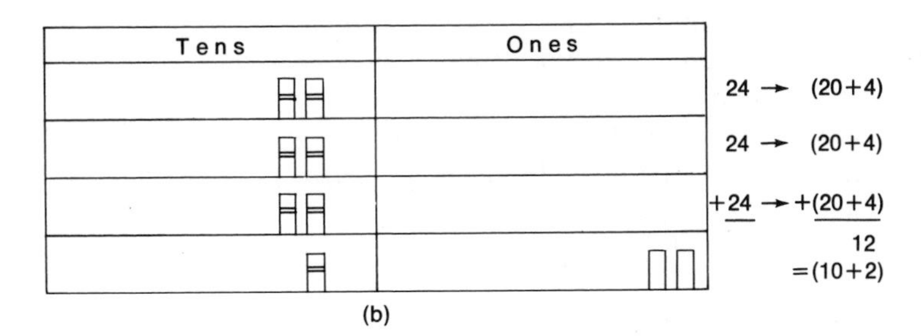


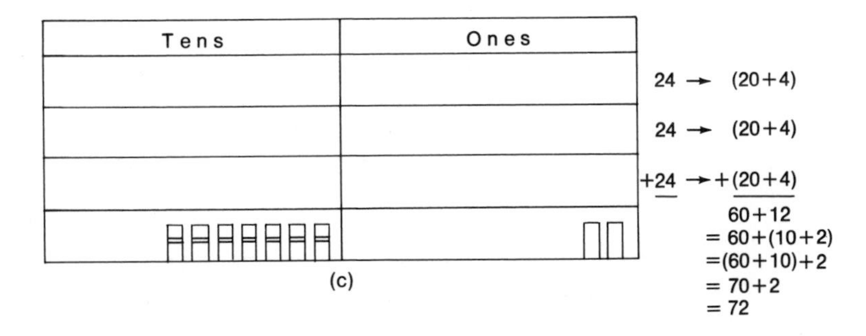


**Figure 10-5** Place value pocket chart used to show steps in completing the multiplication sentence $3 \times 24 = \square$ as repeated addition and multiplication with regrouping

$$
\begin{aligned}
3 \times 24 &= 3 \times (20 + 4) \\
&= (3 \times 20) + (3 \times 4) \\
&= 60 + 12 \\
&= 60 + (10 + 2) \\
&= (60 + 10) + 2 \\
&= 70 + 2 \\
&= 72
\end{aligned}
\qquad
\begin{array}{rcr}
24 & \rightarrow & 20 + \ \ 4 \\
\underline{\times 3} & \rightarrow & \underline{\quad \times 3} \\
 & & 60 + 12 \\
 & & = 60 + (10 + 2) \\
 & & = (60 + 10) + 2 \\
 & & = 70 + 2 \\
 & & = 72
\end{array}
$$

Other examples involving relatively small numbers should be illustrated on the place value devices, with the steps in the sentence and algorithm forms completed as above.

## 2. Arrays

Arrays are useful to illustrate multiplication involving regrouping. Pieces of tagboard or colored railroad board backed with small pieces of magnetic tape or pieces of coarse sandpaper for use on a magnetic or flannel board can demonstrate arrays. To show the meaning of $7 \times 14 = \square$, prepare two arrays, 7 by 10 and 7 by 4. Place the arrays together to show a 7 by 14 array, as in Figure 10-6(a). Then the array can be separated to show the 7 by 10 and 7 by 4 arrays (b). The sentence and algorithm should be rewritten as

$$7 \times (10 + 4) = \square \qquad \text{and} \qquad \begin{array}{r} 10 + 4 \\ \times\ 7 \\ \hline \end{array}$$

The steps to show multiplication with the sentence and algorithm are the same as the ones used with the place value pocket chart. Other arrays should be prepared and used in the same manner.

Naturally, children should eventually learn the standard algorithm for completing this type of multiplication. Instruction might follow this progression:

$$\begin{array}{rcl} 37 & \rightarrow & (30 + 7) \\ \times 2 & \rightarrow & \times 2 \\ \hline & & 60 + 14 \\ & = & 60 + (10 + 4) \\ & = & (60 + 10) + 4 \\ & = & 70 + 4 \\ & = & 74 \end{array} \qquad \begin{array}{r} 37 \\ \times 2 \\ \hline 14 \\ 60 \\ \hline 74 \end{array} \qquad \begin{array}{r} {}^{1}37 \\ \times 2 \\ \hline 74 \end{array}$$

(a) (b) (c)

In (a) the multiplication is completed using the expanded form; in (b) it is completed with both partial products written in a vertical placement; and in (c) it is completed with the standard algorithm.

---

## Self-check of Objective 2

Make up a real-life story problem for the sentence $5 \times 13 = 65$. Use markers, a pocket chart, or a series of drawings to illustrate the solution of your problem. Write the sequence of sentences that should accompany your demonstration with markers, pocket chart, or the series of drawings.

---

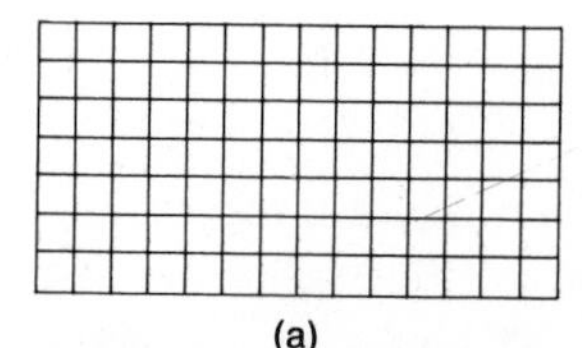

(a)

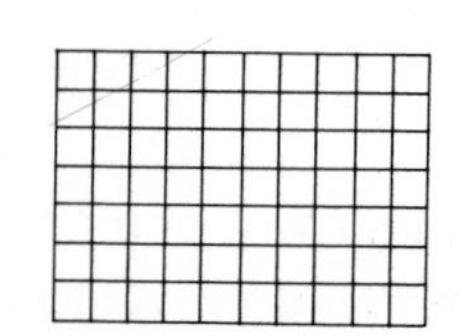

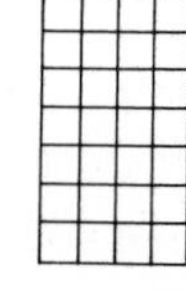

(b)

**Figure 10-6** An array used to show the distributive property used with the sentence $7 \times 14 = \square$. In (a) the array shows the original sentence; and in (b) it is separated to show $7 \times (10 + 4) = (7 \times 10) + (7 \times 4) = 70 + 28 = 98$

## MULTIPLICATION OF TWO NUMBERS LARGER THAN 10

### 1. Sequence of Steps

The commonly used algorithm for multiplying two numbers larger than 10 uses the distributive property of multiplication over addition and the associative property of addition. While these steps are not clear in the algorithm, they can be shown in the sequence of sentences that follow:

$$\begin{array}{r} 53 \\ \times 36 \\ \hline 318 \\ 159\phantom{0} \\ \hline 1908 \end{array}$$

| | |
|---|---|
| $36 \times 53 = 36 \times (50 + 3)$ | Renaming 53 |
| $= (36 \times 50) + (36 \times 3)$ | Distributive property |
| $= [(30 + 6) \times 50] + [(30 + 6) \times 3]$ | Renaming 36 |
| $= (30 \times 50) + (6 \times 50) + (30 \times 3) + (6 \times 3)$ | Distributive property |
| $= 1500 + 300 + 90 + 18$ | Multiplication |
| $= 1500 + 300 + 90 + (10 + 8)$ | Renaming 18 |
| $= 1500 + 300 + (90 + 10) + 8$ | Associative property |
| $= 1500 + 300 + 100 + 8$ | Addition |
| $= 1500 + (300 + 100) + 8$ | Associative property |
| $= 1500 + 400 + 8$ | Addition |
| $= 1908$ | Addition |

Such a long series of steps would not be used with a class of fourth- or fifth-grade children; however, you must understand the complete process so you can demonstrate it to children using simpler examples illustrated with arrays.

### 2. Use of Arrays

Use arrays on a magnetic or flannel board to illustrate examples with smaller numbers such as $12 \times 14 = \square$ (Figure 10-7). Prepare four arrays, 10 by 10, 10 by 4, 2 by 10, and 2 by 4, to display as a single 12 by 14 array (a) along with the sentence $12 \times 14 = \square$. Next, separate the array to show two arrays, as in (b), and rewrite the sentence as $12 \times (10 + 4) = \square$, and then as $(12 \times 10) + (12 \times 4) = \square$. Help children relate each part of the array to the corresponding parts of the sentences. Separate the array again, as in (c), and rewrite the sentence to show $[(10 + 2) \times 10] + [(10 + 2) \times 4] = \square$. Rewrite the sentence once more as $(10 \times 10) + (2 \times 10) + (10 \times 4) + (2 \times 4) = \square$. Again help the children relate each part of the array with the rewritten sentences. Now the children can complete the multiplication and addition. This is a good time to demonstrate to children how parentheses and brackets are used to keep the order of operations clear in mathematical sentences.

### 3. Shortened Forms

When separation of the array and display of the sequence of rewritten sentences have been completed, children should be introduced to the

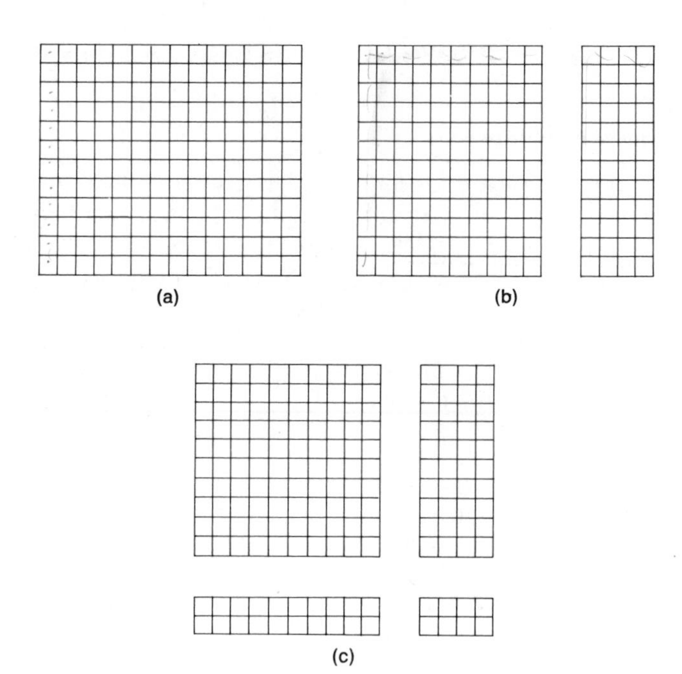


**Figure 10-7** Array used to show the meaning of the sentence 12 × 14 = □. In (a) the 12 by 14 array is shown, in (b) it is separated into arrays of 12 by 10 and 12 by 4, and in (c) into arrays of 10 by 10, 10 by 4, 2 by 10, and 2 by 4

common algorithm form

$$\begin{array}{r} 14 \\ \times 12 \\ \hline \end{array},$$

$$\begin{array}{r} 14 \\ \times 12 \\ \hline 8 \\ 20 \\ 40 \\ 100 \\ \hline 168 \end{array}$$

$$\begin{array}{r} 14 \\ \times 12 \\ \hline 28 \\ 140 \\ \hline 168 \end{array}$$

and relate its parts to the different arrays and parts of the sentence. During their early work with multiplication of two numbers larger than 10, children should see all the partial products written as the algorithm is completed, as in the margin. This makes it easy for them to relate the algorithm form to the arrays and expanded forms in the sentences. Other arrays should be used as children repeat the process with other multiplication sentences and the algorithm form.

After this process and the meaning of the partial products are clear, the writing of the partial products can be shortened, as in the margin. Arrays help children understand the shortened way of doing the multiplication. The array in Figure 10-7 can be used as an example. To discover how the first partial product is found, children should have their attention directed to the 2 by 10 and 2 by 4 arrays. They should note that the total number of squares in the two arrays is twenty-eight. The second partial product is found by using the 10 by 10 and 10 by 4 arrays, which contain a total of 140 squares.

When partial products are indicated in the algorithm in the most sophisticated form, the "0" is not written in the ones place of the second partial product. Children must realize that it is omitted only as a time-saving technique (see example in margin). The "14" is indented in the second partial product to show that it represents 14 tens, so it is placed with the "4" in the tens place.

```
  14
 ×12
  28
 14
 168
```

4. **Regrouping**

Multiplication involving numbers greater than 10 that require regrouping, or "carrying," is taught after children understand how to multiply two such numbers with no regrouping. As the first few examples of multiplication with regrouping are encountered, children should complete them by writing each partial product in the algorithm, as in the margin. Then the combined partial products should be written; the 4 tens from the 48 are added to the 24 tens, giving the first partial product of 288. The carrying of the 2 hundreds in the second partial product should be noted too. "Crutches," as shown in the example in the margin, might be used in the algorithm to help children remember the numbers that are being regrouped. After it has been used, the "4" is marked out to prevent confusion when the "2" is written later. You should also assign frequent "adding-by-endings" exercises that involve the products of two numbers smaller than 10 and the numbers 1 through 8. These experiences should include the use of work sheets and oral exercises, as indicated in Chapter 8, to give children practice with the type of addition required in multiplication with regrouping.

```
   36
  ×48
   48
  240
  240
 1200
```

```
  2
  4 (marked out)
   36
  ×48
  288
 144
 1728
```

Once children understand the process of multiplying two numbers that require regrouping, they can move on to work with larger numbers. The steps in the process are the same as those with numbers expressed by two-digit numerals. You should continually observe children as they work examples involving larger numbers to prevent errors and the formation of incorrect habits.

---

## Self-check of Objective 3

Use markers, such as poker chips or prepared arrays, to illustrate the steps in completing the algorithm for the multiplication sentence 11 × 16 = 176.

---

## ZERO IN MULTIPLICATION

Factors such as 306 and 4,002, especially when they appear in the multiplier, present difficulties to some children. Before multiplying numbers containing 0, multiplication by 10 and its multiples should be mastered. Sets, arrays, the number line, and other aids should be used until children can generalize that when 10 is the multiplier, the product is the multiplicand with a 0 annexed. (An alternate generalization is to think of the process as moving the number one place to the left and placing a 0 in the ones place.) To multiply by a number such as 20 or 30 as in the sentence 20 × 6 = □, children should be able to think "2 × 6 = 12, annex 0 to give 120." Computation with other powers of 10 and their multiples should be taught in a similar manner so that children can generalize about them, too. After they have learned this process, frequent oral exercises reinforce children's ability to use it. Efficient use of this process is important in multiplication, and essential before beginning to learn the division algorithm.

```
  346
 ×209
 3114
6920
72314
```

```
  346
 ×209
 3114
 000
692
72314
```

In multiplying the first problem in the margin above, children should be able to determine the second partial product by multiplying by the 2 hundreds (or 20 tens) in one step rather than two. There is no reason for children to write the work with three partial products as shown in the second algorithm in the margin.

## THE LOW-STRESS MULTIPLICATION ALGORITHM

There are low-stress algorithms for both multiplication and division. Only the one for multiplication is discussed here.[1] First, a process for generating the multiplication and division facts using the low-stress addition algorithm is discussed.

### 1. Multiplication and Division Facts

$$\begin{array}{r} 8 \\ {}_1 8_6 \\ {}_1 8_4 \\ {}_1 8_2 \\ {}_1 8_0 \\ 8_8 \\ {}_1 8_6 \\ {}_1 8_4 \\ {}_1 8_2 \end{array}$$

A process for generating the multiplication and division facts uses the low-stress addition algorithm. The facts for 8 are shown in the margin. To determine the product of 4 × 8, count down four 8s from the top. Note the ones numeral, which is "2." Count the tens from the top down; there are three. The product of 4 × 8 is 32. For the product of 8 × 8, count eight 8s; note the "4." Count the tens; there are six, so 8 × 8 is 64.

The same column is used to generate division facts. To determine how many 8s are in 56, count the tens downward until you have counted five; look at the ones to see if there is a "6" alongside the fifth ten. Since there is, you have reached 56. Now count the 8s upward; there are seven, so 56 ÷ 8 = 7. To determine the answer for 48 ÷ 8, count the tens downward until four are counted. Look at the ones to see if an "8" is opposite the fourth ten. Since there is not, go down one more 8; note the "8" on the ones side. Count the 8s upward; there are six, so 48 ÷ 8 = 6.

[1] Readers who are interested in knowing about the algorithm for division are referred to Chapter 16 in Doyal Nelson, ed., *Measurement in School Mathematics* (Reston, Va.: National Council of Teachers of Mathematics, 1976).

## 2. The Low-Stress Multiplication Algorithm

A special "drop" notation is used to record products in the low-stress multiplication algorithm, as shown in the margin. Products greater than 10 are recorded as in (a). When a product is less than 10, record a "0" in the tens place, as in (b). This notation permits the user to complete the multiplication and record all partial products before doing any adding. This lessens the cognitive load by enabling the student to think of only multiplication during the first part of the computation and addition later.

The process for multiplying a number larger than 100 by one smaller than 10 is shown in the margin. In (a) the product of 7 × 2 is recorded. In (b) the product of 7 × 6 is recorded, while the product of 7 × 4 is shown in (c). The partial products are then added to determine the final product (d). A student can multiply larger numbers very quickly once the multiplication process has been learned. Two examples illustrate the process with numbers as large as millions.

```
 42936     4213846
   ×9           ×4
31825     1001312
 68174     6842264
386424    16855384
```

Barton Hutchings, who directed much of the experimental work during development of the low-stress algorithms, recommends that products be recorded in the left-to-right direction, rather than in the conventional right-to-left direction. "Experience indicates it would be better to do all computations left to right, for two reasons. First, if we assume that the more operations we perform the more likely a mistake becomes, then in working from high place values to low, we shift the greater likelihood of error to the low place values. This should reduce the effects of error even when error frequency is constant. The error-probability structure of conventional algorithms tends to maximize the effects of error. Second, left-to-right operation can offer efficient and simple estimation procedures."[2] The left-to-right process is illustrated with this example:

```
 42369     42369     42369     42369     42369
    ×7        ×7        ×7        ×7        ×7
2         21        212       2124      21246
 8         84        841       8412      84123
                                        296583
```

```
(a)  3     8
    ×7    ×8
     2     6
      1     4

(b)  3     3
    ×3    ×2
     0     0
      9     6
```

```
(a) 462
    × 7
     1
      4

(b) 462
    × 7
     41
      24

(c) 462
    × 7
    241
     824

(d) 462
    × 7
    241
     824
    3234
```

[2] Barton Hutchings, "Low-Stress Algorithms," *Measurement in School Mathematics* (Reston, Va.: National Council of Teachers of Mathematics, 1976), pp. 227–228.

Examples of the algorithm with larger factors and both left-to-right (a) and right-to-left processes (b) are illustrated below. The line beneath the "2" in (a) indicates where the first numerals of the multiplication by ones will be placed.

```
(a)  4264        (b)     9437
      ×36               ×328
    ------             -------
     1011                 7325
     2682                  2246
     2132               1001
      4264                8864
    ------             2102
    153504              7291
                       -------
                       3095336
```

---

## Self-check of Objectives 4 and 5

```
624
×38
```

Demonstrate how the low-stress addition algorithm can be used to generate multiplication and division facts.

Compute the example in the margin by using the low-stress multiplication algorithm. Work it from both the left and the right.

---

## INTRODUCING THE DIVISION ALGORITHM

```
   46
3)138
  12
   18
   18
```

The standard division algorithm for the sentence $138 \div 3 = 46$ is shown in the margin. Before children can perform it with full understanding, they must know and understand certain basic concepts and skills. These include: (1) knowledge of both partitive and measurement division situations, (2) knowledge of the multiplication and division facts, (3) ability to subtract, and (4) ability to multiply and divide by ten, its powers and multiples. The two situations that give rise to division and ways to promote mastery of the facts are considered in Chapter 9, while subtraction is discussed in Chapters 7 and 8.

### 1. Dividing by Numbers Smaller Than 10

Situations that can be used to promote understanding of the division algorithm abound. A problem such as the following is useful to introduce division by a number smaller than 10: "Twenty-eight marbles will be shared equally by six children. How many marbles will each one get?" Initially, children should have manipulative materials to represent this partitive situation. Real marbles, markers on a magnetic board, popsicle sticks, and similar items are useful. First, count 28 objects. Then form six equal-sized groups—there will be 4 objects in each group, with 4 left over. Other, similar partitive situations should be presented.

Once the process has been made clear with manipulative materials, write the algorithm for each situation. Then help children develop a strategy for determining quotients without concrete aids. One strategy is to relate the division to multiplication. For the example $6\overline{)28}$, use these facts:

$6 \times 1 = 6$ $6 \times 2 = 12$ $6 \times 3 = 18$ $6 \times 4 = 24$ $6 \times 5 = 30$

Questions such as, "Will each child get at least 1 marble?" "At least 2 marbles?" "At least 3 marbles?" "At least 4 marbles?" "At least 5 marbles?" help children recognize that 4 is the answer. Point out that $6 \times 4 = 24$ is the sentence with the largest product that is smaller than the dividend. For the example $8\overline{)79}$, use these multiplication facts:

$8 \times 1 = 8$ $8 \times 2 = 16$ $8 \times 3 = 24$ $8 \times 4 = 32$ $8 \times 5 = 40$

$8 \times 6 = 48$ $8 \times 7 = 56$ $8 \times 8 = 64$ $8 \times 9 = 72$ $8 \times 10 = 80$

Here $8 \times 9 = 72$ is the sentence with the largest product that is smaller than the dividend, so 9 is the quotient.

"Think-back" flashcards are helpful for developing children's skills in determining quotients. The card for $8\overline{)79}$ is shown in Figure 10-8. The front of the card (a) shows the division algorithm, without the quotient, while the back (b) shows the basic fact associated with the division. A child is shown $8\overline{)79}$, then thinks back to the related division fact and says, "Seventy-two divided by 8 equals 9." (Avoid the expression "8 goes into 72 9 times.") If a child cannot "think back" to the correct basic fact, reverse the card and have him or her read it. Then show the front of the card again and have the child repeat the statement of the basic fact. Group and individual work with "think-back" cards should be provided frequently until children are skillful in naming quotients.

| $8\overline{)79}$ | $\begin{array}{r}9\\8\overline{)72}\end{array}$ |
|---|---|
| (a) | (b) |

**Figure 10-8** An example of a "think-back" flashcard

## Self-check of Objective 6

Explain orally or in writing how multiplication sentences and "think-back" flashcards are used to help children learn to name quotients when the division algorithm is introduced.

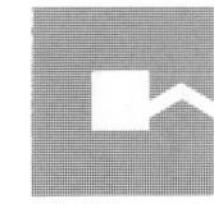

## 2. Multiplication and Division Patterns

Skill in multiplying and dividing by ten, its powers, and its multiples is needed in order to estimate quotients or use divisors greater than 10. Patterns such as the following can be used to develop this skill.

$$
\begin{array}{lll}
1 \times 1 = 1 & 1 \times 10 = 10 & 1 \times 100 = 100 \\
2 \times 1 = 2 & 2 \times 10 = 20 & 2 \times 100 = 200 \\
3 \times 1 = 3 & 3 \times 10 = 30 & 3 \times 100 = 300 \\
\vdots & \vdots & \vdots \\
9 \times 1 = 9 & 9 \times 10 = 90 & 9 \times 100 = 900
\end{array}
$$

$$
\begin{array}{lll}
1 \div 1 = 1 & 10 \div 10 = 1 & 100 \div 100 = 1 \\
2 \div 1 = 2 & 20 \div 10 = 2 & 200 \div 100 = 2 \\
3 \div 1 = 3 & 30 \div 10 = 3 & 300 \div 100 = 3 \\
\vdots & \vdots & \vdots \\
9 \div 1 = 9 & 90 \div 10 = 9 & 900 \div 100 = 9
\end{array}
$$

---

$$
\begin{array}{lll}
2 \times 1 = 2 & 2 \times 10 = 20 & 2 \times 100 = 200 \\
2 \times 2 = 4 & 2 \times 20 = 40 & 2 \times 200 = 400 \\
2 \times 3 = 6 & 2 \times 30 = 60 & 2 \times 300 = 600 \\
\vdots & \vdots & \vdots \\
2 \times 9 = 18 & 2 \times 90 = 180 & 2 \times 900 = 1800
\end{array}
$$

$$
\begin{array}{lll}
2 \div 1 = 2 & 20 \div 10 = 2 & 200 \div 100 = 2 \\
4 \div 2 = 2 & 40 \div 20 = 2 & 400 \div 200 = 2 \\
6 \div 3 = 2 & 60 \div 30 = 2 & 600 \div 300 = 2 \\
\vdots & \vdots & \vdots \\
18 \div 9 = 2 & 180 \div 90 = 2 & 1800 \div 900 = 2
\end{array}
$$

and

$$
\begin{array}{lll}
2 \div 2 = 1 & 20 \div 2 = 10 & 200 \div 2 = 100 \\
4 \div 2 = 2 & 40 \div 2 = 20 & 400 \div 2 = 200 \\
6 \div 2 = 3 & 60 \div 2 = 30 & 600 \div 2 = 300 \\
\vdots & \vdots & \vdots \\
18 \div 2 = 9 & 180 \div 2 = 90 & 1800 \div 2 = 900
\end{array}
$$

Careful development and discussion of these and other patterns help children learn skills for further work with the algorithm. During discussions, note relationships between the sentences in a line; for example, $1 \times 1 = 1$, $1 \times 10 = 10$, and $1 \times 100 = 100$. Also, note relationships between a given line of multiplication sentences and the corresponding line of division sentences, such as $2 \times 9 = 18$, $2 \times 90 = 180$, $2 \times 900 = 1800$, and $18 \div 9 = 2$, $180 \div 90 = 2$, $1800 \div 900 = 2$.

You can prepare work sheets that contain multiplication and division patterns like these for children to complete. Discuss their completed work sheets with them to highlight the relationships. Later, use oral questions to reinforce written work: "If there are three 9s in 27, how many 9s are there in 270?" "If six 8s are 48, what are six 80s?"

### 3. Division with Quotients Greater Than 10

Use a problem situation such as the following to introduce division with quotients greater than 10. "John has forty-eight stamps that he is going to fix for a school display. He will put four stamps on a card. How many cards will he need?" After giving this measurement situation, you might say, "Here is a magnetic board with forty-eight 'stamps' on it. How can we use it to find out how many cards John will need?" The disks should be randomly placed on the board when you first show them. Children with ideas of how to proceed should have a chance to demonstrate and discuss them. This is a measurement-type problem that can be solved by repeated subtraction, and children with good backgrounds developed through earlier informal activities involving similar situations will recognize this. Let them separate the forty-eight disks into groups of four each, and then count the groups to determine the number of cards John needs.

Now discuss the situation to help children associate the disks and the way they separated them with the division sentence for this problem: $48 \div 4 = 12$. The original set (dividend) contained forty-eight disks, which were separated into groups (divisor) of four disks each. The number of groups (quotient) was twelve. You do not need to use the words *dividend, divisor,* and *quotient* at this time unless you think your children are ready for these special terms.

Children will need other problems that can be represented with disks and other manipulative materials before they use the algorithm. Children who understand measurement situations and have the required prerequisite skills will have little difficulty understanding this algorithm:

```
4)48 |
  40 | 10 × 4
  ---+-------
   8 |
   8 |  2 × 4
  ---+-------
     | 12
```

Use a dialogue similar to the following to help children understand this algorithm (your questions are in quotation marks, while children's answers are in parentheses throughout the book): "What does '48' stand for?" (The number of stamps John had.) "What does '4' stand for?" (The number of stamps he planned to put on each card.) "Did John need at least ten cards?" (Yes.) "How do you know?" (Ten cards will hold forty stamps; he had more than forty stamps.) "After he put four stamps on each of ten cards, how many stamps did he have left to put on cards?" (Eight.) "How many cards did he need for eight stamps?" (Two.) "How many cards did John use for his forty-eight stamps?" (Twelve.)

Similar discussions should be carried out as children solve other problems.

You should also give problems that involve partitive situations. Another stamp problem illustrates this. "John had forty-eight stamps to mount in his stamp book. If he puts the same number of stamps on each of four pages, how many stamps will each page have on it?" Once again, a magnetic board with forty-eight disks on it is useful. Those children with ideas about how to proceed should have a chance to discuss them. After the answer has been determined with "stamps" on the magnetic board, present the problem in both sentence and algorithm form: $48 \div 4 = \square$ and $4\overline{)48}$.

## 4. Expanded Notation

Expanded notation will help children learn to place the quotient in the proper place in the algorithm. A pocket chart is helpful for showing the process. For the example $4\overline{)48}$, you might proceed this way: Represent the dividend in the chart, as in Figure 10-9(a). Express the algorithm as $4\overline{)40 + 8}$. Next, separate the 4 bundles of ten markers into 4 groups, each containing ten markers (b). Show this division in the algorithm, as in the margin. Then separate the 8 single markers into 4 groups, each containing two markers (c), and complete the algorithm, as shown also in the margin.

$$\begin{array}{r} 10 \\ 4\overline{)40+8} \end{array}$$

$$\begin{array}{r} 10+2 \\ 4\overline{)40+8} \end{array}$$

Lead children away from reliance on markers and place value aids by asking questions such as, "What must I multiply 4 by to get 40?" or "How many 4s in 40?" and "What do I multiply 4 by to get 8?" or "How many 4s in 8?"

They should then determine the result of separating the four bundles of ten and the 8 ones in the pocket chart. First the four bundles of ten, which represent forty stamps, are separated into four groups, as in (b), to represent the placement of an equal number of stamps on each of the four pages. The division should then be shown in the sentence $40 \div 4 = 10$. Next, the 8 ones are separated to represent the placement of the remaining stamps on the pages (c), and the division in the sentence is completed, $8 \div 4 = 2$. Each of the four pockets of the chart will have one bundle of ten and 2 ones in it. The algorithm should be completed as children review the steps taken with the markers in the pocket chart and with the sentence.

| Tens | Ones |
|---|---|
| | |
| | |
| | |
| | |
| | |

(a)

| Tens | Ones |
|---|---|
| | |
| | |
| | |
| | |
| | |

(b)

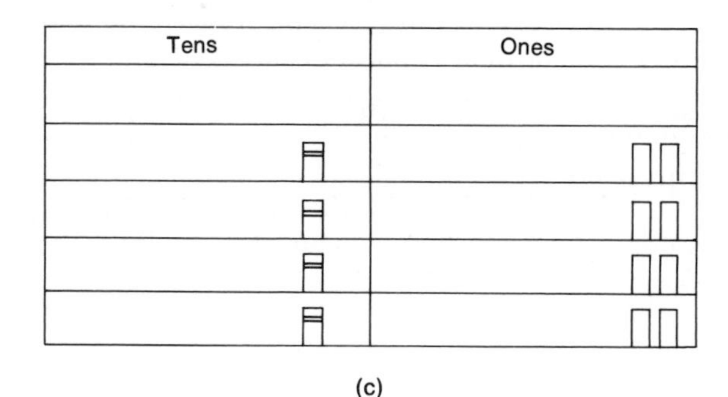


(c)

**Figure 10-9** Place value pocket chart showing the division sentence $48 \div 4 = 12$. In (a) the 48 is represented as $40 + 8$; in (b) the division of 40 by 4 is represented to show the 4 tens; and in (c) the division of 8 by 4 is represented

## Self-check of Objectives 7 and 8

State why it is important for children to know how to multiply by ten and its powers and by multiples of ten before they begin learning the division algorithm. Describe a procedure that can be used to help children learn these skills.

Make up a real-life division situation and demonstrate a procedure that can be used to make a meaningful introduction of the division algorithm using your example. Tell whether yours is a partitive or a measurement situation.

## DIVISION WITH REGROUPING

Regrouping is required when the number in the tens place of the dividend is not a multiple of the divisor.

### 1. Concrete-Manipulative Model

The division $3\overline{)45}$ requires regrouping because 40 (4 tens) is not a multiple of 3. This example can be worked out for a partitive situation with beansticks. Begin by representing 45 with the sticks, as in Figure 10-10(a). Now the beansticks are to be separated into three groups, each the same size. First consider the tens sticks and separate them into three groups, each containing one stick (b). It is not possible to separate the remaining stick without exchanging it for ten loose beans. This exchange is shown in (c). Finally, the 15 single beans are separated into three equal-sized groups, with each group put next to a tens stick (d).

| $3\overline{)45}$ | |
|---|---|
| 30 | $3 \times 10$ |
| 15 | |
| 15 | $3 \times 5$ |
| | 15 |

The steps are illustrated in the algorithm in the margin, where the partial quotients are "stacked" with the divisor at the right.

### 2. Algorithm Form

The algorithm form should be studied next. The form shown above, or some variation of it, is frequently used. This form is preferred because:

1. The parts of the quotient are shown in their complete form; the ten is shown as "10" at the right, rather than as a "1" in the tens place above the dividend.

2. The multiplication is shown in a familiar form, "$3 \times 10$," rather than in the form

$$\begin{array}{r} 1\phantom{5} \\ 3\overline{)45} \\ 3\phantom{5} \end{array}$$

   in which the "1" indicates a ten which is multiplied by 3, the divisor, resulting in a product of 30, which is indicated by a "3" written in the tens place beneath the dividend.

3. The product of 3 and 10 is shown as "30" rather than "3" in the tens place.

Thus the form of the algorithm helps children see the meaning of the steps performed because the different parts of the algorithm are related to the actions with the beansticks used earlier.

A sufficient number of examples help make the meaning of the algorithm clear as the children move from the use of manipulative mate-

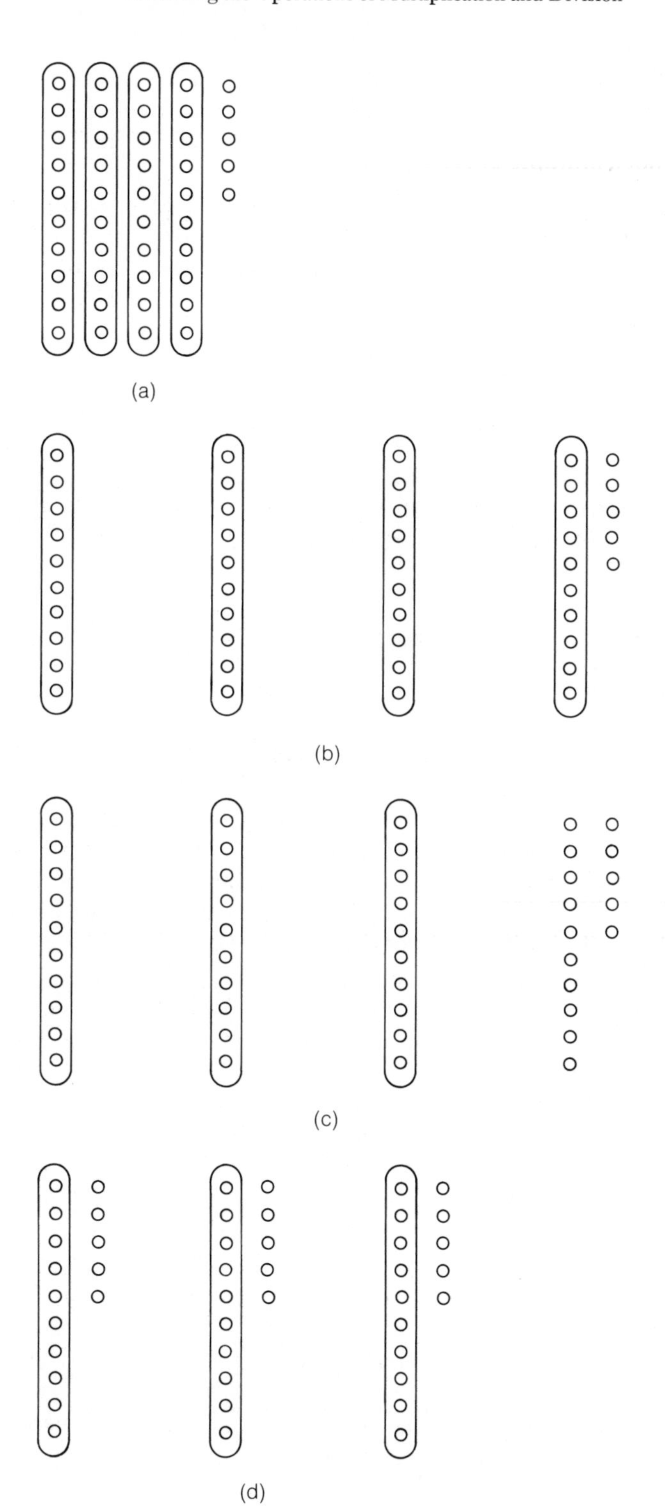


**Figure 10-10** The division 45 ÷ 3 = 15 represented with beansticks

(a) 4)56
40 | 10 × 4
16 |

(b) 4)56
40 | 10 × 4
16 |
16 | 4 × 4
| 14

rials to the abstract algorithm: "When we divide 56 by 4, will there be as many as ten 4s in 56?" (Yes.) "How do we know?" (10 × 4 = 40; 40 is less than 56.) "Will there be as many as twenty 4s in 56?" (No.) "How do we know?" (20 × 4 = 80; 80 is greater than 56.) "Let us show [in (a) in the margin] that 10 times 4 equals 40 and then subtract 40 from 56 to see how much is left to be divided." "What is 56 minus 40?" (16.) "What is 16 divided by 4?" (4.) "We can show that too [in (b)]. Now, what is the answer to 56 divided by 4?" (14.)

---

## Self-check of Objective 9

Make up a real-life problem involving a division situation that requires regrouping. Write an imaginary dialogue that might involve you and several children as you introduce the algorithm with regrouping.

---

## DIVISION WITH REMAINDERS

When the divisor is not a whole-number factor of the dividend, a remainder occurs. The way a remainder is treated depends on the nature of the problem situation giving rise to the division. The following situations show different treatments of remainders.

"There are thirty-two children in our class. If we play a game that requires three equal-sized teams, how many players will there be on each team?" In this situation, which is partitive, there will be ten players on each team. The remainder does not actually become a part of the quotient. The answer is that there will be three teams of ten players each.

"Mother baked twenty-six cookies for Sally and her three friends. If the cookies are given to the children so that each child gets an equal share, how many will each one get?" In this partitive situation there will be two cookies left after twenty-four have been shared by the four children. These can be broken into halves, making four halves, one for each child. In this example it makes sense to divide the remainder. The answer, therefore, can be expressed as 6½ cookies per child.

"It costs 79 cents to buy three cans of dog food. If you buy only one can, what will you pay for it?" In this case the division gives a whole-number quotient of 26, with a remainder of 1. Children will understand that the store is not likely to sell one can of dog food for 26 cents because then the price would become three cans for 78 cents. For this example, the answer is changed to 27 when the price of one item is determined.

"A group of fifty-three children is riding to camp in cars. If each car can carry six children, how many cars will be needed?" In this measurement situation the division results in a quotient of 8. There will be a remainder of five children, which means that eight cars are insufficient. Therefore the answer is nine cars, rather than eight.

"Sixty-eight apples must be put in boxes. If each box can hold eight apples, how many boxes can be filled?" For this measurement situation the quotient is eight boxes with four apples remaining. Since these are not enough to fill another box, the answer is eight full boxes.

Starting with third grade, problem situations featuring these different types should be explored periodically to review appropriate ways to deal with remainders. In practice exercises with division involving remainders, there is sometimes no way to determine what the dividends, divisors, and quotients represent. Then the numeral that represents the remainder can be written at the bottom of the algorithm, as in (a) in the margin. It is also correct to write the numeral after the quotient, as in (b), or, as in (c), where the remainder is recorded as $\frac{4}{9}$.

$$\text{(a)}\quad \begin{array}{r} 7 \\ 9\overline{)67} \\ \underline{63} \\ 4 \end{array}$$

$$\text{(b)}\quad \begin{array}{r} 7\text{r}4 \\ 9\overline{)67} \\ \underline{63} \\ 4 \end{array}$$

$$\text{(c)}\quad \begin{array}{r} 7\frac{4}{9} \\ 9\overline{)67} \\ \underline{63} \\ 4 \end{array}$$

---

## Self-check of Objective 10

Remainders in division are handled differently depending upon the situation from which they arise. Make up at least three real-life story problems in which the remainder is not 0 and is handled differently in each one. Describe how the remainder should be handled for each problem situation.

---

## DIVISION INVOLVING LARGER NUMBERS

A good understanding of division with quotients represented by one-digit numerals, including use of the algorithm and the ability to multiply by ten and powers of ten, is a prerequisite to learning division involving larger numbers; without the proper background, children will encounter many difficulties with this kind of division.

### 1. Two-Digit Divisors and One-Digit Quotients

Children's first work with two-digit divisors is usually done with divisors that are multiples of ten. This is so because patterns can be used to help them develop skill in determining quotients. The patterns illustrated earlier in this chapter lead to this generalization:

____ ones × ____ tens = ____ tens
____ tens ÷ ____ tens = ____ ones

So, for $270 \div 30 = \square$, the 27 tens are divided by 3 tens to determine the quotient—9.

Answers for examples such as these

$30\overline{)123}$ $\quad$ $30\overline{)98}$ $\quad$ $30\overline{)72}$ $\quad$ $30\overline{)131}$

can be determined by using this multiplication pattern:

$1 \times 30 = 30$ $\quad$ $2 \times 30 = 60$ $\quad$ $3 \times 30 = 90$
$4 \times 30 = 120$ $\quad$ $5 \times 30 = 150.$

Since not all divisors are multiples of ten, strategies for estimating quotients for nonmultiples must be learned. A useful strategy is to round the divisor to the nearest ten and use multiples of that number.

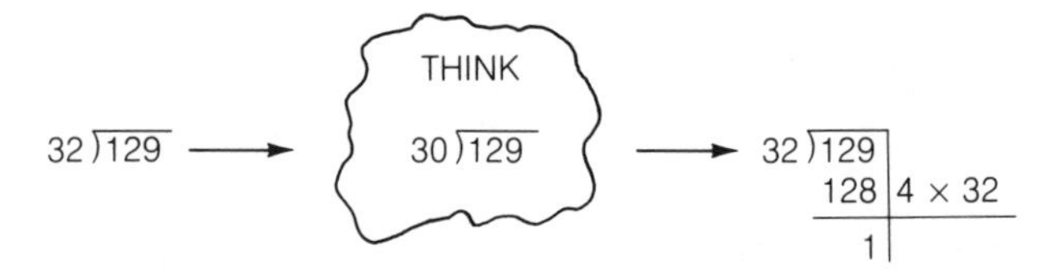


This process will not always yield the correct quotient, as in

THINK

$33\overline{)129}$ → $30\overline{)129}$ → $33\overline{)129}$
132 | 4 × 33

so children must be able to adjust the quotient to the correct one by using a smaller number, in this case 3.

## 2. Two-Digit and Larger Quotients

A situation similar to the following can introduce children to division with larger numbers. "288 marbles are to be packaged in bags of twenty-four marbles each. How many bags will it take to hold the marbles?" A box holding 288 marbles can be used so children can determine the answer by actually putting twenty-four marbles in bags until all the marbles have been bagged. As children bag the marbles, another child can show the repeated subtraction, as in the margin. When they finish, the number of bags should be compared with the number of times 24 served as a subtrahend in the subtraction; the number will be 12 in both cases.

$$\begin{array}{r} 288 \\ -24 \\ \hline 264 \end{array} \quad \begin{array}{r} 264 \\ -24 \\ \hline 240 \end{array} \quad \begin{array}{r} 240 \\ -24 \\ \hline 216 \end{array} \quad \cdots \quad \begin{array}{r} 24 \\ -24 \\ \hline 0 \end{array}$$

Then you or a child should indicate the sentence and the algorithm form for this division, $288 \div 24 = \square$ and $24\overline{)288}$, and write them on the chalkboard. The algorithm form is easier to use with larger numbers than the sentence form, so the children's attention should be directed to it: "Do we know that there are at least ten bags of twenty-four marbles each for the 288 marbles?" (Yes, we counted them.) "Can we tell without counting them?" (We can multiply to find out. [See example (a) in the margin.] Ten times 24 is 240; 240 is less than 288.) "We know there are not as many as twenty bags of marbles because we already know the answer is less than twenty. How would we know that twenty is too many without counting?" (We could multiply 20 × 24, which is 480; 480 is too much.) "After we have filled ten bags, how many marbles are still left to be bagged?" (Forty-eight.) "How do we know?" (We subtracted 240 from 288.) "How many more bags of marbles will there be?" (Two.) [See example (b).] "What is the answer when 288 is divided by 24?" (Twelve.)

```
24)288
   240 | 10 × 24
    48 |
```

(a)

```
24)288
   240 | 10 × 24
    48 |
    48 |  2 × 24
       | 12
```

(b)

If children have a good understanding of the meaning of division, know how to multiply by tens and powers of ten, and know how to use the algorithm with smaller numbers, the first work they do with two-place and larger divisors need not be restricted to examples that have one-place quotients. In the foregoing example, the quotient is larger than 10. Further use of examples with one- and two-digit numerals to represent quotients smaller than 100 will further develop understanding of the algorithm.

### 3. Estimating Quotients

As the size of the numbers used in division increases, skills in determining the quotient must be continuously refined. The following procedure might determine the answer to $36\overline{)4792}$. "Will there be at least ten 36s in 4792?" (Yes.) "How do we know?" (10 × 36 is 360; 360 is less than 4792.) "Are there at least a hundred 36s in 4792?" (Yes. 100 × 36 is 3600; 3600 is less than 4792.) "Are there at least a thousand 36s in 4792?" (No. 1000 × 36 is 36,000; 36,000 is greater than 4792.) "We know that the answer will be more than 100 and less than 1000. Will there be as many as two hundred 36s in 4792?" (No. If 36 is rounded off to 40 to estimate the quotient, 200 × 40 is 8000; 8000 is much too large.) "We know then that our answer will be greater than 100 but less than 200. If we multiply 100 times 36 and subtract the product from 4792 [see example in the margin], how much is left to be divided?" (1192.) "We know that our quotient is going to be less than 200, and we have already used 100, so we know the next quotient figure will have to be less than 100. Can it be 10?" (Yes. 10 × 36 is only 360.) "Can it be 20?" (Yes, 20 × 36 is 720.) "Can it be 30?" (Yes. 30 × 36 is 1080.) "Can it be 40?" (No. 40 × 36 is greater than 1192.) "If we subtract 30 times 36 from 1192 [see example in the margin], how much is left to be divided?" (112.) "How many 36s are there in 112? One? Two? Three? Four?" (Three.) "What is the undivided remainder?" (Four.) "How many 36s are there in 4792?" (133.)

```
36)4792
   3600 | 100 × 36
   1192 |
```

```
36)4792
   3600 | 100 × 36
   1192 |
   1080 |  30 × 36
    112 |
    108 |   3 × 36
      4 | 133
```

When procedures such as the ones just described are first used, they should be developed slowly so that children see the meaning of each step. Each step, including the multiplication used to estimate quotients, should be written on the chalkboard. Children should see how multiplying by ten and its powers helps them first determine the size of a quotient. The multiplication of the divisor by 10, 100, 1000, or their multiples helps the students estimate specific quotient figures.

Those who learn the division algorithm as suggested in the preceding discussion will have little difficulty with the proper placement of zeros in a quotient. To solve the example $42\overline{)12,852}$, they will first determine within what power of ten the answer will be.

$$10 \times 42 = 420$$
$$100 \times 42 = 4200$$
$$1000 \times 42 = 42,000$$

When it is discovered that it will be more than 100 but less than 1000, they determine how many hundreds there will be. That there will be 3 hundreds in the quotient is determined by the following multiplication.

$$100 \times 42 = 4200$$
$$200 \times 42 = 8400$$
$$300 \times 42 = 12,600$$
$$400 \times 42 = 16,800$$

```
42)12,852 |
   12,600 | 300 × 42
      252 |

42)12,852 |
   12,600 | 300 × 42
      252 |
      252 |   6 × 42
          | 306
```

The product of 300 and 42 is then subtracted from 12,852, leaving 252 still to be divided by 42, as in the margin. Because 10 times 42 is 420, there cannot be as many as ten 42s in 252. The undivided parts of the dividend and the divisor are then examined to determine the final quotient figure. Since the product of this figure and 42 equals the dividend, it completes the division: $12,852 \div 42 = 306$, as in the margin. There is little possibility that an incorrect answer such as 36 or 360 will be given, although this is commonly the case when division is learned by "the rules."

## 4. Shortened Forms

Refinements can be made in the division process so that more mature ways are used with the algorithm to reduce the number of steps and lessen the notation that is written as division is done. For example, to help children decide the number of decimal places there will be in a quotient, place a number of division examples on the chalkboard:

$39\overline{)4869}$ $\quad 73\overline{)7421}$ $\quad 86\overline{)6243}$ $\quad 23\overline{)78,552}$

The place value of each quotient figure is revealed by multiplying the divisor by ten and its powers. In the first example, this shows that the answer is more than 100 but less than 1000, so there will be a quotient represented by a three-digit numeral. An "x" or something similar can be marked over the place value position, as shown in the margin. Next, the first quotient figure should be calculated. Since in this example it is known to be some number of hundreds, the exact number of hundreds is found by multiplying by multiples of 100. Once children can readily discover the first quotient figure, they can begin to shorten the steps used with the algorithm. First, the number of decimal places is decided and marked. Next, the first quotient figure is determined. The number of 39s in 4869 is estimated by rounding 39 to 40 and multiplying 40 by multiples of 100. It is more than 100 but less than 200, so "100" is placed in the algorithm, as shown in (a) in the margin. The numeral for the product of 39 and 100 is placed beneath the dividend in the algorithm. The number of 39s in 969 will be about twenty (39 can be rounded to 40 to make the estimate), so "20" is written in the algorithm, as in (b). The product of 20 and 39 is subtracted from 969. Finally, the number of 39s in 189 is estimated by rounding off to 40. The answer, 4, is written in the algorithm, and the numeral for the product of 4 and 39 is written beneath 189. The remainder is then determined. The completed algorithm form is shown in (c).

```
      ×
  39)4869
```

```
          ×
         100
(a) 39)4869
        3900
         969
```

```
         ×20
         100
(b) 39)4869
        3900
         969
         780
         189
```

```
           4
         ×20
         100
(c) 39)4869
        3900
         969
         780
         189
         156
          33
```

Some children will continue to reduce the number of written steps until they put just a "1" in the hundreds place, a "2" in the tens place, and a "4" in the ones place, rather than writing "100," "20," and "4." These children may or may not shorten the way products of the quotient figures and divisor are written or the way the subtraction is done. The only reason zeros might be omitted in the algorithm is to save time. However, there is an increased likelihood that mistakes will be made when zeros are omitted, so don't have children leave them out when they are not ready for this step.

Finally, some children will achieve an understanding of the most mature form of the algorithm. This form is explained for the example in the margin. The divisor, 32, is rounded to 30 to give an easier-to-use divisor. The number of 30s in 79 is two. A "2" is recorded in the hundreds place of the quotient because 79 stands for hundreds. The multiplication of 2 and 32 is completed and the product, 64, is recorded. Sixty-four is subtracted from 79, and the answer, 15, is recorded; the 4 is brought down. Since 32 has been rounded to 30, the second division (that is, 154 divided by 30) gives a quotient of 5. Five is too large, however, and the quotient is reduced to 4. The division is completed by multiplying 4 times 32, recording the 128, subtracting, bringing down the 6, and dividing again to find the number of ones in the quotient. The answer is completed by writing the remainder after the quotient.

```
    248 r10
32)7946
   64
   154
   128
    266
    256
     10
```

## Self-check of Objective 11

The division algorithm is difficult for many children to understand when large numbers are involved. Demonstrate your own understanding of the process by making up a real-life problem similar to the one given here, and write a series of questions and answers that might be a dialogue between you and a group of children learning to use the algorithm with large numbers.

## CHECKING MULTIPLICATION AND DIVISION

The ultimate goals of instruction in computation are to help children know when to use a particular computational process and to compute rapidly and accurately. Teaching procedures such as those in Chapter 9 and in this chapter help children recognize situations out of which multiplication and division computations arise. These procedures should help children develop some degree of speed and accuracy as well. The greatest key to accuracy is a thorough knowledge of the basic facts and an understanding of the common algorithm forms for multiplying and dividing.

Children should establish the habit of estimating answers as a means of determining the reasonableness of their computation. To be able to get close estimates of answers, they must know how to round numbers to powers of ten or their multiples and how to multiply by rounded numbers. Once they do, they can be introduced to the concept of *upper and lower bounds*. For example, to determine the product of 45 and 86, the upper bound—4500—is found by multiplying 50 and 90, and the lower bound—3200—by multiplying 40 and 80. Obviously, the product will lie between 3200 and 4500. Any product that lies outside these bounds cannot be correct.

A more accurate estimate of an answer results by rounding the two factors to the nearest multiple of ten or multiple of a power of ten. Thus to estimate the answer for $49 \times 312 = \square$, the 49 is rounded to 50 and the 312 to 300. The product of $50 \times 300$ can be determined mentally. Children need to have regular practice rounding numbers and multiplying by rounded numbers before they begin to use these skills in estimating answers for multiplication.

Estimation can serve as a rough check of an answer. However, a product may be close to an estimated answer and still be incorrect. Therefore children need to learn other checks to use when complete accuracy is required.

The simplest check is to redo the work. The disadvantage of this procedure is that the work can be repeated without thinking about each step; one can go through the motions without really doing any checking.

This problem can be overcome if the factors are reversed before multiplication is redone. This application of the commutative property usually creates different multiplication combinations, making it necessary to write different partial products. The addition usually involves different combinations of addends, also. Since a child must think about the work as it is done, the probability of an accurate check is enhanced.

Division can be checked by several procedures. None of them is particularly popular with children because each takes about as much time as the original division when the numbers are very large. Therefore the best assurance of accuracy is a good understanding of the process and a good knowledge of the basic facts. A good background for division means fewer errors and less need to check.

Nevertheless, some checking procedures should be learned. The most common one is to multiply the quotient by the divisor and add the remainder, if any, to the product:

```
     248          248
32)7946          ×32
    6400          496
    1546         744
    1280         7936
     266          +10
     256         7946
      10
```

Another procedure is to use the quotient as a divisor and redo the division. If the new quotient is the same as the former divisor and any remainders are the same, the check indicates that the original answer is correct:

```
      66           59
59)3942      66)3942
    3540         3300
     402          642
     354          594
      48           48
```

The minicalculator makes the checking of multiplication and division quite simple.

```
   456
 × 328
  3648
  912
 1368
149568
```

```
   223 r3
17)3794
   34
    39
    34
     54
     51
      3
```

## Self-check of Objective 12

Explain with examples ways to check the accuracy of the answers to the multiplication and division examples in the margin.

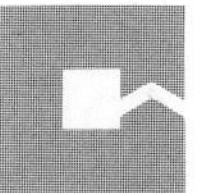

## THE MINICALCULATOR AND MULTIPLICATION AND DIVISION

The minicalculator can be used to help children understand multiplication and division operations and to extend their experiences with these operations beyond the ones usually found in textbooks.

### 1. Repeated Addition and Subtraction

Relationships between multiplication and addition, division and subtraction, and multiplication and division are readily shown on the minicalculator. To show multiplication as repeated addition, use one factor as an addend and add it as many times as indicated by the other factor. For $7 \times 83 = \square$, use 83 as the addend seven times to get the product 581. Compute by multiplying on the machine to verify the answer.

In a similar way, show division as repeated subtraction. Set the dividend into the machine. Subtract the divisor as many times as needed to reach zero, or a number smaller than the divisor. Count each time the "=" key is pressed to get the quotient. When children alternate these two operations with the same set of numbers, they have another way of seeing that multiplication and division are inverse operations.

$3 \times 27$
$6 \times 27$
$9 \times 27$
$12 \times 27$
$15 \times 27$
$18 \times 27$
$21 \times 27$
$24 \times 27$
$27 \times 27$
$30 \times 27$

### 2. Multiplying 27

Complete the products for the examples in the margin. Note the sequence of numbers in the hundreds and ones places of the products, and the sequence of numbers in the tens places. Select any product, say 486 ($18 \times 27$). Rearrange the numerals by putting the "6" before the "48." Is 648 divisible by 27? Put the "4" after "86." Is 864 divisible by 27? Rearrange other products the same ways. Are all of the rearranged numbers divisible by 27?

### 3. Filling the Holes

Multiplication and division examples such as the following can be given children so they can "fill in the holes." Let each child decide how to proceed without your directions.

```
  364        8 2          4 6
  ×9         ×38     39)17784
  ____       ____       156
  364          76       ____
 327        261          2 8
 _____      ____         1 5
 33  4      3136        _____
                          234
                          234
                         _____
```

**4. Russian Peasant Multiplication**

This old multiplication process is made simple by the minicalculator. The process is illustrated with two examples. In the left-hand column the factor is divided by 2 and the quotient is recorded. (If the calculator shows a decimal fraction quotient, only the whole number portion is recorded.) Divide each successive quotient by 2 until "1" is reached. The factor in the right-hand column is multiplied by 2, with its product recorded beneath it. Each successive product is multiplied by 2. Certain rows are marked out. What criterion is used to determine the rows to mark out? What is done with the remaining numbers in the right-hand column to determine the product?

| | | | | |
|---|---|---|---|---|
| ~~48 × 28~~ | | | 37 × | 57 |
| ~~24~~ | ~~56~~ | | ~~18~~ | ~~114~~ |
| ~~12~~ | ~~112~~ | | 9 | 228 |
| ~~6~~ | ~~224~~ | | ~~4~~ | ~~456~~ |
| 3 | 448 | | ~~2~~ | ~~912~~ |
| 1 | 896 | | 1 | 1824 |
| | 1344 | | | 2109 |

**5. Estimating Products and Quotients**

Children should extend their estimation skills by making reasonable estimates for multiplication and division. Give them work sheets you have prepared for this activity, or have them use appropriate textbook pages for practice.

## Self-check of Objective 13

Demonstrate with a minicalculator five activities that can be used to extend children's experiences with multiplication and division.

## Common Pitfalls and Trouble Spots

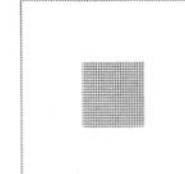

A common pitfall associated with division is the absence of grade-to-grade articulation of procedures for dealing with the algorithm. It is not uncommon for a child to have four different teachers during the years this operation is taught and have each teacher use a different way of thinking about and processing the algorithm. Because division is a difficult operation to learn, it is important that teachers be consistent in their approach to teaching the meaning of the algorithm and ways of processing it. Teachers in grades three through six must agree on their approach so children will not be confused by a year-to-year switch from one approach to another.

There are trouble spots in multiplication and division, just as there are in other areas of mathematics. Two trouble spots are discussed, one for each operation. Keeping numerals properly aligned in the algorithm is a common trouble spot associated with multiplication. Errors like these result from the improper alignment of partial products:

| | |
|---:|---:|
| 29 | 362 |
| ×46 | ×405 |
| 174 | 1810 |
| 116 | 14480 |
| 290 | 16290 |

One way to help children with this problem is to have them turn ruled paper so that the lines are vertical and write the algorithm so each place value position occupies one column (Figure 10-11). Each partial product is written with the first numeral in the column beneath the multiplier for that product.

**Figure 10-11** Multiplication done on vertically lined paper helps keep numerals aligned

| | | | |
|---|---|---|---|
| | | 2 | 9 |
| | × | 4 | 6 |
| | 1 | 7 | 4 |
| 1 | 1 | 6 | |
| 1 | 3 | 3 | 4 |

| | | | | | |
|---|---|---|---|---|---|
| | | | 3 | 6 | 2 |
| | | × | 4 | 0 | 5 |
| | | 1 | 8 | 1 | 0 |
| 1 | 4 | 4 | 8 | 0 | |
| 1 | 4 | 6 | 6 | 1 | 0 |

A common trouble spot in division is the omission of zero(s) in a quotient. This occurs most often when children learn by rote. Children taught through a meaningful approach are not as likely to make this error. A child who omits zeros can also be helped by using ruled paper so that the lines run vertically. The columns help align the numerals representing dividends, products, remainders, and quotients so that absent zeros are likely to be spotted (Figure 10-12).

**Figure 10-12** Vertically ruled paper helps children remember to put zeros in a quotient

| | | | |
|---|---|---|---|
| | 2 | 0 | 7 |
| 4 | )8 | 3 | 0 |
| | 8 | | |
| | | 3 | 0 |
| | | 2 | 8 |
| | | | 2 |

| | | | | | | |
|---|---|---|---|---|---|---|
| | | | | 4 | 3 | 0 |
| 6 | 2 | )2 | 6 | 6 | 6 | 0 |
| | | 2 | 4 | 8 | | |
| | | | 1 | 8 | 6 | |
| | | | 1 | 8 | 6 | |
| | | | | | | 0 |

## Self-check of Objectives 14 and 15

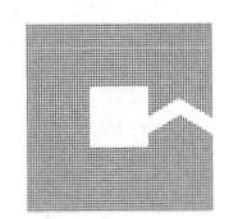

One common pitfall associated with teaching division is discussed. Identify it; then describe its effect on children.

Describe a trouble spot for each of the operations of multiplication and division, and explain a procedure that can be used for overcoming each one.

## SUMMARY

Care must be used when children are taught to use the algorithms for multiplication and division so the algorithms' meanings and applications will be clear. Many of the manipulative materials used to help children learn the meanings of the addition and subtraction algorithms and to introduce multiplication and division can be used to teach the multiplication and division algorithms. These materials include place value devices, markers for magnetic or flannel boards, arrays, and structured materials. Real-life story problems play an important role in making applications of the algorithms meaningful. Division situations frequently result in answers with remainders other than zero. Children should learn to handle remainders according to the nature of the situation, rather than by rule alone. There is a low-stress algorithm for multiplication, and the one for addition can be used to generate multiplication and division facts. There are processes for checking the accuracy of children's work. None are popular, so emphasis should be placed on careful work as the multiplication and division are first done.

The minicalculator gives children the opportunity to better understand multiplication and division algorithms and to extend their work with these operations. In order to avoid confusing students as they progress, teachers need to agree on the procedures they use as they teach these operations. The proper alignment of partial products in multiplication and the omission of zeros in quotients are trouble spots for some children. Having children use vertically lined paper will help them over these trouble spots.

## STUDY QUESTIONS AND ACTIVITIES

1. Make drawings to show how arrays can illustrate the following multiplication sentences on a magnetic board.
   (a) $2 \times 14 = \square$ (b) $13 \times 3 = \square$

2. Make up a problem situation to fit these division sentences: $216 \div 12 = 18$ and $645 \div 15 = 43$. Make one a measurement and the other a partitive situation. Write dialogues similar to those used in this chapter to show how you might discuss these problems with children.

3. Arrange to work with a group of fifth or sixth graders to observe how well they multiply by tens and powers of ten and their multiples. Can they readily multiply such numbers as 23 and 48 by 10, 100, and 1000? Can they readily multiply the same kinds of numbers by multiples of 10, 100, and 1000, such as 20, 200, and 2000? If the children have difficulty doing this kind of multiplication without paper and pencil, what difficulties are they likely to encounter as they do division with larger numbers? What should be done to help the children improve their ability to do this type of multiplication?

4. Work with a group of fifth or sixth graders who are multiplying and dividing large numbers. Do you believe the children have a good understanding of the two algorithms? If they don't, what procedures do you think should be used to help them overcome their weaknesses?

5. Read Ashlock's article (see the reading list for this chapter). What does he mean by "model switching"? What problems are caused by model switching? How can you overcome this common weakness?

6. Heddens and Lazerick say that lies and half-truths are told to children when the "guzinta" approach to division is used (see the reading list). What is the "guzinta" approach? What lies and half-truths does it generate? What alternative to this approach do these authors recommend?

## FOR FURTHER READING

Adkins, Bryce E. "A Rationale for Duplication-Mediation Multiplying," *The Arithmetic Teacher*, XI, No. 4 (April 1964), 251–253. Explains an old multiplication process, frequently called "Russian peasant" multiplication. Some students will enjoy using this procedure to supplement the regular multiplication algorithm.

Alger, Louisa R. "Finger Multiplication," *The Arithmetic Teacher*, XV, No. 4 (April 1968), 341–343. Explains finger multiplication using "cycles." The cycles permit multiplication of a pair of factors to 100. It is a good enrichment exercise for those who are skillful with other algorithm forms.

Ashlock, Robert. "Model Switching: A Consideration in the Teaching of Subtraction and Division of Whole Numbers," *School Science and Mathematics*, LXXVII, No. 4 (April 1977), 327–335. Models are used to help children understand operations and their algorithms. Chips or blocks may be used to represent partitive and measurement division situations. Ashlock cautions teachers to use models consistently so children do not become confused by discrepancies between how a model is used and the problem situation it represents.

Cacha, Frances B. "Understanding Multiplication and Division of Multidigit Numbers," *The Arithmetic Teacher,* XIX, No. 5 (May 1972), 349–354. Illustrates the use of arrays to interpret the meanings of the multiplication and division algorithms. The illustrations can serve as models for transparencies or magnetic board manipulatives.

DeVault, M. Vere. "The Abacus and Multiplication," *The Arithmetic Teacher,* III, No. 2 (March 1956), 65. Describes the classroom abacus as an aid for giving meaning to the process of multiplying two numbers greater than 10.

Hazekamp, Donald W. "Teaching Multiplication and Division Algorithms," *Developing Computational Skills,* 1978 Yearbook of the National Council of Teachers of Mathematics. Reston, Va.: The Council, pp. 96–128. A variety of manipulative materials—abacus, beansticks, base ten blocks, and others—serve as models for developing understanding of the algorithms. Many of the activities are illustrated.

Heddens, James W., and Beth Lazerick. "So 3 'Guzinta' 5 Once: So What!!" *The Arithmetic Teacher,* XXIX, No. 7 (November 1977), 576–578. When the "guzinta" approach to teaching division is used, half-truths and lies are told children. These half-truths and lies are eliminated when a meaning approach described by the authors is used.

Junge, Charlotte W. "Now Try This—In Multiplication," *The Arithmetic Teacher,* XIV, No. 1 (January 1967), 47. Explains a process of multiplication based on separating the multiplier into two or more of its factors and multiplying the multiplicand by one factor, that product by another factor, and so on. The procedure is a good enrichment exercise.

———. "Now Try This—In Multiplication," *The Arithmetic Teacher,* XIV, No. 2 (February 1967), 134–135. Explains several shortcuts used to multiply certain types of whole numbers. Gives three procedures for multiplying a pair of numbers ending in 5.

Pincus, Morris, et al. "If You Don't Know How Children Think, How Can You Help Them?" *The Arithmetic Teacher,* XXII, No. 7 (November 1975), 580–585. Identifies common errors children make while using the addition, subtraction, multiplication, and division algorithms. Describes suggestions for remediating errors.

Shoecraft, Paul. "15 Billion Hamburgers is a Lot of Multiplication and Division," *The Arithmetic Teacher,* XXII, No. 8 (December 1975), 612–613. The total number of McDonald's hamburgers sold at the time the article was written served as the basis for much computation: How many pounds of hamburger? gallons of ketchup? miles high? and so on. Today's numbers will be even larger.

Spitler, Gail. "Multiplying by Eleven—A Place-Value Exploration," *The Arithmetic Teacher,* XXIV, No. 2 (February 1977), 122–124. Shortcuts for multiplying by eleven are easily learned. The process is of little value, however, unless children use their experiences as a basis for making conjectures about why the process works. They also employ many concepts of place value.

Swart, William L. "A Diary of Remedial Instruction in Division—Grade Seven," *The Arithmetic Teacher,* XXII, No. 8 (December 1975), 614–622. Children who fail with a conventional algorithm often will not even attempt to learn to use it when they get older. A new division algorithm and how it was used with two seventh-grade girls who rebelled against the conventional algorithm are described.

Zweng, Marilyn J. "The Fourth Operation Is Not Fundamental," *The Arithmetic Teacher,* XIX, No. 8 (December 1972), 623–627. This article contains a discussion of division. It has good examples of partitive and measurement situations, with sample illustrations.

# 11 Introducing Fractional Numbers—Common Fractions

Upon completion of Chapter 11, you will be able to:

1. Identify orally or in writing at least three different real-world situations that give rise to common fractions, and name an example of a common fraction for each situation.

2. Demonstrate with geometric regions, structured materials, number lines, and markers activities for children who are learning about common fractions.

3. Explain why there are infinitely many fractional numbers between a given pair of fractional numbers, and demonstrate a procedure for helping children develop an intuitive understanding of this fact.

4. Demonstrate activities for children that will enable them to compare fractional numbers.

5. Explain an abstract process that can be used for comparing two or more fractional numbers.

6. Demonstrate at least three systematic procedures for renaming fractional numbers so they are represented by common fractions having the same denominator.

7. Explain what it means to express a fractional number in its simplest form, and demonstrate at least two procedures for helping children to express fractional numbers in simplest form.

8. Rename common fractions as decimal fractions and vice versa, and identify which kinds of common fractions result in terminating decimals and those which result in repeating decimals.

9. Demonstrate how to use a minicalculator to rename common fractions in simplest form and improper fractions as mixed numerals.

10. Identify two pitfalls associated with instruction about common fractions, and explain how each one can be avoided.

11. Identify a trouble spot for children learning about common fractions, and describe orally or in writing a paper-folding activity to help them overcome it.

Key Terms you will encounter in Chapter 11:

| | | |
|---|---|---|
| common fraction | unit fraction | simplest form |
| denominator | equivalent fraction | improper fraction |
| ratio | average | mixed number |
| numerator | common denominator | mixed numeral |
| indicated division | least common multiple (LCM) | terminating decimal |
| rational number | simplifying (reducing) fractions | nonterminating (repeating) decimal |
| region | | |

Children's study of fractional numbers begins as early as kindergarten and continues through the remainder of the elementary school. The fractional numbers treated during the elementary school period are a part of the set of rational numbers; those that can be expressed in the form $a/b$, when $a$ is any whole number and $b$ is any nonzero whole number. It is possible to consider such numbers on a purely abstract basis with the meanings of the numbers and operations developed from undefined terms, postulates, or axioms, and theorems that pertain to them. However, such an approach is not meaningful to elementary school children, so a concrete approach to the study of fractional numbers must be taken.

The most useful approach is one that allows children to intuitively develop a reasoned understanding of fractional numbers. In all grades of elementary school, many concrete representations of fractional numbers help children to develop a clear understanding of the meaning of these numbers and the way operations with them are performed.

Fractional numbers, like whole numbers, can be expressed in a number of ways. Symbolically, they may be expressed in any one of three ways: as common fractions ($\frac{1}{2}$ and $\frac{2}{3}$), as decimal fractions (0.5 and 0.6666 . . .), and as percent (50% and $66\frac{2}{3}$%). Also, any given fractional number has an infinite number of numerals, for example, $1/2 = 2/4 = 3/6 = 4/8 = 5/10 = \cdots$.

## THE MEANING OF COMMON FRACTIONS

Common fractions are numerals used to represent fractional numbers and ratios. Historically, children have experienced considerable difficulty with the meaning of common fractions because there are several situations out of which they arise and these have not always been well understood. If they are to understand common fractions, children must become familiar with each situation and work with them in a mature fashion.

### 1. Unit Subdivided into Equal-Sized Parts

When a unit of measure, such as an inch, is subdivided into smaller equal-sized parts, common fractions can be used to express the meaning of each subunit of measure. For example, when an inch is subdivided into two equal-sized parts, each part is $\frac{1}{2}$ of an inch. Likewise, when an object, such as a cake, is first considered as a whole and is then cut into four parts of equal size, the common fraction ¼ can be used to express the size of each part. The parts of a common fraction indicate the nature of the situation from which the numeral arises. In the numeral ½, the 2 indicates the number of equal-sized parts into which the whole, or unit, has been subdivided. This part of the fraction is the denominator. The 1 indicates the number of parts being considered at the moment, and is called the numerator.

### 2. Set Subdivided into Equal-Sized Groups

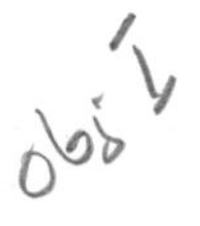

Situations that lead to fractional numbers when a set is subdivided into groups of equal size are clearly related to those that involve division. When a set of twelve objects is subdivided into two equal-sized groups, the mathematical sentence $12 \div 2 = 6$ represents the situation. The 6 in the sentence represents one-half of the original set of twelve. The sentence ½ of 12 = 6 also describes this situation. Obviously, children will often engage in activities that involve parts of sets of discrete wholes. To find ⅙ of 18, for instance, children must think of eighteen objects which are to be subdivided into six groups of equal size. The size of each group relates to the size of the original set in such a way that each is ⅙ of the original set. When a common fraction is used to represent this type of situation, the denominator indicates the number of equal-sized groups into which the set is subdivided, while the numerator indicates how many of the groups are being considered.

### 3. Expression of Ratios

The relationship between a pair of numbers is often expressed as a ratio. Ratios arise from many situations. Examples of common situations are:

1. A comparison made between the number of children in a classroom and the number of textbooks in the same classroom. If each child has six textbooks, the ratio is 1 to 6, and may be expressed as ⅙ or 1:6. Here the numerals in the ratio expression represent the numbers of objects in two completely different sets.

2. A comparison made between the number of blue-covered books in a set and the number of books in the complete set. If there are three blue-covered books in a set of ten books, the ratio is 3 out of 10, and may be expressed as 3/10 or 3:10. Here the numerals in the ratio represent the numbers of objects in a set and one of the groups within it.

3. A comparison made between the length of two objects. If four pieces of dowel rod, each the same length, are laid end-to-end alongside a second rod and have a total length as long as the second rod, the

ratio is 4 to 1, and may be expressed as $\frac{4}{1}$ or 4:1. (The length of one short rod is $\frac{1}{4}$ of the longer rod.) Here the numerals represent the number of shorter objects compared to a single object.

### 4. Indicated Division

Sentences such as $3 \div 4 = \square$ and $17 \div 3 = \square$ indicate that division is to be performed. A physical model of the first sentence is the subdivision of 3 feet of cloth into four equal-sized parts, with the answer, $\frac{3}{4}$, indicating the size of each of the four parts. The second sentence might represent a situation in which 17 cookies are shared equally by 3 children, with $5\frac{2}{3}$ being each child's share.

### 5. Expression of Rational Numbers

At a completely abstract level, it is possible to think of common fractions as representing the elements in the set of nonnegative rational numbers. While children in the elementary school will study rational numbers in this abstract sense only briefly, the development of concepts of fractional numbers and common fractions should be consistent with that of concepts of rational numbers.

---

## Self-check of Objective 1

Common fractions are used when units are subdivided into equal-sized parts, when sets are subdivided into equal-sized groups, when ratios are considered, to represent answers for all division situations, and to give numerals for abstract rational numbers. Give oral or written real-life examples for any three of these situations, and write a common fraction for each example. Identify the meaning of each numerator and denominator in your examples.

---

## INTRODUCING COMMON FRACTIONS

There are those who advocate that common fractions not be taught in elementary school. They contend that the minicalculator and metric system will make these fractions obsolete and that children will not need to learn about them. This view is short-sighted for at least two reasons.

First, the preceding discussion shows that common fractions are used in ways unrelated to those the minicalculator and metric system will replace. Children need to learn these uses because their importance will not diminish.

Second, the amount of time that will be required to change from the customary system to the metric system of measure is undetermined. Until the transition is complete, children will need to know how common fractions are used in the customary system.

The study of fractional numbers, represented by common fractions, should begin with a variety of shapes and markers for children to subdivide into equal-sized parts and with structured materials, such as Unifix cubes and Cuisenaire rods. Develop each concept carefully so that children understand the meanings of the parts of the whole. While it is not essential that the equal-sized parts be the same shape, it is easier for young children to understand the meanings of common fractions when the parts are congruent.

## 1. Using Regions

The concept of ½ can be introduced by giving each child several pieces of colored construction paper cut into shapes such as the ones shown in Figure 11-1(a), but not marked with dotted lines. Let the children experiment with shapes to see how many ways they can fold them to make two parts that are the same size. After time for exploration, have children show and discuss the results of their work. During the discussion, tell children how to describe the parts of the whole by using the words *one-half, a half,* and *halves.* A similar lesson can be used to introduce the concept of one-fourth (b), but not for thirds. You must prepare shapes with broken lines indicating thirds in advance (see Figure 11-2) because accurate models of thirds are difficult to fold. (You can see that this is true by trying to fold a model of some shapes to show thirds.)

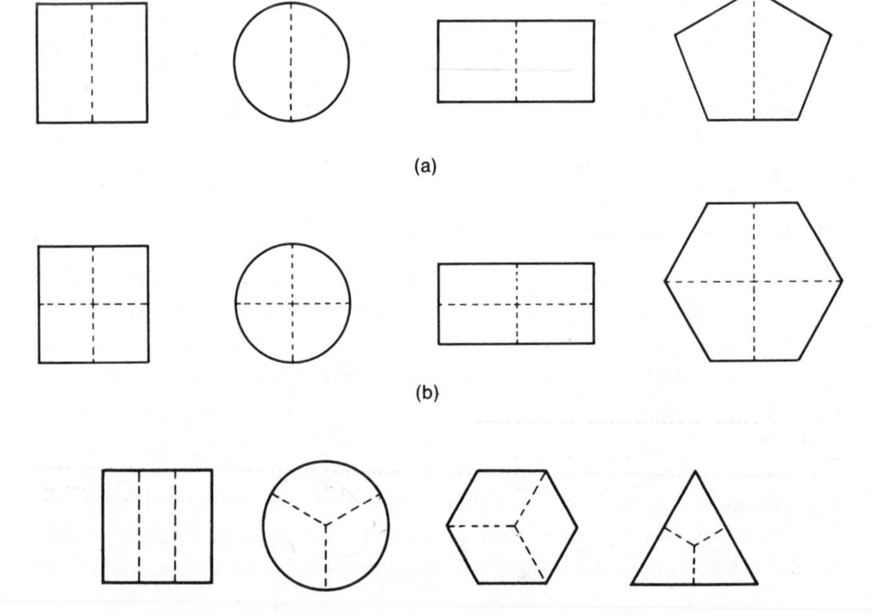


**Figure 11-1** Geometric regions used to represent (a) halves and (b) fourths

**Figure 11-2** Common geometric shapes used to represent thirds

Have children subdivide other regions into equal-sized parts until the meanings of the subdivisions are clear. The number of subdivisions need not be confined to only two, three, and four as long as models are used and common fraction numerals are not introduced too quickly. Regions subdivided into five, six, seven, and eight parts are easily understood when shapes are actually separated into equal-sized parts.

From time to time, work sheets similar to those in Figure 11-3 should be completed by the children. In (a) they color one of the three equal-sized parts, while in (b) they mark an X over each drawing that does not show two equal-sized parts. Workbooks for first and second graders contain a limited number of practice exercises of these types; you can duplicate additional work sheets if they are needed.

Once the meanings of the shapes and their parts are clear, introduce common fractions as numerals for fractional numbers. Relate the numerals to shapes that have been cut into two, three, or more parts. At first, use only one part of each shape to show the related unit fraction. (A unit fraction is one with a numerator of 1.) Also, refrain from using *denominator* and *numerator* too early. However, do help children understand that the bottom part of the numeral tells the number of equal-sized parts into which a shape has been cut, while the top part tells the number of parts being considered at the moment. Later, introduce numerators greater than 1 as children consider $2/3$ of a shape cut into three equal-sized parts or $5/6$ of one cut into six equal-sized parts.

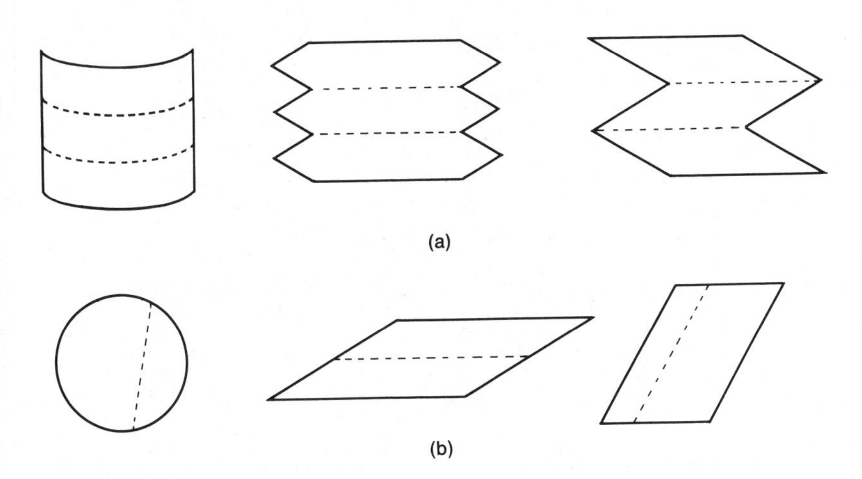


**Figure 11-3** Samples of items to use on work sheets. The shapes in (a) are to be colored to show one of the three equal-sized parts and in (b) are to be crossed out if the two parts are not equal-sized parts of the unit region

## 2. Using Structured Materials

Cuisenaire rods and Unifix cubes are structured materials children can use to explore the way common fractions express the meaning of ratios. Children in third and higher grades can carry out their explorations under your direction or by following instructions on problem cards. Directions for several cards are given here to suggest activities with Cuisenaire rods. Unifix cube activities are similar, except that children must make rods by

EXPLORING COMMON FRACTIONS (1)

1. Get a set of rods from the shelf.

2. Write the answers to these questions on your answer sheet:
   a. How many white rods do you use to make a train as long as one red rod?
   b. How many white rods do you use to make a train as long as one purple rod?
   c. How many white rods do you use to make a train as long as one black rod?
   d. How many red rods do you use to make a train as long as one purple rod?
   e. How many red rods do you use to make a train as long as one brown rod?
   f. How many yellow rods do you use to make a train as long as one orange rod?

**Figure 11-4** Problem cards for exploring the meaning of common fractions

(2)

   g. How many white rods do you use to make a train as long as one green rod?
   h. How many red rods do you use to make a train as long as one dark green rod?

3. Use a white rod and a red rod to make this pattern:

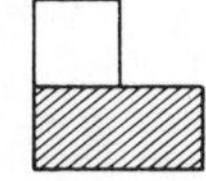

If you call the red rod 1, the white rod is 1/2. The white rod is one-half as long as the red rod.

connecting cubes. Your instructions can tell them the colors to use and how long to make the different rods.

(3)

4. Use a white rod and a purple rod to make this pattern:

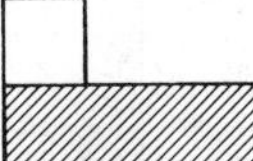

<u>If you call the purple rod 1, the white rod is</u> ___ .
<u>The white rod is</u> ___ <u>as long as the purple rod</u>.

5. Use rods to make these patterns:
   a. white, green
   b. white, yellow
   c. white, black
   d. white, brown
   e. white, orange
   f. white, purple

(4)

If the colored rod in each pattern in step 5 is called 1, what is the white rod called? Write your answers on your response sheet.

6. Make these patterns with your rods:
   a. red, purple
   b. red, dark green
   c. red, brown
   d. red, orange

If the rod that is not red is called 1, what is the red rod called? Write your answers on your response sheet.

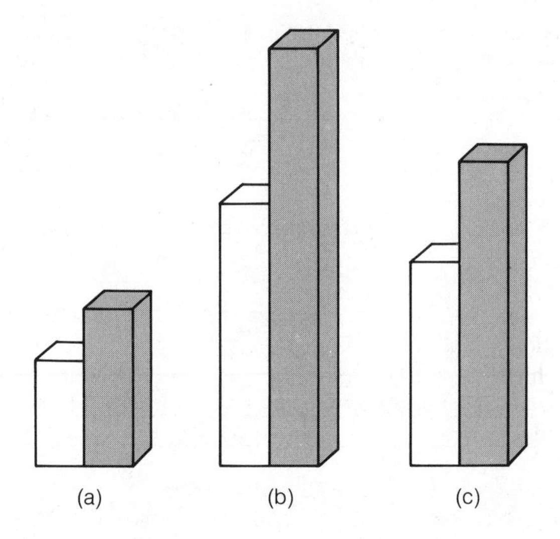


**Figure 11-5** Cuisenaire rods help children understand common fractions. In (a) the red rod is $\frac{2}{3}$ as long as the green rod, in (b) the yellow rod is $\frac{5}{8}$ as long as the brown rod; and in (c) the purple rod is $\frac{4}{6}$, or $\frac{2}{3}$, as long as the dark green rod

The activities on these problem cards deal with unit fractions, but Cuisenaire rods and Unifix cubes can be used for work with other fractions also. The patterns in Figure 11-5 illustrate several nonunit fractions. In (a) the relationship between a red and a green rod is $\frac{2}{3}$, in (b) the relationship between a yellow and a brown rod is $\frac{5}{8}$, while in (c) the relationship between a purple and a dark green rod is $\frac{4}{6}$.

The activities just described deal with the relationships between pairs of rods or cubes and are useful for introducing the ratio concept, although in the beginning the term *ratio* need not be used.

### 3. Using a Number Line

Number lines are also useful for illustrating the meaning of common fractions (Figure 11-6). Keep activities simple in the primary grades. Use a number line with the distance between whole-number points great enough so that other points can be marked between them without crowding as in (a). Then show a second line with the same unit distances. Mark a point midway between each pair of original points and compare the new segments with those on the first line. Children will see that there are two new segments for each old one. In other words, each new segment is one-half of the length of each original segment. Label these new points, as in (b), and discuss the numerals' meanings. Have children use two different ways of counting these new segments: one-half, two-halves, three-halves, . . . , and one-half, one, one and one-half, . . . ; and two ways of writing the numerals: $\frac{1}{2}$, $\frac{2}{2}$, $\frac{3}{2}$, . . . , and $\frac{1}{2}$, 1, $1\frac{1}{2}$, . . . .

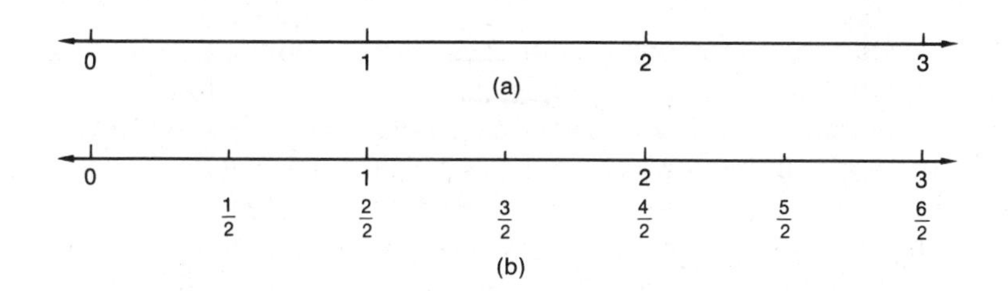


**Figure 11-6** Number lines marked to help children develop concept of halves

Number lines marked with segments showing thirds, fourths, fifths, and so on, should be used to extend children's understanding of these common fractions.

When children make measurements, many of the instruments they use have a form of number line on them. For instance, while they use a ruler such as the one in Figure 11-7(a), the scale is in inch units. In (b) the ruler is marked in half-inch units. Other measures, such as measuring cups, have scales with units and parts of units marked on them.

## 4. Using Sets of Objects

The introduction of the concept of a fractional part of a set can be made only after children have a good grasp of whole numbers and their use as names for the number of objects in a set. The first work must be handled carefully so that this fraction concept is developed meaningfully.

One way to introduce the concept is with disks or other small markers. Give each child a container of markers and say, "Count out a set of eight markers." After each child has a set of eight, instruct the children to separate their sets into two groups so each one has the same number of markers in it (Figure 11-8). During discussion of the activity, help chil-

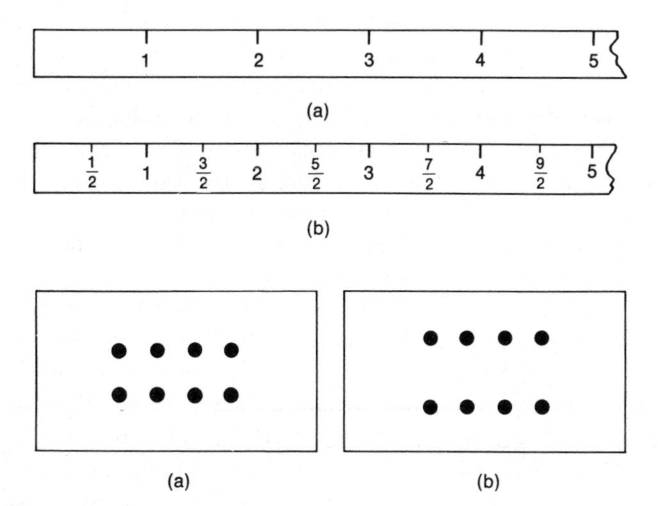


**Figure 11-7** Rulers marked by (a) inches and (b) inches and half inches

**Figure 11-8** A child's work space showing (a) a set of eight markers and (b) the two sets that represent the two halves of the original set

dren relate the present work with their earlier activities with parts of regions, structured materials, and the number line, if they have used all of these materials, so they can associate separation of the set into two groups of equal size with the way regions, rods, and number lines are separated or compared to show one-half. This association will help them realize that when half of a set is to be found, it is necessary to form two groups of the same size, in the same sense that a shape or segment is subdivided into two parts of the same size, or two shorter rods are put together to make a train as long as another rod. Children find one-half of the set as they separate the markers into two equal-sized groups and then count the objects in one group. The fact that the answer is a whole number rather than a fractional number is sometimes troublesome, so it is necessary to repeat activities with different-sized sets of markers until the concept is clear. In the beginning, always use sets that can be subdivided into equal-sized groups without any objects left over.

Discussion of this concept is continued in the next chapter in the section dealing with multiplication of fractional numbers.

---

## Self-check of Objective 2

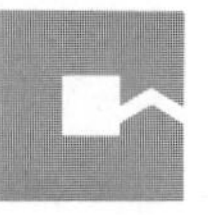

Select at least three of these materials and demonstrate the meaning of two-thirds of a unit, how two-thirds expresses a ratio, and the meaning of two-thirds of a set: unit regions, number lines, structured materials, and markers.

---

### COMPARING FRACTIONAL NUMBERS

As children study whole numbers, they learn to compare numbers so that they can tell when one number is greater than, less than, or equal to another. They use sets, number lines, and other means of comparing numbers until the idea is clear. Generally, they learn that one number is greater than another if it is to the right of the other on the number line. At the same time, they learn that every whole number has an immediate predecessor and an immediate successor. It is always possible to determine the predecessor of a whole number greater than 0 by subtracting 1 from it. The successor of a number is determined by adding 1 to it. As children study fractional numbers, they learn that it is possible to order them in sequence according to size also. At the same time, they learn that it is not possible to determine an immediate predecessor or successor for a given fractional number. An infinite number of fractional numbers lie between any pair, so fractional numbers do not have immediate predecessors or successors. It is always possible to determine a fractional number midway between a pair of fractional numbers by adding the numbers and determining their average. A number between $\frac{5}{8}$ and $\frac{3}{4}$ is $\frac{11}{16}$, which is determined by expressing $3/4$ as $6/8$ and adding it to $5/8$; $5/8 + 6/8 = 11/8$, $11/8 \div 2 = 11/16$.

It is possible to compare the sizes of two fractional numbers by multiplying the numerator of the first by the denominator of the second and the denominator of the first by the numerator of the second and comparing the two products. For the two fractional numbers $a/b$ and $c/d$, when $a \times d = b \times c$, the two numbers are equivalent. The fractional number $a/b$ is greater than $c/d$ when $a \times d > b \times c$; the fractional number $a/b$ is less than $c/d$ when $a \times d < b \times c$.

It should be clear that such abstract approaches for ordering and comparing numbers are unsuitable for children. Initial experiences should come through investigations with models of various kinds. Later, abstract procedures can be used with mathematically mature children in the higher grades of the elementary school.

## 1. Using Congruent Geometric Shapes

Children can use congruent geometric shapes cut from white construction paper to compare the sizes of fractional numbers. Give the children paper cut in circles, squares, or other shapes that they can fold to show halves, fourths, and eighths. When the shapes have been folded, have the children color half of one, a fourth of another, and an eighth of the remaining one and compare the colored portions to see which is larger (Figure 11-9). The expression $\frac{1}{2} > \frac{1}{4} > \frac{1}{8}$ indicates the relationship among these three numbers. Shapes marked to show thirds, sixths, ninths, and twelfths, and halves, fifths, and tenths can be used to compare the fractional numbers they represent.

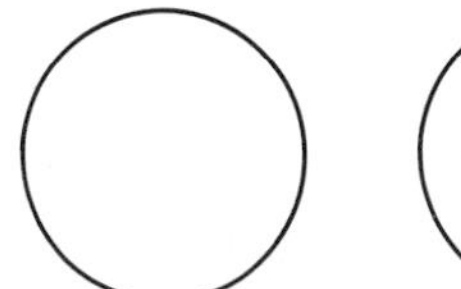
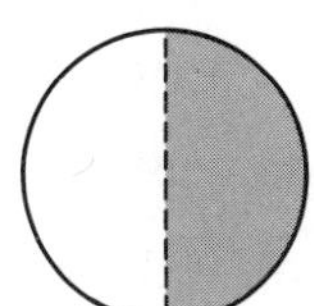
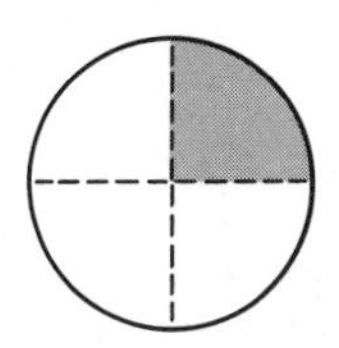
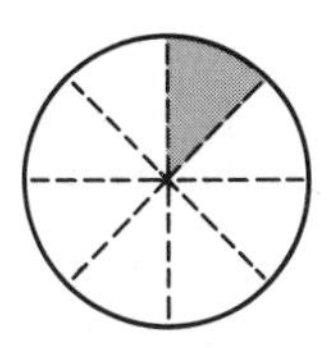

**Figure 11-9** Geometric shapes can be used to show that $\frac{1}{2} > \frac{1}{4} > \frac{1}{8}$

Congruent shapes are also useful for investigations in which children compare nonunit fractions. Models folded and shaded like those in Figure 11-10 help them see that $\frac{3}{4} > \frac{2}{3} > \frac{1}{2}$. As long as the shapes are congruent, children can compare and order any fractional numbers that can be conveniently represented with shapes.

## 2. Using Fraction Strips

Fraction strips should also be used for comparing fractional numbers. A set of strips can be cut from colored railroad board and fixed with magnets

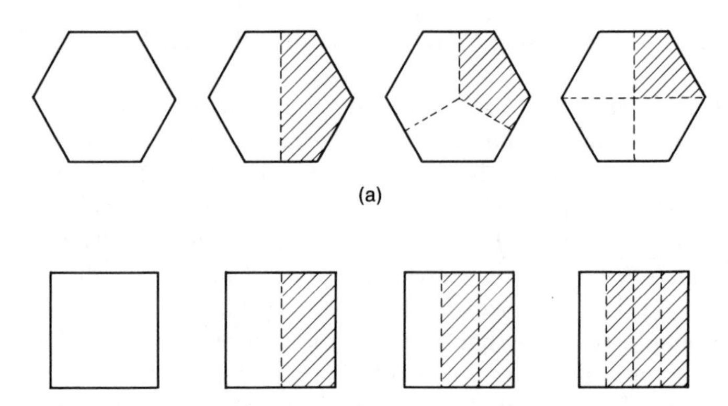


**Figure 11-10** Geometric shapes used to compare the sizes of (a) $1/2$, $1/3$, and $1/4$, and (b) $1/2$, $2/3$, and $3/4$

on the back for manipulation on a magnetic board. One useful set is pictured in Figure 11-11. Devices similar to fraction strips can also be purchased. Questions such as the following can be put on problem cards or given orally to guide children's work:

1. How many $1/2$ strips are as long as the 1 strip? How many $1/3$ strips are as long as the 1 strip? Which is longer, a $1/2$ strip or a $1/3$ strip?
2. What is the shortest fraction strip in this set? Which strips are longer than this strip?
3. Use the fraction strips to put these common fractions in order, beginning with the largest and ending with the smallest: $1/8$, $1/2$, $1/3$, $1/6$, $1/4$.
4. Which is longer, the $1/2$ strip or two of the $1/3$ strips?
5. Which is longer, two of the $1/6$ strips or one of the $1/4$ strips?
6. Which is shorter, two of the $1/2$ strips or two of the $1/8$ strips?
7. Use strips to put these common fractions in order, beginning with the smallest and ending with the largest: $1/8$, $2/3$, $2/4$, $3/6$, $4/12$.

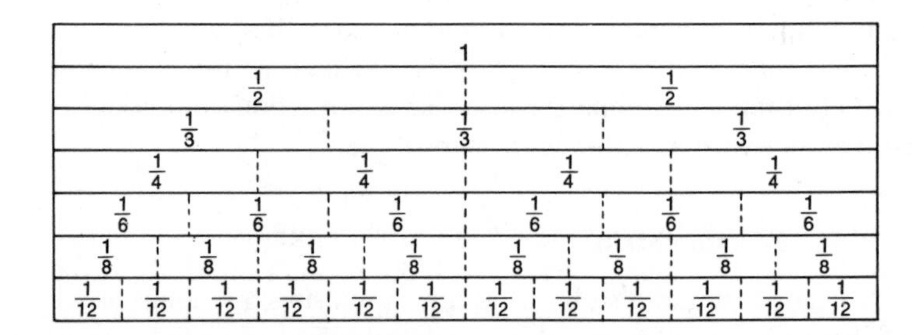


**Figure 11-11** A set of fraction strips for comparing fractions

COMPARING FRACTIONAL NUMBERS

1. If the purple rod is a unit, what are the names of the other rods? Some answers have been written in for you. Write the others on your response sheet. Then answer the questions at the bottom of this card.

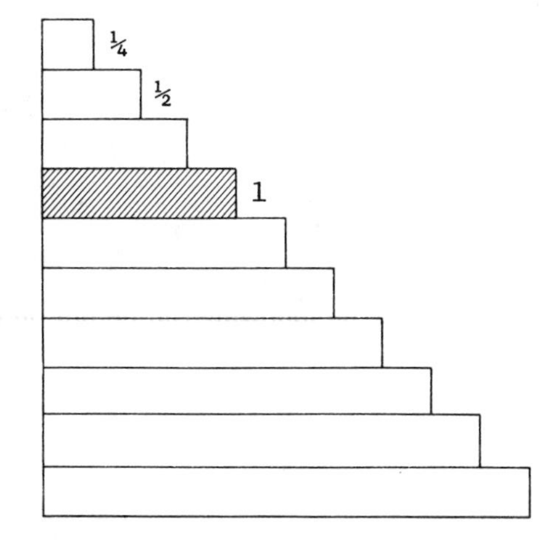


Which is larger,
3/4 or 1/2? 5/4 or 5/2? 7/4 or 3/2?

**Figure 11-12** Sample problem card for using structured materials to compare fractional numbers

### 3. Using Structured Materials

Cuisenaire rods or Unifix cubes can be used for comparing fractional numbers. A problem card such as the one shown in Figure 11-12 can be used to guide children's work.

### 4. Using the Number Line

Use the number line extensively with children in grades four through six. There is no better device for comparing fractional numbers and helping children to see that there are infinitely many numbers between any pair of fractional numbers. The following activity is one children can do to develop an intuitive understanding of the infinity of fractional numbers.

Make a set of number lines marked on a large sheet of paper, as in Figure 11-13(a). Mark the top line with points and numerals to show the whole numbers 0 and 1. Mark only the points for whole numbers with

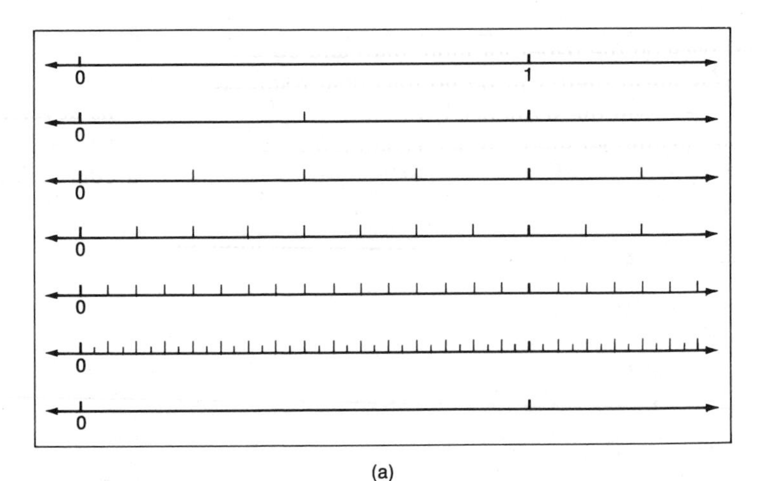


(a)

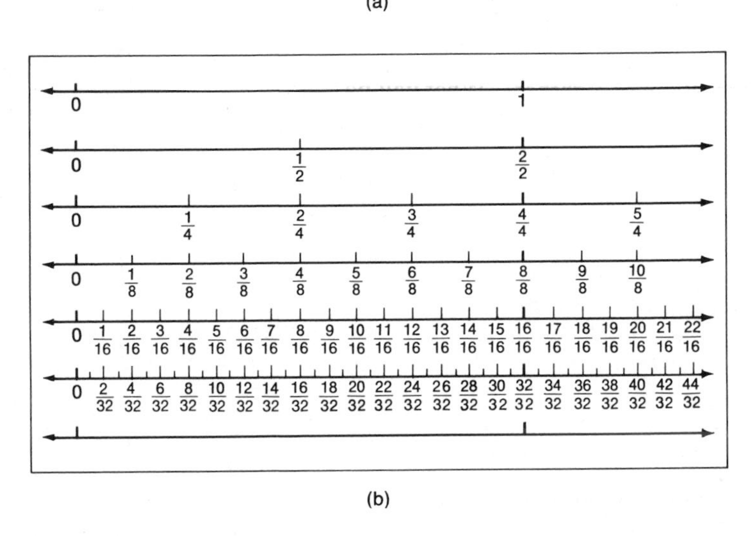


(b)

**Figure 11-13** Set of number lines (a) ready to be developed by children and (b) after points on the lines have been identified and marked

heavy marks on the other lines; mark the in-between points lightly so they can be accurately located later as the activity develops (a).

Begin the activity by directing the children's attention to the top line to have them note the unit segments along it. Then go to the next line and mark the point between the unit points. Have children identify, label, and count the marks for $\frac{1}{2}$ and $\frac{2}{2}$. Go to the next line and mark points on it; have the children identify, label, and count these points: $\frac{1}{4}$, $\frac{2}{4}$, $\frac{3}{4}$, and $\frac{4}{4}$.

Continue until the line that shows thirty-seconds has been marked and counted. Your questions and comments will extend children's thinking so that they develop an intuitive understanding that the subdivisions of the line could go on forever. "If we follow this pattern, what will be the name of the first point on the line at the bottom of the paper?" "I have run out of space on the paper for more lines and on the lines for more points. Does this mean there can be no fractional numbers between 0 and $\frac{1}{64}$, $\frac{1}{128}$, . . .?" "Will the sequence ever end? Why not?" When this development of the number lines has been completed, the chart will look like (b).

An interesting extension of this activity is to mark a curved line beginning at 1, passing through $\frac{1}{2}$, then $\frac{1}{4}$, $\frac{1}{8}$, $\frac{1}{16}$, and $\frac{1}{32}$. Ask, "Will this line ever reach 0?" Children will recognize that while the line approaches 0, it will theoretically never reach it. (The limitation of the size of the chart will make the curved line actually touch the 0 point somewhere below the last line, even though theoretically it will not.)

Children in fifth and sixth grades should explore the number lines to learn how to average numbers to determine fractional numbers that lie midway between any pair of fractional numbers. For example, when they compare the lines that show fourths and eighths, they can see that on the fourths line a point midway between $\frac{2}{4}$ and $\frac{3}{4}$ can be located that is in the same location as $\frac{5}{8}$ on the eighths line. "What will $\frac{2}{4}$ and $\frac{3}{4}$ be called if we express them as eighths?" "What will be the sum if we add the $\frac{4}{8}$ and $\frac{6}{8}$?" "Can you figure out what we can do to the $\frac{10}{8}$ to determine the fraction midway between $\frac{4}{8}$ and $\frac{6}{8}$?" (We divide by 2.) Further exploration and similar questioning should enable fifth or sixth graders to generalize that by adding a pair of fractional numbers and dividing their sum by 2, they can find a number that lies midway between any pair of fractional numbers.

Number lines have many other uses:

1. During discussion periods, children can use number lines to indicate fractional numbers that are equivalent and to compare two or more fractional numbers for the less than and greater than relationships. Children can use their number lines as they complete work sheets on which the $=$, $<$, and $>$ symbols have been omitted between numerals. Pairs of fractional numbers such as $\frac{1}{2}$ and $\frac{7}{16}$, $\frac{14}{32}$ and $\frac{7}{16}$, and $\frac{17}{32}$ and $\frac{8}{16}$ should be compared.

2. Children can count by fractional numbers. Steps along a number line can be counted as "$\frac{1}{16}$, $\frac{2}{16}$, $\frac{3}{16}$, . . . ." As they count beyond 1, children should name points as $\frac{17}{16}$, $\frac{18}{16}$, . . . and as $1\frac{1}{8}$, $1\frac{2}{8}$, $1\frac{3}{8}$, . . . .

3. Children can learn that any point on a line has many names. As they examine their number line chart, they will see that $\frac{1}{2}$, $\frac{2}{4}$, $\frac{4}{8}$, $\frac{8}{16}$, . . . all name a point that is a given distance from 0. They should begin to develop an intuitive understanding of the abstract concept of equivalence classes of fractional numbers. While this concept is not developed extensively in the elementary school, children can see that all common fractions that name a given point

on the line are names for the same number, or are equivalent. They will also learn that there is only one way a fractional number can be expressed in simplest form. All the other fractional numbers in a given equivalence class can be renamed in simpler form. Intuitively, children will begin to understand that there is a set of fractional numbers that equals $\frac{1}{2}$, a set that equals $\frac{1}{4}$, and so on.

4. Children can use their number line chart to discover ways to rename fractional numbers. You can guide them by asking questions. "We see that $\frac{1}{2}$ and $\frac{2}{4}$ are both names for a given point on the number line. What can we do with $\frac{1}{2}$ to rename it as $\frac{2}{4}$?" (We can multiply both numerator and denominator by 2.) "We see that $\frac{6}{8}$ and $\frac{3}{4}$ are both names for a given point on the line. What can we do with $\frac{6}{8}$ to rename it as $\frac{3}{4}$?" (We can divide both numerator and denominator by 2.) After discussion of a number of similar questions, children should develop the general procedure for renaming fractional numbers by multiplying or dividing both numerator and denominator by the same number. The multiplication of numerator and denominator by the same number is mathematically the same as multiplying the fractional number by 1. The numeral $\frac{2}{2}$ is another name for 1, so $\frac{1}{2} \times \frac{2}{2} = \frac{1}{2} \times 1$. The value of the fractional number named by $\frac{1}{2}$ is unchanged by the multiplication. The division of the numerator and denominator by the same number is mathematically the same as dividing the fractional number by 1. Since $\frac{6}{8} \div \frac{2}{2} = \frac{6}{8} \div 1$, the value of the fractional number $\frac{6}{8}$ is not changed by the division.

### 5. Using Abstract Procedures

Children who have a mature understanding of the procedures for comparing numbers that have already been discussed, who understand the meaning of common fractions, and who can multiply fractional numbers in ways developed in the next chapter, can use the multiplication process explained on page 295 to compare fractions. As you explain the process, give children practice pages containing pairs of common fractions and help them relate the abstract process to earlier work with models. This is a useful process for dealing with ratios and proportions, and is discussed in this connection in Chapter 12.

## Self-check of Objectives 3, 4, and 5

Give an oral or written explanation of why there are infinitely many fractional numbers between any given pair of fractional numbers. Demonstrate with a sequence of number lines a children's activity that will help them discover this fact.

Activities for comparing fractional numbers with unit regions, structural materials, fraction strips, and number lines are described. Demonstrate with at least two activities how children can show that these relations are true: $1/5 < 1/4 < 1/3 < 1/2$ and $6/7 > 4/5 > 2/3$.

The abstract process for comparing fractional numbers described on page 295 can be used with mature elementary school children. Explain how this process can be used to compare each of these pairs of numbers and name the greater fractional number in each pair: $4/15$ and $5/16$, $21/36$ and $19/35$.

## SYSTEMATIC PROCEDURES FOR NAMING EQUIVALENT FRACTIONS

Before children can become proficient in adding and subtracting fractional numbers using common fractions, they must learn to express common fractions so that they have the same denominators. The process for doing this, commonly referred to as *changing to a common denominator*, must be developed carefully so children will understand both the necessity for doing it, and how to do it meaningfully. Children should learn that when they use two equivalent common fractions to rename fractional numbers, the value of each number remains unchanged. Children who have had many experiences with various learning aids will have little difficulty with numbers such as $1/2$, $1/4$, $1/8$, $1/3$, and $1/6$. They will "just know" that $1/2$ can be renamed as $2/4$ or $4/8$, $1/4$ as $2/8$, and $1/3$ as $2/6$. However, they should learn systematic procedures for dealing with less familiar numbers.

### 1. Using Multiples

In some cases the denominator of one of the fractions is a multiple of the other. When this is the case, as with $2/5$ and $3/20$, the larger of the two denominators is a common denominator. The following thought process can be used. "Twenty is a multiple of 5. By what number do we multiply 5 to get 20?" (4.) "To rename $2/5$ with a denominator of 20, multiply both the numerator, 2, and the denominator, 5, by 4. Two-fifths equals $8/20$."

When neither of the denominators is the common denominator, a different procedure must be used. The easiest procedure if denominators are small—for example, $1/4$ and $1/5$, $1/4$ and $1/6$—is to use successive multiples of the larger denominator until a common denominator has been determined: for $\frac{1}{4}$ and $\frac{1}{5}$, multiples of 5 would be used: Twenty is the first multiple of 5 that is also a multiple of 4, so this is the common denominator. To find the least common denominator for $\frac{1}{4}$ and $\frac{1}{6}$, 12 is the first multiple of 6 that is also a multiple of 4, so it is their least common multiple (LCM). To determine the number by which both numerator and denominator of the fractional number $\frac{1}{4}$ should be multiplied to rename it as twelfths, children should think, "By what number is 4 multiplied to give a product of 12?" The answer, 3, is the number by which both terms of the fraction are multiplied.

$$\frac{1 \times 3}{4 \times 3} = \frac{3}{12}.$$

The same procedure should be used to rename $\frac{1}{6}$ as twelfths.

## 2. Using Equivalence Classes

Another procedure for finding common denominators is to list the first several numbers in the equivalence class of each number. To add $\frac{1}{6}$ and $\frac{3}{8}$, the following numerals would be listed:

$$\frac{1}{6}, \frac{2}{12}, \frac{3}{18}, \frac{4}{24}, \frac{5}{30}, \frac{6}{36}, \frac{7}{42}, \frac{8}{48}, \ldots$$

$$\frac{3}{8}, \frac{6}{16}, \frac{9}{24}, \frac{12}{32}, \frac{15}{40}, \frac{18}{48}, \frac{21}{56}, \ldots$$

Guide children to see that the first list names some fractional numbers with the same denominators as some in the second listing. These should be indicated: $\frac{1}{6} = \frac{4}{24} = \frac{8}{48}$ and $\frac{3}{8} = \frac{9}{24} = \frac{18}{48}$. Children can use both pairs of numbers, $\frac{4}{24}$ and $\frac{9}{24}$ and $\frac{8}{48}$ and $\frac{18}{48}$, to complete the addition.

$$\begin{array}{rcrcr} \frac{1}{6} & \rightarrow & \frac{4}{24} & \rightarrow & \frac{8}{48} \\ + \frac{3}{8} & \rightarrow & + \frac{9}{24} & \rightarrow & + \frac{18}{48} \\ \hline & & \frac{13}{24} & & \frac{26}{48} \end{array}$$

After the sums in the algorithms have been determined, children should note that the answer cannot be simplified when the denominator is 24 (in this example), but can be when it is 48.

### 3. Using Prime Factorizations

Mature children can determine the LCM for fractional numbers by using the prime factorization of their denominators. Processes for factoring numbers are discussed in Chapter 16. For work with fractional numbers, factoring is applied as follows: Determine the prime factorization of the denominators of the two numbers. For example,

$$\frac{4}{15} = \frac{4}{3 \times 5}$$

$$\frac{5}{12} = \frac{5}{2 \times 2 \times 3}$$

Next, form the union of the sets of factors used in the prime factorization of the denominators: $2 \times 2 \times 3 \times 5$. This becomes the denominator that is used in renaming the fractional numbers.

$$\frac{4}{15} = \frac{4 \times 2 \times 2}{2 \times 2 \times 3 \times 5}$$

$$\frac{5}{12} = \frac{5 \times 5}{2 \times 2 \times 3 \times 5}$$

The number by which each numerator is multiplied is found by noting the factors by which the original denominator is multiplied to yield the new denominator: for $4/15$, the factors $2 \times 2$ are used so the numerator is multiplied by 4; for $5/12$, the factor 5 is used so the numerator is multiplied by 5.

$$\frac{4}{15} \rightarrow \frac{4}{3 \times 5} \rightarrow \frac{4 \times 2 \times 2}{2 \times 2 \times 3 \times 5} \rightarrow \frac{16}{60}$$

$$+\frac{5}{12} \rightarrow +\frac{5}{2 \times 2 \times 3} \rightarrow +\frac{5 \times 5}{2 \times 2 \times 3 \times 5} \rightarrow +\frac{25}{60}$$

$$\frac{41}{60}$$

---

## Self-check of Objective 6

Use any two of these procedures to show how children can learn to rename fractional numbers so they are represented by common fractions having like denominators: multiples, equivalence classes, and prime factorizations.

---

## RENAMING FRACTIONAL NUMBERS IN SIMPLEST FORM

A fractional number is expressed as a common fraction in its simplest form when the numbers represented by the numerator and denominator have no common factor, or are relatively prime. There are times when the original answers to problems involving fractional numbers are not in their simplest form and it is desirable to simplify them. Thus after the answer to a subtraction such as $5/8$ minus $1/8$ has been named as $4/8$, it may need to be renamed as $1/2$.

### 1. Using Fraction Strips and Number Lines

Fraction strips and number lines are particularly useful for developing the concept of simplest terms. (The process of renaming a fraction in simplest terms may be more familiar to you as "reducing a fraction to its lowest terms.") During early work with their learning aids, children observe that $1/2 = 2/4 = 4/8$ and $1/3 = 2/6 = 3/9$. At this time they see a fraction such as $1/2$, which is in its simplest form, renamed as a fraction in higher terms. You now want them to learn that the inverse process is also possible, so that a fraction in higher terms can be simplified.

You can help children recognize the inverse nature of these two processes by having them think "both ways" as they use their learning aids. Thus when they use a set of fraction strips, they should rename $1/2$ as $2/4$ and $2/4$ as $1/2$. The fraction strips in Figure 11-11 illustrate how you can do this. The strips show that although six of the twelfth strips, four of the eighth strips, and other sets of strips are all like one of the half strips in that they are half as long as the unit strip, it is the half strip that requires the fewest subdivisions to represent one-half of the unit. At the same time, children will see that $4/8$ is a simpler expression than $6/12$.

Children will not know the simplest form for all fractional numbers, so they need to learn systematic procedures for simplifying all fractions.

### 2. Examining Numerator and Denominator

One procedure that is commonly used is to examine the numerator and denominator to determine the largest number by which both may be divided. Once a common divisor is found, both numerator and denominator can be divided by it. This procedure is satisfactory as long as the numbers represented by the numerators and denominators are reasonably small and the greatest common divisor can be readily determined, as in $6/12$ or $9/15$. However, it may fail to be useful when the numerators and denominators are larger numbers.

### 3. Using Prime Factorizations

With larger numbers in the numerator and denominator, a better procedure is to find the prime factorization of each one. An example, $24/36$, will be treated in the same way that children should proceed.

Rewrite each term using its prime factorization:

$$\frac{24}{36} = \frac{2 \times 2 \times 2 \times 3}{2 \times 2 \times 3 \times 3}$$

Determine the greatest common factor (GCF) by noting all factors common to both numerator and denominator. (The GCF is made up of the factors in the intersection of the two factorizations. This may be shown with a Venn diagram, to help children visualize it. See Figure 16-10.) Both numerator and denominator of the fraction should be divided by this common factor to rename it in its simplest form:

$$\frac{24}{36} = \frac{24 \div 12}{36 \div 12} = \frac{2}{3}$$

### 4. Simplifying Answers Greater Than 1

Sometimes addition or one of the other operations with fractional numbers results in an answer that is greater than 1. Children should note that these answers can be left as common fractions or expressed as a combination of whole and fractional numbers. It has been common practice to call the common fractions with numerators greater than their denominators improper fractions, and a combination of whole and fractional numbers mixed numbers.

All of the operations with fractional numbers should be introduced with learning aids. Examples will involve situations where answers greater than 1 are found. Then children should learn how to express the answers both as common fractions and combinations of whole and common fraction numerals. The addition sentence $2/3 + 2/3 = \square$ is illustrated with several learning aids in Figure 11-14. In (a), the two addends are represented by circular regions, and the sum is shown as both $4/3$ and $1\,1/3$. The same sentence is illustrated with fraction strips in (b) and a number line in (c). All three examples stress that $1\,1/3$ is a simplification of $4/3$. The series of steps

$$\frac{4}{3} = \frac{3+1}{3} = \frac{3}{3} + \frac{1}{3} = 1 + \frac{1}{3} = 1\frac{1}{3}$$

shows the full process for simplifying this answer. One point you should emphasize is that $1 + 1/3$ and $1\,1/3$ are both names for the same number.

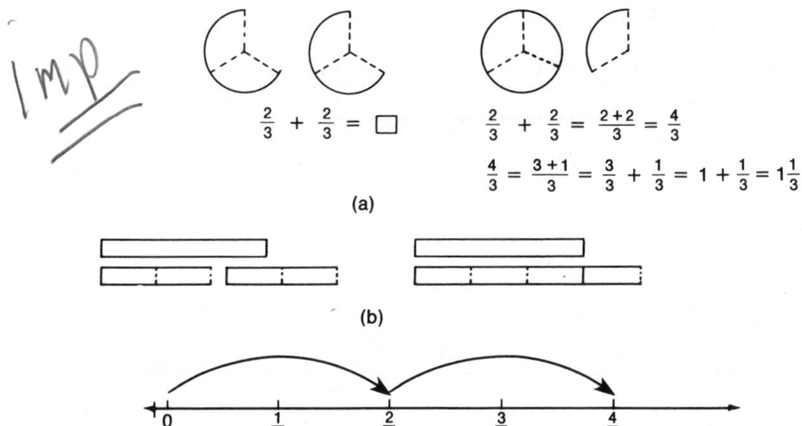


**Figure 11-14** The addition sentence $\frac{2}{3} + \frac{2}{3} = \square$ represented with (a) geometric shapes, (b) fraction strips, and (c) number line

This point is often not stressed, and many children do not understand this important fact. It is also more mathematically precise to refer to numerals such as $1\frac{1}{3}$ as "mixed numerals," rather than as "mixed numbers."

---

## Self-check of Objective 7

Give an oral or written explanation of what is meant by the expression "The common fraction $\frac{2}{3}$ represents the fractional number two-thirds in its simplest form." Demonstrate with at least two different materials or abstract processes some activities children can use to learn to simplify fractional numbers.

---

## RENAMING COMMON FRACTIONS AS DECIMAL FRACTIONS

As is true of all mathematics, some skills and facts dealing with fractional numbers are easier to learn and remember than others. And, of course, there are some children who grasp and remember ideas about these numbers more readily than others. As children work with common and decimal fractions, some will learn quickly that $\frac{1}{2}$ can be renamed as 0.5, and that 0.75 is another name for $\frac{3}{4}$. Such knowledge will come more slowly for others. And there are some children who may need to continue to work with manipulative materials throughout their study of these numbers. These children may never get more than a rudimentary understanding of common and decimal fractions in the elementary school.

## 1. Renaming Common Fractions as Decimal Fractions

$$\begin{array}{r} 0.5 \\ 2\overline{)1.0} \\ \underline{1\,0} \end{array}$$

$$\begin{array}{r} 0.333\ldots \\ 3\overline{)1.000} \\ \underline{9} \\ 10 \\ \underline{9} \\ 10 \\ \underline{9} \end{array}$$

Mature children at the higher levels of the elementary school should be given assistance in learning systematic procedures for renaming fractional numbers in different forms. The decimal fraction equivalent of a common fraction can be found by dividing the numerator of the common fraction by its denominator, as shown in the margin: ½ is found to be the equivalent of 0.5 by dividing 1 by 2; ⅓ to be 0.333 . . . by dividing 1 by 3. When fractional numbers are renamed, the result is a decimal fraction that is either a terminating or repeating decimal. For example, ½ results in a terminating decimal because the division is complete when the quotient, 0.5, has been identified; and ⅓ results in a repeating decimal because no matter to how many decimal places the division is extended, the quotient is always 3 with a remainder of 1. Repeating decimals are also called nonterminating decimals.

After children have studied common fractions and have learned to rename them as decimals, help them find the rule for determining which common fractions result in terminating decimals and which ones result in decimals that repeat. A chart such as the one in Figure 11-15 will aid their search. Have them examine the common fractions on the left to discover a property of their denominators that does not apply to the denominators of the common fractions on the right. You may need to raise such questions

| Terminating Decimals | Repeating Decimals |
|---|---|
| $\frac{1}{2} = 0.5$ | $\frac{1}{3} = 0.333\ldots$ |
| $\frac{1}{4} = 0.25$ | $\frac{2}{3} = 0.666\ldots$ |
| $\frac{3}{4} = 0.75$ | $\frac{1}{6} = 0.1666\ldots$ |
| $\frac{1}{5} = 0.2$ | $\frac{5}{6} = 0.8333\ldots$ |
| $\frac{2}{5} = 0.4$ | $\frac{1}{7} = 0.142857142857\ldots$ |
| $\frac{3}{5} = 0.6$ | $\frac{2}{7} = 0.285714285714\ldots$ |
| $\frac{4}{5} = 0.8$ | $\frac{3}{7} = 0.428571428571\ldots$ |
| $\frac{1}{8} = 0.125$ | $\frac{4}{7} = 0.571428571428\ldots$ |
| $\frac{3}{8} = 0.375$ | $\frac{5}{7} = 0.714285714285\ldots$ |
| $\frac{5}{8} = 0.625$ | $\frac{6}{7} = 0.857142857142\ldots$ |
| $\frac{7}{8} = 0.875$ | $\frac{1}{9} = 0.111\ldots$ |
| $\frac{1}{10} = 0.1$ | $\frac{2}{9} = 0.222\ldots$ |
| $\frac{3}{10} = 0.3$ | $\frac{3}{9} = 0.333\ldots$ |
| $\frac{7}{10} = 0.7$ | $\frac{4}{9} = 0.444\ldots$ |
| $\frac{9}{10} = 0.9$ | $\frac{5}{9} = 0.555\ldots$ |
| $\frac{1}{20} = 0.05$ | $\frac{6}{9} = 0.666\ldots$ |

**Figure 11-15** Classroom chart listing some terminating and repeating decimals

as these: "If you write the prime factorization for the denominator of each fraction on the left, what prime numbers do you get? What prime numbers do you get when you write the prime factorizations of the denominators on the right?" The children will see that when the denominators of fractions giving terminating decimals are factored, the prime numbers are always 2s, 5s, or a combination of 2s and 5s. When denominators of fractions giving repeating decimals are factored, there *may* be 2s and 5s, but there is *always* at least one prime number other than 2 or 5. Once children are familiar with the rule, give them examples such as $5/16$, $7/32$, $8/15$, $4/11$, $3/50$, and $21/40$ and have them tell which kind of decimal fraction they will get if they divide the numerator by the denominator.

## 2. Renaming Decimal Fractions as Common Fractions

Mathematically mature children should also learn to rename decimal fractions as common fractions. Fractional numbers expressed as terminating decimal fractions can be readily renamed as common fractions: 0.5 is equivalent to $5/10$, which in simplest form is $1/2$; 0.75 as $75/100$, simplified to $3/4$; and 0.125 as $125/1000$, simplified to $1/8$.

A fractional number expressed by a repeating decimal cannot be renamed as a common fraction in such a direct manner. A different procedure must be used. Use $n$ to represent the repeating fraction 0.121212 . . . in an equation: $n = 0.121212. \ldots$ Then multiply both terms of the equation by 100. One hundred is used in this case because the pattern of repetition is a block of two digits; if the pattern consists of one digit, multiply by 10; if the pattern consists of three digits, multiply by 1,000, and so on. Subtract the first equation from the second, as shown in the margin. Finally, divide both terms of the new equation by 99 to determine the value of $n$ expressed as a common fraction. The common fraction $12/99$ is equivalent to the repeating decimal 0.121212. . . . When expressed in its simplest form, $12/99$ becomes $4/33$. Children should verify their work by expressing $4/33$ as a repeating decimal by dividing 4 by 33.

$$\begin{array}{rl} 100n = & 12.121212\ldots \\ -\quad n = & 0.121212\ldots \\ \hline 99n = & 12.000000\ldots \end{array}$$

An ellipsis (. . .) is used in each repeating decimal on the chart in Figure 11-15. It is one of several ways to show repeating decimals and is convenient for decimals such as 0.333 . . . and 0.1666 . . . , where only one numeral is repeated. It is less convenient for a decimal such as 0.142857142857. . . , because so many numerals are repeated. For decimals of this type a simpler, alternate procedure may be used. At least two are generally acceptable: $0.\dot{1}\dot{4}\dot{2}\dot{8}\dot{5}\dot{7}$ and $0.\overline{142857}$.

## Self-check of Objective 8

Rename each of these common fractions as a decimal fraction: $8/75$, $3/50$, $4/15$, and $27/200$. What distinguishes the denominators of common fractions that result in terminating decimals from those that result in repeating decimals?

## THE MINICALCULATOR AND COMMON FRACTIONS

Computation with common fractions cannot be done directly with a minicalculator; they must be renamed as decimal fractions first. The renaming of common fractions is done on a minicalculator simply by dividing the numerator by the denominator. Once children learn to do this, they can then perform operations with common fractions on a handheld calculator by renaming each one as a decimal fraction. They can also use the machine and the chart in Figure 11-15 to simplify common fractions and to rename improper fractions as mixed numerals.

### 1. Simplifying Common Fractions

A common fraction is in its simplest form when the numbers represented by the numerator and denominator have no common factor, or are relatively prime. To simplify $6/20$ with a minicalculator, divide 6 by 20. Read the answer—0.3—on the machine's display and then locate it on the chart. The chart shows that 0.3 is equivalent to $3/10$, so $6/20$ is expressed in simplest form as $3/10$.

### 2. Renaming Improper Fractions

To rename $42/8$ as a mixed numeral, divide 42 by 8. Note the answer on the display—5.25. The whole number is already identified—it is 5. Now locate 0.25 on the chart. The chart shows that it is equivalent to $1/4$. Combine the 5 and the $1/4$ to rename $42/8$ as $5\,1/4$.

## Self-check of Objective 9

Demonstrate how to use a minicalculator and the chart in Figure 11-15 to simplify these common fractions—$8/10$, $15/20$, and $36/40$—and to rename these improper fractions as mixed numerals—$69/4$, $33/7$, and $47/8$.

## Common Pitfalls and Trouble Spots

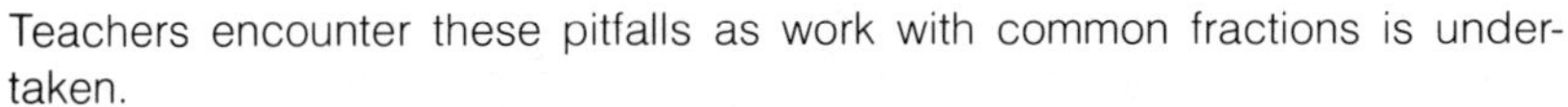

Teachers encounter these pitfalls as work with common fractions is undertaken.

Common fractions are frequently introduced before children are cognitively ready for them. A child who is not yet conserving quantity is not ready to deal with common fractions. Even when suitable manipulative materials are used, the nonconserver will not grasp the meaning of the fractions the models represent. Delay the introduction of common fractions until Piagetian-type tests reveal that a child is able to conserve quantity.

Another pitfall teachers often encounter in the introductory work with common fractions is that they frequently emphasize the "part-of-a-whole" use of common fractions and neglect other uses. Children's first work with common fractions usually deals with the "part-of-a-whole" meaning modeled by geometric regions, such as circles and squares. This is properly so, because this meaning is the simplest one and is the easiest for children to understand. However, unless children have equal opportunities to use sets of objects and markers and pictorial representations, which teach the "part-of-a-set," ratio, and indicated division meanings too, they will have a limited understanding of common fractions. Your awareness of these meanings and how to deal with them in the concrete and semiconcrete modes is the best assurance that children will have a broad understanding of common fractions.

Finding it difficult to rename a pair of common fractions so they have a common denominator is a trouble spot many children encounter. Naturally, these children will have problems when they add and subtract fractional numbers represented by common fractions. The following activity focuses on this problem.

To rename $\frac{1}{3}$ and $\frac{1}{4}$ so they have a common denominator, begin with two equal-sized pieces of paper, as in Figure 11-16(a). Have the children fold one so that it shows thirds and the other so it shows fourths (b). (Premarked "fold lines" will assure that the folds are made accurately.) Instruct the children to shade one part of each paper—that is, one third or one fourth. Next, have the children experiment with their paper models to determine how to fold them so each one has the same number of parts. The models in (c) and (d) show ways the folds can be made. The models in (d) are preferred because each part is congruent with the other. Once the folds have been made, discuss the models to highlight these facts:

1. Each model is folded to show 12 congruent parts. The number of the shaded part of the thirds model increased by a factor of 4; that is, there are four times as many shaded parts. The number of the shaded part of the fourths model increased by a factor of 3; there are three times as many shaded parts.

2. The fraction $\frac{1}{3}$ is equivalent to $\frac{4}{12}$, while the fraction $\frac{1}{4}$ is equivalent to $\frac{3}{12}$.

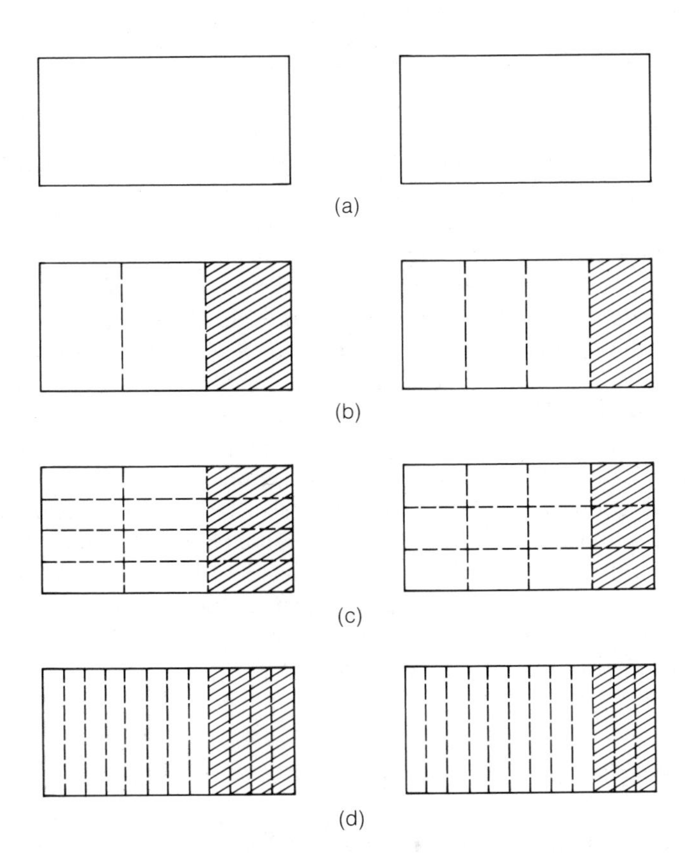


**Figure 11-16** Paper folding can be used to rename fractions with common denominators

## Self-check of Objectives 10 and 11

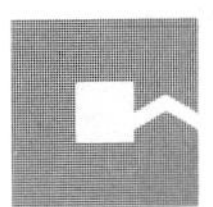

Two common pitfalls associated with instruction about common fractions are discussed. Identify them, and describe orally or in writing a way each one can be avoided.

Demonstrate with pieces of paper or a series of pictures how a paper-folding activity can be used to help children learn to name these fractions with common denominators—⅔, ⅜.

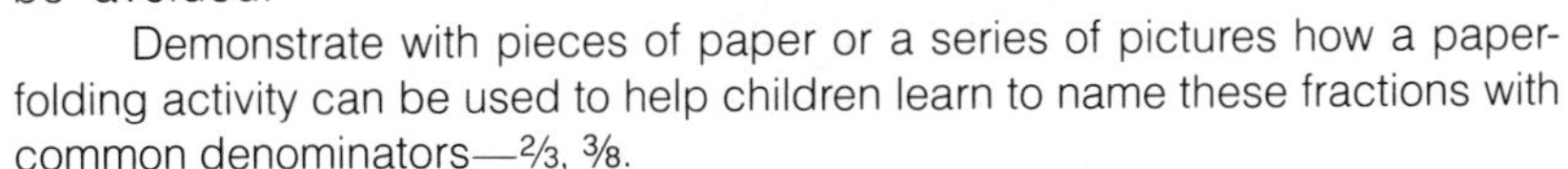

## SUMMARY

Common fractions have several meanings, depending upon the situation in which they are used. Children need experience with concrete and semiconcrete models that represent all of the meanings. Geometric regions, fraction strips, structured materials, markers, and number lines are used during investigative activities designed to make the common fraction representation of fractional numbers meaningful. Children learn that common fractions are used to express parts of a unit or set of objects that has been subdivided into equal-sized parts or subsets, to express ratios, to indicate division, and to express abstract fractional numbers. These same materials, along with certain abstract procedures, help children learn to compare fractional numbers, to rename them as common fractions in their simplest form, and to rename two or more fractional numbers so they are expressed by common fractions having the same denominator. Children with a good understanding of common and decimal fractions can learn to rename common fractions as decimals, and vice versa, and distinguish between terminating and repeating decimal fractions.

The minicalculator has limited use with common fractions because operations cannot be performed with these numerals. A calculator can be used with a chart to simplify common fractions and rename improper fractions as mixed numerals. Two pitfalls associated with instruction about common fractions are (1) introduction of common fractions before children are cognitively ready for them and (2) neglect of some of the uses of common fractions. The renaming of a pair of fractions so they have a common denominator is a trouble spot for many children. A paper-folding activity will help children learn to do this properly.

## STUDY QUESTIONS AND ACTIVITIES

1. Design and prepare a set of teaching aids for a flannel or magnetic board to help children learn about halves, thirds, fourths, sixths, and eighths of a unit. Make another set for helping children learn about halves, thirds, fourths, sixths, and eighths of a set of objects. Explain briefly how children may use your materials. Prepare problem cards if you think they might be useful with one or both sets of materials.

2. Describe three different ways children can compare these common fractions: $\frac{2}{3}$, $\frac{3}{6}$, and $\frac{5}{8}$. Include illustrations of any manipulative materials you would have children use. Order your descriptions from most concrete to least concrete.

3. Several of the articles summarized in this chapter's reading list describe activities with games and manipulative materials that contribute to understanding common fractions. Add some of them to your collection of teaching-learning aids.

4. Read Usiskin's article (see the reading list) and summarize his argument that the need for common fractions will not diminish in the future.

5. Write the common fraction equivalent for each of these decimal fractions:
(a) 0.25 (b) 0.323232 . . . (c) 0.143143143 . . . (d) 0.32

## FOR FURTHER READING

Brown, Christopher N. "Fractions on Grid Paper," *The Arithmetic Teacher,* XXVII, No. 5 (January 1979), 8–10. Describes and illustrates grid paper activities to develop fundamental concepts of common fractions and visual recognition of parts and wholes, and to give students experiences with equivalent fractions.

Bruni, James V., and Helene J. Silverman. "An Introduction to Fractions," *The Arithmetic Teacher,* XXII, No. 7 (November 1975), 538–545. Fractions are introduced through a series of games and related tables. Geometric regions cut from oaktag, outlines of regions, and spinners serve as the games' playing pieces.

———. "Using Rectangles and Squares to Develop Fraction Concepts," *The Arithmetic Teacher,* XXIV, No. 2 (February 1977), 96–106. Explains and illustrates a variety of activities, including paper folding and cutting, coloring regions, and a fraction game.

Kennedy, Leonard M. *Models for Mathematics in the Elementary School.* Belmont, Calif.: Wadsworth Publishing Company, 1967. Pages 99–130 are devoted to a discussion of learning aids for study of the common fraction representation of fractional numbers.

Molenoski, Marie. "Fracto," *The Arithmetic Teacher,* XXI, No. 4 (April 1974), 321–322. Fracto is a game that reinforces children's knowledge of the equivalence of certain common fractions, decimal fractions, and percents. The game has a teacher-made 52-card deck and resembles rummy.

Moulton, J. Paul. "A Working Model for Rational Numbers," *The Arithmetic Teacher,* XXII, No. 4 (April 1975), 328–332. The model is a balance beam scaled to show common fractions with denominators of 1, 2, 3, . . . , 20 and large paper clips. It shows equivalence relations, and addition and subtraction with common fractions.

Souviney, Randall J. "Seeing Through Fractions," *Learning,* V, No. 7 (March 1977), 66–67. Regions cut from colored transparent acetate serve first as models for developing the meaning of common fractions. Later, they are used to give meaning to addition, subtraction, and multiplication with common fractions.

Usiskin, Zalman P. "The Future of Fractions," *The Arithmetic Teacher,* XXVII, No. 5 (January 1979), 18–20. Usiskin refutes the argument that common fractions will soon become obsolete by exploring uses of fractions that will persist regardless of the import of either the minicalculator or metric system.

# 12 Operations with Fractional Numbers Represented by Common Fractions

Upon completion of Chapter 12, you will be able to:

1. Demonstrate with unit regions, fraction strips, structured materials, and number line activities involving addition of fractional numbers expressed as common fractions where denominators are the same, where denominators are different, and where there are mixed numerals.

2. Demonstrate ways to show that addition of fractional numbers is both commutative and associative, that there is an identity element for addition of these numbers, and that closure applies to addition in the set of fractional numbers.

3. Demonstrate with appropriate materials activities involving subtraction of fractional numbers expressed as common fractions where denominators are the same, where denominators are different, and where there are mixed numerals.

4. Illustrate situations where the repeated addition and array interpretations of multiplication apply to fractional numbers, and tell why the Cartesian product interpretation is not useful for multiplication with fractional numbers.

5. Use manipulative and other materials to represent the meanings of various types of multiplication sentences involving fractional numbers.

6. Explain how to show that commutativity and associativity apply to multiplication of fractional numbers, and give a definition of the identity element and reciprocal.

7. Give examples of real-life situations illustrating partitive and measurement division situations involving fractional numbers, and illustrate each situation with learning aids.

8. Explain the steps for completing division of fractional numbers by both the common denominator and the invert-and-multiply processes.

9. Explain orally or in writing how children should determine whether or not an answer should be simplified (reduced to lowest terms) in situations involving any of the four operations with fractional numbers.

10. Explain what cancellation means, and describe how children should learn to do it.

11. Demonstrate with appropriate materials some procedures that can be used to help children understand ratios and use them in real-life situations.

12. Describe a major pitfall associated with teaching children to compute with fractional numbers represented by common fractions, and explain how it can be avoided.

13. Identify common errors children commit as they compute with common fractions.

Key Terms you will encounter in Chapter 12:

like denominators
unlike denominators
least common multiple (LCM)
lowest common denominator (LCD)
mixed numeral
commutative property for addition
identity element for addition
associative property for addition
closure property for addition
number plane
commutative property for multiplication
associative property for multiplication
identity element for multiplication
reciprocal
common denominator process
invert-and-multiply process
cancellation
ratio
rate pair
proportion

By the time children begin to perform the operations of addition, subtraction, multiplication, and division with fractional numbers, they have already developed some skill in performing these operations with whole numbers. Armed with this knowledge, teachers frequently introduce these operations with fractional numbers by using a series of rules: "If two fractions have the same denominator, their sum can be determined by adding the numerators and placing the sum over the denominator." "If two fractions have different denominators, they must be changed to their least common denominator before they can be added." Rules for renaming fractional numbers with a least common denominator are then listed so children can proceed with the addition. However, teaching these operations by rules alone means that many children will have little or no understanding of what they are doing or why they are doing it.

Procedures that help children visualize operations with fractional numbers are extensions of those used with whole numbers. As children learn to add whole numbers, they learn the meaning of the operation in terms of joining sets. Objects, markers, number lines, and other devices are used so children can discover the meaning of the operation. Children should learn to add fractional numbers in much the same way. Again physical models help extend children's understanding of addition and its properties to include fractional numbers.

## ADDITION WITH COMMON FRACTIONS

Children's initial experiences with addition should be with familiar situations that are easily represented by physical models.

### 1. Adding When Denominators Are the Same

The setting in which children work should be well stocked with manipulative materials: geometric regions left whole and cut into fractional parts, sets of fraction strips, Cuisenaire rods, and fractional number lines. Present a problem, for example: "Last night Jim practiced his piano lesson for ¼ hour before dinner and ¼ hour after dinner. What part of an hour did he practice altogether?" Let each child determine the answer in his or her own way. Many will give the answer immediately, while others may need to use one of the learning aids. After the answer has been given, someone who used each aid should discuss its use. For example, "I used fourths of the circle to represent each of the two fourth hours Jim practiced (Figure 12-1). When I put the two fourths together, I saw that one-fourth and one-fourth are the same as one-half."

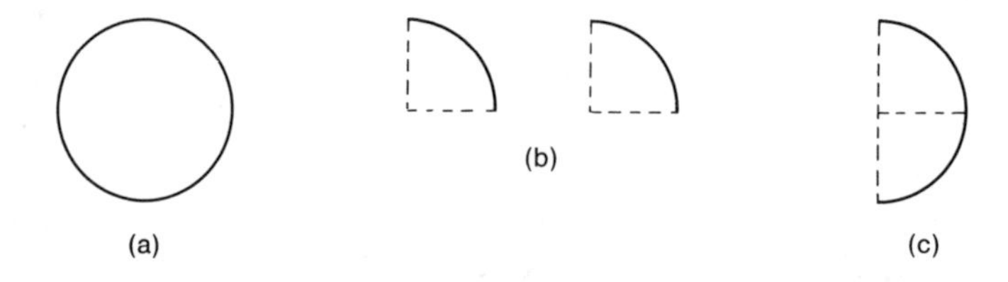


**Figure 12-1** Geometric shapes used to represent the addition sentence ¼ + ¼ = 2/4; (a) shows the unit region; in (b) the two shapes represent the addends; and the two shapes are joined to represent the sum in (c)

"I used two pieces that show fourths on our fraction strips (Figure 12-2). They showed me that two fourths are the same as one-half."

**Figure 12-2** Fraction strips used to represent the addition sentence ¼ + ¼ = 2/4

"I used Cuisenaire rods (Figure 12-3). I know that a red rod is one-fourth when the brown rod is one, so I know that two red rods are one-half."

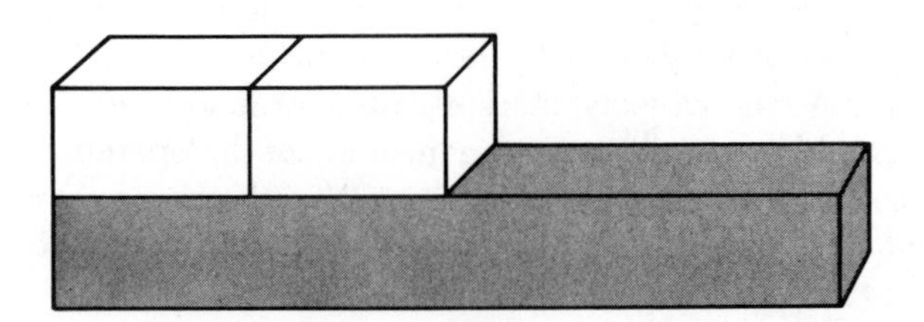

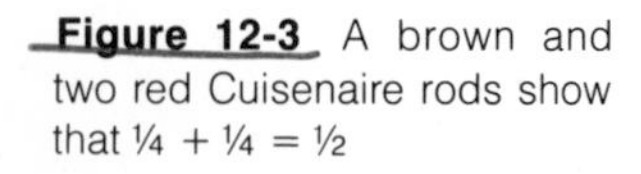

**Figure 12-3** A brown and two red Cuisenaire rods show that ¼ + ¼ = ½

"I used our number line (Figure 12-4). When I took jumps that were each $\frac{1}{4}$-unit long, I stopped at $\frac{2}{4}$. That's the same as $\frac{1}{2}$."

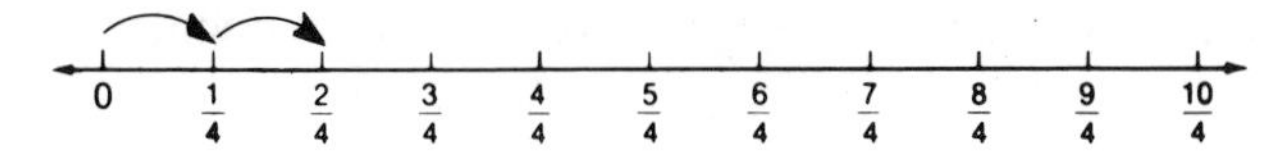


**Figure 12-4** Jumps on a number line showing the addition sentence $\frac{1}{4} + \frac{1}{4} = \frac{2}{4}$

(In each illustration, the unit is shown along with the fractional parts so children can see the relationships between them.)

Some children will say, "I just know that $\frac{1}{4}$ and $\frac{1}{4}$ is $\frac{1}{2}$." Few, if any, will have used the mathematical sentence $\frac{1}{4} + \frac{1}{4} = \frac{2}{4}$ (or, $\frac{1}{4} + \frac{1}{4} = \frac{1}{2}$) to express what they did to determine the answer. However, after children have determined the answers for several problem situations, display the addition sentence for each one and call attention to them: "What have we done in each of these sentences to determine the answer?"

$$\frac{1}{4} + \frac{1}{4} = \frac{2}{4}$$

$$\frac{1}{3} + \frac{1}{3} = \frac{2}{3}$$

$$\frac{1}{2} + \frac{1}{2} = \frac{2}{2}$$

$$\frac{1}{4} + \frac{2}{4} = \frac{3}{4}$$

Guide the children to see that the sum is found by adding the numerators of the fractional numbers. It will be helpful to children to see the sentences rewritten as

$$\frac{1+1}{4} = \frac{2}{4}, \quad \frac{1+1}{3} = \frac{2}{3}, \quad \frac{1+1}{2} = \frac{2}{2}, \quad \text{and} \quad \frac{1+2}{4} = \frac{3}{4}$$

This emphasizes the fact that the numerators are added. If children have difficulty understanding the process, the examples can be further simplified:

| 1 fourth | 1 third | 1 half | 1 fourth |
|---|---|---|---|
| +1 fourth | +1 third | +1 half | +2 fourths |

Problems with answers greater than 1 can also be examined by children during laboratory activities. "Josie is making a flag and will need $\frac{2}{3}$ of a yard of red material and $\frac{2}{3}$ of a yard of white material for it. How much material will she use for the flag?" Learning aids such as those in Figure 11-14 can make the meaning clear to children.

## 2. Adding When Denominators Are Different

The hierarchy of skills and understandings required as background for addition of fractional numbers expressed by common fractions with unlike denominators includes:

1. Understanding the meaning of addition.
2. Mastery of the basic addition facts.
3. Understanding of common fractions and physical models that represent them.
4. Understanding of addition using fractions with like denominators.
5. Understanding that by using equivalent fractions, fractional numbers can be renamed so they have common denominators.

Children with these understandings and skills are ready for an addition situation such as the following: "While making some clothes for her sister, Suzy used ½ yard of material for a skirt and ¼ yard for a blouse. What part of a yard of material did she use for the complete outfit?" Some children will probably "just know" that the answer is ¾ of a yard. Other children will probably be uncertain. As children explore ways to find the answer, they will note that the addition process is different from adding two fractional numbers with the same denominator. Ask the children to express the sentence for this situation: ½ + ¼ = □, and to examine it to find a procedure for finding the answer. Children who "just know" can probably rationalize the process: they thought of ½ and ²⁄₄ and then added ²⁄₄ and ¼ to get ¾. This process should be verified with fraction strips, as shown in Figure 12-5, and with number lines. In (a), the sentence is

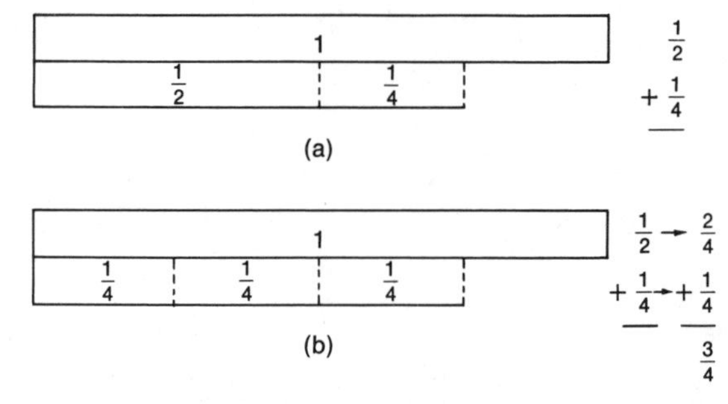


**Figure 12-5** Fraction strips used to show the meaning of (a) ½ + ¼ = □ and (b) changing ½ + ¼ to ²⁄₄ + ¼

written in the vertical form alongside fraction strips that represent the two fractional numbers. Children will see that the two pieces of fraction strips in (a) are equivalent to ¾ of the unit strip. They should then note that in order to express the answer represented by fraction strips in simplest form, it is necessary to exchange the fraction strip that represents ½ for two pieces each representing ¼. The algorithm is rewritten to express the ½ as ²⁄₄ (b). The answer can now be determined: ²⁄₄ and ¼ are added.

The addition sentence $\frac{1}{2} + \frac{1}{3} = \square$ is illustrated with number lines in Figure 12-6. Children will note that it is necessary to rename both fractional numbers so they have common denominators before the addition can take place. To do this, they should study the number lines to locate the line with points that correspond to points on the lines for both thirds and halves. The number lines indicate that a number line showing sixths has points that correspond to points on both halves and thirds lines. The addition for $\frac{1}{2} + \frac{1}{3} = \square$ is shown when $\frac{1}{2}$ is expressed as $\frac{3}{6}$ and $\frac{1}{3}$ as $\frac{2}{6}$.

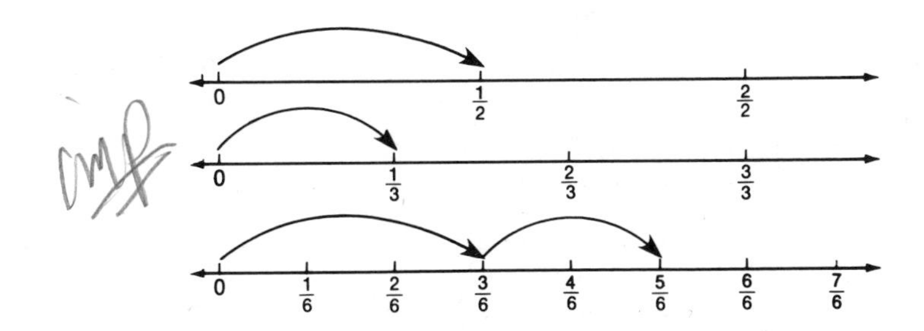


**Figure 12-6** Number lines show that $\frac{1}{2} + \frac{1}{3} = \square$ is renamed as $\frac{3}{6} + \frac{2}{6} = \square$ to determine the sum. $\frac{5}{6}$

Ultimately, children who understand least common multiples, or lowest common denominators, and processes for renaming fractions as equivalent fractions, will use mature processes for adding fractional numbers. Processes for finding common denominators are discussed on pages 301 to 303. As children learn these processes, they should also learn to use them when adding fractions with unlike denominators. By the time they complete elementary school, many children will handle examples such as

$$\begin{array}{r} \frac{3}{4} \\ +\frac{5}{6} \\ \hline \end{array} \qquad \begin{array}{r} \frac{3}{8} \\ +\frac{5}{12} \\ \hline \end{array} \qquad \begin{array}{r} \frac{7}{12} \\ +\frac{11}{15} \\ \hline \end{array}$$

with little difficulty.

## 3. Adding with Mixed Numerals

Historically, the expression "adding mixed numbers" has described addition situations in which both whole and fractional numbers are involved. An example of this addition is

$$\begin{array}{r} 1\frac{1}{2} \\ +1\frac{1}{4} \\ \hline \end{array}$$

In order to maintain the proper distinction between "number" and "numeral," the expression "addition with mixed numerals" is a better description of addition of this kind.

Children can represent the addition 1½ + 1¼ = □ with geometric regions, as in Figure 12-7. In (a), the two addends are illustrated. In (b), the shape showing ½ of a unit is exchanged for two pieces showing fourths, and the algorithm is changed. Finally, the regions are joined and the addition is completed (c).

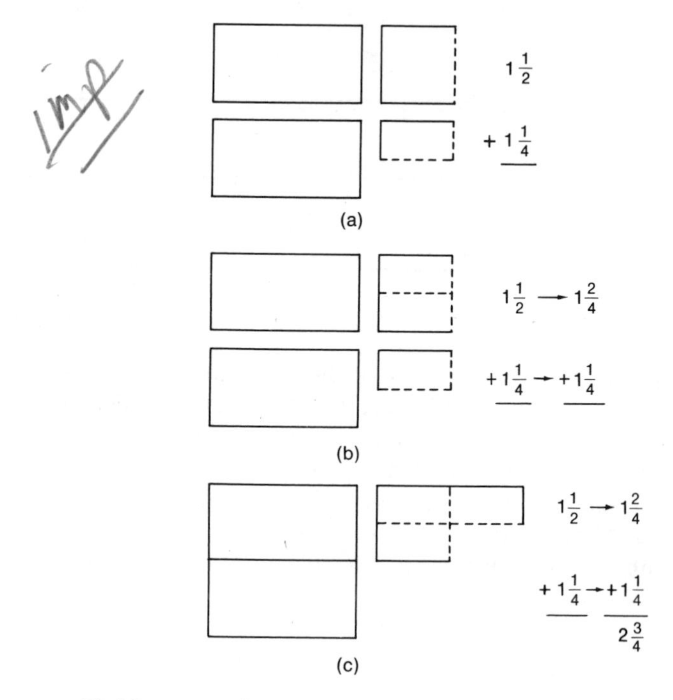


**Figure 12-7** Rectangular regions used to represent the addition sentence 1½ + 1¼ = 2¾

Children can also use a sentence form of the algorithm and number lines for addition of this type. The sentence 1⅓ + 1⅔ = □ is illustrated in Figure 12-8. The following steps represent the addition: the sentence is rewritten as (1 + ⅓) + (1 + ⅔) = □; by applying the commutative and associative properties of addition, the sentence is rewritten as (1 + 1) + (⅓ + ⅔) = □; the sentence can now easily be represented on the number line and the addition completed.

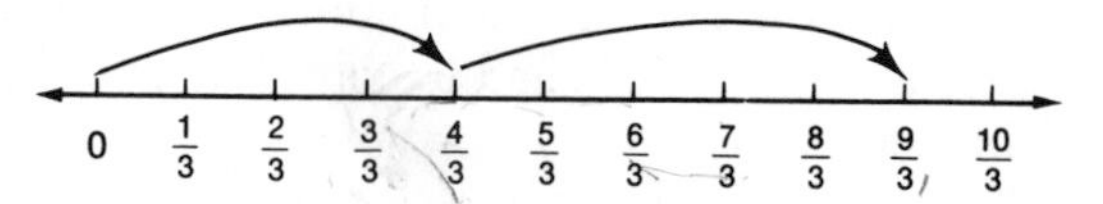


**Figure 12-8** The addition sentence 1⅓ + 1⅔ = □ represented on the number line

A third way to add with mixed numerals is to express them as improper fractions. The addition involves only the numerators. The process for 1⅙ + 1⅓ = □ is shown with fraction strips in Figure 12-9.

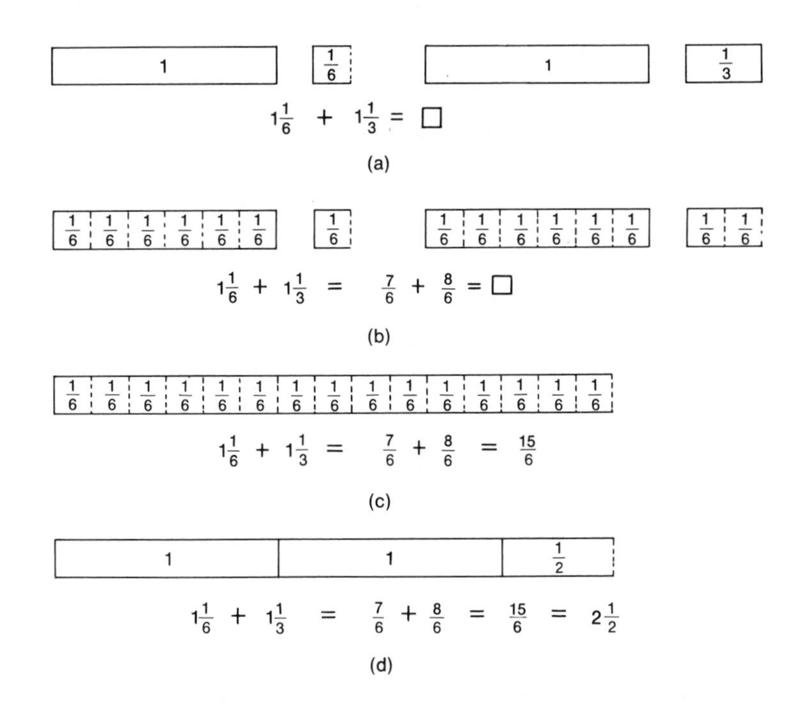


**Figure 12-9** The addition sentence $1\frac{1}{6} + 1\frac{1}{3} = \square$ represented with fraction strips

## Self-check of Objective 1

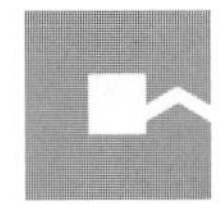

Use at least two different types of learning aids to represent each of these addition sentences: $2/5 + 1/5 = 3/5$, $1/3 + 1/6 = 3/6$, $1/4 + 1/6 = 5/12$, and $1\,1/4 + 1\,1/3 = 2\,7/12$. (Use a series of illustrations for each example if learning aids are not available.)

## THE PROPERTIES OF ADDITION APPLIED TO FRACTIONAL NUMBERS

Children should extend their understanding of the addition properties to include the properties' uses with fractional numbers.

### 1. Commutative Property

As children learn to add fractional numbers, you can help them see how the commutative property applies by asking questions that encourage thinking about the property: "We have seen how certain rules, or properties, sometimes simplify our work by letting us use shortcuts. One of these properties is the commutative property for addition. Who will give me an example of this property using whole numbers?" "Here is a number line (Figure 12-10). I want someone to use it to show that addition with fractional numbers is also commutative. Use the line to show that $\frac{1}{6} + \frac{5}{6} = \frac{5}{6} + \frac{1}{6}$."

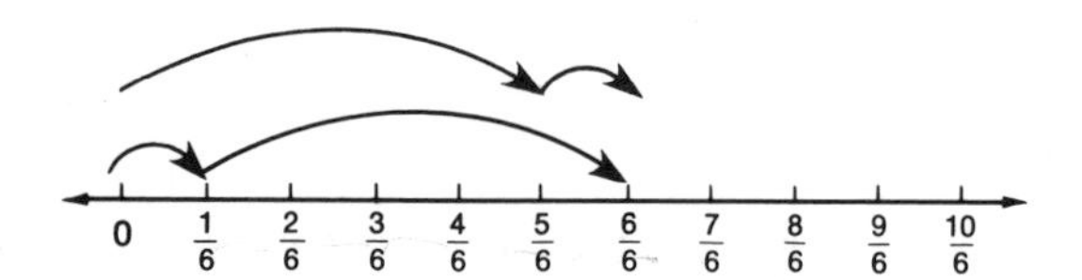


**Figure 12-10** The commutative property demonstrated on a number line for the sentence $\frac{1}{6} + \frac{5}{6} = \frac{5}{6} + \frac{1}{6}$

## 2. Identity Element

The identity element for addition of whole numbers is zero. Before children extend this property to fractional numbers, they should understand that when zero is a numerator, the number represented by the common fraction is equal to zero.

Since a fractional number can represent division, one way to help children understand the use of zero as a numerator is to relate fractional numbers to division. Sentences such as $0 \div 2 = \square$ and $0 \div 16 = \square$ should be used as examples to help children see that $\frac{0}{2}$ and $\frac{0}{16}$ can be used as other names for zero.

To develop an intuitive understanding of the use of zero as numerator, children should note sequences of fractional numbers represented on number lines: $\frac{1}{2}, \frac{2}{2}, \frac{3}{2}, \frac{4}{2}, \ldots; \frac{1}{12}, \frac{2}{12}, \frac{3}{12}, \frac{4}{12}, \ldots$. You should ask questions as children study their number line chart. "When we count the halves along the line on the chart, what is the number of halves that are counted at this point?" (The 0 point is indicated.) "What is another name that we could use to identify this point?" (The name, zero-halves, should be given and its numeral, $\frac{0}{2}$, written on the chalkboard.) "What name would we use for the same point if we count along our twelfths line?"

Numbers with zero as a numerator represent the identity element for addition of fractional numbers. This can be demonstrated easily with the number lines.

## 3. Associative Property

An activity such as the following can be used to extend understanding of the associative property to fractional numbers. "I want one of you to go to the magnetic board and use the fraction strips to show how we can use the associative property of addition with fractional numbers. I've put strips there to show the sentence $\frac{1}{2} + \frac{1}{3} + \frac{1}{4} = \square$ [Figure 12-11(a)]. Look at

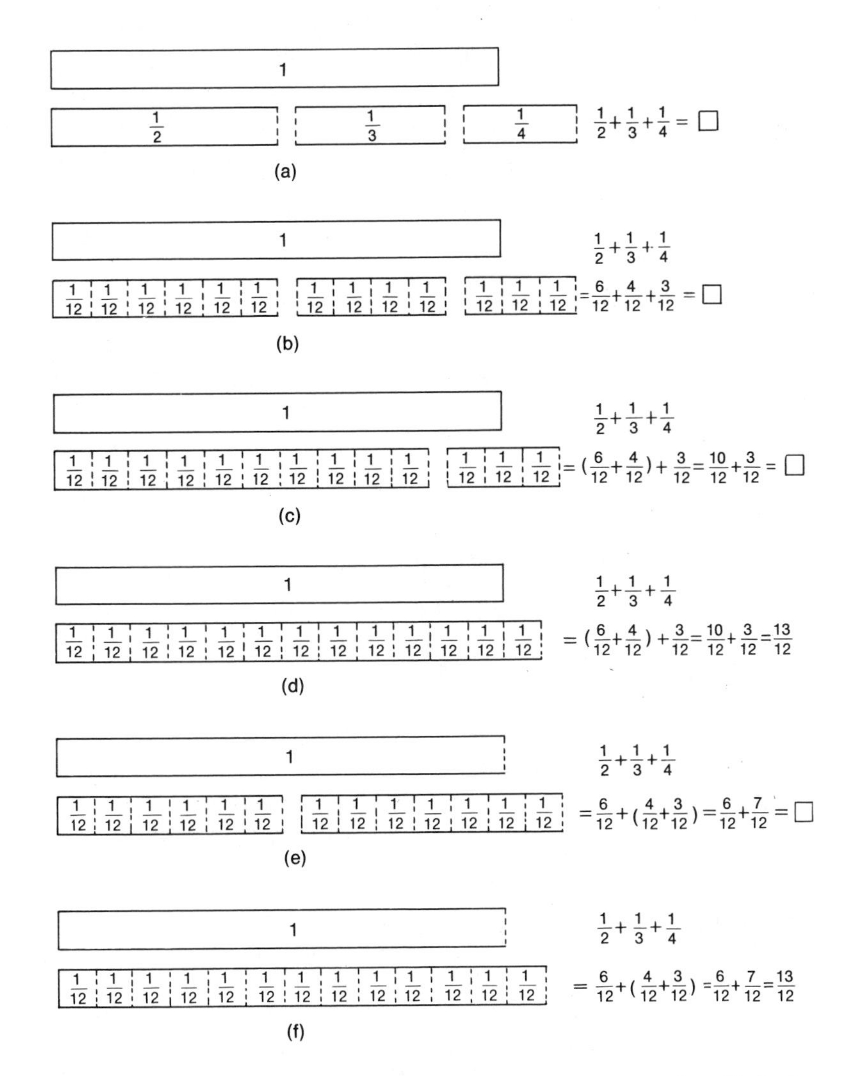


**Figure 12-11** The associative property demonstrated with fraction strips for the sentence $(\frac{1}{2} + \frac{1}{3}) + \frac{1}{4} = \frac{1}{2} + (\frac{1}{3} + \frac{1}{4})$

the strips and the addition sentence and tell me what we should do first. Yes, we should use our strips to help us rename our fractions so they have the same denominators." The children change the fraction strips and rewrite the sentence, as shown in (b). The strips that represent $\frac{6}{12}$ and $\frac{4}{12}$ are joined and the numbers added (c). Finally, the remaining strips are joined and the addition is completed (d). In (e) the strips for $\frac{4}{12}$ and $\frac{3}{12}$ are joined first, followed by the joining of strips showing $\frac{6}{12}$ (f).

## 4. Closure Property

During discussions of these properties, you can bring in the idea that closure extends to addition of fractional numbers. An addition table for fractional numbers is not useful for this, however, because there are an infinite number of addends that can be put on it. However, children who are capable of understanding the three properties already discussed will accept the fact that for every pair of fractional number addends there is a fractional number that is their sum.

## Self-check of Objective 2

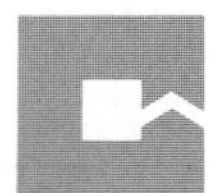

Use manipulative materials or any other suitable materials to show how children's understanding of commutativity, associativity, and the identity element for addition applies in the set of positive fractional numbers. (Use drawings if suitable aids are not available.)

### SUBTRACTION WITH COMMON FRACTIONS

Subtraction of fractional numbers arises from the same types of situations as subtraction of whole numbers. Learning aids can illustrate these different types of situations.

1. Subtraction is used to determine the fractional part that remains after a part is removed. "If there is ¾ of a pie in a pan and Billy eats a piece that is ¼ of the original pie, how much pie is left?" This situation is represented with circular regions in Figure 12-12.

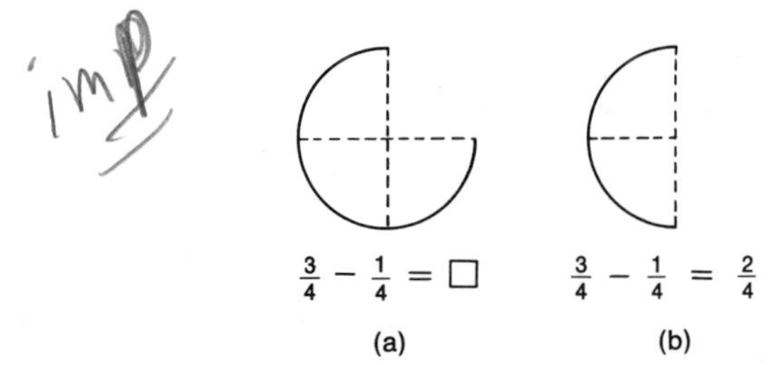


**Figure 12-12** Circular regions used to represent the subtraction sentence $3/4 - 1/4 = 2/4$: (a) shows ¾ of the unit; (b) shows $2/4$

2. Subtraction is used to determine the difference in size of two fractional numbers. "It is ¾ of a mile from Ted's house to school and ⅞ of a mile from Larry's house to school. How much farther is school from Larry's house than from Ted's?" A number line illustrates this subtraction in Figure 12-13.

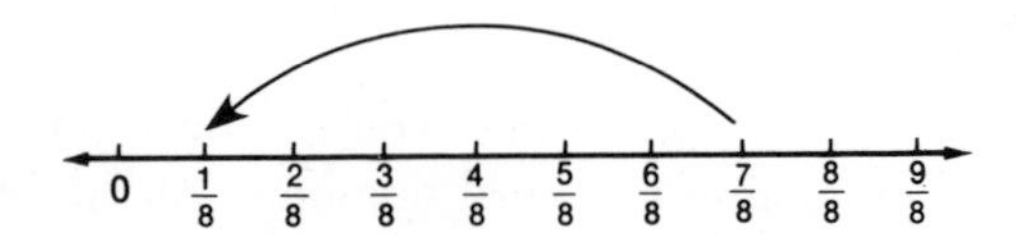


**Figure 12-13** Number line used to represent the subtraction sentence $7/8 - 3/4 = 1/8$

3. Subtraction is used to determine how large a fractional number is to be added to a second one of known size to yield a third of known size. "Judy is working on a new track layout for her model railroad. She will have a total of 8½ feet of track in the new layout. If she has put down 3½ feet of the new track, how many more feet of track will she put down?" The meaning of this situation is best illustrated with a number line (Figure 12-14).

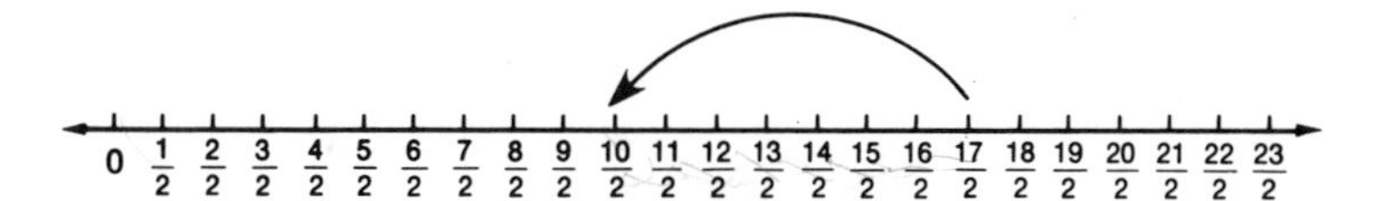


**Figure 12-14** Number line used to represent the subtraction sentence $8\frac{1}{2} - 3\frac{1}{2} = 5$

4. Subtraction is used to determine the missing addend when a sum and one addend are known. "Jack has a mixture of two chemicals that weighs $\frac{13}{16}$ of an ounce. He has $\frac{7}{16}$ of an ounce of sulfur in the mixture. How much of the mixture is made up of the other chemical?" Fraction strips are used to illustrate this situation in Figure 12-15.

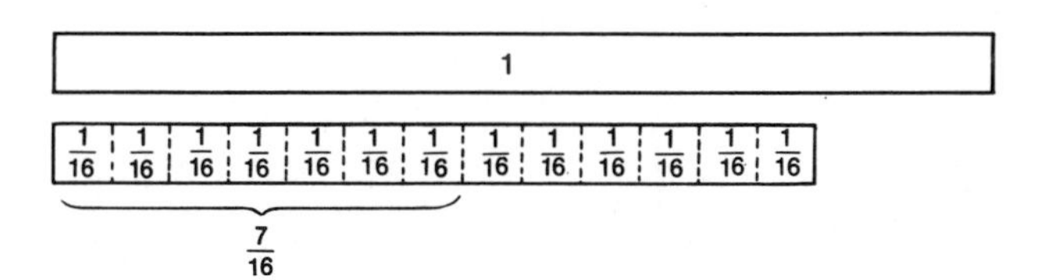


**Figure 12-15** Fraction strips used to illustrate the subtraction sentence $\frac{13}{16} - \frac{7}{16} = \frac{6}{16}$

Children already know that subtraction and addition are inverse operations, so they learn to add and subtract fractional numbers simultaneously. You should use the same care to introduce subtraction through situations that are relevant to them. Children's first subtraction should be with fractions having like denominators; examples 1 and 4 illustrate suitable situations and appropriate learning aids. Subtraction with fractions having unlike denominators, as in example 2, should be introduced along with addition with unlike denominators. In this way the need for renaming fractions so as to result in common denominators will be clear. The renaming process is the same for either operation.

Subtraction with mixed numerals is troublesome for some children, especially when the common fraction part of the minuend is smaller than the common fraction part of the subtrahend. A child may rename the subtrahend as in the margin. Here the child has used the same procedure for renaming the mixed numeral as he or she used earlier for regrouping when subtracting whole numbers, renaming it as $\frac{11}{4}$, rather than $\frac{5}{4}$. To avoid this type of error, begin instruction with this kind of subtraction with situations such as the one in example 3. This example is quite simple and is easily illustrated.

| | |
|---|---|
| $6\frac{1}{4}$ | $5\frac{11}{4}$ |
| $-3\frac{3}{4}$ | $-3\frac{3}{4}$ |

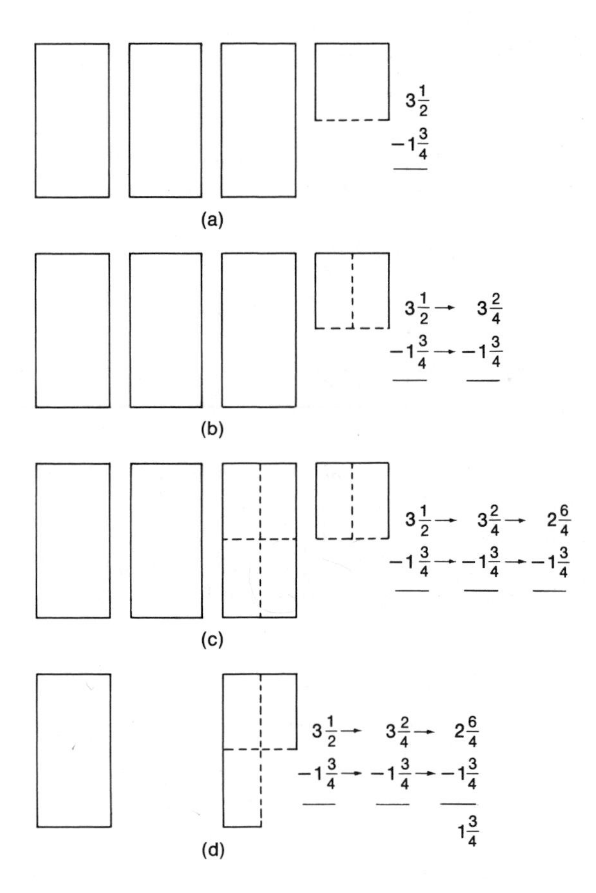


**Figure 12-16** Rectangular regions used to illustrate the subtraction sentence 3½ – 1¾ = 1¾

Once the students are familiar with the algorithm, introduce an example such as 3½ – 1¾ = □. Use a story situation to illustrate it. "Billy was in charge of the cake sale at his school's bazaar. At 8:00 he counted 3½ cakes on the table. By 8:30 he had sold 1¾ more cakes. How many cakes were left at 8:30?" Begin with 3½ regions, as in Figure 12-16(a). Before 1¾ of the model can be removed, it is necessary to rename the ½ as fourths (b), and exchange one unit for 4/4 (c). There are now two units and six fourths. Now remove 1¾, leaving 1¾ (d). Change the algorithm to indicate the actions each time changes are made in the model. Repeat the process with other story situations and models until it is clear to children and they can perform the work without reference to models.

## Self-check of Objective 3

Use suitable learning aids to represent each of these subtraction sentences: 5/8 – 2/8 = 3/8, 7/8 – 3/4 = 1/8, 3 5/6 – 2 1/6 = 1 4/6, and 2¼ – 1½ = ¾.

## INTERPRETATIONS OF MULTIPLICATION WITH COMMON FRACTIONS

The operations of multiplication and division of fractional numbers are deceptively easy for teachers to teach and children to perform, but their meanings are elusive. Children can be taught rules for performing these two operations and, as long as they remember the rules, they can multiply or divide pairs of fractional numbers with ease. However, if children learn to perform these operations using only rules, they probably will understand very little of the meaning behind them. If you adopt the philosophy of this book, you will use learning aids, and not simply the rules, as you explore with children the meanings of the operations.

Before discussing procedures, the meaning of multiplication with fractional numbers will be considered. Chapter 9 presents three interpretations of multiplication with whole numbers. These explanations fail in part or totally as explanations of the meaning of multiplication with fractional numbers.

### 1. Using Repeated Addition

The repeated addition interpretation is suitable only when a fractional number is multiplied by a whole number. For example, $5 \times \frac{1}{2} = \square$ and $6 \times \frac{2}{3} = \square$ may be interpreted as $\frac{1}{2} + \frac{1}{2} + \frac{1}{2} + \frac{1}{2} + \frac{1}{2} = \square$ and $\frac{2}{3} + \frac{2}{3} + \frac{2}{3} + \frac{2}{3} + \frac{2}{3} + \frac{2}{3} = \square$, respectively. This is consistent with the repeated addition interpretation of whole-number multiplication, because a fractional number is used as an addend a given number of times. Repeated addition cannot be used to interpret multiplication such as $\frac{1}{2} \times 5 = \square$ and $\frac{1}{3} \times \frac{5}{6} = \square$, however, because it is not reasonable to think of using 5 as an addend $\frac{1}{2}$ of a time or $\frac{5}{6}$ as an addend $\frac{1}{3}$ of a time.

### 2. Using Arrays

The array concept of multiplication with whole numbers can be extended to examples such as $\frac{1}{4} \times 12 = \square$ and $\frac{2}{3} \times \frac{3}{4} = \square$. The array in Figure 12-17 illustrates these sentences. In (a), the sentence $\frac{1}{4} \times 12 = \square$ is interpreted. The twelve disks are arranged in four columns because the 4 in the denominator indicates that the array is to be separated into four parts. Three of the disks, or one-fourth of the total, are encircled. (Note that this situation is exactly like $12 \div 4 = 3$, where a set of twelve disks is separated into four equal-sized groups, each containing three disks.)

The array in (b) interprets $\frac{2}{3} \times \frac{3}{4} = \square$. The number of disks used for the array is determined by multiplying 3 times 4, the numbers represented by the denominators. The 3 and 4 in the denominators also suggest that the disks be arranged in a 3 by 4 array. Three-fourths of the array is shaded and bracketed to show how much of the array is being considered. Two-thirds of the array is then indicated by bracketing and shading two of

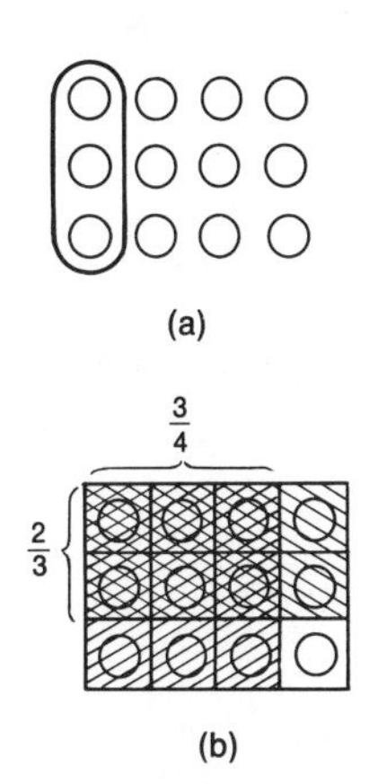


**Figure 12-17** An array used to represent the multiplication (a) $\frac{1}{4} \times 12 = 3$ and (b) $\frac{2}{3} \times \frac{3}{4} = \frac{6}{12}$

the three rows. The product of $\frac{2}{3}$ times $\frac{3}{4}$, which is $\frac{6}{12}$, is represented by the six disks that are both bracketed and doubly shaded. When $\frac{6}{12}$ is simplified, it becomes $\frac{1}{2}$, and indicates that the six disks are half of the twelve disks.

When you work with examples such as these, proceed slowly and use many different arrays to clarify why multiplication is used in what appear to be division situations. Children must realize that they are finding a fractional part of a group, unit, or another fractional part, rather than the total number of objects in an array, as is the case when arrays are used with multiplication of whole numbers.

### 3. Using Cartesian Products

The Cartesian product interpretation of multiplication of whole numbers is not useful for interpreting multiplication of fractional numbers.

## LEARNING TO MULTIPLY WITH COMMON FRACTIONS

The following procedures can be used to help children learn to multiply fractional numbers using common fractions.

### 1. Multiplication of Fractional Numbers by Whole Numbers

Multiplication of fractional numbers by whole numbers is usually used for beginning work because the examples are easily related to the repeated addition interpretation of multiplication. When a group of children is ready, use a realistic situation. "Sarah practices the piano $\frac{3}{4}$ of an hour each day. What is the total amount of time she practices in a week?"

Children can interpret this repeated addition situation with geometric regions, fraction strips, number lines, or Cuisenaire rods. Circular regions are illustrated in Figure 12-18, and a number line is shown in Figure 12-19. In Figure 12-20, Cuisenaire rods, each representing three-fourths of a unit, are placed end to end (a). Then they are interpreted in terms of units to show the answer as $5\frac{1}{4}$ (b). Fraction strips are used in the same way.

### 2. Multiplication of a Whole Number by a Fractional Number

After children are able to multiply examples such as $7 \times \frac{3}{4} = \square$, you can use the commutative property for multiplication to show that the product of $\frac{3}{4} \times 7$ will be the same. However, to use the commutative property at this point will eliminate the children's opportunity to develop an understanding of the kinds of situations from which such multiplication arises. A sentence such as $\frac{3}{4} \times 7 = \square$ is most frequently interpreted as finding $\frac{3}{4}$ of

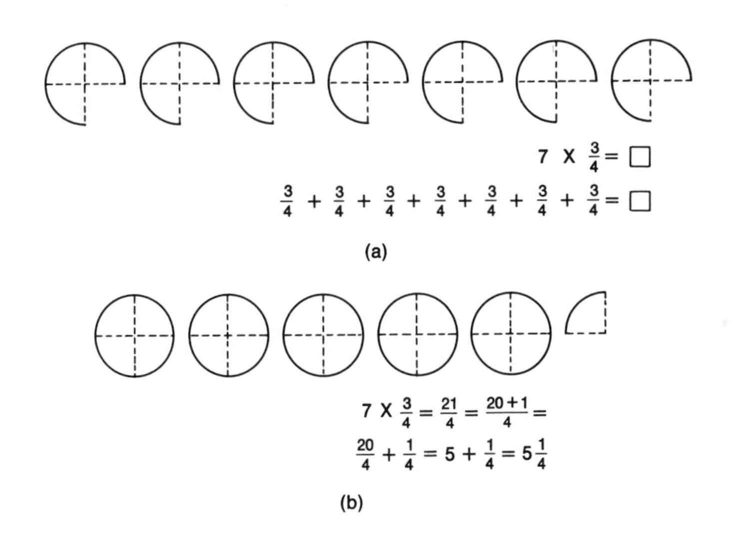

7 X 3/4 = □
3/4 + 3/4 + 3/4 + 3/4 + 3/4 + 3/4 + 3/4 = □
(a)
7 X 3/4 = 21/4 = (20+1)/4 =
20/4 + 1/4 = 5 + 1/4 = 5 1/4
(b)


**Figure 12-18** The multiplication sentence 7 × ¾ = □ represented as repeated addition with circular regions

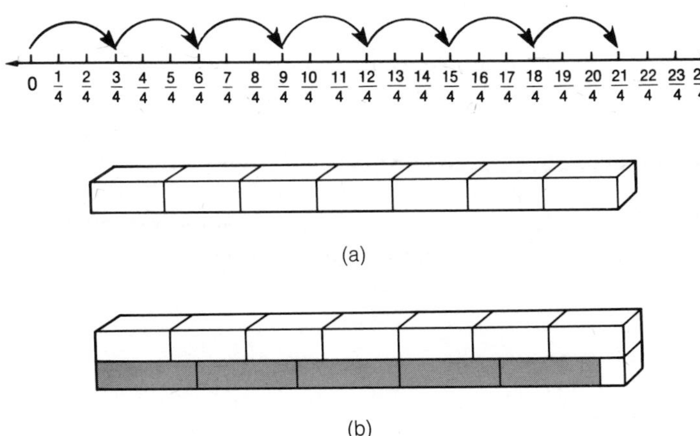


**Figure 12-19** The multiplication sentence 7 × ¾ = □ represented as repeated addition on a number line

**Figure 12-20** The multiplication sentence 7 × ¾ = □ represented with Cuisenaire rods

7. Using "of" as an interpretation of the multiplication sign is disturbing to some mathematicians because, strictly speaking, it is nonmathematical. However, careful use of it will help children understand that multiplication is used to find part of a whole number, such as ⅔ of 18, and part of a fractional number, such as ⅔ of ¾.

Many practical situations give rise to the need for finding a fractional part of a whole number. When one cube of butter weighs ¼ of a pound, the weight can be expressed as ¼ of 16 ounces, or 4 ounces. One-third of a foot is the same as ⅓ of 12 inches, or 4 inches; ½ of a dollar is ½ of 100 pennies, or 50 cents. Frequent encounters with these and similar applications enhance children's appreciation of the need for knowing how to perform such multiplication.

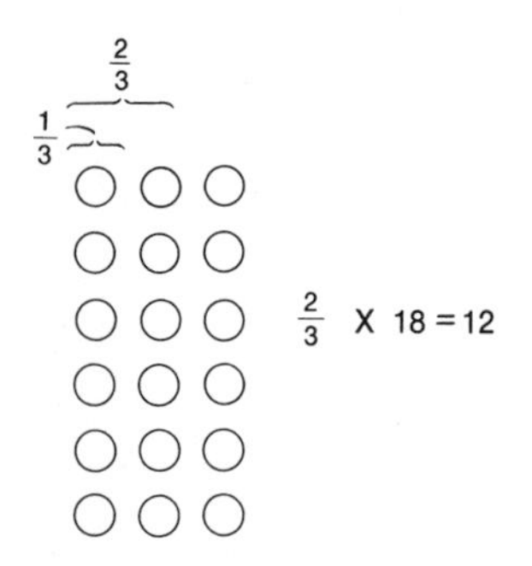


**Figure 12-21** An array used to represent ⅔ of 18, or the multiplication sentence ⅔ × 18 = 12

Activities with learning aids further children's understanding of this multiplication. To find ⅔ of 18, use an array (Figure 12-21) and guide children's thinking with questions such as these: "How many disks are in the set that makes up this array?" (Eighteen, indicated by the second factor of the sentence.) "What will be the number of rows (or columns) in the array?" (Three, as suggested by the denominator of the fractional number.) "How many markers are there in each row (or column) of the array?" (Six.) "How many markers are there in one-third of the set of eighteen?" (Six.) "How many markers are there in two-thirds of the set?" (Twelve, determined by 2 × 6 = 12.) After children have used arrays many times, they will learn that the multiplication of a pair of numbers, such as ⅔ and 18, can be completed by either of two procedures:

$$\frac{2 \times 18}{3} = \frac{36}{3} = 12 \quad \text{or} \quad \frac{18}{3} \times 2 = 6 \times 2 = 12$$

Structured materials and graphs also provide the means for extending multiplication of whole numbers by fractional numbers. Children using Unifix cubes, for example, can assemble a rod of three cubes, one of which is green [Figure 12-22(a)], and see that the green cube is one-third of the number of cubes in the entire rod. Then by doubling the length of the rod and the number of green cubes in it, they can see that ⅓ of 6 is 2 (b). By extension, the meaning of ⅓ of other numbers can be shown (c and d).

Graphs of multiplication of whole numbers by fractional numbers are useful for mature children to construct and examine. Such graphs are a

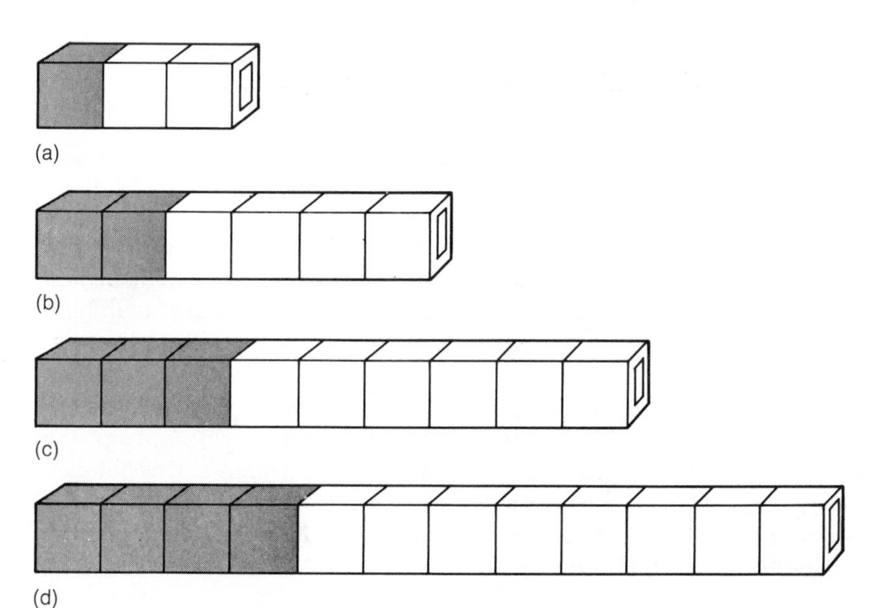


**Figure 12-22** Unifix cubes illustrate the meaning of (a) ⅓ of 3, (b) ⅓ of 6, (c), ⅓ of 9, and (d) ⅓ of 12

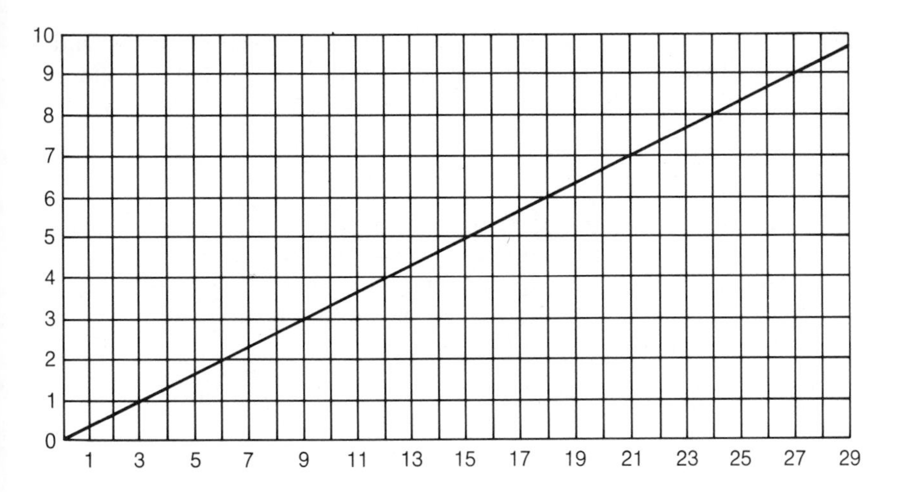


**Figure 12-23** The graph illustrating ⅓ of the numbers from 0 through 29

good way of showing the "fraction-of-a-number" concept for a given fraction and several numbers. (The magnitude of the numbers children use is limited only by the size of the graph.) In Figure 12-23, the graph of ⅓ is illustrated, while in Figure 12-24, the graphs of several fractional numbers are shown on the same grid.

## 3. Multiplication of Fractional Numbers by Fractional Numbers

A paper-folding learning center can be used to give children an interesting approach to multiplication of one fractional number by another. Problem

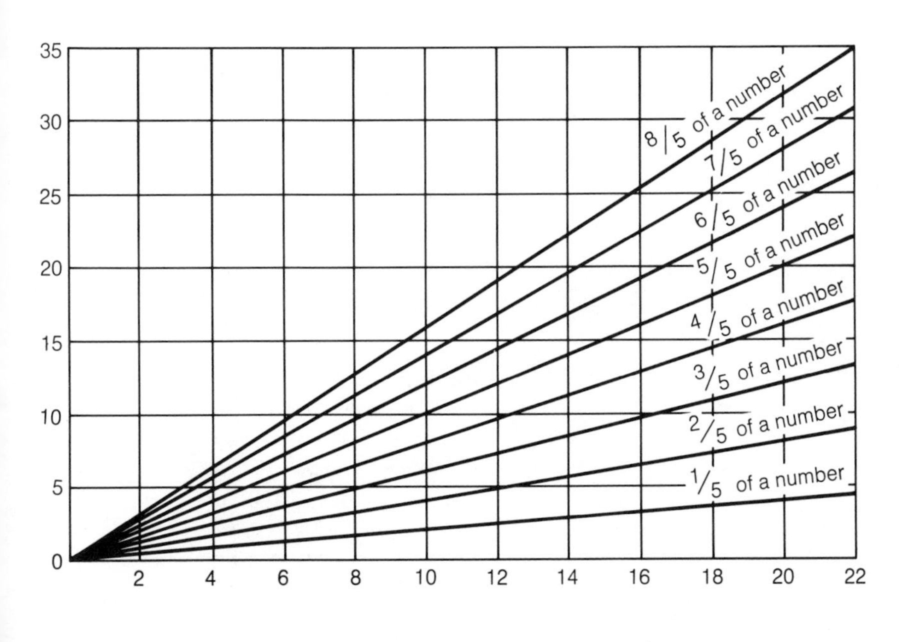


**Figure 12-24** The graphs for ⅕ through 8/5 of the numbers 0 through 22

cards give directions, and paper cut in the shapes of squares, circles, and rectangles provides the materials for investigations. A manila envelope

**Figure 12-25** Sample paper-folding problem cards

FINDING PARTS OF PARTS (1)

1. Get a square of paper from your envelope. Fold it in half by putting two opposite edges together and making a crease in it, like this:

The paper on each side of the crease is ___ of the square piece of paper.

(2)

2. Fold the paper one more time so that it looks like this:

Now, the paper is folded so there are ___ sections, or parts. Each part is ___ of the square I began with.

3. After your first fold, each section was ___ of the square. When you made the second fold, each section was ___ of the square. This shows that 1/2 of 1/2 is ___ of the whole piece of paper.

(3)

4. Take out the sheet of paper that looks like this:

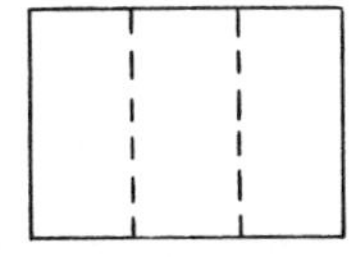

Fold on the dotted lines. Each section of the paper is ___ of the whole rectangle.

5. Fold the paper so that you fold the thirds in half. What is the size of each section now? This shows that 1/2 of 1/3 is ___ of the whole piece of paper.

containing a set of shapes and a response sheet for each person makes it easy for a child to begin work. Sample cards are shown in Figure 12-25.

Prepare other cards for activities with thirds, fourths, sixths, and eighths of fractions, and for nonunit fractions like two-thirds and two- and three-fourths of units.

When children have completed the center activities, discuss their work with them to help them summarize their findings. Overhead transparencies are useful for a review of the paper-folding activities. The transparencies for the sentence $\frac{2}{3} \times \frac{3}{4} = \frac{6}{12}$ are illustrated in Figure 12-26. In (a) a unit square is shown. In (b) $\frac{3}{4}$ of the unit is represented by shading on an overlay, while in (c) another overlay represents $\frac{2}{3}$ of the unit. Finally, $\frac{2}{3}$ times $\frac{3}{4}$ is illustrated when both overlays are placed over the unit region at the same time. The answer, $\frac{6}{12}$, is represented by the six doubly shaded sections.

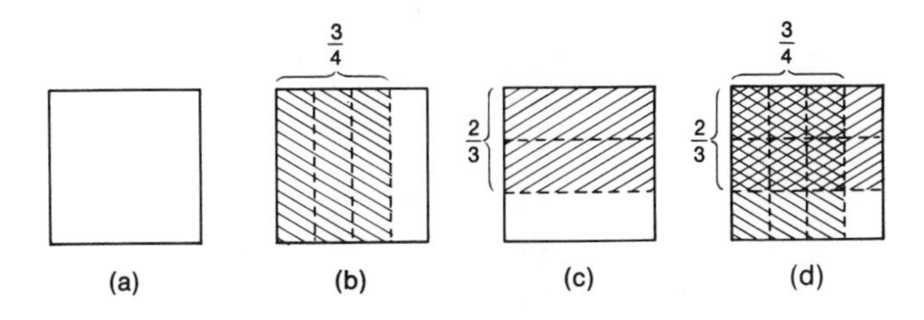


**Figure 12-26** The multiplication sentence $\frac{2}{3} \times \frac{3}{4} = \frac{6}{12}$ illustrated with a series of overhead transparencies

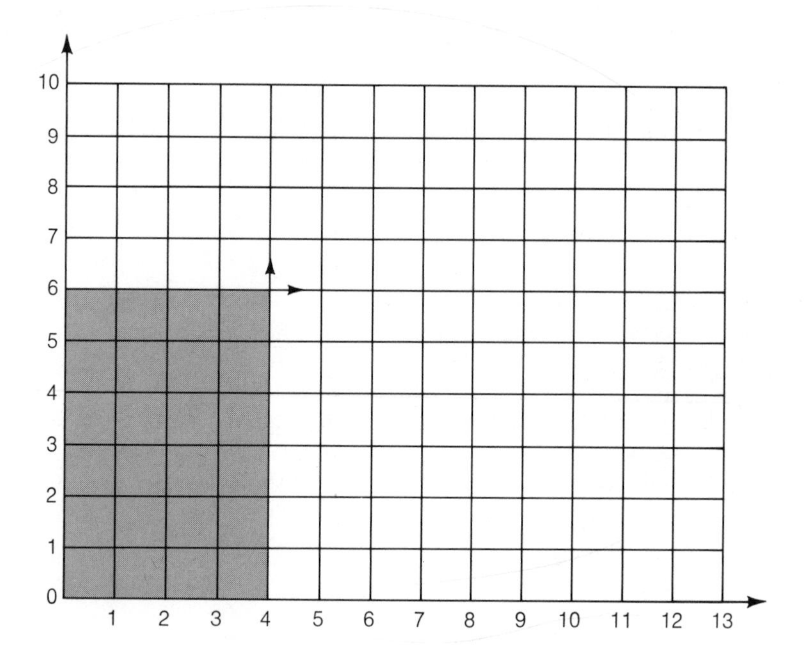


**Figure 12-27** The multiplication sentence $4 \times 6 = 24$ illustrated on the number plane

Another way to illustrate multiplication of two fractional numbers is on the number plane, or graph of whole numbers. Begin with an illustration of multiplication of two counting numbers, such as $4 \times 6 = 24$, which is illustrated in Figure 12-27. Count the shaded squares bounded by the

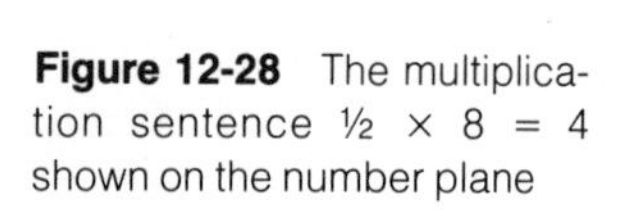
**Figure 12-28** The multiplication sentence $\frac{1}{2} \times 8 = 4$ shown on the number plane

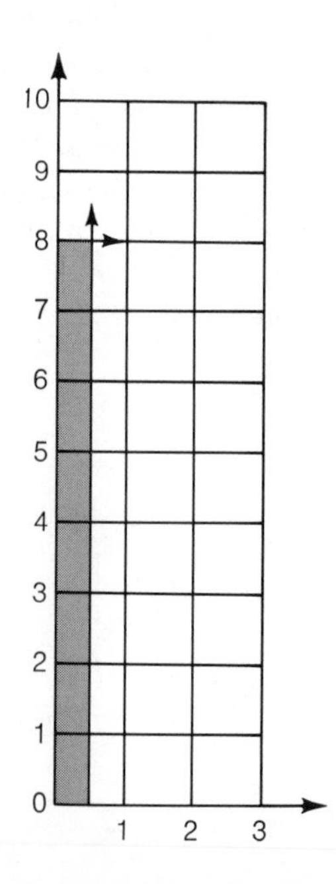

horizontal and vertical number lines and the two arrows to find the answer. (Note the similarity of the pattern of squares on the number plane and the arrays of squares used to show multiplication in Chapter 9.) Next, show an example of multiplication of a whole number by a fractional number, as illustrated in Figure 12-28. Finally, illustrate a sentence that involves two fractional numbers. The sentence $\frac{3}{4} \times \frac{1}{2} = \frac{3}{8}$ is shown in Figure 12-29. The horizontal arrow goes along the horizontal line that represents $\frac{1}{2}$ and beyond the $\frac{3}{4}$ mark, while the vertical arrow goes along the vertical line that represents $\frac{3}{4}$ and past the $\frac{1}{2}$ mark. The shaded portion of the grid is $\frac{3}{8}$ of one square unit on the grid.

As children work with their various models to gain an understanding of multiplication of fractional numbers, their attention should be directed so they will see the relationship between the fractions' numerators and the product and the denominators and the product. Eventually, they should generalize that to multiply fractional numbers, they multiply the numbers represented by numerators and the numbers represented by denominators to get the product. Thus the answer for the sentence $\frac{3}{4} \times \frac{7}{8} = \square$ is found by multiplying 3 times 7 to name the numerator of the product and 4 times 8 to name its denominator.

## MULTIPLICATION INVOLVING MIXED NUMERALS

An extension of multiplication with pairs of whole numbers and pairs of fractional numbers is made when multiplication is done with what are commonly referred to as "mixed numbers." As explained earlier, the phrase "multiplication of mixed numbers" is being replaced by the expression "multiplication with mixed numerals" in order to maintain a proper distinction between number and numeral. An example of this multiplication is $3\frac{2}{3} \times 2\frac{1}{2} = \square$.

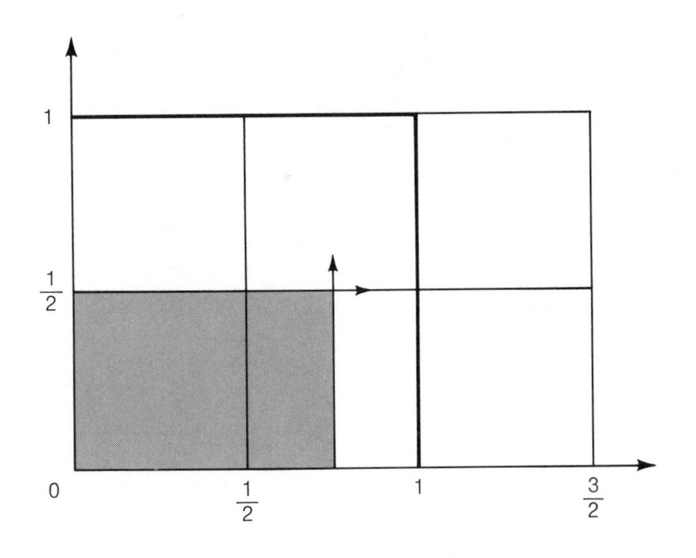


**Figure 12-29** The multiplication sentence $\frac{3}{4} \times \frac{1}{2} = \frac{3}{8}$ illustrated on the number plane

Illustrations that show this multiplication with markers, regions, and other devices can become quite cumbersome, so work with it should be postponed until children can handle it at an abstract level. This means that some children may not work with it until after they leave elementary school.

The number plane does provide one model that can be used during introductory lessons (Figure 12-30). Reading an answer on the number plane for an example such as $3\frac{2}{3} \times 2\frac{1}{2} = \square$ is not as easy as it is with simpler fractional numbers. You will need to guide children's work carefully so they read answers correctly. In Figure 12-30 the answer is found by counting the whole units and those parts of units that are bounded by the number lines and arrows. There are six whole units, three ½ units, two ⅔ units, and one ⅓ unit. When these units are represented as common fractions and added, we get $6 + \frac{3}{2} + \frac{4}{3} + \frac{1}{3} = 9\frac{1}{6}$.

The distributive property of multiplication over addition can be extended to fractional numbers and used to explain multiplication with mixed numerals. Use the number plane model along with number sentences so children can see how this is done. For the model in Figure 12-30, draw children's attention to this sequence of sentences as they examine the model:

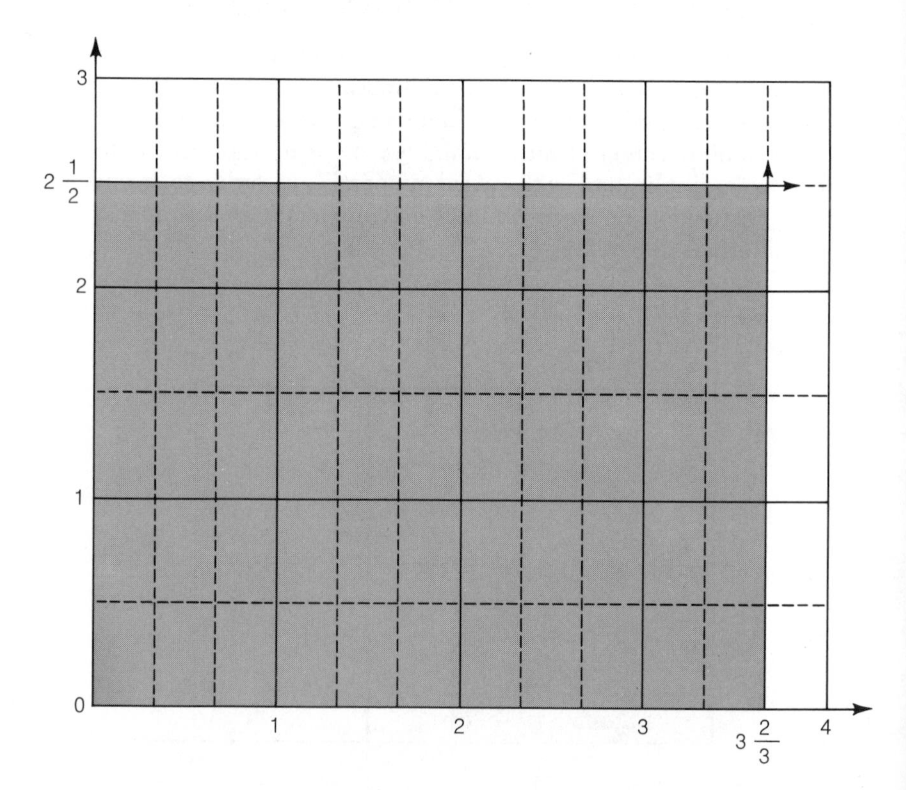


**Figure 12-30** The multiplication sentence $3\frac{2}{3} \times 2\frac{1}{2} = \frac{55}{6}$, or $9\frac{1}{6}$, illustrated on the number plane

$3\frac{2}{3} \times 2\frac{1}{2} =$

a. $3\frac{2}{3} \times 2\frac{1}{2} =$
b. $3\frac{2}{3} \times (2 + \frac{1}{2}) =$
c. $(3\frac{2}{3} \times 2) + (3\frac{2}{3} \times \frac{1}{2}) =$
d. $[(3 + \frac{2}{3}) \times 2] + [(3 + \frac{2}{3}) \times \frac{1}{2}] =$
e. $(3 \times 2) + (\frac{2}{3} \times 2) + (3 \times \frac{1}{2}) + (\frac{2}{3} \times \frac{1}{2}) =$
f. $6 + \frac{4}{3} + \frac{3}{2} + \frac{2}{6} =$
g. $6 + \frac{8}{6} + \frac{9}{6} + \frac{2}{6} =$
h. $6 + \frac{19}{6} =$
i. $9\frac{1}{6}$

Beginning with four multiplications in (e), and continuing through (i), children should see that for each part of the sentence there is a matching part in the model.

Another way to do this multiplication is to change each mixed numeral to a common fraction greater than 1. Thus $3\frac{2}{3} \times 2\frac{1}{2} = \square$ becomes $\frac{11}{3} \times \frac{5}{2} = \frac{55}{6} = 9\frac{1}{6}$. This, too, can be related to the number plane shown in Figure 12-30. Count the thirds across the bottom; there are eleven. Then count the halves up the left side; there are five. Each unit square is separated into six parts by the broken lines. Altogether, 11 times 5, or 55, of these parts are shaded. This is a procedure most children who attempt this multiplication will eventually use.

---

## Self-check of Objectives 4 and 5

Make up real-life problem situations showing how each of the repeated addition and array interpretations apply to multiplication with fractional numbers. Use learning aids or picture sequences to illustrate each example and write each situation's number sentences. Explain why the Cartesian product interpretation is not useful for multiplication with fractional numbers.

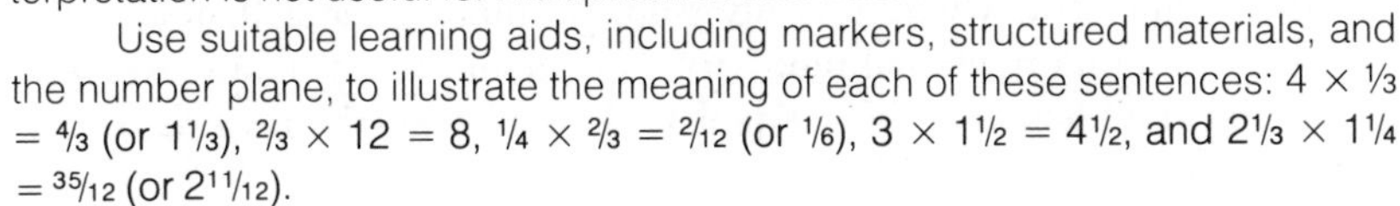

Use suitable learning aids, including markers, structured materials, and the number plane, to illustrate the meaning of each of these sentences: $4 \times \frac{1}{3} = \frac{4}{3}$ (or $1\frac{1}{3}$), $\frac{2}{3} \times 12 = 8$, $\frac{1}{4} \times \frac{2}{3} = \frac{2}{12}$ (or $\frac{1}{6}$), $3 \times 1\frac{1}{2} = 4\frac{1}{2}$, and $2\frac{1}{3} \times 1\frac{1}{4} = \frac{35}{12}$ (or $2\frac{11}{12}$).

---

## THE PROPERTIES OF MULTIPLICATION APPLIED TO FRACTIONAL NUMBERS

The properties of multiplication can be extended to include fractional numbers as children work with examples such as the ones already discussed.

## 1. Commutative Property

Models and algorithms will verify the commutative property. Children can see that $\frac{1}{2} \times \frac{1}{4} = \frac{1}{4} \times \frac{1}{2}$ by first finding one-half of a fourth and then finding one-fourth of a half. In each case, they will see that the answer is one-eighth of the original unit. They can also verify this property by completing a sentence such as $\frac{5}{7} \times \frac{7}{8} = \square$, and then reversing the factors and finding the answer for $\frac{7}{8} \times \frac{5}{7} = \square$. Their experiences with models and sentences will lead them to generalize that for any multiplication represented by common fractions, $a/b \times c/d = c/d \times a/b$.

## 2. Associative Property

There must be at least three factors for use of the associative property for multiplication to be necessary. Sets of cubes and arrays illustrate the validity of this property with whole numbers, but simple models to verify it for fractional numbers are not easy to make. Therefore it is best to postpone its use with fractional numbers until children have a mature enough understanding of multiplication to work with it at an abstract level. Begin with a sentence such as $\frac{1}{2} \times \frac{3}{4} \times \frac{2}{3} = \square$. Have children multiply the $\frac{1}{2}$ times $\frac{3}{4}$ to get $\frac{3}{8}$. Then have them multiply this product by $\frac{2}{3}$ to name the final product, $\frac{6}{24}$, or $\frac{1}{4}$. To show the associative property, have them multiply a second time, beginning with $\frac{3}{4}$ times $\frac{2}{3}$, and then multiply the product, $\frac{6}{12}$, by $\frac{1}{2}$ to get the final product. In each case, the product is the same, regardless of which pair of factors was multiplied first.

## 3. Identity Element

Understanding the role of the identity element for multiplication in operations with fractional numbers is particularly important. Multiplication by 1 in the form of $a/a$, where $a$ is any whole number greater than 0, is the basis for renaming fractional numbers. You should assign many exercises in naming fractional numbers equal to 1 and using them as factors in multiplication.

## 4. Reciprocals

The set of rational numbers, of which the fractional numbers studied by children in elementary grades are a subset, includes a property of multiplication that does not exist in the set of whole numbers. This is the reciprocal property of multiplication. In general, this property states that every number in the set of rational numbers, except 0, has a number by which it can be multiplied to yield a product of 1. For every number $a$, there exists a reciprocal, $1/a$. The product of $a$ and $1/a$ is 1. Another name for a reciprocal is a multiplicative inverse.

Children can learn about reciprocals through examples such as these:

$$\frac{3}{4} \times \square = \frac{12}{12},$$

$$\frac{6}{7} \times \square = \frac{42}{42},$$

$$\frac{2}{3} \times \square = \frac{6}{6}.$$

"By what number must we multiply $3/4$ to obtain the product $12/12$? By what do we multiply $6/7$ to obtain the product $42/42$? By what do we multiply $2/3$ to obtain the product $6/6$?" After the second factor for each sentence has been determined, the questions should continue. "What do you notice about each product in these sentences?" (They all are names for 1.) "What do you notice about the two factors in each example?" (The numerator and denominator of the second are the reverse of the numerator and denominator of the first.)

"What do you think is the product of each of the following sentences?"

$$\frac{4}{15} \times \frac{15}{4} = \square$$

$$\frac{9}{7} \times \frac{7}{9} = \square$$

$$2 \times \frac{1}{2} = \square$$

After children understand how to determine a reciprocal, help them compose a definition of it. For example, "A number and its reciprocal yield a product of 1." "The reciprocal of a number is the number by which the original number is multiplied to yield a product of 1."

The idea should be pursued one step further with more mature children: "What is the quotient of the sentence $1 \div 3/4 = \square$?" Help children determine the answer to this and similar examples that involve 1 as the dividend by having them see that the sentence $1 \div 3/4 = \square$ can be rewritten as $\square \times 3/4 = 1$. Children should readily see that the quotient in division of this type is always the reciprocal of the number that is the divisor.

## Self-check of Objective 6

Explain orally or in writing how children can see that multiplication of fractional numbers is both commutative and associative. Define *reciprocal* and give an example; give an example of a fractional number identity element for multiplication.

## DIVISION WITH FRACTIONAL NUMBERS

Two processes can be used to divide fractional numbers. One of these is based on the idea that dividing one number by another is the same as multiplying the number by the reciprocal of the divisor. Thus $c \div b = a$ and $c \times 1/b = a$. This is the basis for the invert-and-multiply method of dividing with fractional numbers. The other process requires both fractional numbers to have the same denominator. Division is done by dividing the numerator of the dividend by the numerator of the divisor. Division of the denominators is unnecessary because their division always yields a quotient of one. Thus

$$\frac{a}{b} \div \frac{c}{b} = \frac{a \div c}{b \div b} = \frac{a \div c}{1} = a \div c.$$

There is no agreement as to which is best. There have been many articles published in journals such as *The Arithmetic Teacher* to support one or the other of the processes. Most writers discuss the invert-and-multiply procedure, but even these do not agree on how the process should be rationalized for children. They do not even agree on whether or not the process can and should be rationalized for children. Other writers recommend the common denominator procedure. This book's discussion will not attempt to settle the issue, but will suggest those procedures that encourage a discovery approach which develops fifth and sixth graders' understanding of division with fractional numbers.

### 1. Situations That Illustrate Division with Common Fractions

Children should relate division with fractional numbers to division with whole numbers. Their understanding of both partitive and measurement situations should be extended to include fractional numbers; as before, early experiences should include work with physical models to aid in visualizing the situations and the role of division in their solution. An exploratory lesson should begin with a review of both situations used with whole numbers. The examples need not involve large numbers. For example, "There are fifteen markers on the magnetic board. If we put them into groups that each have three markers, how many groups will there be?" (Review the idea that this is a measurement situation and that the answer can be determined by repeated subtraction.)

"There are eighteen markers on the board. If we separate them into three groups of the same size, how many markers will there be in each group?" (Review the idea that this is a partitive situation and that the answer can be determined by creating three groups and distributing the eighteen markers equally among them.)

"If I have a piece of lumber 3 feet long and I cut it into pieces each ½ foot long, how many pieces will I have?" Help children relate this measurement situation to a comparable one that involves whole numbers, and then use rectangular regions (Figure 12-31), fraction strips, and the number line to visualize the meaning of the situation. Also, help children see that the answer can be determined by repeated subtraction. Using repeated subtraction here reinforces children's understanding of the measurement situation, and, at the same time, emphasizes the value of an efficient process.

"If I have three-fourths of a cake and cut it into pieces each one-fourth of the original cake, how many pieces will there be? What is the open sentence for this situation?" Children should think of the sentence $\frac{3}{4} \div \frac{1}{4} = \square$ and then use various aids to see that a piece that is three-fourths of a unit consists of three pieces one-quarter unit in size. This situation is illustrated on number lines in Figure 12-32. Again, the repeated subtraction should be completed. The quotients in both these situations are whole numbers that tell how many of something are created by division. This is in agreement with the interpretation of measurement situations involving whole numbers in which quotients tell how many groups are created by separating sets into equal-sized groups.

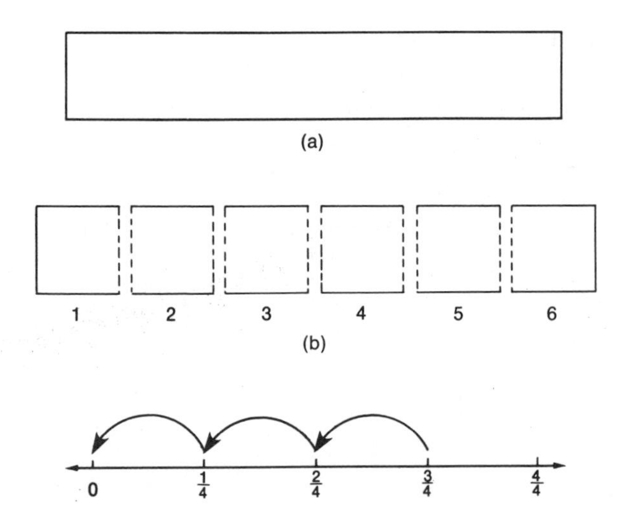


**Figure 12-31** A rectangular region used to represent the division sentence $3 \div \frac{1}{2} = 6$. In (a) a 3-foot board is represented; and in (b) the board is separated into pieces each ½ foot long to represent the repeated subtraction concept of division

**Figure 12-32** Number line used to represent the division sentence $\frac{3}{4} \div \frac{1}{4} = 3$

"There is half a pie in Jack's refrigerator. If he cuts it into three equal-sized pieces, what part of the whole pie will each piece be?" The children should have time to explore the meaning of the situation and determine the answer. They should be guided to see that in a situation of this type, the answer represents the size of a fractional part of a whole. This is a partitive situation in which something is separated into a given number of equal-sized pieces. The division algorithm is used to determine the size of each piece (Figure 12-33).

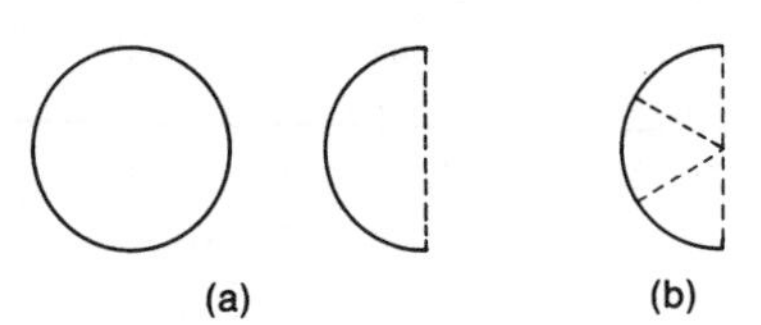


**Figure 12-33** A circular region used to represent the division sentence $\frac{1}{2} \div 3 = \frac{1}{6}$. In (a), the unit region and one-half of it are represented; in (b), the half is shown divided into three congruent parts

## 2. The Common Denominator Process

Encourage exploratory activities such as those suggested here until you are certain the children understand measurement and partitive concepts used with fractional numbers. Then a process for dividing them will be more easily understood. Even though most textbook series introduce the invert-and-multiply process first, this book recommends that the common denominator process be used first because this method can be more readily rationalized for children than the invert-and-multiply method.

Use problems similar to those in the exploratory lessons so children can use learning aids to visualize the meaning of this process. To determine the quotient for the sentence $2\frac{1}{2} \div \frac{1}{2} = \square$, children can use fraction strips or rectangular regions such as those in Figure 12-34. The dividend is represented by two and one-half unit regions (a). Then the two-unit regions are represented as four-halves to show a total of five-halves, and the sentence is rewritten as $\frac{5}{2} \div \frac{1}{2} = \square$ (b). Children should be guided to see that the quotient can be determined by dividing 5 by 1.

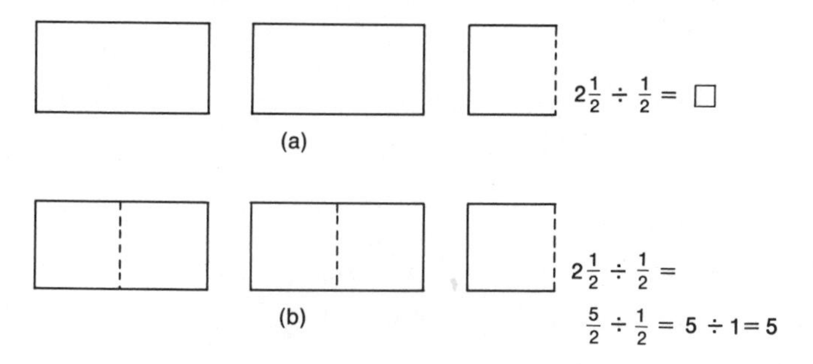

$2\frac{1}{2} \div \frac{1}{2} = \square$
(a)
$2\frac{1}{2} \div \frac{1}{2} =$
$\frac{5}{2} \div \frac{1}{2} = 5 \div 1 = 5$
(b)


**Figure 12-34** Rectangular regions used to help children rationalize the common denominator method of division for the sentence $2\frac{1}{2} \div \frac{1}{2} = 5$

The process should be repeated for other division sentences with learning aids as children extend their understanding of the common denominator process to include division of fractional numbers with unlike denominators, such as $3/4 \div 1/8 = \square$, and pairs of numbers that yield a mixed numeral quotient, such as $2 \div 3/4 = \square$. In the first case, models help children to see that when $3/4$ is renamed as $6/8$, the division involves only the numerators:

$$\frac{6}{8} \div \frac{1}{8} = 6 \div 1 = 6.$$

In the case of $2 \div 3/4 = \square$, use two-unit regions to show that

$$\frac{8}{4} \div \frac{3}{4} = 8 \div 3 = 2\frac{2}{3}.$$

After the quotients have been found, their meanings should be interpreted. The sentence $6/8 \div 1/8 = 6$ can be considered in this way: In a piece that is six-eighths of a unit in size, how many pieces each one-eighth in size are there? (Six.) The sentence $2 \div 3/4 = 2\,2/3$ can be interpreted as: In two whole pieces, there are how many pieces three-fourths of a unit in size? (Two, plus two-thirds of another.)

Physical models for representing the division of a fractional number by a whole number using the common denominator process are cumbersome to use. Therefore this type of division should be introduced after children understand the process for whole numbers divided by fractional numbers. For the sentence $1/2 \div 4 = \square$, children should see that the quotient can be determined by renaming the 4 as a fraction that has the same denominator as the dividend. The sentence $1/2 \div 4 = \square$ thus becomes $1/2 \div 8/2 = 1 \div 8 = 1/8$.

### 3. Invert-and-Multiply Process

Children who are introduced to the common denominator method of division can also learn the invert-and-multiply method. It is efficient for dividing many pairs of numbers that are more difficult to divide with the common denominator method.

Four ideas must be clear before children will understand the invert-and-multiply process:

1. A common fraction can be used to indicate division: $a \div b = a/b$.
2. Division by 1 always yields a quotient that is the same as the dividend: $a \div 1 = a$.
3. A number and its reciprocal yield a product of one: $a \times 1/a = 1$.

4. If both numerator and denominator are multiplied by the same number, the value of the fraction is not changed:

$$\frac{a}{b} = \frac{a \times c}{b \times c}$$

Children's understanding of the invert-and-multiply process should be developed through a study of all the steps in the process, even though they later learn that it is a shortcut procedure based on several principles of mathematics, the steps for which are omitted when the mature process is used. For the open sentence $2/3 \div 1/3 = \square$, the steps are

$$\frac{2}{3} \div \frac{1}{3} = \frac{2/3}{1/3} = \frac{2/3 \times 3/1}{1/3 \times 3/1} = \frac{2/3 \times 3/1}{1} = \frac{2}{3} \times \frac{3}{1} = \frac{6}{3} = 2.$$

First, children should recognize that

$$\frac{2}{3} \div \frac{1}{3} = \frac{2/3}{1/3}$$

Next, they should see that when they multiply $1/3$ by its reciprocal, the denominator of the compound fraction is renamed as 1. This eliminates any further need for consideration of the denominator because division by 1 always yields a quotient that is equal to the dividend. Children should note that when the denominator is multiplied by its reciprocal, the numerator must also be multiplied by this same number. Therefore the numerator becomes $2/3 \times 3/1$. Thus the original sentence, $2/3 \div 1/3 = \square$, has become $2/3 \times 3/1 = \square$. They see that the quotient is found by multiplying the dividend by the reciprocal of its divisor.

Guide children through the process with other examples before you help them state the generalization: "To divide fractional numbers, multiply the dividend by the reciprocal of the divisor." This definition of the process is preferable to the commonly used, but mathematically meaningless, expression, "To divide fractional numbers, invert and multiply."

---

## Self-check of Objectives 7 and 8

Describe a real-life partitive situation for this division sentence: $1/3 \div 4 = 1/12$. Illustrate your example with an appropriate learning aid. Do the same for this sentence, using a measurement situation: $3 \div 1/4 = 12$.

Write the sequence of steps that illustrate the completion of the division sentence $3/4 \div 1/2 = 1\frac{1}{2}$ by the common denominator method. Show the sequence of steps for completing the same sentence by the invert-and-multiply method.

---

## SIMPLIFICATION OF ANSWERS AND CANCELLATION

### 1. Simplification of Answers

For many years children have been taught to automatically simplify ("reduce to lowest terms") the answers to any problem using fractional numbers whenever such simplification is possible. They have been told to do this regardless of the situation out of which the computation has arisen. (The processes for simplifying fractions are discussed in Chapter 11.) However, it should be recalled that in Chapter 10 the discussion of division advises that remainders should be expressed according to the situation and not automatically written in any one way according to a rule. The same advice applies for the answers to problems dealing with fractional numbers.

There are many times when an answer should not be simplified, even though it might be renamed in a simpler form. For example, if a number of pieces of fine jewelry wire have been measured to the nearest sixteenth of an inch and their total length is $^{14}/_{16}$, it would not be correct to rename $^{14}/_{16}$ as $^{7}/_{8}$. To do so would suggest that the pieces of wire were measured to the nearest eighth rather than to the nearest sixteenth of an inch. On the other hand, in the example on page 316 about Jim practicing on the piano, the answer $^{2}/_{4}$ can be renamed as $^{1}/_{2}$ because it is common to speak of one-half of an hour rather than two quarter hours. You should help children assess each problem situation carefully to determine the best way to express the answer rather than give them a rule to automatically follow.

### 2. Cancellation

Another process children often learn by rule is "cancellation." The example

$$\frac{\overset{1}{\cancel{2}}}{\underset{1}{\cancel{3}}} \times \frac{\overset{3}{\cancel{9}}}{\underset{5}{\cancel{10}}} = \frac{3}{5}$$

will help you recall this process. Cancellation is a valid process, and children should learn to use it while they work with fractional numbers. However, instead of learning it simply as cancellation, they should understand its meaning.

The process is actually one of simplifying the common fractions *before* the multiplication is done rather than *after*. Children can use their knowledge of factoring and the identity element for multiplication to make cancellation meaningful. For example, begin with an example like (a) in the margin. Rewrite the sentence as in (b). Next, factor the com-

(a) $\frac{2}{3} \times \frac{9}{10} = \square$

(b) $\frac{2 \times 9}{3 \times 10} = \square$

(c) $\frac{2 \times (3 \times 3)}{3 \times (2 \times 5)} = \square$

(d) $\frac{2 \times 3 \times 3}{2 \times 3 \times 5} = \square$

(e) $\frac{2}{2} \times \frac{3}{3} \times \frac{3}{5} = \square$

(f) $1 \times 1 \times \frac{3}{5} = \square$

posite numbers as in (c). Application of the commutative property permits rewriting the sentence as in (d). In (e), the sentence is written in a way that emphasizes fractions that are names for one. Finally, two of the fractions are rewritten as ones and the multiplication is completed. Eventually, children will complete the process without showing all of the steps. Before you encourage them to use cancellation, be certain they understand its rationale. Have them show all of the steps for examples such as these:

$$\frac{4}{5} \times \frac{3}{4} = \square,$$

$$\frac{3}{4} \times \frac{8}{9} = \square,$$

$$\frac{5}{12} \times \frac{3}{10} = \square.$$

Then have them use cancellation to complete the multiplication for each.

## Self-check of Objectives 9 and 10

Explain some of the factors that need to be considered when determining whether or not a common-fraction answer should be simplified (reduced to lowest terms).

Explain the meaning of cancellation. List the sequence of steps that illustrate why cancellation can be used in this multiplication example: $6/10 \times 12/15 = \square$.

## RATIOS

One of the uses of common fractions discussed at the beginning of Chapter 11 is that of indicating ratios. When common fractions are used to indicate ratios, they are interpreted differently than when they represent parts of a whole or group. For example, the numerator of a common fraction that represents a ratio may tell the number of elements in one of two sets, while the denominator tells the number of elements in the other set. Ratios cannot be added, subtracted, multiplied, and divided like fractional numbers, either. Consequently you must use care when common fractions are introduced as ways of representing ratios so that their meaning is clear.

### 1. Rate Pairs and Tables

You do not need to delay introduction of ratios until children have mastered all the other uses of common fractions. In fact, many children who have only begun work with common fractions as representations of fractional numbers can be introduced to the idea of ratios. A good way to make the introduction is through work with rate pairs. A simple example shows how this can be done. "I bought two pencils for 5 cents. Will one of you use these play coins and these pencils to illustrate this for me?" Five pennies (or one nickel) and two pencils should be displayed by a child. "How much money will I need to buy four pencils at this price?" Through questions like these, you can help children understand many-to-many correspondence and the ordered pairs that describe this and similar situations. Show a simple table, such as the one in Figure 12-35, to record and organize the data.

| Number of Pencils | 2 | 4 | 6 | | | | |
|---|---|---|---|---|---|---|---|
| Cost of Pencils | 5¢ | 10¢ | 15¢ | | | | |

**Figure 12-35** Example of table used to organize ordered pairs from a problem situation

Help third and fourth graders develop tables for rate pairs that arise from varied problem situations. A few are listed below:

1. Comparing miles walked and hours of time.
2. Comparing the number of bricks and feet of fence.
3. Comparing pints and quarts (or any other related measures).
4. Comparing books and number of children.
5. Comparing apples and boxes of apples.

Many fifth and sixth graders will be ready to use common fractions to represent rate pairs. By using situations similar to those for which tables were developed earlier, they can learn to express rate pairs as both ordered pairs $(a,b)$, and as common fractions, $a/b$. The word *ratio* should also be introduced at this time. Children should be able to determine and list the set of common fractions that represents the 1-to-2 ratio in the following situation. "Jack and Bob are going to share a paper route. Jack has agreed to deliver the papers one day out of three; Bob is going to deliver them two days out of three. You have learned how to use tables to keep track of numbers in situations of this kind. If we make a table for this situation, what will be the first entry?" A table should be drawn and the first entry made (Figure 12-36). "If Jack has delivered the papers for three

| Number of Days for Jack | 1 | 2 | 3 | 4 | 5 | 6 | 7 |
|---|---|---|---|---|---|---|---|
| Number of Days for Bob | 2 | 4 | 6 | 8 | 10 | 12 | 14 |

**Figure 12-36** Example of table used to organize data arising from a realistic situation

days, how many days of delivering will Bob's share be?" Guide the children's thinking as they complete the rest of the table and interpret the data in it.

Children should also be guided to observe that the data in this table can be represented by ordered pairs: (1, 2), (2, 4), (3, 6), . . . . They should also be helped to observe that common fractions can be used. "The first ordered pair can be written as $\frac{1}{2}$. What does the numerator stand for? What does the denominator stand for?" The ordered pairs should be listed as common fractions as their meanings are discussed: $\frac{1}{2}$, $\frac{2}{4}$, $\frac{3}{6}$, $\frac{4}{8}$, $\frac{5}{10}$, . . . . "When we read a common fraction used in this way, we say '1 to 2, 2 to 4,' and so on. Why are they read this way rather than as 'one-half, two-fourths,' and so on?" (Because they are being used to express the comparison of two sets rather than the meaning of fractional numbers that tell about parts of a whole or a group.) "Common fractions used in this way express ratios. A ratio is a way of comparing two sets. All the fractions we have listed here express a 1-to-2 ratio. What do you notice about all the numerals in this set?" (All of them can be expressed in their simplest form as "$\frac{1}{2}$.") "What do you notice about the numbers that are represented by the first numerals in each of the ordered pairs?" (They are in sequence by ones.) "What do you notice about the numbers that are represented by the second numerals in each ordered pair?" (They are in sequence by twos.)

## 2. Proportions

"If we know that our ordered pairs have a ratio of 1 to 2, we can determine the missing part of any ordered pair that is equivalent to $\frac{1}{2}$. Suppose Jack has delivered papers for ten days. How many days will Bob have delivered them?" Children should be able to determine the answer, 20, by noting that Bob's share of the delivery days is always twice as much as Jack's share. "If Bob has delivered the papers thirty times, how many times will Jack have delivered them?" With your guidance, children should organize the problem as a proportion:

$$\frac{1}{2} = \frac{n}{30}.$$

Help them note that they can solve the proportion by thinking, "By what number is the denominator, 2, multiplied to yield a product of 30?" The numerator, 1, must be multiplied by the same number. You should then help the children see that this process is the same as the one used to rename fractional numbers to create common denominators. Children's understanding of ratio and proportion will grow by completing many exercises similar to this one.

Some children will learn a process that helps find the missing term in any expression of proportion. In general, two ordered pairs expressed as common fractions represent the same ratio when the product of the numerator of the first and the denominator of the second is the same as the product of the denominator of the first and numerator of the second: that is, $a/b = c/d$ when $a \times d = b \times c$. Thus the ratio is the same for $3/4$ and $6/8$ because $3 \times 8 = 4 \times 6$.

The missing term in any proportion can be determined by "cross multiplying" and dividing; to find the value of $n$ in $6/7 = n/42$, 6 is multiplied by 42 and 7 by $n$; then the product of 6 and 42 is divided by 7. This process can be taught as children use simple proportions. They should begin with a proportion such as $3/4 = n/8$. "You know a way to determine the value of $n$ in this proportion. What is its value?" "We are going to learn another way to determine the missing term in proportions. You will need to use common denominators as we do this new work, so let's review what we know about common denominators." (Review the concept of common denominators.)

"Look at the proportion $3/4 = n/8$. We know that 6 can be used to replace $n$. Now let's see how we can determine the value of $n$ in another way. First, we should rename the two common fractions so they have a common denominator that is the product of 4 and 8. What will the new denominator for each be?" (32.) "By what do we multiply the terms of $3/4$ to rename it with the denominator 32?" (Each term is multiplied by 8.) "By what do we multiply each term of $n/8$ to rename it with a denominator 32?" (Each term is multiplied by 4.) Rewrite the proportion to show the renaming of the common fractions:

$$\frac{3 \times 8}{4 \times 8} = \frac{n \times 4}{8 \times 4}.$$

The proportion should again be rewritten to show the result of the multiplication:

$$\frac{24}{32} = \frac{4n}{32}$$

"We know that when two common fractions are equal, they have the same numerators and denominators. In the above fractions, we have the same denominators. What must we do to get the same numerators?" (We must determine the value of $n$ so we will know the factor to use with 4 to give the product 24.) "We know that we can determine a missing factor if we divide the product by a known factor. What is the missing factor in this example?" ($24 \div 4 = 6$.)

Have the children follow this procedure with a sufficient number of different proportions so they can generalize the procedure and name it "cross multiplication." For the proportion $4/7 = n/35$, children should see the steps as $4 \times 35 = 7 \times n$, $140 = 7n$, $n = 140 \div 7$, $n = 20$.

Proportions are used frequently to solve problems. Ratios used with percent and proportions are discussed in the next chapter. Chapter 4 also discusses proportions used in problem solving.

---

## Self-check of Objective 11

Several examples of situations involving ratios are described or listed. List two or three of these, and name at least two similar but different examples. Prepare a table that reports data for one of your situations. Use cross multiplication to determine whether these are expressions of proportions: $^{15}/_{35} = {}^{45}/_{105}$ and $^{18}/_{39} = {}^{72}/_{120}$.

---

## PRACTICING OPERATIONS WITH COMMON FRACTIONS

Once children have been introduced to processes of renaming common fractions, finding their common denominators, and performing operations with them, they need opportunities to practice their newly acquired skills. Most of the games used for practice with whole numbers can be adapted for fractional numbers. In addition, there are games devoted to common fractions that should be a part of the materials available to your children.

Many commercial games are available at low cost. *The Arithmetic Teacher* is a source of many games. For example, see "Make a Whole—A Game Using Simple Fractions," by Joann Rode, in the February 1971 issue.[1] *Games for Individualizing Mathematics Learning*[2] has games on pages 79 to 95. Another useful set of books for fraction games is *Let's Play Games in Mathematics,*[3] volumes 4 through 7.

There are also educational films, filmloops, and filmstrips dealing with fractional numbers that should be included in mathematics laboratories and classroom learning centers. The National Council of Teachers of Mathematics film series distributed by the Silver Burdett Company contains seventeen films that deal with all phases of fractional numbers represented by common fractions.[4]

[1] *The Arithmetic Teacher* is a journal published eight times a year by the National Council of Teachers of Mathematics, 1906 Association Drive, Reston, Va. 22091.

[2] Leonard M. Kennedy and Ruth L. Michon, *Games for Individualizing Mathematics Learning* (Columbus, Ohio: Charles E. Merrill Books, Inc., 1973).

[3] George L. Henderson, *Let's Play Games in Mathematics* (Skokie, Ill.: National Textbook Company, 1970).

[4] *Elementary Mathematics for Teachers and Students* (Morristown, N.J.: Silver Burdett Company, 1970).

## Common Pitfalls and Trouble Spots

One pitfall associated with teaching operations with common fractions is the failure to relate what children already know about whole numbers when they begin to work with common fractions. Too often, common fractions and operations with them are treated as entirely new. If you prepare a hierarchy of the concepts and skills for adding, subtracting, multiplying, or dividing with common fractions, you will include in each one subordinate concepts and skills learned during work with whole numbers. Knowledge of these concepts and skills should serve as the foundation upon which common fractions are built, so that children can see that operations with fractional numbers are an extension of operations with whole numbers, rather than something new.

Many children make errors while computing with common fractions. The following errors are ones that often occur in spite of well-sequenced and properly paced lessons, and require special attention:

1. Addition
   a. Adding both the numerators and denominators:

      $7/8 + 3/8 = 10/16, \qquad 5/6 + 4/8 = 9/14.$

   b. Failure to rename one fraction when there are unlike denominators:

      $2/3 + 1/6 = 3/6, \qquad 4/5 + 3/10 = 7/10.$

2. Subtraction
   a. Subtracting both numerators and denominators:

      $3/8 - 1/4 = 2/4, \qquad 2/5 - 1/3 = 1/2.$

   b. Failure to subtract whole numbers when there are mixed numerals:

$$\begin{array}{rl} 4\,1/3 \rightarrow & 4/12 \\ -2\,1/4 \rightarrow & \underline{3/12} \\ & 1/12, \end{array} \qquad \begin{array}{rl} 5\,5/6 \rightarrow & 10/12 \\ -2\,1/3 \rightarrow & \underline{4/12} \\ & 6/12. \end{array}$$

3. Multiplication
   a. Renaming with common denominators, multiplying numerators only:

      $3/4 \times 1/8 = 6/8 \times 1/8 = 6/8.$
      $5/6 \times 2/3 = 5/6 \times 4/6 = 20/6.$

   b. Inverting and multiplying:

      $2/3 \times 1/4 = 2/3 \times 4/1 = 8/3.$
      $5/6 \times 2/3 = 5/6 \times 3/2 = 15/12.$

4. Division
   a. Dividing both numerator and denominator (disregarding order when divisor number is larger than dividend number):

$4/6 \div 1/3 = 4/2$, $1/2 \div 1/4 = 1/2$.

b. Failure to invert:

$1/2 \div 1/4 = 1/2 \times 1/4 = 1/8$.
$3/8 \div 3/4 = 3/8 \times 3/4 = 9/32$.

These erroneous ways of computing with common fractions are applied systematically, so they become habitual unless detected and corrected early. The guidelines for correcting errors listed at the end of Chapter 8 should be used to establish a remedial program for any children who make these and similar errors with common fractions.

---

## Self-check of Objectives 12 and 13

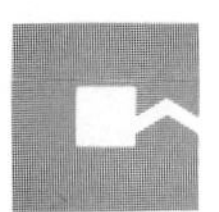

One pitfall associated with teaching computation with common fractions is discussed. Describe it orally or in writing, and tell how it can be avoided.

Examples of common errors committed by children as they compute with common fractions are given. Explain how a child would compute each of these examples if that child used one of the erroneous ways of computing:

$1/4 + 1/5 = \square$ $5/8 - 1/2 = \square$ $4/5 \times 1/3 = \square$ $5/8 - 1/4 = \square$

---

## SUMMARY

Operations with fractional numbers expressed as common fractions require the same careful development that is made for operations with whole numbers. Even though children are older when operations with fractional numbers are developed, it is still necessary that the processes be developed through carefully sequenced activities with appropriate learning aids, rather than by rules alone. Real-life problems can illustrate situations that give rise to addition, subtraction, multiplication, and division with fractional numbers. Geometric regions, fraction strips, structured materials, markers, and number lines should accompany the real-life problems to give further meaning to the operations. References to similarities that exist between the operations with whole numbers and those with fractional numbers should be stressed whenever it is appropriate.

There are procedures for developing children's understanding of when to simplify (reduce to lowest terms) the answers to problems involving common fractions and to show the rationale behind cancellation when multiplying fractional numbers. Uses of common fractions to express ratios can be introduced early in the program and be developed through simple examples at first, with more complex examples introduced as children mature. The common error of presenting material about computing with common fractions without relating it to computing with whole numbers is not likely to occur if you are aware of the hierarchy of skills for each operation with common fractions. There are common errors committed by children as they compute with common fractions. The errors must be detected and corrected early so they do not become habitual.

## STUDY QUESTIONS AND ACTIVITIES

1. Word problems are used frequently in this chapter to make the four operations with fractional numbers expressed as common fractions more meaningful. Write word problems that might give rise to the following mathematical sentences.

   (a) $3/8 + 5/8 = \square$ (b) $1\frac{1}{2} + 1/4 = \square$ (c) $7/8 - 3/8 = \square$

   (d) $2 \times 3/8 = \square$ (e) $2/3 \times 18 = \square$ (f) $6 \div 1/2 = \square$

   (g) $1/3 \div 2 = \square$

2. Use simple drawings to show how a child might use learning aids to represent each of the sentences in activity 1. When a sequence of illustrations is used, indicate their order by labeling them (a), (b), (c), . . . .

3. Prepare a set of teaching aids (perhaps for a magnetic board or an overhead projector) that illustrates the following multiplication sentences.

   (a) $1/3 \times 12 = \square$ (b) $5/6 \times 24 = \square$ (c) $6 \times 2/3 = \square$
   (d) $2/3 \times 5/7 = \square$

4. Use the common denominator method of dividing to determine the quotient for the following division sentences. Next, use the invert-and-multiply process. In the examples used here, either process is relatively easy to use. Under what other circumstances would the invert-and-multiply method be superior to the common denominator method?

   (a) $3/8 \div 1/4 = \square$ (b) $8/12 \div 1/3 = \square$ (c) $12/16 \div 3/8 = \square$

5. Examine a modern mathematics textbook series to determine at which grade level children are introduced to the concept of ratio. What types of situations are used to introduce this concept? When is the concept of proportion introduced? What situations illustrate this concept?

## FOR FURTHER READING

Bray, Claud J. "To Invert or Not to Invert," *The Arithmetic Teacher,* X, No. 5 (May 1963), 274–276. The author discusses the invert-and-multiply method of dividing fractional numbers. He indicates the essential background of mathematical principles for understanding the process.

Chabe, Alexander M. "Rationalizing 'Inverting and Multiplying,' " *The Arithmetic Teacher,* X, No. 5 (May 1963), 272–273. The author suggests that after children have an understanding of the relationships among the four fundamental processes of arithmetic and the concept of reciprocal, they are ready for a rationalization of the invert-and-multiply algorithm for dividing fractional numbers.

Easterly, Nancy J., and Roy M. Bennett. "Teaching Fractions with Dominoes," *School Science and Mathematics,* LXXVII, No. 2 (February 1977), 117–121. A teacher-made set of 28 fraction dominoes becomes a game for adding with common fractions. A description of the dominoes, rules, and a simulated game are included.

Elashhab, Gamal A. "Division of Fractions—Discovery and Verification," *School Science and Mathematics,* LXXVIII, No. 2 (February 1978), 159–162. A model using arrays of objects or Xs can be used to give added meaning to the invert-and-multiply algorithm for dividing with common fractions.

Ellerbruck, Lawrence W., and Joseph W. Payne. "A Teaching Sequence from Initial Fraction Concepts through the Addition of Unlike Fractions," *Developing Computational Skills,* 1978 Yearbook of the National Council of Teachers of Mathematics. Reston, Va.: The Council, pp. 129–147. The sequence includes initial work with common fractions using real objects and diagrams leading to oral names and fraction symbols. The concepts are extended to include work with equivalent fractions, comparing, and renaming. Addition of like and unlike fractions is described briefly.

Freeman, William W. K. "Mrs. Murphy's Pies—An Introduction to Division by Fractions," *The Arithmetic Teacher,* XIV, No. 4 (April 1967), 310–311. Relates an amusing story to introduce fifth and sixth graders to division of fractional numbers.

Green, George F., Jr. "A Model for Teaching Multiplication of Fractional Numbers," *The Arithmetic Teacher,* XX, No. 1 (January 1973), 5–9. Green lists five characteristics of a satisfactory model for teaching a mathematical concept. Then he identifies a model for teaching multiplication with fractional numbers that has the five characteristics.

Heddens, James W., and Michael Hynes. "Division of Fractional Numbers," *The Arithmetic Teacher,* XVI, No. 2 (February 1969), 99–103. Discusses diagrams and number lines that give meaning to the division processes involving fractional numbers represented as common fractions. An algorithm form is associated with the learning aids.

Jencks, Stanley M., and Donald M. Peck. "Symbolism and the World of Objects," *The Arithmetic Teacher,* XXII, No. 5 (May 1975), 370–371. Children frequently fail to grasp the meaning of operations with numbers because symbols have no meaning to them. Egg cartons provide one means of making addition with common fractions meaningful.

Johnson, Harry C. "Division with Fractions—Levels of Meaning," *The Arithmetic Teacher,* XII, No. 5 (May 1965), 362–368. Presents several approaches to the teaching of division of fractional numbers, including rules presented by texts and teachers, intuitive discovery, and uses of mathematically meaningful or structurally logical approaches. The author concludes that a combination of approaches is appropriate.

Morgenstern, Frances B., and Morris Pincus. "Division with Fractional Numbers: Invert and Multiply?" *School Science and Mathematics,* LXXV, No. 7 (November 1975), 644–648. Reviews the difficulty children have understanding the invert-and-multiply process. Uses geometric regions as models for giving meaning to division with common fractions.

Olberg, Robert. "Visual Aid for Multiplication and Division of Fractions," *The Arithmetic Teacher,* XIV, No. 1 (January 1967), 44–46. Discusses and illustrates the use of rectangular regions as aids in understanding the meanings of multiplication and division with fractional numbers.

Phillips, E. Ray. "Rational Number Division: It's Not So Difficult," *School Science and Mathematics*, LXXVI, No. 5 (May/June 1976), 408–414. The effective use of regions and discrete objects gives meaning to division with common fractions. Examples are amply illustrated.

Thompson, Charles. "Teaching Division of Fractions with Understanding," *The Arithmetic Teacher*, XXVII, No. 5 (January 1979), 24–27. Illustrates and explains a set of materials and sequence of steps for teaching the common denominator method of dividing with common fractions.

Thornton, Carol A. "A Glance at the Power of Patterns," *The Arithmetic Teacher*, XXIV, No. 2 (February 1977), 154–157. Patterns can be developed and recorded to aid children who are learning to multiply and divide with common fractions. (Some uses of patterns in multiplying with decimal fractions and negative numbers and making conversions within the metric system are also described.)

Zweng, Marilyn J. "The Fourth Operation Is Not Fundamental—Fractional Numbers and Problem Solving," *The Arithmetic Teacher*, XXII, No. 1 (January 1975), 28–32. Zweng presents a case for doing away with division with common fractions in the elementary school program. She does believe that the situations that are commonly used to present such division need to be presented, but that they can be solved by methods other than division.

# 13 Fractional Numbers—Decimals and Percent

Upon completion of Chapter 13, you will be able to:

1. Demonstrate the uses of at least two different learning aids for activities dealing with the meaning of decimal fractions.

2. Demonstrate a procedure for extending children's understanding of place value notation to include positions to the right of the ones place.

3. Write two different expanded numeral forms for representing fractional numbers expressed as decimal fractions.

4. Describe real-life situations involving addition and subtraction with decimal fractions, and demonstrate learning aids children can use to develop an understanding of this type of addition and subtraction.

5. Make up story problems for multiplication sentences such as $4 \times 0.3 = 1.2$, $0.5 \times 10 = 5$, and $0.2 \times 0.4 = 0.08$, and use suitable learning aids to show the meaning of each sentence.

6. Explain two procedures that can be used to determine the number of places to the right of the ones place in the product of multiplication involving decimal fractions.

7. Demonstrate with appropriate learning aids activities that extend partitive and measurement division situations to include fractional numbers represented by decimal fractions.

8. Explain two ways of determining where to put the decimal point in quotients for division involving decimal fractions.

9. Define percent, and describe a real-life situation involving it.

10. Distinguish between percent (rate), base, and percentage, and give an example of each in a real-life situation.

11. Describe materials that are suitable for children's investigations into the meaning of percent.

12. Rename common and decimal fractions as percents, and rename percents as common and decimal fractions.

13. Use the proportion and at least one other method for solving percent problems.

14. Describe four activities with the minicalculator dealing with decimal fractions and percent.

15. Identify two pitfalls commonly associated with teaching decimal fractions and percent, and describe how to avoid them.

16. Describe a common problem children encounter as they add and subtract *ragged decimals,* and explain orally or in writing two ways to help them overcome it.

Key Terms you will encounter in Chapter 13:

- fractional number
- common fraction
- decimal fraction
- percent
- tenth
- hundredth
- decimal point
- expanded notation
- exponent
- partitive division
- measurement division
- ratio
- rate
- base
- percentage

The need for knowledge of decimal fractions will increase as uses of the minicalculator and metric system increase. It is probable that decimal fractions will be introduced earlier in the elementary school program and that more time will be devoted to them in the future than presently. This chapter discusses materials and procedures for helping children learn about decimal fractions and percent.

## INTRODUCING DECIMAL FRACTIONS

Decimal fractions are one of three ways to represent fractional numbers. Studying them should be related to what has been learned about common fractions and the Hindu-Arabic numeration system. Models for decimal fractions should be similar to those used to learn about common fractions so that relationships between the two ways of representing fractional numbers can be emphasized. In many instances, middle-grade children can learn about common and decimal fractions simultaneously, using the same models. This approach has two benefits. First, children learn that common fractions and decimal fractions both represent fractional numbers rather than viewing them as being unrelated, as is often the case when they are studied separately. Second, it saves time because many of the same concrete and semiconcrete materials can be used simultaneously to develop an understanding of both types of fractions.

You should also make clear that decimal fractions are an extension of the already familiar base ten numeration system.

Once it is clear that children are ready to learn about decimal fractions, familiar concrete aids should be used for developing their understanding. Cuisenaire rods or Dienes multibase blocks are useful, if they are available. Children who are accustomed to these materials will know that any particular piece can be named as a unit. Therefore they can use a tens flat from a Cuisenaire or Dienes set as a unit. Ten rods cover a flat, so each rod is a tenth of the flat; while 100 small unit squares cover a flat and each one is a hundredth of a flat.

Squared paper with centimeter squares can be cut to make an activities kit. A kit, which each child can make, consists of several squares that are ten units along each side, at least ten strips that are ten units long, and 100 small unit squares. If squared paper is not available, you can draw a kit's pattern on duplicating masters and make copies of the parts for each child. Figures 13-1 and 13-5 show the parts for one kit.

## 1. Tenths

When a group of children are ready to begin investigating decimal fractions, give each child a kit of structured or squared-paper materials. Discuss the idea that each large square is a unit or whole, and review past work, if any, where the children have subdivided units into halves, thirds, fourths, and other fractional parts. Next, have the children take ten rods or strips and put them on top of a unit square and lay them beside another unit square (Figure 13-1). Have the children count the rods or strips covering the unit and discuss how many there are. Point out that each piece is one-tenth of the unit; write both the numeral $^1/_{10}$ and 0.1 and tell the children that they each represent a single rod or strip. Help children understand the meaning of 0.2, 0.3, 0.4, . . . , 1.0 in the same way. Be sure children pay particular attention to 1.0 so they will see that it represents a unit region that has been subdivided into ten congruent parts. The 0 in the tenths place of 1.0 is necessary because it shows the subdivision of the unit into ten congruent parts.

Following an introduction to the decimal notation of tenths, children should be helped to find uses of tenths represented by decimal fractions. Application of the decimal fraction notation in the metric system should be emphasized. Other applications that are familiar to many children are division of the mile into tenths on automobile odometers, a gallon into tenths on gasoline pumps, and a pound into hundredths on butchers' scales. Odometers for classroom use can be obtained from automobiles in wrecking yards. A classroom display model of an odometer can be made from masonite or heavy oaktag paper.[1]

Some teachers use United States money as an example of decimal notation. However, there are sound arguments against its use during introductory lessons. Many children do not think of a dollar as the unit of

[1]See Leonard M. Kennedy, *Models for Mathematics in the Elementary School* (Belmont, Calif.: Wadsworth Publishing Company, Inc., 1967), pp. 131–133. See also pp. 133–140 for descriptions of a set of transparent overlays for decimal fractions.

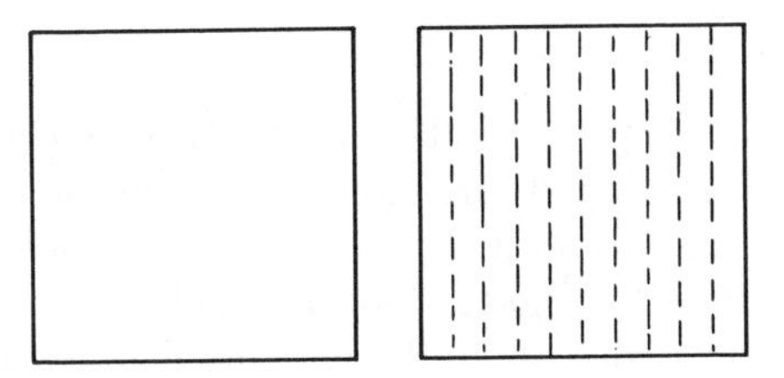

Figure 13-1 Unit region and region separated into ten congruent parts

which pennies and dimes are fractional parts, but instead that the penny is the unit from which other coins and bills of greater value are built. Also, any discussion of the different coins — pennies, nickels, dimes, quarters, and half dollars — requires that hundredths as well as tenths be discussed. Children should understand tenths and their decimal notation before they begin to study hundredths and their notation.

Other learning aids can be used before children begin their study of tenths. For example:

1. *Geometric shapes.* Shapes other than squares can be separated into ten congruent parts. The parts should be identified and represented by decimal fraction notation (Figure 13-2).

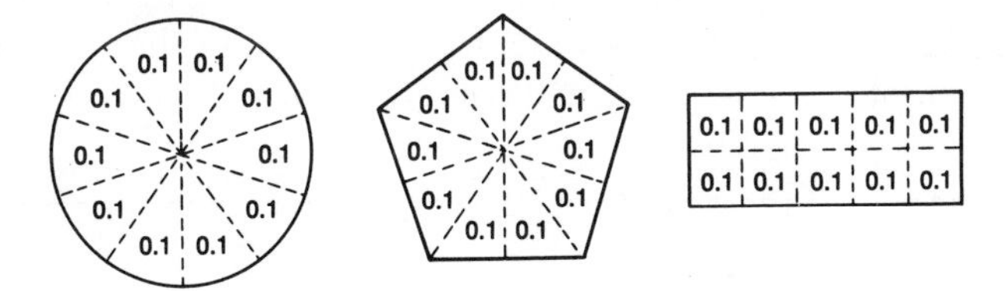


Figure 13-2 Geometric shapes separated into ten congruent parts to illustrate tenths

2. *Fraction strips.* Use a unit strip and a strip separated into ten congruent parts. Children will note that decimal fraction notation represents tenths (Figure 13-3).

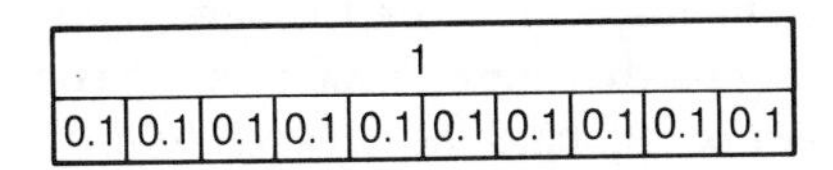


Figure 13-3 Fraction strips show the decimal fraction representation of tenths

3. *Number lines.* You can use number lines to develop understanding of decimal fractions. Begin with a line that shows only unit segments. On a second line show each unit marked into ten congruent segments. The children should identify and name each point on the line as its meaning is discussed (Figure 13-4).

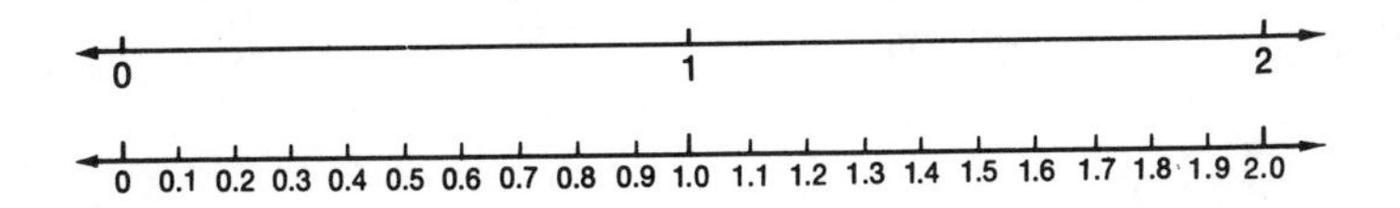


Figure 13-4 Number lines used to illustrate units and tenths

## 2. Hundredths

Children's understanding of the decimal fraction representation of hundredths should be developed through extension of the activities used for tenths. A unit region should first be displayed, separated into ten, and then 100 congruent parts (Figure 13-5). Have children count some of the parts by hundredths. Then show the decimal notation for the parts the children counted: 0.01, 0.02, 0.03, . . . .

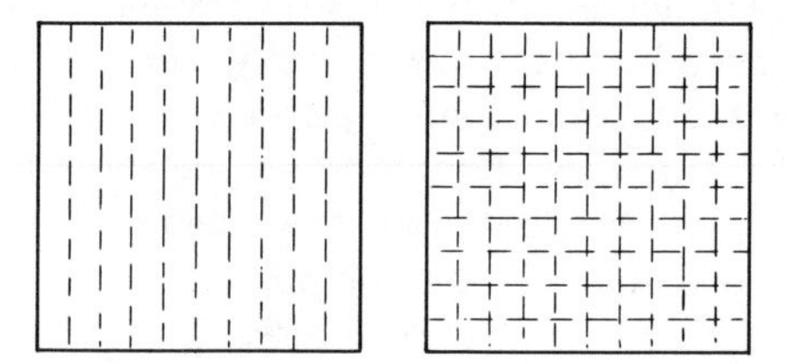

**Figure 13-5** Unit regions separated into 10 and 100 congruent parts

A number line with units separated into 100 congruent segments also helps develop an understanding of hundredths (Figure 13-6). Even though it may be impractical to write the numeral for each point on the line, children can recite their names as they indicate numbers in sequence. Fraction strips separated into 100 congruent parts are also useful learning aids. The fact that ten-hundredths are equivalent to one-tenth should be emphasized as these materials are used.

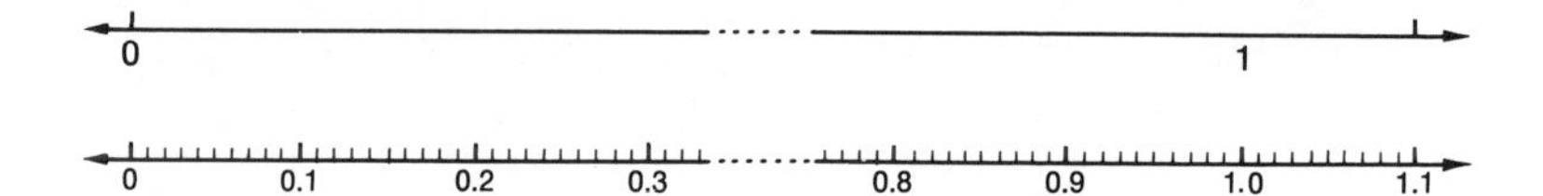


**Figure 13-6** Number line with a unit segment separated into 100 congruent parts to illustrate hundredths

## 3. Other Decimal Fractions

Learning aids for representing fractional numbers smaller than hundredths are difficult to make in sizes that are practical for children to manipulate. A unit region that shows 1000 or more congruent parts can be made on a large sheet of tagboard for children to use during discussion of decimal fractions smaller than hundredths. It is not unreasonable for children to *imagine* that such a unit has been cut into 1000, or 10,000, or even 100,000 parts as they deal with the smaller decimals.

---

## Self-check of Objective 1

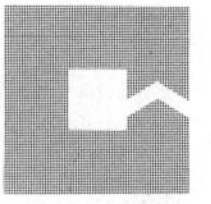

Select any two learning aids and demonstrate how they are used to show the meaning of decimal fractions.

---

## EXTENDING PLACE VALUE TO INCLUDE FRACTIONAL NUMBERS

Kits of structured or squared-paper materials provide a basis for developing the concept of decimal fractions as numerals for fractional numbers. By using these kits, children will learn that $\frac{2}{10}$ and 0.2, $\frac{13}{100}$ and 0.13, and $1\frac{3}{10}$ and 1.3 are pairs of numerals for naming three different numbers. Children's work must go beyond the simple fractions studied through activities with manipulative materials so that they can see how decimal fractions fit into the Hindu-Arabic numeration system. The Hindu-Arabic system is a decimal system because each place value position is a 10 or power of 10. Each place value position has ten times the value of the position to its immediate right. Or, looking from the other direction, each position has a value that is one-tenth of the one to its immediate left. Children must see how decimal fractions provide place positions to the right of the ones place to accommodate numbers smaller than one.

### 1. Using Place Value Pocket Charts

Place value pocket charts are useful aids for extending this concept. The number 21.4 is represented in the pocket chart shown in Figure 13-7. The two bundles of ten markers represent twenty unit squares, the single marker in the ones place represents one unit square, and the four markers in the tenths position represent four-tenths of another unit. Another place value position can be added to the right of the tenths place to represent hundredths on the chart. Once children are familiar with the extended chart, have them complete two types of exercises: (1) Show them pocket charts (or pictures of charts) such as the one in Figure 13-7, and have them write the decimal fraction for the number represented on each chart; (2) Give them decimal fractions and have them represent each one with markers in a chart, or by drawing pictures of charts.

### 2. Using an Abacus

When an abacus is used, a rod other than the one on the extreme right represents the ones place. Put a small tag with a dot on it to the right of the rod used to indicate the ones place. The abacus in Figure 13-8 represents the number 234.062. Use exercises such as the ones mentioned

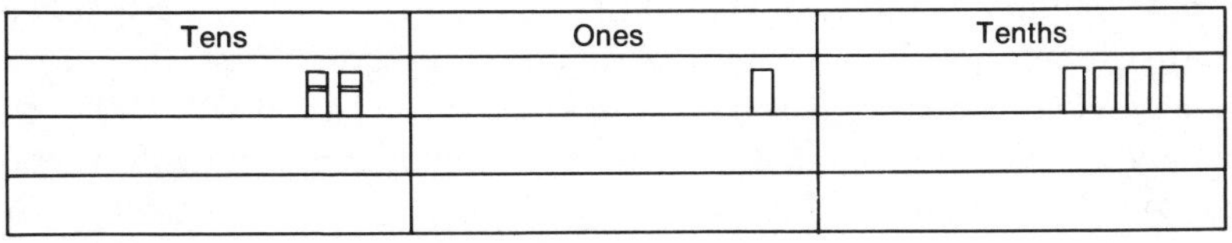

| Tens | Ones | Tenths |
|---|---|---|
| | | |
| | | |
| | | |

**Figure 13-7** Place value pocket chart used to represent tens, ones, and tenths, in this case, the number 21.4

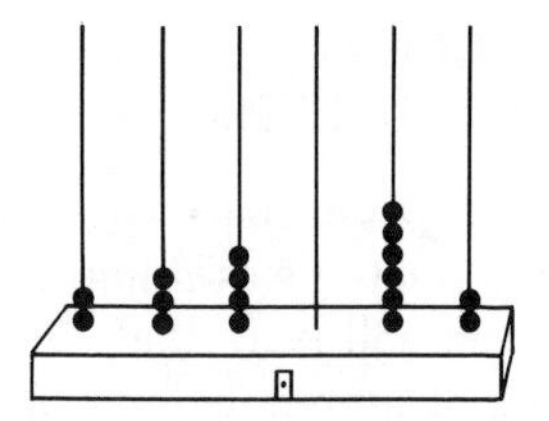

**Figure 13-8** The number 234.062 represented on an open-end abacus

above to reinforce children's understanding of smaller place value positions represented on an abacus.

Help children summarize ideas about place value by guiding their development of a chart such as the one in Figure 13-9. Their summary should emphasize that the ones place, not the decimal point, is the point of symmetry for place value in the Hindu-Arabic numeration system. Also, each place value position has a value ten times that of the position to its right and one-tenth that of the position to its left, regardless of where the positions are with reference to the ones place. The chart clearly shows that the ones place is central to the system; it stands alone, while positions to the right and left of it are connected by lines to show pairs of related positions, such as tens and tenths. The decimal point is a convenient way to indicate the ones place when a numeral is written. Once the chart has

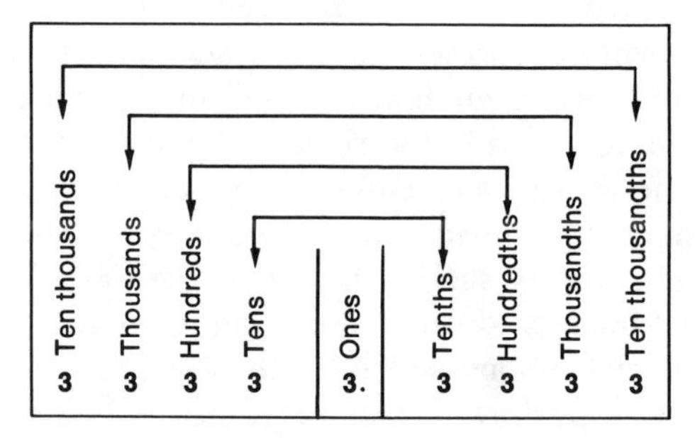


**Figure 13-9** Place value positions of a number indicated on a chart

been developed, children should refer to it frequently as they study the relationships that exist among the various positions. Children can speculate about other ways they might indicate the ones place in numerals as they imagine that the decimal point has not been "invented." Such speculation should help them reinforce their understanding of the ones place having a central position in the system. (Children might come up with notations such as those that follow: 23④12, 23 $\underline{4}$ 12, 23 4 $\dot{1}\dot{2}$, or 234·12.)

As children work with whole numbers, they learn to use expanded notation to interpret the meaning of different numbers. First, they learn that a whole number, such as 6382, can be expressed as 6000 + 300 + 80 + 2. By the end of elementary school, they should refine the interpretations to include these:

$$(6 \times 1000) + (3 \times 100) + (8 \times 10) + 2.$$
$$(6 \times 10 \times 10 \times 10) + (3 \times 10 \times 10) + (8 \times 10) + 2.$$
$$(6 \times 10^3) + (3 \times 10^2) + (8 \times 10) + 2.$$

They should also learn to use expanded notation with decimal fractions.

Initial exercises, such as the following ones, should involve filling in the blanks.

2.36 = ____ ones ____ tenths ____ hundredths.
0.46 = ____ ones ____ tenths ____ hundredths.
2.482 = ____ ones ____ tenths ____ hundredths ____ thousandths.

Later exercises should stress more abstract symbolism:

$4.68 = 4 + (\_\_\_\_ \times 0.1) + (\_\_\_\_ \times 0.01).$
$2.43 = 2 + (4 \times \_\_\_\_) + (3 \times \_\_\_\_).$
$2.006 = 2 + (\_\_\_\_ \times 0.1) + (\_\_\_\_ \times 0.01) + (\_\_\_\_ \times 0.001).$

Finally, by the end of grade six, some children will be ready to use exponential notation to represent decimal fractions:

$$343.68 = (3 \times 10^2) + (4 \times 10) + 3 + (6 \times 1/10) + (8 \times 1/10^2)$$
$$241.32 = (2 \times 10^2) + (4 \times 10) + 1 + (3 \times 1/10) + (2 \times 1/10^2)$$

The following form may be used by children who have a mature enough understanding of negative numbers to make the use of exponents in this notation clear.

$$343.68 = (3 \times 10^2) + (4 \times 10^1) + (3 \times 10^0) + (6 \times 10^{-1}) + (8 \times 10^{-2})$$

## WRITING DECIMAL FRACTIONS FROM DICTATION

It has been commonplace for teachers to dictate numerals containing decimal fractions for children to write. The practice is of marginal value, and the way it was done in the past is no longer recommended. Business requirements have changed so that the reading and copying of numerals are seldom done and the skill is of little importance. When numbers containing decimals are dictated, one such as 346.62 should be read as "three, four, six, point, six, two" and not as "three hundred forty-six and sixty-two hundredths." If you believe your children should practice writing dictated numbers, use the first way of reading them rather than the second.

## Self-check of Objectives 2 and 3

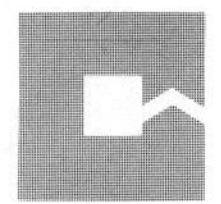

Use a place value pocket chart or an abacus to show how each of these numbers can be represented: 23.4, 36.50, 42.031, and 40.36.

Write two different expanded numerals for each of these decimals: 246.48, 304.06, and 0.342.

## ADDING AND SUBTRACTING WITH DECIMAL FRACTIONS

Children who have developed a good understanding of fractional numbers and their decimal fraction representations will have little difficulty adding and subtracting them. Fractional numbers expressed with decimals are added and subtracted just like whole numbers expressed with Hindu-Arabic numerals.

### 1. Adding with Decimal Fractions

Begin addition and subtraction work with familiar situations. For example, "Judy uses her bicycle to do errands. She keeps track of the distance from place to place with the metric odometer on her bicycle. One day she rode 0.7 of a kilometer from home to a hardware store, then 0.3 of a kilometer from there to a paint store, and finally 0.8 of a kilometer from the paint store back home. How many kilometers did she ride?"

Children should first determine the mathematical sentence that represents the situation: $0.7 + 0.3 + 0.8 = \square$. Next, they should use learning aids such as fraction strips and the number line to determine the answer. In Figure 13-10, fraction strips represent the sentence with sets of seven, three, and eight one-tenth-unit pieces (a). Then the three sets of fractional parts are joined (b). Finally, ten of the tenth pieces are exchanged for a unit strip, and the answer is indicated by one unit strip and eight of the tenth strips (c). On the number line, Figure 13-11, each addend is indicated by one of three arrows, the sum by the numeral at the head of the last arrow.

Addition with hundredths can be illustrated in similar ways. In Figure 13-12 the addition sentence $0.26 + 0.49 = \square$ is represented with parts of a unit region. First, twenty-six of the 100 congruent parts are shaded to represent the first addend; then forty-nine parts of another square are shaded to represent the second addend. When the two sets of shaded parts are considered together, they indicate the sum, 0.75.

Addition with decimal fractions can be represented with place value pocket charts and an abacus, also. These devices are used with decimal fractions in the same way as with whole numbers. Procedures with whole numbers are described in Chapter 8.

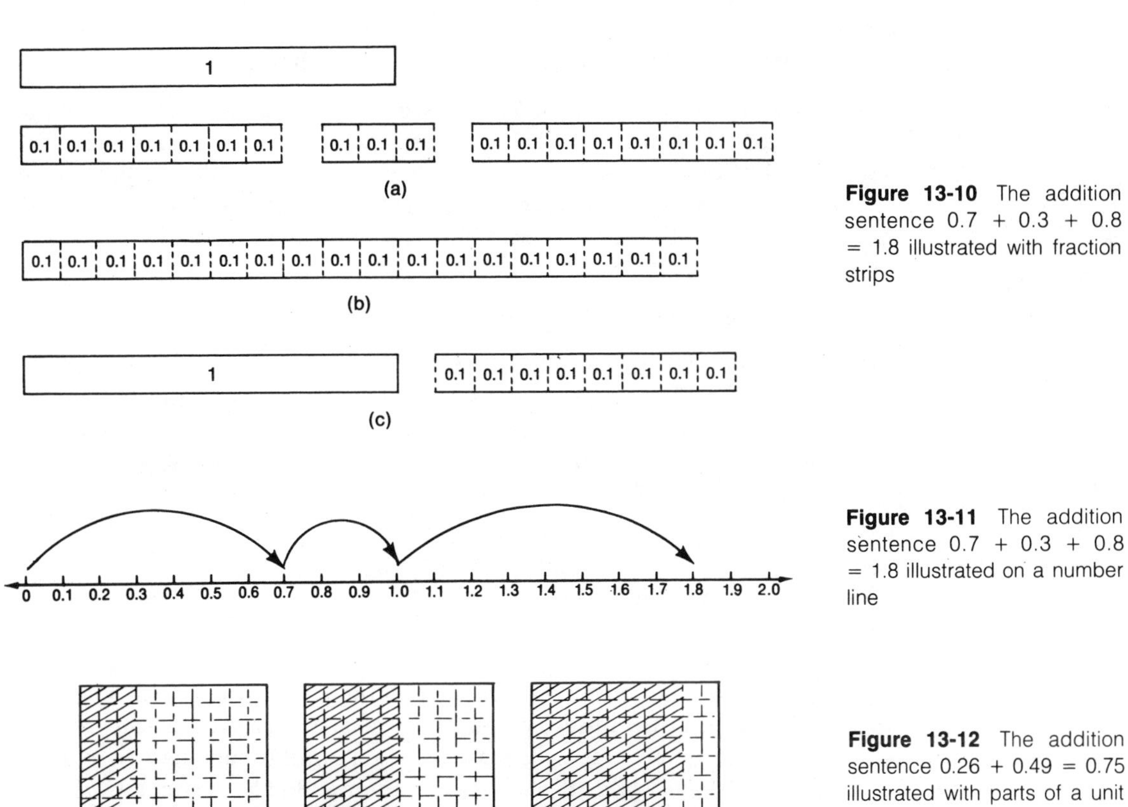


**Figure 13-10** The addition sentence 0.7 + 0.3 + 0.8 = 1.8 illustrated with fraction strips

**Figure 13-11** The addition sentence 0.7 + 0.3 + 0.8 = 1.8 illustrated on a number line

**Figure 13-12** The addition sentence 0.26 + 0.49 = 0.75 illustrated with parts of a unit region

## 2. Subtracting with Decimal Fractions

Two subtraction algorithms for whole numbers are discussed in Chapter 8. Either one of these forms can be used with decimal fractions. Children who have mastered the decomposition method should use it with decimal fractions. The example in the margin illustrates the process. Children who have learned the equal addition method should use it with decimal fractions, too.

```
  1 1
 63.24
-12.69
 50.55
```

## Self-check of Objective 4

Describe orally or in writing real-life problem situations for each of these sentences: 0.4 + 0.5 + 0.7 = 1.6 and 1.4 − 0.8 = 0.6. Demonstrate with suitable learning aids how children can learn the meaning of the algorithms for these two sentences. (Use illustrations if the learning aids are unavailable.)

## MULTIPLYING WITH DECIMAL FRACTIONS

Because fractional numbers expressed as decimal fractions are multiplied using the same algorithm form as whole numbers, you should guard against this process becoming mechanical for your children. Introductory work with such multiplication should be presented within a familiar context relevant to other aspects of the children's lives.

### 1. Multiplication Situations

"Jack is repairing his railroad track layout. He has found that he will need six pieces of wire, each 0.6 of a meter long. How much of the wire will he use altogether?" This situation is interpreted as repeated addition; it is illustrated on the number line in Figure 13-13, where each unit represents one meter.

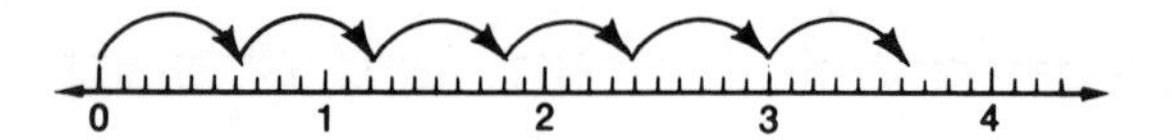


**Figure 13-13** The multiplication sentence $6 \times 0.6 = 3.6$ illustrated on a number line

"Gold leaf is sold in sheets 10 centimeters square. Sarah has 0.7 of a full sheet. If she uses 0.3 of this piece, what part of an entire sheet will she use?" A square unit is best for representing this situation, as in Figure 13-14. The 0.7 of a full sheet can be represented by shading seven of the unit's ten congruent parts (a). Then 0.3 of the 0.7 can be represented by separating each of the tenths into ten congruent parts and shading three of the parts in each tenth (b). The answer is represented by 21 of the parts that are shaded twice. Each of the 21 parts represents one of 100 congruent parts into which the unit has been separated, so the product is 0.21.

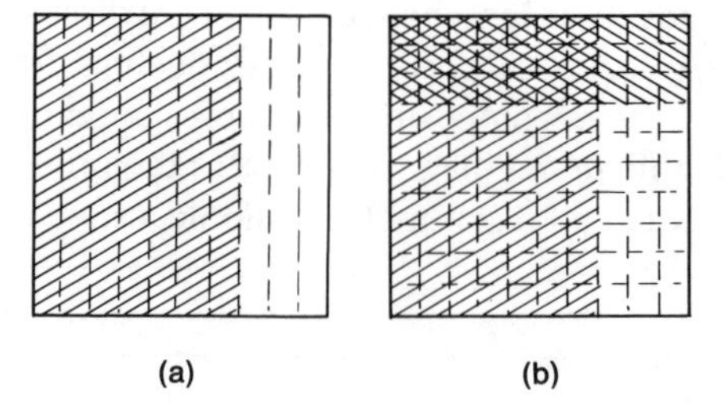


**Figure 13-14** The multiplication sentence $0.3 \times 0.7 = 0.21$ illustrated with parts of a unit region

The multiplication sentence $0.2 \times 4 = \square$ is illustrated in Figure 13-15. A piece of adding machine tape four meters long is shown in (a). It is separated into ten congruent parts in (b). Two-tenths of the four meters are shown in (c), where 0.8 of one meter is shaded.

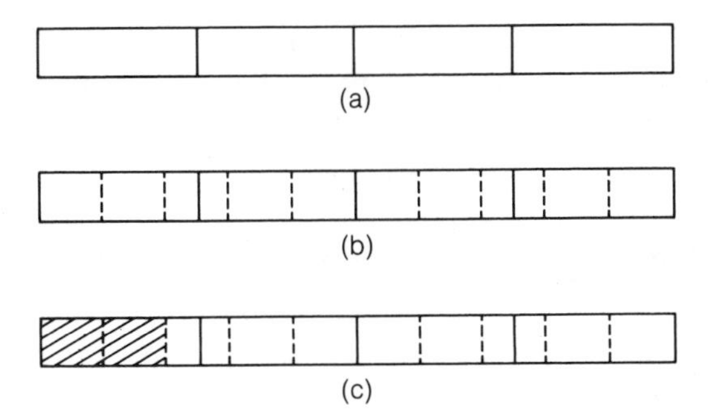


**Figure 13-15** The multiplication sentence 0.2 × 4 = 0.8 illustrated with a strip of adding machine tape

## 2. Placement of the Decimal Point in a Product

After visualizing the meaning of multiplication of decimals with the help of learning aids, children should learn where to place the decimal point in a product. Eventually, they should be guided to generalize that the number of decimal places in a product equals the sum of decimal places to the right of the ones place in the factors.

Estimation is a useful way to learn how to put decimal places in a product. In the sentence 0.9 × 2.1 = □, since 0.9 is almost 1 and the 2.1 is only slightly larger than 2, the product will be about 2. After the multiplication has been completed, children can use the estimated answer to determine that the 1 in the product will be in the ones place, so the answer is 1.89. The sentence 11.2 × 23.4 = □ can be estimated by thinking: 10 times 25 equals 250, so 11.2 times 23.4 will be about 250. The product of the sentence should then be determined to be 262.08.

After children understand how to determine the placement of the decimal point, give them practice with exercises similar to the following ones. Have them underline the numeral that names the correct product for each pair of factors.

| | | | | |
|---|---|---|---|---|
| 2.4 × 6.8 = | 16.32 | 1.632 | 163.2 | 1632 |
| 10.6 × 3.68 = | 3900.8 | 390.08 | 39.008 | 3.9008 |
| 9 × 2.98 = | 2682 | 268.2 | 2.682 | 26.82 |

After children have completed exercises such as these, ask questions to help them discover the generalization about determining the number of decimal places in a product. "How many decimal places are there following the one in the first factor of this sentence?" "How many are there in the second factor?" "What is the sum of the number of places to the right of the ones in both factors?" "How many decimal places are there to the right of the ones place in the product?" "Look at the next sentence. What do you notice about the number of decimal places to the right of the ones place in both of its factors and in its product?" "Now look at the remainder of the sentences. What do you notice about them?" "Can anyone tell how to determine where to put the decimal point in a product when the factors are decimal fractions?"

---

## Self-check of Objectives 5 and 6

Describe orally or in writing a real-life problem situation for each of these sentences: $4 \times 0.3 = 1.2$, $0.5 \times 10 = 5$, and $0.2 \times 0.4 = 0.08$. Demonstrate with suitable learning aids ways children can learn the meanings of these sentences. (Use illustrations if the learning aids are unavailable.)

Explain how the two procedures described in the preceding paragraphs can be used to determine the number of decimal places to the right of the ones place in the product of each of these multiplication examples: $0.8 \times 0.9 = \square$, $3.1 \times 15.14 = \square$, and $11 \times 4.92 = \square$.

---

## DIVIDING WITH DECIMAL FRACTIONS

Division involving decimal fractions determines either the size of each group when a set is divided into a given number of groups or the number of groups when a set is subdivided into groups of a given size, just as it does in division of whole numbers. The first of these is a partitive situation; the second, a measurement situation. The algorithm forms used for division of whole numbers are also used for division of fractional numbers expressed as decimal fractions. The concepts and algorithms are the same for division of both kinds of numbers. The new element is the treatment of the decimal point in the quotient. A logical explanation, not a memorized rule, is needed to help children learn to place it properly.

### 1. Division Situations

The first experiences can use either a partitive or a measurement situation. For example: "A rope is 16.4 meters long. If Jane cuts it into four pieces of equal length, how long will each piece be?" The solution to this partitive situation can be represented with several of the learning aids. In Figure 13-16 it is represented with a pocket chart. The dividend is shown by the markers in (a). The 1 ten is then exchanged for 10 ones and the 16 ones are separated into four groups (b). Finally, the four tenths are separated to become a part of the four groups. The quotient is represented by the markers in one of the four groups shown in the pocket chart in (c).

The same sentence can be represented with unit regions; sixteen of the unit regions and four tenths of another unit region are partitioned into four groups, each containing four of the unit regions and one of the tenths pieces. Fraction strips and number lines can also be used.

| Tens | Ones | Tenths |
|---|---|---|
| | | |
| | | |
| | | |
| | | |
| | | |

(a)

| Tens | Ones | Tenths |
|---|---|---|
| | | |
| | | |
| | | |
| | | |
| | | |

(b)

| Tens | Ones | Tenths |
|---|---|---|
| | | |
| | | |
| | | |
| | | |
| | | |

(c)

**Figure 13-16** The division sentence 16.4 ÷ 4 = 4.1 illustrated with markers in a set of pocket charts

"A relay team is to run a 27.5-kilometer cross-country race. Each member of the team will run 5.5 kilometers. How many runners are there on the team?" A number line is effective to represent this situation, as in Figure 13-17. The quotient is represented by the five jumps, each 5.5 units long, that extend from 27.5 to 0. This sentence can be represented with unit regions, fraction strips, and place value pocket charts.

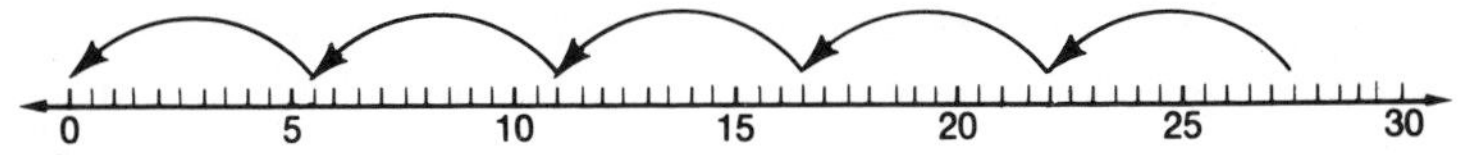


**Figure 13-17** The division sentence 27.5 ÷ 5.5 = 5 illustrated on a number line

## 2. Algorithms

At the same time children determine quotients using their learning aids, they should also learn to use algorithms for determining the answers. The vertical algorithm form for the sentence 16.4 ÷ 4 = □ is illustrated in the margin. You should direct children to think, "If 16 ones are separated into

```
4)16.4 |
  16.0 | 4 × 4
  ----
   0.4 |
   0.4 | 4 × 0.1
       |     4.1
```

four groups of equal size, what is the size of each group?" The $4 \times 4$ at the right in the algorithm indicates that 4 names the size of each of the four groups. "What part of the 16.4 is still undivided?" The 0.4 in the algorithm represents the undivided part after 16.0 is subtracted from 16.4. "If four tenths are separated into four groups, what is the size of each group?" The $4 \times 0.1$ at the right indicates this size. The answer, 4.1, is indicated at the bottom of the algorithm.

For the sentence $27.5 \div 5.5 = \square$, children should be helped to recognize that repeated subtraction can determine the quotient:

$$\begin{array}{r} 27.5 \\ -\ 5.5 \\ \hline 22.0 \end{array} \qquad \begin{array}{r} 22.0 \\ -\ 5.5 \\ \hline 16.5 \end{array} \qquad \begin{array}{r} 16.5 \\ -\ 5.5 \\ \hline 11.0 \end{array} \qquad \begin{array}{r} 11.0 \\ -\ 5.5 \\ \hline 5.5 \end{array} \qquad \begin{array}{r} 5.5 \\ -\ 5.5 \\ \hline \end{array}$$

After solving the problem with repeated subtraction, they should use the algorithm. Guide their thinking by asking, "If 27.5 is separated into groups that are each 5.5 in size, how many groups will there be?"

If children think of partitive and measurement situations when they use the algorithms, they will understand how to interpret quotients so placement of the decimal point develops logically. If the sentence $6.3 \div 3 = \square$ arises from a partitive situation, the quotient indicates the size of each of three groups. If it is the result of a measurement situation, the quotient indicates that there are two sets of three and one-tenth of another set of three in 6.3.

### 3. Placement of Decimal Point in Quotient

Understanding the following ideas helps children learn how to place the decimal point in quotients.

1. Estimation is a way to determine the number of decimal places in a quotient. To divide a pair of numbers as in the sentence $2.36 \div 0.9 = \square$, children should think "2.36 is a little larger than 2 and 0.9 is a little smaller than 1. Since $2 \div 1 = 2$, 2.36 divided by 0.9 will be about 2." When the quotient is determined, children should place the decimal point so that 2 is in the ones place. An exercise such as the following one provides practice in estimating. Each sentence becomes a true statement only when the decimal point is correctly placed.

$$69.3 \div 3 = 231.$$
$$811.8 \div 22 = 369.$$
$$20.74 \div 3.4 = 61.$$

2. Children who know that common fractions can be used to express division can rewrite a decimal fraction, such as $0.3\overline{)6.3}$, as a

common fraction: $^{6.3}/_{0.3}$. When both numerator and denominator are multiplied by 10, it becomes $^{63}/_{3}$. This can then be rewritten in algorithm form $3\overline{)63}$, so the division can be completed. This is the mathematical justification for the "caret" method of placing the decimal point in a quotient. If both divisor and dividend are multiplied by 10 or a power of 10 so the divisor is changed to a whole number, the decimal point in the quotient can be placed immediately above the decimal point in the changed dividend. This is illustrated in the margin; a caret (∧) indicates the new position of the decimal point in the divisor and dividend. Children should use this procedure after its rationale has been made clear to them.

```
          4.98
6.3.)  31.3.74
       252
        617
        567
         504
         504
```

Eventually, children should be guided to make this generalization: The number of decimal places to the right of the ones place in a quotient is the difference between the number of places to the right of the ones place in the dividend and the number there are in the divisor. Once they have stated this generalization, have children compare it with the generalization about how to determine the number of decimal places in a product. They should observe that the operations indicated by the two generalizations are the inverse of each other.

## Self-check of Objectives 7 and 8

Make up a story problem for partitive division involving this sentence: 1.2 ÷ 4 = 0.3. Use an appropriate learning aid to illustrate the sentence's meaning. Describe a measurement situation for this sentence: 4 ÷ 0.5 = 8. Use an appropriate learning aid to illustrate this sentence's meaning.

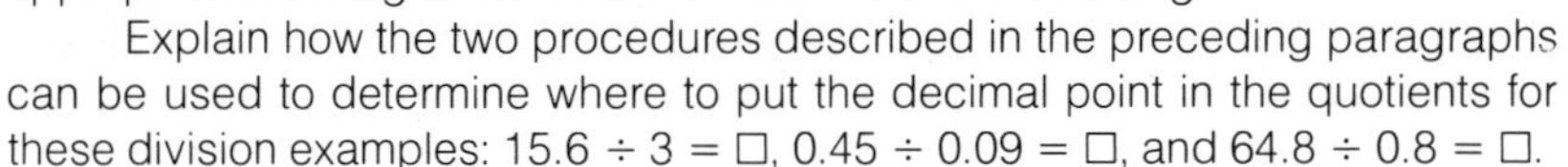

Explain how the two procedures described in the preceding paragraphs can be used to determine where to put the decimal point in the quotients for these division examples: 15.6 ÷ 3 = □, 0.45 ÷ 0.09 = □, and 64.8 ÷ 0.8 = □.

## PERCENT

In a sense, activities with percent in the sixth grade are a capstone for the work with comparisons that began in the first year of school. A part of children's first activities in mathematics is concerned with comparing sets to determine which one is larger or smaller, or whether they are equivalent; and comparing the length of two objects to find out which is longer or shorter, or whether they are the same length. Later, children count discrete objects and measure continuous objects and use subtraction to compare the sizes of sets or measures of objects. As their knowledge of mathematics increases, they use common and decimal fractions for making comparisons. Work with percent gives children yet another way of dealing with comparisons.

## 1. Meaning of Percent

When percent is used, the comparison is expressed as a ratio between some number and 100. For example, when 15 is used as a percent, it is an expression of the ratio between the numbers 15 and 100; it is symbolized as 15%. The symbol % expresses a denominator of 100; the word *percent* names the symbol and means "per hundred," or "out of 100."

In practice, percent is used in various ways, and its applications in business, government, science, industry, and other aspects of our lives are numerous. For example, a store reports that 16% of its sales are made in its garden department. This indicates that for every 100 dollars of sales, 16 are in the garden department. The complete report will contain information that tells the portion of sales made in the store's other departments, with the total amount of sales equaling 100%.

Some states use a sales tax as a means of raising revenue for part of their annual budget. A tax of 5% means that for every 100 cents (1 dollar) of the purchase price of an article, a tax of 5 cents (0.05 of a dollar) must be collected. This situation provides the setting for a discussion of terms used in connection with percent and consideration of a point of confusion that often arises during the study of percent.

For a number of years the state of California used a 5% sales tax as a means of collecting a part of the money needed for its budget. When a $10.00 sale was made, 50¢ was collected for the sales tax. Three words are used to describe the different numerals in this situation. The "5%" is the rate, "$10.00" is the base, and "50¢" is the percentage.

When children encounter a statement such as the following, which might appear in a social studies textbook, they need to know how to interpret it so it will have meaning: "The imports of grain are 36% of the country's imports. The total value of all imports for one year was $236,000,000." In this statement, 36% is the rate and $236,000,000 is the base. Children must analyze the statement by thinking somewhat as follows: "I know that 36% represents a rate of $36 out of every $100. This means that about ⅓ of all the money value of the imports is for grain. One-third of $236,000,000 is about $80,000,000. The country imports about $80,000,000 worth of grain." $80,000,000 is the estimated percentage of grain among all imports.

*Percent* and *percentage* are often confused, and many children (and adults, as well) are uncertain about the distinction between them. Because confusion is likely, you must make it clear that percent indicates the *rate* (of sales in a store's department, or of sales tax), while percentage indicates the *quantity,* or *amount* (of sales, or of tax). Thus a base and a percentage always represent numbers that refer to the same thing, for example, numbers of dollars of sales made or the weights of different ingredients in a mixture, while percent is the rate by which the percentage compares with the base. The potential for confusion is great enough to justify delayed use of the word *percentage* during elementary school, leaving it for teachers in higher grades to introduce. Its omission will not interfere with children's understanding of the basic ideas involved.

The California state sales tax situation provides the basis for discussion of another potential point of confusion. On the first of July, 1973, the sales tax rate was changed from 5% to 6%. This was a very unpopular change because at the time the state's treasury had a budget surplus of nearly 600 million dollars. Later, the legislature passed and the governor signed into a law a bill that rescinded the increase for 6 months beginning October 1, 1973. Was the *rate* (percent) of decrease the same as the rate of increase? In both cases, the *amount* was the same. The increase was 1¢ for every dollar spent, and the decrease was also 1¢ for every dollar spent. However, the rate of decrease was not the same as the rate of increase. The number line graphs in Figure 13-18 show why the rates in this increase-decrease situation are different, even though the amounts are the same. In (a), the original sales tax, which was 5¢ for every dollar spent (5%), is represented by the first line and is shown as 100%, since 5¢ was the whole amount of tax for each dollar before the increase. The second line in (a) shows the sales tax after it was changed to 6%. The amount of increase in the length of this line is the same as the length of one of the five parts of the original line. One of five is 20% of the whole, so the rate of increase was 20%.

In (b), the new sales tax is represented by the top line. It now represents 100%, since the rate was 6¢ for every dollar spent. The amount of decrease in the length of the second line is the same as the length of one of the six parts of the top line. One of six is 16⅔% of the whole, so the rate of decrease was 16⅔%.

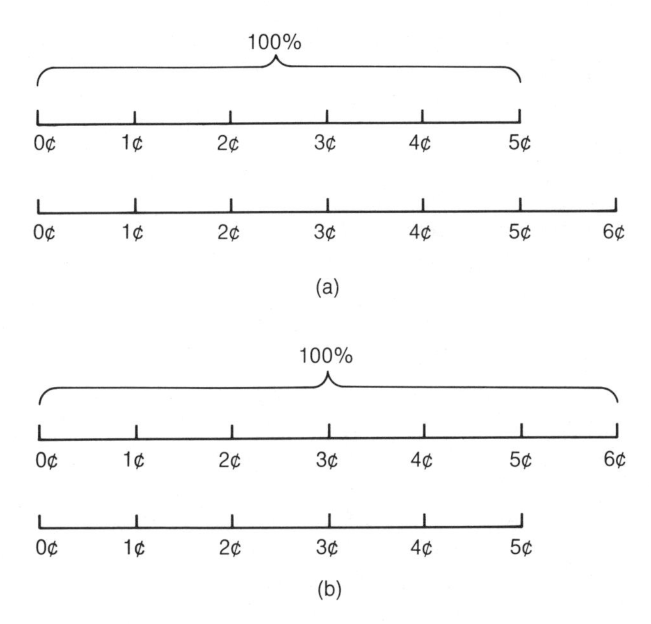


**Figure 13-18** Number lines show that (a) when a sales tax is increased from 5¢ to 6¢ the rate of increase is 20%, while (b) for a decrease from 6¢ to 5¢ the rate of decrease is 16⅔%

## Self-check of Objectives 9 and 10

Give an oral or written definition of the meaning of percent. Describe a real-life situation that involves it.

Describe a real-life situation in which percent (rate), percentage, and base are used. Identify each term in your example.

### 2. Introducing Percent

Before children begin work with percent, they must have a clear understanding of fractional numbers and decimal fraction representations of them because of the relationship between decimal hundredths and percent. There are many elementary school children who fail to understand their work with percent because they lack the necessary background. Don't start activities with youngsters whose backgrounds are inadequate, even if it means delaying their work until after their elementary school years.

You can introduce percent to children who are ready through activities with learning aids used for other topics. The key point to make during early work is that *percent* means "per 100" or "out of 100," so materials that show a unit, or whole, that can be readily subdivided into 100 parts or easily compared with that many parts are necessary. Especially helpful are magnetic materials, such as sets of poker chips with magnets attached, a hundreds board (such as the one in Figure 8-2), structured materials, and square-paper decimal fraction kits.

In most instances you will introduce percent to those children who are mathematically mature, and not to the entire class at once. Therefore problem cards are useful for providing instruction during children's early work. The activities offered on the cards shown on the following pages, which are written for Cuisenaire rods, can be adapted for work with squared-paper or other materials. They can also be used for teacher-directed activities. Give individual or pairs of children a flat, five orange rods, and 100 white cubes, a response sheet, and set of problem cards.

When children have completed activities such as the ones on the problem cards, discuss the meaning of the word *percent* and the % symbol. Follow-up activities, such as naming the percent of blue, red, and white chips in sets of 100 poker chips, are useful. (After children have used poker chips for three or four examples, you can give them pictured sets for practice.) Other activities include naming the percents of numbers blocked off on a hundreds board and naming the percents of shaded or colored squares in 10 by 10 arrays. A variation of these activities that children enjoy is one in which they make colored designs in 100-square sections of squared paper. Each child can mark off his or her own blocks of

100 squares. The blocks do not need to be 10 by 10 arrays, but can be any arrangement that contains 100 squares. Have each child leave enough space between blocks to write the percents for the different colors used in each of the designs.

## Self-check of Objective 11

Describe at least two learning aids children can use to learn the meaning of percent. Explain why problem card activities are well suited for children who are learning about percent.

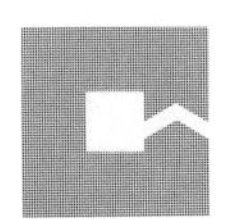

### 3. Extending Understanding of Percent

The problem cards in Figure 13-19 contain activities that acquaint children with the idea that percent is an extension of the common and decimal fraction forms of representing comparisons. Subsequent work with poker chips, hundreds boards, and other manipulative and pictorial materials should continue to emphasize the relationships among common and decimal fractions and percents. Once the relationships are clear, children are ready to move to more abstract levels of work. They can begin to name the equivalents for different fractions that are given in one or another of the three numerals. An activity sheet that contains several sections can be prepared. Each section has expressions of one type, which are to be changed to another form (Figure 13-20).

Not all percents arise from situations in which 100 is conveniently present as the quantity against which another number is compared, as in the introductory situations given above. Therefore, once the basic concept is well established, children's work should focus on activities that deal with groups other than 100. For example, if you begin with a set of ten chips, four of which are blue and six white, what percent of the entire set is the group of blue chips? Have children recall their earlier work with ratios as they deal with this and similar situations. A chart of equivalent ratios will help (Figure 13-21). If necessary, use blue and white chips on a magnetic board to build up the sets from ten, to twenty, and on to 100. Children can use the table to see that 4 is 40% of 10, 8 is 40% of 20, and so on, up to 40 is 40% of 100.

Percent is used to compare two sets as well as a group within a set. Children should become familiar with both uses. In Figure 13-22 a set of blue chips is compared with a set of white ones. The question now is to compare the two sets to determine what percent the set of blue chips is compared with the set of white chips. Again begin with sets smaller than 100 (a). Enlarge the sets until a set of forty blue chips is compared with a set of 100 white chips, if necessary. Many children will realize without

**Figure 13-19** Examples of problem cards and response sheet for activities to help students understand percent

PERCENT (1)

Use your flat, orange rods, and white cubes to do these activities:

1. Lay your flat on your desk. Put 5 orange rods side by side on top of it. Write a statement on your response sheet using a common fraction to tell how much of the flat is covered by the 5 rods. Write a decimal fraction that tells how much of the flat is covered by the rods.

2. Cover the rest of the flat with white cubes. Count the white cubes, and then write a statement on your response sheet telling how many there are. Write another statement using a common fraction to tell what part of your flat is covered by the white cubes. Write a decimal fraction that tells how much of the flat is covered by white cubes.

3. Cover your flat completely with white cubes. Write a statement that tells how many white cubes cover one flat.

4. Some sets of white cubes are listed below. Use one set at a time and put the cubes on your flat, then use both a common and a decimal fraction to tell the part of the flat that is covered by each set. The picture on the next card shows how to do the first one.

| | |
|---|---|
| a. 6 cubes | e. 63 cubes |
| b. 15 cubes | f. 94 cubes |
| c. 23 cubes | g. 36 cubes |
| d. 45 cubes | h. 1 cube |

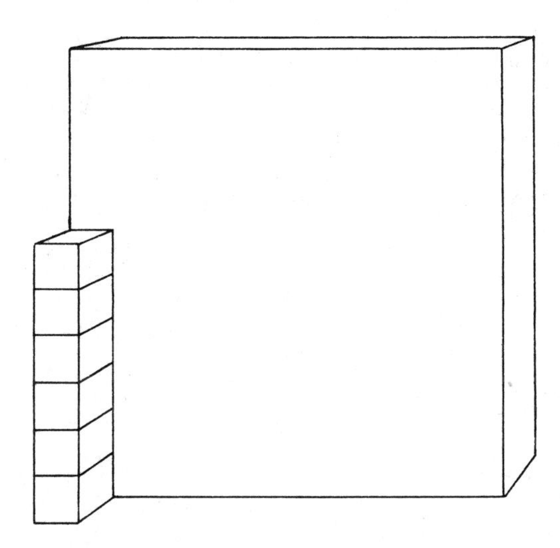

Six cubes cover 6/100, or 0.06, of the flat.

5. So far you have used common and decimal fractions to tell what part of the flat is covered by rods and cubes. There is another way you can write your answers. You will learn about PERCENT so you can use this new way.

   In problem 1, you put 5 orange rods on the flat. Your statement with a common fraction could have used either 1/2 or 5/10 to tell how much of the flat was covered by orange rods. Your decimal answer could have been either 0.5 or 0.50.

(3)

In problem 2, you put 50 white cubes on the rest of the flat. Your statement with common fractions is correct if you used 1/2, 5/10, or 50/100 in it. Your decimal fraction can be either 0.5 or 0.50.

Another way to tell how many orange rods or white cubes cover half of the flat is with percent. Write 50 percent on your response sheet in the space you left blank earlier.

6. In problem 3, you used 100 white cubes to cover the flat. These cubes cover 100 percent of the flat. One hundred percent means that all of the flat was covered by white cubes. What do you think 50 percent means? Write your answer on your response sheet.

7. In problem 4, you used eight different sets of cubes and wrote common and decimal fractions to tell what part of the flat was covered by each set. The answer for the set of 6 cubes was written as 6/100 and 0.06. Another way to answer is to say that the cubes cover 6 percent of the flat.

   Use the rest of the spaces in problem 4 to tell what percent of the flat is covered by each of the sets of cubes.

PERCENT RESPONSE SHEET

Name ______________________

1. ______________________________________________

______________________________________________

______________________ Decimal fraction __________

2. ______________________________________________

______________________________________________

______________________________________________

______________________ Decimal fraction __________

__________________ (Leave this line blank for now.)

3. ______________________________________________

______________________________________________

4. a. ___ ___ e. ___ ___

b. ___ ___ f. ___ ___

c. ___ ___ g. ___ ___

d. ___ ___ h. ___ ___

(Leave these lines blank for now.)

________ ________

________ ________

________ ________

________ ________

______________________________________________

______________________________________________

______________________________________________

1. Rename each decimal fraction as a percent:
   a. 0.50 = ___ c. 0.14 = ___ e. 0.79 = ___
   b. 0.03 = ___ d. 0.19 = ___ f. 0.83 = ___

2. Rename each common fraction as a decimal:
   a. 1/4 = ___ c. 1/2 = ___ e. 2/10 = ___
   b. 3/4 = ___ d. 1/10 = ___ f. 9/10 = ___

3. Rename each common fraction as a percent:
   a. 1/4 = ___ c. 1/2 = ___ e. 2/10 = ___
   b. 3/4 = ___ d. 1/10 = ___ f. 9/10 = ___

4. Rename each percent as a decimal fraction:
   a. 13% = ___ c. 69% = ___ e. 4% = ___
   b. 45% = ___ d. 21% = ___ f. 1% = ___

**Figure 13-20** Sample activity sheet for renaming common and decimal fractions and percents

| 4 | 8 | 12 | 16 | 20 | 24 | 28 | 32 | 36 | 40 |
|---|---|---|---|---|---|---|---|---|---|
| 10 | 20 | 30 | 40 | 50 | 60 | 70 | 80 | 90 | 100 |

**Figure 13-21** Chart of equivalent ratios for showing what percent 4 is of 10

using all of the chips that the ratio remains the same after seeing no more than a second set, as shown in (b). A table will be useful in this situation, too. Use whatever additional activities are necessary to make certain that children can find what percent one number is of another in situations where the comparison is made with both 100 and numbers other than 100. When numbers other than 100 are used, they should be those for which the equivalent ratio containing 100 can be easily determined. A ratio of 4 out of 10 is suitable, while a ratio of 7 out of 13 is not. In the elementary school, percents less than 1 and greater than 100 are seldom used.

## Self-check of Objective 12

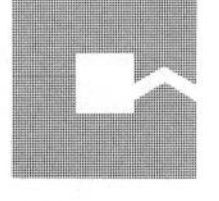

Rename each of these common and decimal fractions as a percent: $^{31}/_{100}$, $^{3}/_{100}$, $^{6}/_{50}$, $^{2}/_{6}$, 0.51, and 0.05. Rename each of these percents as both a common and a decimal fraction: 75%, 83%, 90%, and 6%.

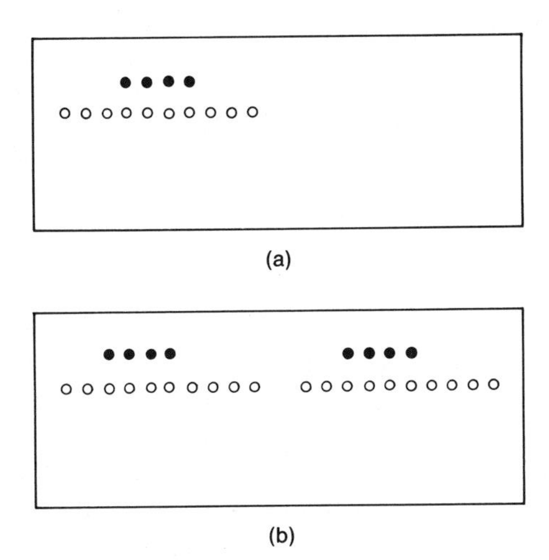


**Figure 13-22** Markers used to build the ratio of 40 to 100 to help children understand one of the uses of percent

## 4. Working Percent Problems

Children who successfully complete the activities already discussed are ready for percent problems. A few elementary school children will understand these problems and will learn to work with them in different situations; others will work only briefly with them; and some will not work with them at all. Three methods of solving percent problems are used: (1) the *case* method, (2) the *unitary analysis* method, and (3) the *proportion* method. These methods will be discussed in the order named here. The proportion method is being used with greater frequency than in the past, and every teacher should be familiar with it.

A common situation will be used to explain each of these methods. It can be easily illustrated with colored poker chips on a magnetic board or a 100-hole pegboard and golf tees. The situation is this: "There is a parking lot containing fifty cars. Eighteen of the cars are foreign-made. Eighteen is 36% of 50."

**a. The Case Method.** This method is based on three rules, or formulas, one for each of the three possible situations that can arise from a percent problem. In one situation (case 1), we know that there are fifty cars in the lot and that 36% of them are foreign. The question is: "How many foreign cars are in the lot?" The formula for finding the answer is $36\% \times 50 = \square$. It is solved by renaming 36% as 0.36 and multiplying 0.36 and 50. In a case 2 situation, we know the lot has fifty cars and that eighteen of them are foreign-made. The question is: "What percent of the cars are foreign-made?" and the formula is $18 \div 50 = \square$. It is solved by using division, as in the margin. For a case 3 situation, we know that there are eighteen foreign-made cars in the lot and that they are 36% of the total. The question is: "Eighteen is 36% of what number?" The formula is

```
     0.36
50) 18.00
    15 0
     3 00
     3 00
```

$$0.36_{\wedge})\overline{18.00_{\wedge}}$$
0 50
18 00

$18 \div 36\% = \square$. It is also solved by division, as shown in the margin. Note that 36% is renamed as a decimal fraction before it is used as the divisor and that carets (∧) are used. Work of this sort requires a high level of understanding. The level of maturity required for understanding the case method is one reason why it is delayed beyond the elementary years for most children.

b. **The Unitary Analysis Method.** This method is based on the idea that given a percent problem, we can simplify it if we first determine what 1% is, then use multiplication or division to determine the total percent. For a case 1 situation, the problem is to find 1% of 50, then multiply the answer by 36. One percent of 50 is ½, or 0.5. Thirty-six times ½ is 18, the number of foreign-made cars in the lot. For a case 2 situation, the thought process is: "Eighteen is some percent of 50. If I know the number that is 1% of 50, I can divide 18 by it to find out what percent 18 is of 50." One percent of 50 is ½ or 0.5. Eighteen divided by ½ is 36, which is the percent of foreign-made cars in the lot. In a case 3 situation, we know that 18 is 36% of some number. Now the thought process is: "If 18 is 36% of a number, I can find the number if I know what part of 18 1% of the number is. The answer is that number multiplied by 100." Divide 18 by 36, and multiply the answer, ½, by 100. The answer, 50, is the total number of cars in the lot. This method also requires considerable maturity before it can be understood, so teaching it is often postponed beyond the elementary school years.

c. **The Proportion Method.** This method has gained favor in recent years because it is easier for most children to learn and use. It is based on the idea that a single expression can be used to show each of the three types of percent problems. Children must understand the meanings of percent (rate), percentage, and base, and how to express these as a proportion. The proportion expression is

$$\frac{\text{rate}}{100} = \frac{\text{percentage}}{\text{base}}.$$

Children will have encountered other proportion expressions if they have completed activities such as those described on pages 348 to 350, where tables are developed and markers used to illustrate proportion situations. The parking lot will again provide the setting for discussion of the proportion method.

When the percent and total number of cars are known, as in a case 1 problem, fill in the proportion expression with the two known terms:

$$\frac{36}{100} = \frac{\text{percentage}}{50},$$

and solve for the unknown term.

When the number of foreign-made cars and the total number of cars are known (case 2) fill in the expression with the two known terms and solve for the unknown term:

$$\frac{\text{percent}}{100} = \frac{18}{50}.$$

In the third situation, the percent and percentage are known and the base is to be determined, so the expression becomes

$$\frac{36}{100} = \frac{18}{\text{base}}.$$

In the examples used here, the solution is not difficult. For $^{36}/_{100} = n/50$, children will know that 100 is divided by 2 to give 50, so they know that 36 divided by 2 will give the answer, 18. For $n/100 = {}^{18}/_{50}$, they will know that 2 times 50 is 100, so the answer can be found by multiplying 2 times 18. And, for $^{36}/_{100} = 18/n$, 36 is divided by 2 to give 18, so 100 is divided by 2 to give 50.

Not all proportion problems can be solved as easily as these. Therefore children should become familiar with the procedures explained in Chapter 12, so they can use the more mature methods to solve all types of percent problems, as well as the simple ones used here.

---

## Self-check of Objective 13

Show how the proportion and at least one other method are used to solve each of these percent problems: 25% of 160 = □, □ of 160 = 40, and 25% of □ = 40.

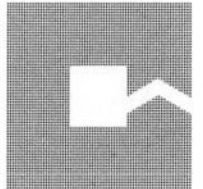

---

## THE MINICALCULATOR, DECIMALS, AND PERCENT

The minicalculator offers many opportunities for children to extend their understanding of decimal fractions and to solve problems involving percent. The examples that follow are samples of the type of work children can do with their machines in these areas of study.

### 1. Reading Words for Decimal Fractions

Children can check their skill in translating decimals expressed in words with activities such as this one:

Enter each of the following decimals in your calculator. Press the ⊞ key after each entry. Your work is correct if the display matches the numeral at the bottom of each list of words.

| | |
|---|---|
| two and five tenths | sixty-one and forty hundredths |
| four and three tenths | seventy-five and sixty-seven hundredths |
| six and nine tenths | four and twenty-one hundredths |
| five and seven tenths | ninety-six and nine hundredths |
| 19.4 | 237.37 |

### 2. Placing Decimal Points

Estimate where to put the decimal point in each product or quotient. Check your answers by computing each example on your calculator.

$$\begin{array}{r}21.3\\ \times 4.8\\ \hline 10224\end{array}\qquad \begin{array}{r}0.78\\ \times 4.3\\ \hline 3354\end{array}\qquad \begin{array}{r}0.789\\ \times 26.3\\ \hline 207507\end{array}\qquad \begin{array}{r}3.621\\ \times 0.4\\ \hline 14484\end{array}$$

$$0.3\overset{263}{\overline{)7.89}}\qquad 3.6\overset{654}{\overline{)235.44}}\qquad 9\overset{0051}{\overline{)0.459}}\qquad 0.27\overset{632}{\overline{)170.64}}$$

### 3. Checking Products

Below are 22 multiplication combinations. Put a check mark in front of each combination you think has a product of 3.6. Check your work by computing each one on your calculator.

| | |
|---|---|
| 3.6 × 1 | 0.4 × 9 |
| 0.12 × 30 | 1.8 × 2 |
| 1.0 × 36 | 0.04 × 0.9 |
| 9 × 0.04 | 0.3 × 12 |
| 120 × 0.03 | 6 × 0.6 |
| 18 × 0.02 | 40 × 0.09 |
| 3 × 1.2 | 60 × 0.6 |
| 2 × 1.8 | 0.05 × 720 |
| 4 × 0.9 | 18 × 0.2 |
| 0.6 × 0.6 | 36 × 0.1 |
| 72 × 0.5 | 360 × 0.01 |

### 4. Solving Percent Problems

The minicalculator simplifies the process of computing answers for percent problems. It is especially useful with the proportion method because each step can be completed in sequence, with the answer appearing on the display at the conclusion. Have children practice with examples similar to the parking lot situation used earlier in this chapter. When you make up your own examples, be sure that the numbers you use are ones that will come out with whole-number and nonfraction-percent answers. Here is one to get you started.

"John completed all but 22 of the 200 mathematics problems he was assigned. What percent of the problems did he complete?"

"Juanita completed 98% of the 200 mathematics problems she was assigned. How many problems did she finish? How many did she have left to complete?"

"Susie completed 180, or 90%, of the problems she was assigned. How many problems was she assigned?"

## Self-check of Objective 14

Describe orally or in writing four minicalculator activities that deal with decimal fractions and percent.

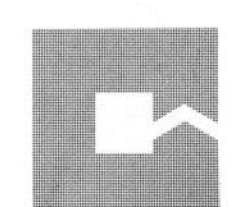

## Common Pitfalls and Trouble Spots

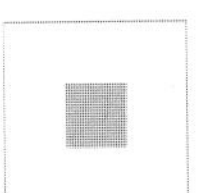

Two common pitfalls associated with decimal fractions and percent are discussed.

Many teachers instruct children to add, subtract, multiply, and divide with decimal fractions without giving them opportunities to learn the fractions' meanings first. This is because all four operations can be performed with fractional numbers represented as decimal fractions with the same algorithms that are used for whole numbers. Even though many children will use the algorithms correctly and learn rules for placing the decimal point in answers, they will not have the knowledge they need to interpret answers properly in real-life situations. Children must progress slowly, beginning with concrete models, and learn about decimal fractions themselves before they are introduced to operations with them. Only through a carefully sequenced series of activities will they develop the understanding they need in order to know what they are doing when they compute with decimal fractions.

The second common pitfall is that instruction about percent is frequently confined to the materials included in a textbook series or a district mathematics program. When this is the case, children have only limited work with the concept. One way to broaden children's experiences with percent is to have them use it in curriculum areas other than mathematics. For example, data gathered during a sixth-grade science project can often be reported using percent. As you plan science and social studies projects, look for ways children can use percent to report information

$$\begin{array}{r} 30.061 \\ 1.23 \\ 600.2 \\ +\ \ 3.908 \\ \hline \end{array}$$
(a)

$$\begin{array}{l} 30.061 \\ 1.23 \\ 600.2 \\ +3.908 \\ \hline \end{array}$$
(b)

*Ragged decimals* that children encounter in addition and subtraction are frequently troublesome. The addends in example (a) in the margin are sometimes called ragged decimals because the number of decimal places to the right of the ones place varies from addend to addend. When given these addends in a sentence—30.061 + 1.23 + 600.2 + 3.908 = □, many children will record them as shown in (b). Two procedures can be used to help children avoid this error.

First, relate the decimal fractions to common fractions. Children who know that each addend in the example can be renamed as a common fraction can use that knowledge to see the correct way to write the algorithm. They will know that $30.061 = 30^{61}/_{1000}$, $1.23 = 1^{23}/_{100}$, $600.2 = 600^{2}/_{10}$, and 3.908 =

$$\begin{array}{r} 30.061 \\ 1.230 \\ 600.200 \\ +\ \ 3.908 \\ \hline \end{array}$$
(c)

$3^{908}/_{1000}$, and that these common fractions cannot be added until they are renamed with common denominators. So, before the decimal fractions are added, they can be expressed as in the algorithm (c). Now the decimal fractions are written with each place value position aligned and the decimal points one under the other.

Second, use estimation to determine an approximate answer. An estimated answer for the example is 635, to the nearest whole number. A child who makes this estimate will realize that the algorithm with the improperly aligned place value positions cannot possibly yield the correct sum.

Both of these procedures work equally well for subtraction and addition.

Once the reason for aligning numerals by place value positions is clear, children can learn the rule that addition and subtraction algorithms must be written with the decimal points in a line.

---

## Self-check of Objectives 15 and 16

Two pitfalls associated with teaching decimal fractions and percent have been discussed. Identify each one, and describe how you can avoid it.

Describe orally or in writing two ways children can develop their understanding of the need to align decimal points in a row when adding and subtracting with decimal fractions.

---

## SUMMARY

The fractional number system includes all of the positive and negative integers, and numbers between integers that can be expressed in the form $a/b$, where $a$ is any integer and $b$ is any integer other than 0. Fractional numbers can be expressed with common fractions, decimal fractions, and percent. Fractional numbers greater than 0 are included in the elementary curriculum. Ways to introduce decimal and percent representations are presented in this chapter, along with ways of teaching operations with decimal fractions.

Work with decimal fractions and percent extends the understanding of fractional numbers and ways to express them. Decimal fractions express numbers smaller than 1 that have a denominator of ten or a power of ten. The denominator is not written, but is indicated by how many numerals there are to the right of the ones place in a decimal numeral. Structured materials, unit-region kits, and number lines are aids children use to learn the meaning of decimal fractions. Once the meaning of decimal fractions is understood and children can write decimal numerals accurately, they are ready to learn to add, subtract, multiply, and divide fractional numbers expressed as decimals. The algorithms for these operations are like those for whole numbers. Real-life situations and work with learning aids will help children extend their use of the algorithms to include decimals.

Carefully developed procedures for determining where to put the decimal points in answers should be used rather than teacher-dictated rules.

Percent is used to express a ratio between some number and 100. Carefully sequenced activities will help mature elementary school children understand the meaning of percent. Special attention needs to be given to the meanings of percent (rate), base, and percentage. There are three methods for solving percent problems: (1) the case method, (2) the unitary analysis method, and (3) the proportion method. The proportion method has gained favor in recent years because it is easier for most children to understand and use.

The minicalculator offers opportunities for children to extend their understanding of decimal fractions and percent. Children can use it for computing problems and for playing games dealing with both of these ways of expressing fractional numbers. Two pitfalls to avoid while working with decimal fractions and percent are (1) having children perform operations using algorithms without first knowing the meaning of the fractions they are using, and (2) failure to use percent in other curriculum areas once it is learned. Some children have trouble putting numerals in proper alignment when adding and subtracting with decimal fractions. Having them relate the decimals to common fractions and estimate answers before computing will help many children overcome this error.

## STUDY QUESTIONS AND ACTIVITIES

1. Decimal fractions are sometimes called an extension of the system for recording whole numbers. How is this true?

2. Represent each of these numbers on pictures of a classroom abacus.
(a) 23.492 (b) 306.004 (c) 911.01

3. Use fraction strips and number lines to illustrate each of these addition and subtraction sentences.
(a) $0.4 + 0.8 = \square$ (b) $1.2 + 1.9 = \square$
(c) $2.3 - 1.8 = \square$ (d) $1.7 - 0.9 = \square$

4. Design a set of transparent overlays for illustrating the meaning of each of these multiplication sentences.
(a) $0.3 \times 0.4 = \square$ (b) $3 \times 0.6 = \square$ (c) $0.3 \times 2 = \square$

5. Use a process of estimation to find correct placement of the decimal point in the following products and quotients. Why do children need to develop facility with rounding off numbers to estimate the placement of decimal points?

(a) $\begin{array}{r} 4.8 \\ \times\ 2.3 \\ \hline 1104 \end{array}$ (b) $\begin{array}{r} 36.2 \\ \times\ 1.08 \\ \hline 39096 \end{array}$ (c) $\begin{array}{r} 48.36 \\ \times\ 51.28 \\ \hline 24799008 \end{array}$ (d) $3.2\overline{)21.824}$ = 682 (e) $0.33\overline{)9.966}$ = 302

6. Why is the estimation process not well suited for determining placement of the decimal point in a product and quotient such as the following ones? What procedures for placing the decimal point in examples such as these should children learn?

(a)
$$\begin{array}{r} 0.341 \\ \times\ 0.682 \\ \hline 232562 \end{array}$$

(b)
$$\begin{array}{r} 2 \\ 0.036\overline{)0.0072} \end{array}$$

## FOR FURTHER READING

Amstutz, Mildred Gilston, "Let's 'Place' the Decimal Point, not 'Move' It," *The Arithmetic Teacher*, X, No. 4 (April 1963), 205–207. The author argues to use meaningful procedures to help children learn to determine where to place the decimal point during division with decimals.

Broussard, Vernon. "Using the Subtraction Method in Dividing Decimal Fractions," *The Arithmetic Teacher*, X, No. 5 (May 1963), 288–289. Broussard explains how to use the vertical algorithm for dividing with decimal fractions. He also gives the rationale for changing all decimal fraction divisors to whole numbers before dividing with them.

Cole, Blaine L., and Henry S. Weissenfluh. "An Analysis of Teaching Percentage," *The Arithmetic Teacher*, XXI, No. 3 (March 1974), 226–228. Describes the relationship between the ratio method and the case method of solving percent problems. Explains the ratio method and gives a brief outline of the background for understanding it.

Firl, Donald H. "Fractions, Decimals, and Their Futures," *The Arithmetic Teacher*, XXIV, No. 3 (March 1977), 238–240. The author predicts an increasing emphasis on decimal fractions because of the advent of the minicalculator and greater use of the metric system. While the role of common fractions will diminish, they cannot be eliminated entirely. Even after total metrication, there will be a continuing need for certain common fractions.

Fisher, John W. "Deci-Deck," *The Arithmetic Teacher*, XXII, No. 2 (February 1975), 149. The deci-deck card game provides practice in naming equivalent common fractions, decimal fractions, and percent. Several card games are possible with the deck.

Mueller, Francis J. "The Neglected Role of the Decimal Point," *The Arithmetic Teacher*, V, No. 2 (March 1958), 87–88. Discusses the role of the decimal point as indicator of the ones place in a numeral and the symmetric nature of the place value scheme of the Hindu-Arabic numeration system.

Rappaport, David. "Percentage—Noun or Adjective?" *The Arithmetic Teacher*, VIII, No. 1 (January 1961), 25–26. Rappaport indicates the nature of some of the confusion that results from improper use of the word *percentage*.

Schmalz, Rosemary, S.P. "A Visual Approach to Decimals," *The Arithmetic Teacher*, XXV, No. 8 (May 1978), 22–25. Dienes multibase arithmetic blocks, with a cardboard shape ten times the area of a flat, serve as a model for decimal representation of fractional numbers. Procedures for their use are described.

# 14 Measures and the Processes of Measuring

Upon completion of Chapter 14, you will be able to:

1. Explain briefly the Metric Conversion Act of 1975.
2. Explain the meaning of measurement.
3. Trace briefly the historical development of the metric system of measure.
4. List the seven basic and two supplementary units of the International System of Units (SI).
5. Identify at least two characteristics of the metric system.
6. List at least four advantages of the metric system over the English (customary) system.
7. Give a rationale for beginning instruction with metric units, and identify the role of the English (customary) system in the program.
8. Explain the role of exploratory activities in the measurement program and differentiate between nonstandard and standard units.
9. Identify the grade levels at which standard units for these measures should be introduced: length, weight, capacity, volume, area, time, temperature, angle.
10. Identify and demonstrate activities for building a preoperational-level foundation for measurement.
11. Describe activities for introducing underlying concepts of measurement using nonstandard units.
12. Define the meaning of linear measure, and demonstrate procedures children can use to understand it and to measure lengths directly and by estimation.
13. Identify the commonly used units of weight (mass) measure, and name two types of scales children should learn to use.
14. Define the meaning of area measure, and describe materials and activities for learning how to measure the areas of squares and rectangles, parallelograms, and triangles.
15. Describe the meaning of volume measure, and demonstrate a procedure children can use to understand it and to determine how to find the volumes of space figures such as rectangular and triangular prisms, and cylinders.
16. Demonstrate a procedure for helping children learn to measure angles.
17. Describe activities to help children learn to tell time and use money.
18. Summarize the important measurement concepts included in the elementary school mathematics program.
19. Identify two common pitfalls associated with teaching measurement, and describe how each one can be avoided.
20. Identify two trouble spots experienced by children as they learn about measurement, and describe ways to overcome each one.

Key Terms you will encounter in Chapter 14:

| | |
|---|---|
| Metric Conversion Act of 1975 | hectometer |
| English (customary) system | kilometer |
| metric system | liter |
| length | kilogram |
| area | are |
| capacity | hectare |
| weight (mass) | International System of Units (SI) |
| temperature | gram |
| meter | milligram |
| direct measure | milliliter |
| indirect measure | Celsius |
| iteration | conservation of length |
| approximate | conservation of area |
| decimeter | nonstandard unit |
| centimeter | trundle wheel |
| millimeter | volume |
| dekameter | rectangular solid |

On December 23, 1975, President Gerald Ford signed the Metric Conversion Act of 1975. With this act (Public Law 94–168) he set in motion the process of converting the United States from the English (customary) system to the metric system of measurement. President Ford emphasized at the time he signed the bill that the conversion is completely voluntary, with the government providing direction and coordination through the United States Metric Board. This board—made up of men and women representing business, education, labor, industry, government, and the engineering and scientific professions—has no compulsory powers, but develops and conducts a broad program of planning, coordination, and public education.

Since the enactment of Public Law 94–168, conversion has moved slowly, with the most notable changes taking place in the private sector. Producers of many consumer products are marketing goods that have been metricated. Companies already converted partially or entirely include Black and Decker; Chrysler, Ford, and General Motors; IBM;

Coca-Cola and Seven-Up; Penney's; Butterick and Simplicity; Mead Paper; and many food processors.

(A worthwhile activity for children in the intermediate grades is to collect and display consumer goods, or pictures of them, that have metric dimensions or are packaged in metric containers.)

There are indications, however, that the metrication process will not proceed without interruption. The Federal Highway Administration cancelled its plan to change all speed limit signs to kilometers per hour by July 1, 1978 because of public opposition. Also, the National Weather Service planned to have temperature, precipitation, windspeed, distance, pressure, and all other weather elements reported in metric units by January 1, 1979. Following hearings at which public opposition was expressed, the Service's weather conversion schedule was revised to allow for a longer period of public awareness activities and a later implementation date. The General Accounting Office (GAO) of the United States government has indicated that benefits may not be as great as anticipated, and has recommended that immediate conversion not be undertaken.

The above incidents, the GAO recommendation, and frequent newspaper editorials and letters show that there are those who strongly oppose the idea of changing to the metric system. It is not likely, therefore, that total conversion will occur in the immediate future. Children in today's schools need to develop an understanding of both the metric and customary systems, know common units of measure for both, and become skillful users of instruments for determining measures of length, area, capacity (volume), weight (mass), and temperature in metric and customary units.

This chapter includes a discussion of the meaning of measurement, a brief history of the metric system, the metric system's characteristics, and examples of activities for teaching about measures and measuring processes. The metric system serves as the example for each activity. However, the customary system is taught in identical ways, so its units could be substituted wherever metric units are used.

## WHAT IS MEASUREMENT?

The distinction between discrete and continuous objects is made in Chapter 5, where children's early mathematical experiences are discussed. Discrete objects are those which can be counted, whereas continuous objects are measured. Measurement is the process of attaching numbers to the physical qualities of length, area, volume, weight (mass), or temperature. (Time is also measured, but it lacks a physical quality.) For each physical quality there is one or more units that refer to it. For example, the meter[1] is the basic unit for referring to an object's length. Units larger and smaller than the meter are used for finding measurements of objects for which the meter is inappropriate.

## 1. Two Types of Measuring Processes

Two types of measuring processes exist—direct and indirect. The processes for determining most measures of length and capacity are direct, made by applying the appropriate unit directly to the object being measured. This direct process is referred to as *iteration*. Iteration is illustrated by the use of a meter stick to measure a room's length. The length is measured by placing the zero end of the stick against one wall, marking a spot on the floor at the other end, moving the stick to put the zero end at the mark, and repeating the process until the opposite wall is reached. All the while, the number of times the stick is moved is counted. This number is the room's length expressed in meters.

Weight (mass), temperature, and time cannot be measured in the direct fashion just described. The characteristics, or properties of weight, temperature, and time require an instrument that measures each one by translating the measurable property into numbers indirectly. Thus one type of weather thermometer has a number scale which is aligned with a tube containing a liquid. The liquid expands, or goes up, when it is warmed as the surrounding air becomes hotter, and it contracts (goes down) when it is cooled as the air becomes colder. The temperature at a given time is determined indirectly by reading the numeral on the scale that indicates the height of the column of liquid at that time.

There are indirect processes for determining length (height, width, distance, and so on). These processes are too technical for most elementary school children and are not considered in this book.

## 2. Measurement Is Approximate

While the counting of discrete objects is exact, all measurement is approximate. The approximate nature of measurement stems from the fact that for any particular unit of measure there is, theoretically at least, one that is more precise. If a city block is measured with a meter stick it might be found to be 100 meters long. However, if a unit one-tenth of a meter—a decimeter—is used to measure the same block, it might be found to be 998 decimeters long, or slightly less than 100 meters. Even more precise measurements could be made using centimeters or millimeters, although in most instances this would be impractical and unnecessary.

[1]"Meter" is the accepted spelling in the United States for the basic unit of length, while "liter" is the spelling for the basic unit of capacity (volume). Children should also learn the alternate spellings for these words—"metre" and "litre."

## Self-check of Objectives 1 and 2

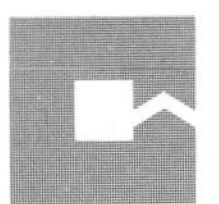

Give a brief oral or written explanation of the Metric Conversion Act of 1975.
Give a brief oral or written explanation of the meaning of measurement.

## THE METRIC SYSTEM

The metric system is assuming a more prominent role in the United States, and children must learn to understand and use it. This section gives a brief overview of the history of this system, discusses its key characteristics, and presents its advantages over the English (customary) system.

### 1. Brief History of the Metric System

The earliest proposal for a worldwide decimalized system of measure was made in France in the late 1600s, but a century passed before serious attention was given to such a system. Then, in 1790, the National Assembly of France asked the French Academy of Sciences to devise a system having an invariable standard for all measures. The academy derived its unit of length from a portion of the earth's circumference. This standard, which is called a meter, was one ten-millionth of the distance from the equator to the north pole along the meridian running near Dunkirk in France and Barcelona in Spain and was represented by a prototype measure made of platinum.

Other units were derived from the meter (m). The units of linear measure are decimal parts or multiples of the meter based on powers of ten. A *decimeter* (dm) is $\frac{1}{10}$ of a meter, a *centimeter* (cm) is $\frac{1}{100}$ of a meter, and a *millimeter* (mm) is $\frac{1}{1000}$ of a meter. A *dekameter* (dam) is 10 meters, a *hectometer* (hm) is 100 meters, and a *kilometer* (km) is 1000 meters.[2]

Basic units for capacity (volume), weight (mass), and area were established at the time the base unit for linear measure was established. A *liter* (L),[3] the basic unit for capacity, is the volume of one cubic decimeter ($dm^3$). A *kilogram* (kg) is the basic unit of weight (mass) and was originally determined to be one cubic decimeter of water at 4°C, the temperature at which water is at its greatest density. Today's international prototype of

[2] There are smaller subdivisions and larger multiples of the metric system's basic units, but they are seldom used outside the field of science and are not included in the elementary school curriculum.

[3] A capital L is used in the United States as the symbol for liter. This avoids the possibility of confusing the symbol with the numeral 1. The use of the script ℓ has been discarded because of its unavailability on typewriters and other printing devices.

the kilogram is a platinum-iridium solid maintained at Sèvres, France. For measuring land, the basic unit is the *are* (a), which is 100 square meters ($m^2$). A *hectare* (ha) is 100 ares. Volume is also measured in cubic meters ($m^3$) and cubic centimeters ($cm^3$).

In 1960 the Eleventh General Conference of Weights and Measures adopted the International System of Units (SI). There are seven basic units and two supplementary units, as shown in Tables 14-1 and 14-2. The SI system is accepted worldwide with nearly full agreement among nations as to the units, their values, names, and symbols. An exception to full agreement is the use of L as the symbol for liter in the U.S., whereas l is the SI symbol.

One detail of interest coming from the 1960 conference was a redefinition of the meter. While it is approximately the same length as the original meter, it is now defined as 1 650 763.73[4] wavelengths in vacuum

**Table 14-1** Base Units in the International System of Units (SI)

| *Quality to be Measured* | *Unit* | *Symbol* |
|---|---|---|
| Length | meter | m |
| Weight (mass) | kilogram | kg |
| Time | second | s |
| Electric current | ampere | A |
| Thermodynamic temperature | kelvin | K* |
| Amount of substance | mole | mol** |
| Luminous intensity | candela | cd |

*Degrees Celsius is an acceptable alternate unit for measuring temperature.

**Some listings contain only six units, omitting the mole. See M. J. B. Jones, *A Guide to Metrication* (Oxford, England: Pergamon Press Ltd., 1969), p. 15.

**Table 14-2** Supplementary SI Units

| *Quality to be Measured* | *Unit* | *Symbol* |
|---|---|---|
| Plane angle | radian | rad |
| Solid angle | steradian | sr |

[4]The comma is omitted from numerals for large and small numbers in SI. Large numbers appear as indicated here, while small ones appear as 0.362 45. This practice is followed because the comma serves as a decimal point in some countries.

of the orange-red line of the spectrum of the krypton-86 atom. This standard was adopted because it is unvarying and indestructible, and can be reproduced in any laboratory possessing the proper equipment.

The SI system is highly technical and only the parts dealing with length, mass, and temperature and certain non-SI metric units dealing with time, plane angles, area, and volume are studied by elementary school children.

## 2. Characteristics of the Metric System

One significant characteristic of the metric system has already been mentioned—that is the fact that measures of capacity, weight, and area are all based on the meter.

Another important feature is that the metric system is a decimal system. This means that for each type of measure there is an interrelationship of units based on multiplication or division by 10. This scheme is essentially the same as the Hindu-Arabic numeration system, a fact that should be capitalized on to help children develop their understanding of the metric system. Table 14-3 shows the relationship between the metric system and the Hindu-Arabic numeration system and certain portions of our monetary system. This table could be extended in both directions to show both larger and smaller units. It will not be common practice to use the larger and smaller units, however, so it is not expected that children will become familiar with them. In fact, only certain of the measures shown in Table 14-3 will be commonly used. These are: meter, kilometer, centimeter, and millimeter; gram, kilogram, and milligram; and liter and milliliter. These, along with certain units for area—square meter and square centimeter; volume—cubic meter and cubic centimeter; time—year, day, hour, minute, and second; temperature—degrees Celsius; and plane angles—degrees, are the units elementary school children will deal with during their study of the metric system.

## 3. Advantages of the Metric System

The following advantages of the metric system are reported in "1 . . . to get ready"[5]:

1. The metric system is simple and logically planned and is a decimal system like our numeration and monetary systems.

[5] "1 . . . to get ready" is a selected bibliography on metrication issued in April 1973 by a Joint Committee of the American Association of School Librarians and the National Council of Teachers of Mathematics. Copies are available from the AASL, 50 East Huron Street, Chicago, Ill. 60611.

**Table 14-3** Relationship between Parts of the Metric System and the Hindu-Arabic Numeration System and Pertinent Parts of Our Monetary System

| | *1000x* ($10^3$) | *100x* ($10^2$) | *10x* ($10^1$) | *Base* | $\frac{1}{10}x$ ($10^{-1}$) | $\frac{1}{100}x$ ($10^{-2}$) | $\frac{1}{1000}x$ ($10^{-3}$) |
|---|---|---|---|---|---|---|---|
| Hindu-Arabic system | thousands | hundreds | tens | ones | tenths | hundredths | thousandths |
| Monetary system | $1000.00 | $100.00 | $10.00 | $1.00 | $0.10 (10¢) | $0.01 (1¢) | $0.001 (mill) |
| Metric prefix | kilo- | hecto- | deka- | | deci- | centi- | milli- |
| Length | kilometer (km) | hectometer (hm) | dekameter (dam) | meter (m) | decimeter (dm) | centimeter (cm) | millimeter (mm) |
| Capacity | kiloliter (kL) | hectoliter (hL) | dekaliter (daL) | liter (L) | deciliter (dL) | centiliter (cL) | milliliter (mL) |
| Weight (mass) | kilogram* (kg) | hectogram (hg) | dekagram (dag) | gram (g) | decigram (dg) | centigram (cg) | milligram (mg) |

*The kilogram is the basic unit for weight. Therefore it is a contradiction of the idea that the base units stand without a prefix.

2. The basic unit, the meter, is always reproducible from natural phenomena.
3. A single set of prefixes is used to designate all subdivisions and multiples of units for all types of measures. This simplifies the process of converting from one unit to another.
4. Uses of common fractions will be reduced, which means that less time will be required for children to understand the common fractions they will need. The time gained can be devoted to other important topics.
5. Familiarity with the meter, gram, and liter and their subdivisions and multiples will be sufficient for most persons. Extension of the study of the International System of Units can be made by those persons who need a deeper understanding of the system.
6. The metric system is the universal measuring language.

---

## Self-check of Objectives 3, 4, 5, and 6

Give a brief oral or written statement in which you trace the history of the metric system.

The International System of Units (SI) has seven basic and two supplementary units. List these units, and identify whether each is basic or supplementary.

Two characteristics of the metric system are described. Give an oral or written explanation of them.

The AASL/NCTM publication quoted lists six advantages of the metric system over the English (customary) system. List at least four of these advantages.

---

## THE ENGLISH (CUSTOMARY) SYSTEM

For the foreseeable future children will live in a society that uses both metric and customary units. Until meat and fresh produce, oil and gasoline, ribbon and yard goods, and other consumer products are available in metric units exclusively, children will need to be familiar with the common customary units.

It is generally recommended that children's first work with standard units of measure should be with metric units. Instruction that begins with the metric system will stress its importance and indicate that its use is increasing. Once children are familiar with the common metric units for length, they should then engage in activities that will acquaint them with the inch, foot, yard, and mile. Once they are familiar with metric units for capacity and weight, they can learn about cups, pints, quarts, gallons, ounces, and pounds. Generally, these activities will come in the intermediate grades.

The time spent on learning the commonly used customary units need not be extensive. Children who understand the basic concepts of measure and who have learned to use metric measuring instruments properly can learn the names of the customary units and their measuring instruments in a few hours of carefully planned activities.

It is not recommended that children have experiences converting measurements from customary units to metric units and vice versa. Beyond learning that a meter is slightly longer than a yard, a liter is a little more than a quart, and a kilogram is about 2.2 pounds, there is little to be gained from having children compare units between the two systems. Too heavy doses of conversions tend to weaken children's interest in measurement and may lead to a negative attitude toward it.

The remaining sections of this chapter are focused on the metric system, except for the part dealing with money. Teachers who want to introduce the customary units first or exclusively can use the information from these sections as a guide for planning the proper sequence and appropriate activities by substituting customary units for their metric counterparts.

---

## Self-check of Objective 7

It is recommended that instruction begin with the metric system. Give a brief oral or written statement explaining why this is so. Include an explanation of the role of the English system in your statement.

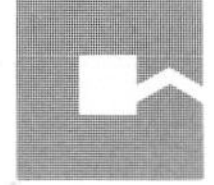

---

## TEACHING MEASURES AND MEASURING

Piaget's research and its general implications for teaching mathematics are reviewed in Chapter 2. The section of that chapter dealing with this research should be reviewed at this point because the contents of that discussion are important in the planning of measurement activities. That discussion is now extended to introduce additional information dealing with children and conservation.

To study children's ability to conserve length, a pair of rods or sticks of equal length can be used. Initially, place the rods side by side on a table with their ends an equal distance from one edge as in Figure 14-1(a). Ask the child if the rods are of equal length. A child who understands the question will usually agree that they are. Next, move one stick so it is farther from the table's edge (b), and repeat the question. A child who is not a conserver of length will say that the sticks no longer have equal lengths. A conserver recognizes that the length of either rod is not altered

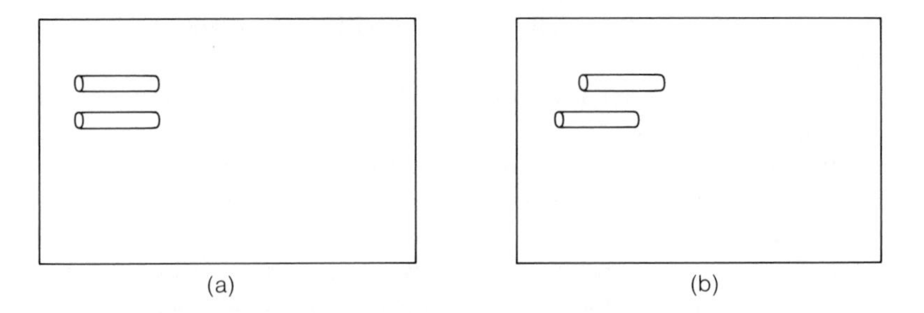

**Figure 14-1** A test for determining a child's ability to conserve length

by changing the position of one of them. Generally, children are unable to conserve length until they are about eight years of age.

A similar test can be used to determine whether a child is a conserver of area. A pair of "fields," represented by equal-sized pieces of green paper, and "buildings," represented by smaller equal-sized pieces of brown paper, are used. Place the fields side by side with a building in place on each, as in Figure 14-2(a). Ask the child if one building covers the same amount of land as the other. Next, place an additional building on each field, as shown in (b), and rephrase the question. A nonconserver will say that the buildings on one field cover more land than those on the other. The conserver recognizes that an equal number of equal-sized buildings cover the same amount of land (area) regardless of their positions. Children generally do not conserve area until around eight years of age.

Neither of these tests alone is sufficient to determine whether a child is a conserver of area or length. Several tests using similar materials

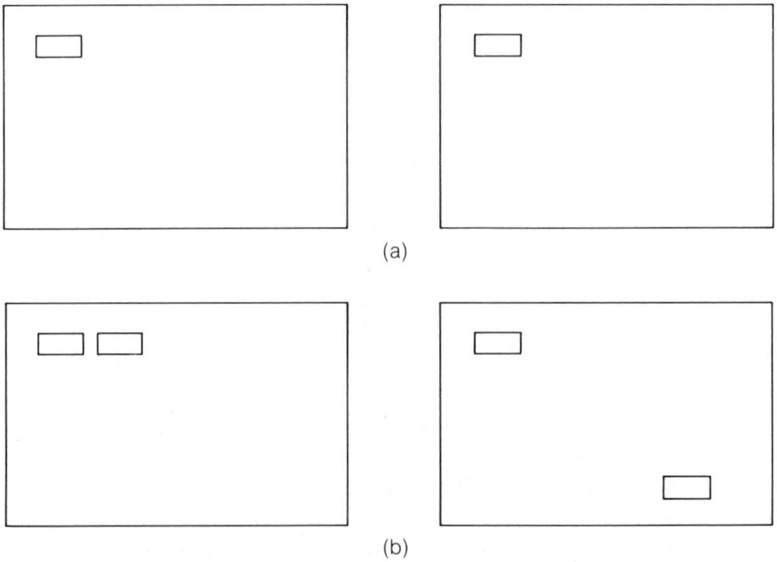

(a)

(b)

**Figure 14-2** A test for determining a child's ability to conserve area

and techniques should be used before a firm decision about a child's status can be reached.[6]

Children's reactions to these and similar tests give curriculum planners and teachers a basis for establishing the proper scope and sequence of activities dealing with measurement. It is generally agreed that children's early experiences should be exploratory, with first activities being playlike in nature. Nonstandard units, such as cans and jars, acorns, paper clips, and 3 by 5 cards, are used to introduce the underlying concepts of measurement once it is determined that children are conservers. Finally, standard units are introduced. Figure 14-3 illustrates a suggested scope and sequence for measurement in elementary schools.

The scope-and-sequence chart in Figure 14-3 cannot be applied directly as a means of establishing a program for your children. Variations in children's levels of cognitive development, particularly with reference

| Activities | Measurement | Grade Levels K 1 2 3 4 5 6 |
|---|---|---|
| *Exploratory* | length | |
| pouring | weight | |
| matching | capacity | |
| balancing | (L, mL) | |
| ordering | time | |
| sorting | area | |
| comparing | temperature | |
| *Nonstandard Units* | length | |
| pouring | weight | |
| matching | capacity | |
| balancing | (L, mL) | |
| ordering | volume | |
| sorting | ($m^3$, $cm^3$) | |
| comparing | time | |
| measuring | area | |
| estimating | temperature | |
| recording | angle | |
| *Standard Units* | length | |
| measuring | weight | |
| estimating | capacity | |
| recording | (L, mL) | |
| choosing | volume | |
| converting | ($m^3$, $cm^3$) | |
| describing | time | |
| applying | area | |
| | temperature | |
| | angle | |

**Figure 14-3** A suggested scope and sequence for measurement in elementary schools. Adapted from Mathematics Education Task Force, *An Introduction to the SI Metric System* (Sacramento: California State Department of Education, 1975), p. 15

[6] For additional discussion of conservation of length and area, and examples of other tests, see Richard W. Copeland's *Math Activities for Children* (Columbus, Ohio: Charles E. Merrill Publishing Company, 1979), Ch. 4.

to conservation, their abilities to comprehend mathematical concepts, and previous experiences with measurement must be considered as you plan their activities.

At all levels many of the children's activities with measurement can be placed in a laboratory setting. From the kindergartner's free play at a sand or water table to the sixth grader's work with the sophisticated notion of volume expressed as cubic centimeters, children need opportunities to investigate the underlying concepts through carefully planned activities involving appropriate manipulative materials. Learning centers with children working singly, in pairs, or in groups up to four, depending on the nature of the activity and availability of apparatus, can provide the setting for much work. Commercial or teacher-made activity cards can provide directions. Even young children can follow carefully prepared pictorial or simply worded rebus-type instructions displayed on colorful cards. There are a number of commercially prepared laboratory manuals for older children.

There are, however, certain types of measures that must be learned through teacher-directed lessons. Measures of time and temperature are examples. The concepts of time and temperature and uses of clocks, watches, and thermometers cannot be "discovered" by children; they must be introduced and developed through carefully sequenced lessons directed by a teacher. It is also important for children that they have opportunities to discuss their laboratory work with their teacher so they can summarize and organize their findings.

The activities that follow are representative of those you can use to help children develop their understanding of the metric system. More activities are presented in materials listed in the metric section of Appendix C.

---

## Self-check of Objectives 8 and 9

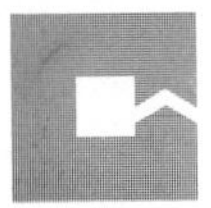

Explain orally or in writing the three-stage progression of activities for a measurement program. Include a statement that indicates why this progression is beneficial to children.

These types of measures are included in the elementary school program: length, capacity, area, time, temperature, weight. List each one, and write the grade level that is recommended for its introduction.

---

### EXPLORATORY ACTIVITIES

Children who are still in the preoperational stage of development and who are not yet conservers should not be given activities dealing with standard units of measure. Rather, they need a variety of activities that build a foundation for later work. In addition to those in the readiness section of Chapter 5, the following types of activities should be made available to young children.

*One.* The "Rice Game"[7] provides background for understanding capacity. There are glass or plastic containers of various sizes, colored rubber bands glued to different levels on the containers, 2 to 6 pounds of rice, a funnel, a scoop, and a large plastic tub to hold the rice. Children fill each container to the rubber band level. Some children may be able to count the number of scoops required to fill each jar or bottle, record the number for each one on small squares of paper, and order the bottles from least to most amount of rice.

*Two.* Balance beams help children understand the meaning of weight. Balance beams with a platform or container at each end of the crossbar can be bought, or you can make one from wood, with two metal pie tins suspended at each end of the crossbar by string or small metal chains (Figure 14-4). First, let children spend time playing with the balance beam so they will become acquainted with it. They will learn how to make it balance by putting something in one pan, then putting objects in the other one until the crossbar is horizontal. They will also see the bar go down on one end when they put an object in one of the pans that is heavier than the object in the other pan. Once children are familiar with the balance beam, they can use it to compare pairs of objects. Put pairs of small objects—toy cars, dolls, blocks, boxes—on a table and let children use the balance beam to find out which is heavier and which is lighter. Provide a place for children to set the heavier object of each pair and another place for the lighter object.

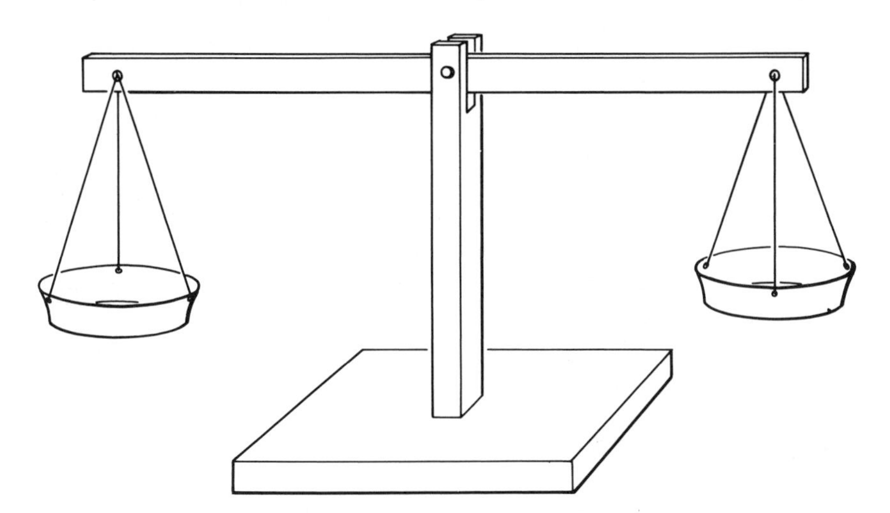

**Figure 14-4** Teacher-made balance beam

[7] Mary Baratta-Lorton, *Workjobs* (Menlo Park, Calif.: Addison-Wesley Publishing Company, 1972), pp. 22–23.

*Three.* An extension of activities in which children compare the length or height of two things is provided when they order quantities according to height, length, and size. The idea of length is developed when children use objects that can be matched when placed side by side. For example, two Cuisenaire rods can be put alongside a rod that is longer than either one to make a train the length of the longer rod. When they do this, children see the equivalence of lengths. Later, they can order the rods to make a stairstep pattern that increases or decreases in length from left to right. (Dowel rods cut to different lengths, different-sized wooden or plastic blocks, different-sized squares cut from colored railroad board, and other commercial and teacher-made materials can be used for ordering activities.)

*Four.* A teacher-directed activity can give children an experience dealing with the passage of time. Use four or five cans alike both in shape and size. Punch a hole in the bottom of each—a small hole in one, a large hole in another, and in-between sizes in the others. Paint each can a different color. Fill a plastic tub with water or clean, fine-grained sand. Seat a group of children around the tub where they can see the action. Have a child select a can, fill it, and hold it up so the water or sand drains through the hole. Have the children watch carefully as each can is filled and drained. Talk about what they have seen. "Did that can drain fast or slowly?" "Which can drained more slowly, the red one or the blue one?" "What color is the can that drained fastest?" "Can you put the cans in order from fastest to slowest?" "What can you tell me about the size of the hole and how fast or slowly a can drains?"

*Five.* The outdoors on a warm, sunny day provides a good setting for conducting an investigation and discussion dealing with temperature. Choose a location where there is a sidewalk, some grass, and a patch of bare ground, and select a time when part of each is in sunlight and in shade. Have the children remove their shoes and socks and walk on the sunny sidewalk. Talk about how it feels. Have them walk on the shady sidewalk. Compare the feel of it with that of the sunny walk. Have the children walk on the sunny grass, then the shady grass, the sunny ground, then the shady ground. Discuss the differences in temperature, if any, between each spot and the others. Can the children order the six locations from warmest to coolest? If so, cards numbered 1 through 6 might be used to indicate the order.

---

## Self-check of Objective 10

Describe materials and activities for building preoperational-level children's backgrounds for capacity, weight, linear, time, and temperature measures.

---

## NONSTANDARD UNITS

Children's understanding of measure begins as they engage in activities such as those already described. Continue with these activities until you see evidence that some children are able to conserve length, volume, and so on. After simple tests show which children are able to conserve length, for example, group them for their first work with nonstandard units.

### 1. Purposes of the Activities

Activities with nonstandard units bridge the gap between exploratory work and the introduction of standard units. Two goals of these activities are to help children recognize the need for a uniform set of measures and to lay the groundwork for their appreciation of the simplicity of the metric system.

### 2. Suggested Activities

The activities that follow are examples of those dealing with nonstandard units for the various types of measures.

*One.* For their first work with linear measure children should use such things as 3 by 5 cards, new pencils, and chalkboard erasers as non-standard units. Show them how to put cards end to end to find the length and width of their desks or tables in "file-card" units. Stress the importance of not leaving gaps or having cards overlap and of counting the cards carefully. Problem cards can give directions once children know how to use the units properly. You can prepare cards with instructions similar to those in Figure 14-5.

Later, children should use one card, pencil, or body parts for measuring. When a single unit, such as a file card, is used, children use the process of iteration; that is, one end of the card should be put at one end of the object that is being measured and a light pencil mark should be made on the object at the card's other end. Then the card is moved so the end that was away from the mark is at the mark; another mark is put at the opposite end, and so on. A record of the number of times the card is put down must be kept so the object's measure can be determined. Useful body parts are the foot, span (distance from end of thumb to tip of little finger on outspread hand), and cubit (distance from elbow to tip of middle finger).

Paces can be used as units of measure, also. One problem card in Figure 14-5 describes an activity using paces.

*Two.* A simple extension of the exploratory activities with a balance beam provides experiences with nonstandard units for measuring weight. Objects such as walnuts, wooden blocks, small bags filled with equal amounts of sand or rice, marbles, and like objects are used. Problem cards with open sentences such as these can provide directions: "How many balance 1 ?" "How many walnuts balance one bag?"

MEASURING WITH PENCILS

Use new pencils from the box to answer these questions.

1. How long is the reading table?
2. How wide is the reading table?
3. Can you figure out a way to find out how high the table is? If you can, tell how high it is.

Write each of your answers in a complete sentence.

USING YOUR PACE TO MEASURE

One pace is the distance you walk when you take a normal step. You will use your paces to measure the classroom.

1. How many of your paces do you think you will take when you cross the room the short way?
2. How many of your paces do you think you will take when you cross the room the long way?
3. Walk across the room the short way and count your paces. How many paces did you take? Which was more, your estimate or the number of paces you took?
4. Walk across the room the long way and count your paces. How many paces did you take? Which was more, your estimate or the number of paces you took?
5. Compare your numbers with a classmate's. Did you take the same number of steps each way as your classmate? If you did not, why do you think your answers are different?

**Figure 14-5** Problem cards for activities with nonstandard units

*Three.* Water- and sand-table activities can also be extended to include nonstandard units. Use a set of cans selected so that one has four times the capacity of another, the third has five times the capacity of the fourth, and so on. Use several cans of each size. Paint all cans with the same capacity a given color. Use problem cards with sentences such as these: "______ blue cans fill one red can?" "How many green cans can you fill from one yellow can?"

*Four.* Children's first work with volume measure can be done with common wooden or plastic blocks and their storage containers. As children fill the containers, direct their attention to how the blocks fill the box without empty spaces. Have them count the rows of blocks; the blocks in each row. "How many rows in the bottom layer?" "How many blocks in each row?" "How many blocks in all?"

It is also beneficial if two containers of different shapes but capable of holding the same number of blocks without empty spaces are available. Children will see that two differently shaped containers have the same volume. (Hold off any discussion of the formal meaning of volume or the introduction of a formula for determining volume for the present.)

*Five.* Glass sand-filled egg timers can be used as nonstandard time-measuring devices. Get as many different timers as possible. Children can observe that the sand transfers more quickly in one than another; and that the sand in two differently sized timers will transfer in the same amount of time.

*Six.* A number of different shapes can be investigated as possible units for area measure. Various units—circles, triangles, rectangles, diamonds, squares, and others—should be cut from cardboard. Groups of three or four children should work with a handful of congruent shapes. Each group should try to arrange its shapes to cover a rectangular region, such as a book cover, to determine its measurement in terms of the shape they are using. Guide the children's thinking by asking about each shape's suitability to cover an area. "Is there a unit that can cover the book without leaving gaps or overlapping? Will circles cover it?" (Figure 14-6.) "How about triangles?" (Figure 14-7.) The children will discover that the square unit is the most convenient, and its exclusive use will begin.

At this point, everyone should receive square and rectangular regions and a grid marked on a sheet of clear plastic or on a piece of

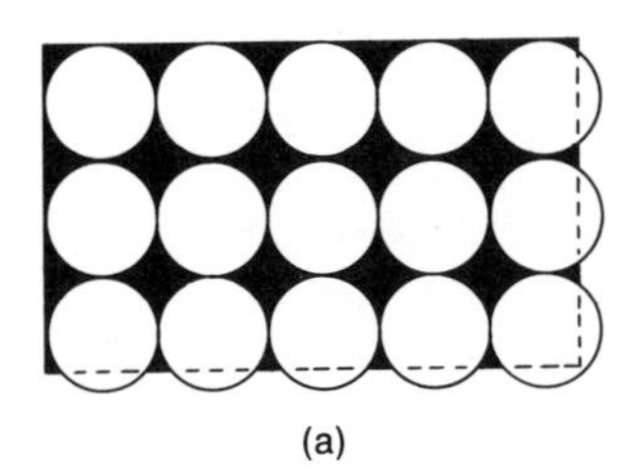
(a)

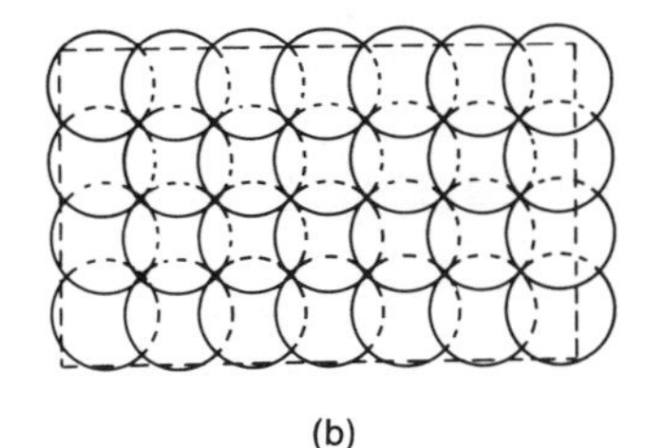
(b)

**Figure 14-6** When circles are used as a unit of measure of a region, they either (a) leave gaps or (b) overlap

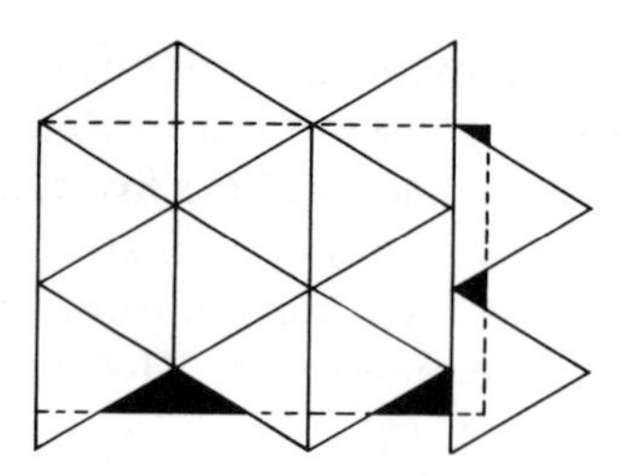

**Figure 14-7** Triangular shapes used as units for finding the area of a rectangular region

high-quality tissue. To determine a region's measure, childen should lay the grid over it and count the units within the region's bounds, as in Figure 14-8.

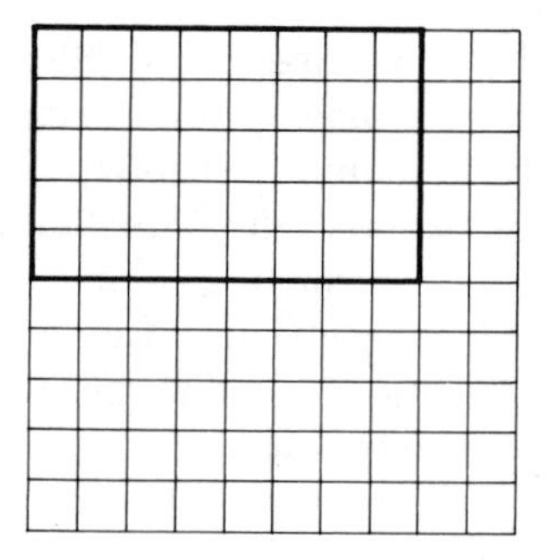

**Figure 14-8** Grid used to measure area. The measure can be determined by counting the units of the grid inside the bounds of the region

*Seven.* You must understand the meaning of degree, as used to measure an angle, before you can instruct children in its use. It can be explained in this manner: Imagine a ray that is moved around its end point so that it completes one full revolution (Figure 14-9). If the ray is stopped 360 times, with all stops being an equal distance apart, the distance from one stop to the next has a measurement of 1 degree. A protractor is used for measuring angles.

Children can begin their study of angle measure by making their own protractors, using nonstandard units. They can cut units from cardboard or mark them on clear plastic or tissue paper. While a given child's protractor will have uniform units, a group of children's protractors will have different-sized units. As they use their protractors to measure angles, children will discover that smaller units are more precise and therefore more useful than larger ones (Figure 14-10).

## Self-check of Objective 11

Nonstandard units for length, weight, capacity, volume, time, area, and angle have been described. Name at least one nonstandard unit for each type of measure, and explain its use by children.

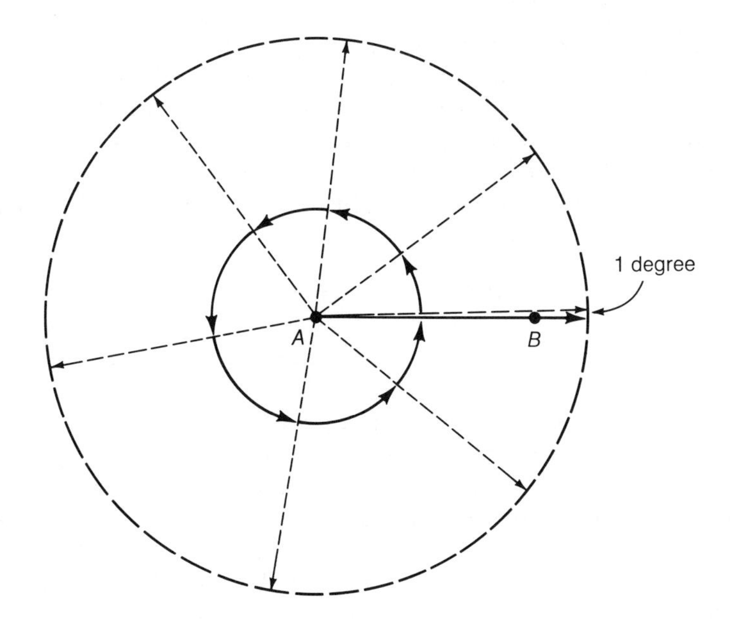


**Figure 14-9** Ray *AB* is swept through one revolution to illustrate the meaning of 1 degree

## STANDARD UNITS OF MEASURE

The progression from working with nonstandard to standard units goes smoothly for children experienced in working in a laboratory setting. Instead of file cards or pencils as units of linear measure, they use centimeter and meter rulers. Sets of kilogram weights replace walnuts and marbles, and liter containers replace cans.

### 1. Linear Measures

The commonly used units of linear measure are the meter, centimeter, millimeter, and kilometer. By the time children complete the sixth grade, they should demonstrate their knowledge of these units by being able to show a reasonable approximation of each and use devices for measuring meters, centimeters, and millimeters accurately. They should also be able to name the other units of linear measure—decimeter, dekameter, and hectometer—and identify the relationships between each of them and the other units.

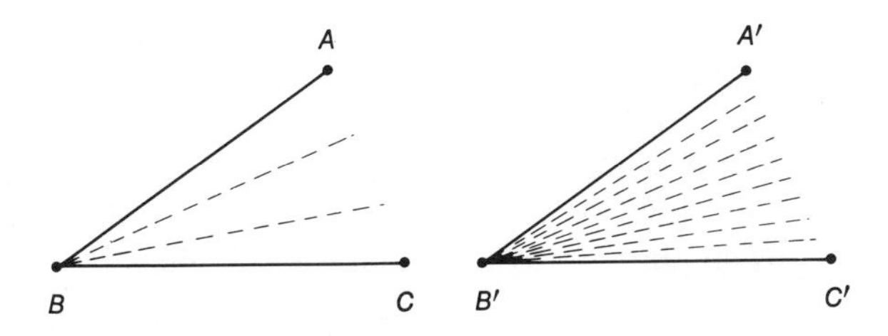


**Figure 14-10** The measure of the angle *ABC* is not as precise as the measure of angle *A′B′C′* because a larger unit of measure is used

Children will generally begin work with the centimeter. When introduced to this unit, children should use rulers marked off only in units, that is, the units should not be marked with numerals (Figure 14-11). Children are forced to count units as they use these rulers; thus their understanding of the meaning of linear measure is reinforced as they work. If such rulers cannot be bought, they can be easily duplicated on tagboard.

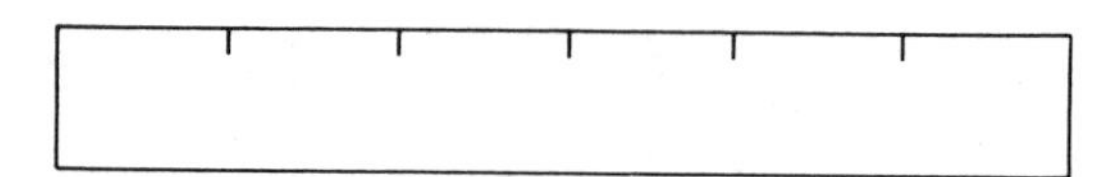

**Figure 14-11** A ruler for use by youngsters; only unit segments are marked

At first, objects whose lengths are a whole number of centimeters should be measured. This can pose a problem because many familiar objects do not possess this dimension. You can locate appropriate articles for children to measure by looking around the classroom and your home. For instance, a role of 35 millimeter film is packaged in a box that is 4 centimeters wide and 6 centimeters long. A Parker ballpoint pen is 1 centimeter thick and 13 centimeters long. You can also cut sticks, sharpen pencils, and trim paper edges so they have centimeter dimensions.

As children's activities continue, you will want them to measure things that have not been preselected. As they do, they will encounter objects whose dimensions are not a whole number of centimeters. Children will note that one pencil is a little more than 4 centimeters long, while another is a little less than 4 centimeters long; Figure 14-12 is an example of this. Even so, each pencil is 4 centimeters long, to the nearest whole unit.

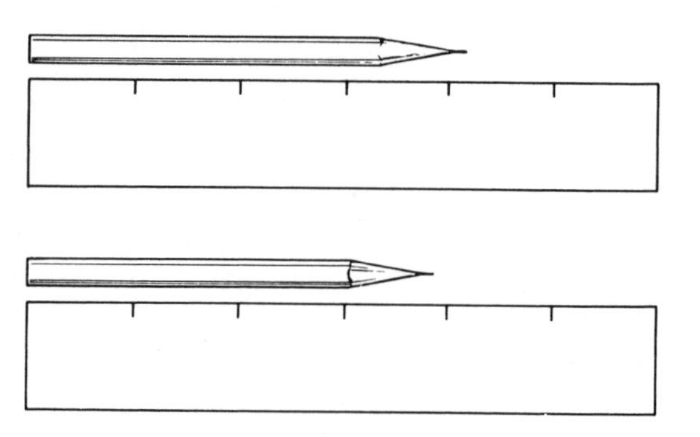

**Figure 14-12** Unit rulers used to measure the length of two pencils. Each pencil has a measurement of 4 units

When the end of an object falls near the midpoint between two units, children must make a judgment about the measurement to be recorded as its length (Figure 14-13). It should be clear that different individuals may record either of two different measurements for this pencil. The measurement might be recorded as 5 centimeters by one child and 6 centimeters by another. Both of them are considered correct.

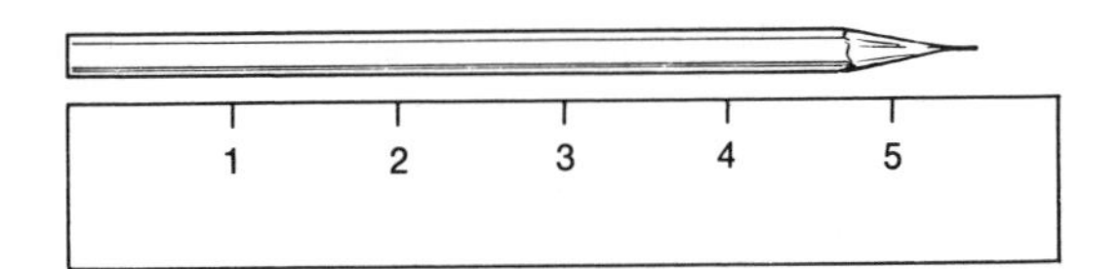


**Figure 14-13** The pencil might be recorded as having a measurement of 5 units by some children and 6 units by others

The ruler in Figure 14-13 has units indicated by numerals. Now a measure can be determined by noting the numeral rather than counting the units.

Work with the meter follows the centimeter. The first meter stick should be one that has no subdivisions. Inexpensive pine molding from a lumber yard cut into meter lengths is a convenient source of such sticks. Have children measure the width, length, and height of their classroom, the distance from the classroom door to the cafeteria door, the length of the sidewalk from curb to front door, and so on.

A wheel having a circumference of 1 meter attached to a handle so it rolls easily across flat surfaces is handy for measuring longer distances. Such a device is called a trundle wheel and can be purchased or the parts cut from plywood by a teacher and assembled by children.

Eventually, children should work with meter sticks subdivided into decimeters and centimeters. Metric measuring tapes and millimeter rulers should also be available. Now children will measure all sorts of objects, including themselves. Measurements will be recorded in different ways, depending upon children's maturity. For instance, a child's height might be recorded as 148 cm or 1.48 m. Mixed units are not used to record measurements. The above height measurement would not be recorded as 1 m 48 cm.

As their understanding of the metric system matures, children should be guided toward recognition of the fact that a given measurement can be converted from one unit to another by shifting the decimal point in one direction or the other. Moving the decimal point one position to the left is the same as dividing by 10; moving it two positions to the left is the same as dividing by 100, and so on. Moving the decimal point to the right one position is the same as multiplying by 10, two positions is the same as multiplying by 100, and so on. To change 148 centimeters to meters, divide 148 by 100 (there are 100 centimeters in 1 meter). The answer is 1.48; hence 148 cm = 1.48 m. To change 3.26 meters to millimeters, multiply 3.26 by 1000. The answer is 3260, so 3.26 m = 3260 mm. Devices for making conversions quickly and easily are available.[8]

One objective of activities dealing with linear measure is that children be able to show reasonable approximations of the meter, decimeter, millimeter, and kilometer. One way to help them do this is to give them opportunities to estimate lengths and distances as well as measure them.

[8]The "Metric Genie" is one such device. It is available from Metric Genie Company, P.O. Box 305, Corte Madera, Calif. 94925.

Identify a list of objects to be measured. Have each child first record an estimate of the length of each object, then have the child measure the objects and compare the two sets of figures.

Another way to meet this objective is to help children establish references based on familiar objects. A dime is approximately 1 millimeter thick. The width of the nail on a person's little finger is about 1 centimeter. The chalk rail in many classrooms is about 1 meter above the floor.

The best way for children to gain a "feeling" for the kilometer is for them to walk the distance several times. Locate landmarks that are about a kilometer from a given point at school—a child's house, a store, a playground, a bridge. As children walk from the point at school to a point at each landmark, have each child count his or her paces. By doing this several times, each child can determine the average number of paces he or she takes while walking a kilometer.

---

## Self-check of Objective 12

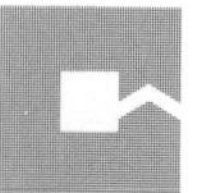

Describe materials and activities that will help children to understand linear measure and to make accurate measurements and reasonable estimates of length.

---

### 2. Weight (Mass) Measures

Technically, weight and mass are not the same. Weight is the force exerted on an object because of gravity, while mass is the quantity of matter of which the object is composed. Thus, a given object will weigh less on the moon, where the force of gravity is less, than it will weigh on Earth, where gravity's effect is greater. It will have the same mass either place because the quantity of matter remains the same. Even though this technical difference exists, it is common practice to consider weight and mass to be the same. Therefore the term that is commonly used—weight—will be referred to in the remaining part of this chapter.

The basic unit of weight in the metric system is the kilogram. As pointed out earlier, this unit contradicts the notion that basic units have no prefixes. The commonly used units are:

1 metric ton = 1000 kilograms
1 kilogram = 1000 grams
1 gram = 1000 milligrams

By the time children complete the elementary school, they should be able to name these as commonly used units and identify familiar objects that weigh about 1 kilogram or 1 gram. They should also be able to name the less commonly used units, that is, hectogram, dekagram, decigram, and centigram, and state the relationships between each unit and those that are larger or smaller.

Two types of scales should now be available—balance scales and spring scales. In order to weigh objects on a balance scale, a set of weights is needed. A commercial set may contain these weights: one 1-kilogram, two 500-gram, two 250-gram, two 100-gram, two 50-gram, four 10-gram, two 5-gram, and five 1-gram. You can duplicate most of these by filling cloth bags with the appropriate amount of sand or rice. Lead fish weights can be used for the smaller ones. Both kitchen-type spring scales, weighing to 5 kilograms, and platform (bathroom) spring scales are needed.

Children's experiences will be more meaningful if they weigh objects commonly sold by weight. Provide them with fruit and vegetables, loose candy, nuts in the shell, and similar items, both loose and packaged (without the weight indicated). Naturally, they should weigh themselves several times during the year.

Once children are acquainted with the two types of scales, many of their experiences can be gained through learning-center activities. Each learning center should have a particular type of scale and problem cards to direct the children's work. There is no need for all children to deal with weight measures simultaneously; some can work at the centers while others are with you, or are engaged in other independent activities. Do discuss with the children plans for their work at the centers and prepare a schedule so there will be no confusion.

A time for the children to discuss their center work with you after it is completed should also be included in the schedule. You need the opportunity to introduce the less commonly used units and direct the children's thinking about the relationships among the various units. It is now that you can have children fill in tables and charts that summarize the major features of the metric system. Table 14-3 on page 397 is an example of one that fifth and sixth graders can complete with profit during discussions with you.

---

## Self-check of Objective 13

Three commonly used units and two types of scales are discussed. Name the units and identify the two types of scales.

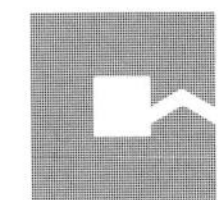

---

### 3. Capacity Measures

The liter is the basic unit of capacity measure in the metric system. It is a derived unit based on a part of the meter—the centimeter. A cube with inside dimensions of 1 decimeter has a volume of 1 cubic decimeter ($dm^3$)

or 1000 cubic centimeters ($cm^3$); this is 1 liter. Any container having a volume of 1 cubic decimeter has a capacity of 1 liter. Plastic containers of various sizes are inexpensive and can be purchased for classroom investigations. The metric containers in which soft drinks, wine, medicines, and similar products are sold should not be overlooked. By choosing carefully, you can accumulate a collection of graduated containers for your children.

Elementary school children should be able to identify the liter and milliliter as common units of capacity measure. They should use appropriate containers to measure a given quantity of water, for example, 250 mL. They should also be able to name the less commonly used units and state their relationships one to the other.

Once a set of containers has been collected, there should be sufficient opportunities for children to pour the contents of one measure into another measure to compare them. Instructions for activities can be written on problem cards or taped on a cassette recorder.[9]

## 4. Area Measure

Children's experiences with nonstandard units of area measure will show them that a square unit is the most convenient to use. Once they realize this and can use collections of congruent squares and plastic or tissue grids to determine areas of book covers, desk tops, and such objects, you will introduce them to standard units—the square centimeter for small things and the square meter for larger ones. The goals of instruction will be to enable children to name these as commonly used units, and to be able to measure the area directly be using square centimeter grids, and indirectly by measuring the edges of regions with centimeter and meter rulers and applying a formula.

Not all regions can be measured in an exact number of square centimeter or meter units. A square centimeter grid can be used to give an estimation of the measure of such a region's interior. The situation is illustrated in Figure 14-14. First, children should count the units entirely within the bounds of the region—there are 32; then they should count those entirely or partially within the region's bounds—there are 45. The region's measure will be greater than 32 but less than 45; a good estimate is 38.

The next experiences should enable children to state the general formula for finding the measure of any rectangular region using square units: $A = l \times w$ ($A$ represents the measure of the area; $l$ and $w$ represent the measure of length and width, respectively). As they use their

[9] See Leonard M. Kennedy. *Experiences for Teaching Children Mathematics* (Belmont, Calif.: Wadsworth Publishing Company, Inc., 1973), pp. 166, 169–171, for an example of taped instructions for primary-grade children.

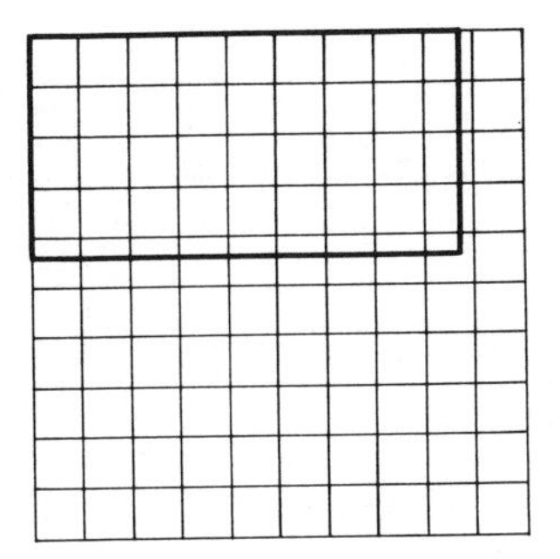

**Figure 14-14** The grid is used to estimate a region's area

grids to measure regions, children should be guided to note that each time they measure a region by counting, they are counting the units in an array. By now the children should know that the product of an $a$ by $b$ array is the product of $a$ times $b$. If a 1-centimeter unit is used to measure the length and width of a rectangular region, the two measures indicate the array of square units for that region. Figure 14-8 shows a 5 by 8 array; the measurement in square centimeters is 40.

Once children understand how to find the area of rectangular regions, they should determine formulas for the areas of regions bounded by parallelograms (Figure 14-15). Each child should receive a model of a parallelogram and its interior cut from paper (a). If they cut one end off the model (b) and put it at the other end, they have a rectangular shape (c). The area of the interior of the parallelogram is seen to be equivalent to that of a rectangle having the same height and length, so the formula used to find these two areas is the same. (This formula is sometimes stated as $A = b \times h$. $A$ represents the area measure, $b$ represents the base, and $h$ represents the height.)

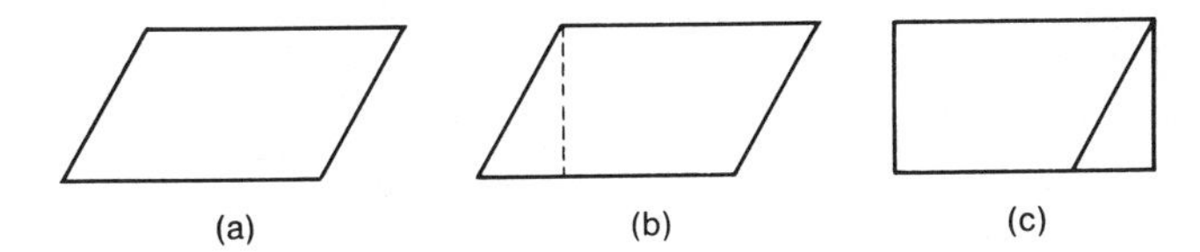


**Figure 14-15** The measure of area bounded by a parallelogram is determined in the same way as the area of a rectangular region

Similarly, children should explore with triangular-shaped regions. Two congruent triangular regions can be placed side by side to form a parallelogram-shaped region (Figure 14-16). The measure of this area is determined by the formula $A = b \times h$. Since the area for only one triangular region is to be determined, the product of $b \times h$ must be divided by 2; this is most often expressed by the formula

$$A = \frac{b \times h}{2}.$$

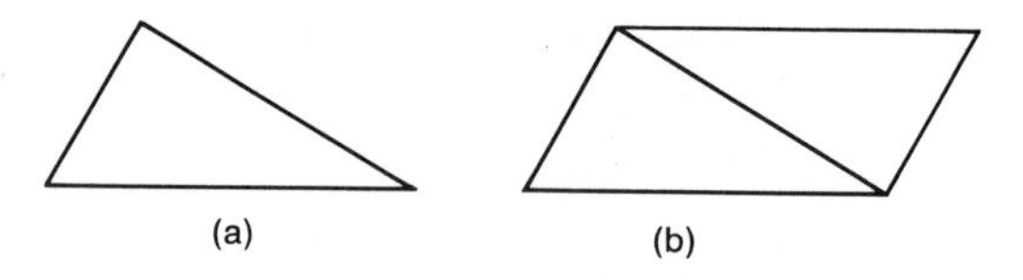


**Figure 14-16** Two congruent triangular regions can form the shape of a parallelogram to help children derive the formula for determining their areas

## Self-check of Objective 14

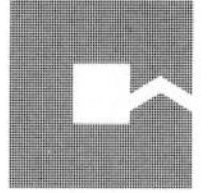

Describe materials and activities that will help children understand the meaning of area measurement for each of these plane figures: square and rectangle, parallelogram, and triangle.

### 5. Volume Measures

Earlier experiences with blocks will convince children that a cube is a satisfactory measure for determining the volume of containers, such as boxes. Children in the fifth or sixth grade who are ready for the work can extend their study of volume to include cubic centimeters and cubic meters. They will work primarily with rectangular solids (boxes) as in Figure 14-17; some may work with geometric solids such as those illustrated in Figure 14-18. Those who work with volume measures should be able to name cubic meters and cubic centimeters as commonly used units; determine the volumes of certain small containers by direct measurement, that is, fill them with centimeter cubes and count the cubes; and measure the edges of containers and determine the volume by using a formula by the time they complete the sixth grade.

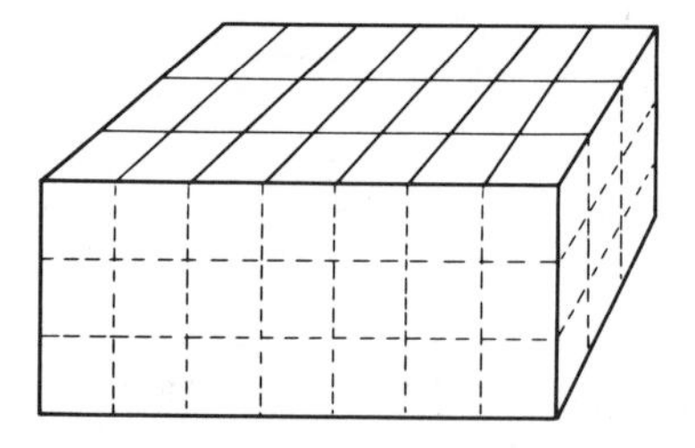

**Figure 14-17** A rectangular solid (box) filled with cubes helps children learn the formula for finding its volume

After numerous experiences with direct measurement, children should discover that indirect measurement can also determine the volume of the solid. The length, width, and height of the rectangular solid can be represented by the elements of the formula $V = l \times w \times h$. They should also see that the measures of length and width can be used to determine the measure of the area of the solid's base. Its volume can then be determined by multiplying the base's area measure times the height's measure:

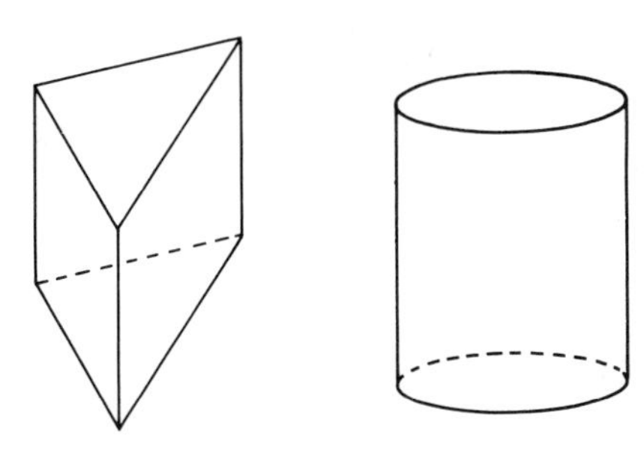

**Figure 14-18** Children can find the volume of simple geometric solids by using the formula $V = b \times h$

$V = A \times h$. This formula can be used by children to determine the volume of triangular prisms and cylinders, also. They should first measure each base's area, and multiply it by the measure of the height (Figure 14-18).

---

## Self-check of Objective 15

Tell what is meant by the volume of a geometric solid. Describe investigations for children that will help them develop their understanding of the concept and the formulas for finding the volumes of such figures as rectangular and triangular prisms (boxes), and cylinders.

---

### 6. Angle Measure

Not all elementary school children will study angle measures; there will be some who are not ready for it. The goals for those for whom the work is appropriate are that they be able to explain the meaning of angle measure, name the degree as the unit of measure, and use the protractor accurately.

After children understand the basis for measuring angles, introduce the standard protractor by using a large demonstration protractor or a clear plastic one that can be projected on a screen by an overhead projector. As children observe the demonstration or projected protractor, they should discuss the meaning of the marks and numerals on it. Then instruct them in its use by measuring several angles drawn on the board or projected on the screen. Children should realize how important it is to be careful in measuring with a protractor; the point at its center must be on the vertex of the angle, and the base of the protractor must be placed along one ray of the angle (Figure 14-19). Each child should then be given a protractor with which to measure angles. Each measurement should be recorded to the nearest whole degree. The typical protractor that children use does not permit measurements any more precise than this.

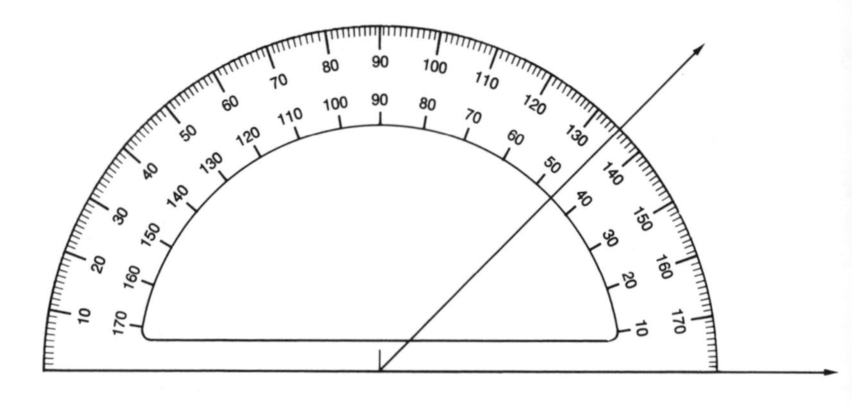


**Figure 14-19** The correct alignment of a protractor to measure an angle

## Self-check of Objective 16

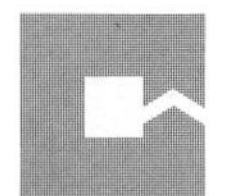

Demonstrate a procedure that can be used to help children learn to measure angles.

### 7. Measuring Temperature

Children in the second and third grades will learn to use thermometers by reading them and recording temperatures during daily activities connected with weather study. In the higher grades there are opportunities for children to use thermometers as they engage in science activities. By the end of the elementary school, children should read temperatures indicated on a Celsius thermometer, know that the freezing point of water is 0°C and the boiling point is 100°C, and identify some temperatures associated closely with their lives—normal body temperature is approximately 37°C, an outdoor temperature of 10°C indicates a cool day, a temperature of 30°C indicates a warm day, and 40°C indicates a hot day.

Primary-grade children should use large, easy to read alcohol thermometers. You can make a demonstration thermometer by printing the scale on a large piece of stiff cardboard or masonite painted a light color. The alcohol is represented by a strip of red cloth, made by sewing a strip of red cloth to a strip of white cloth, inserting the ends into slits at the top and bottom of the scale, and sewing the two ends together so the band is taut (Figure 14-20). After one child has read the temperature from the real

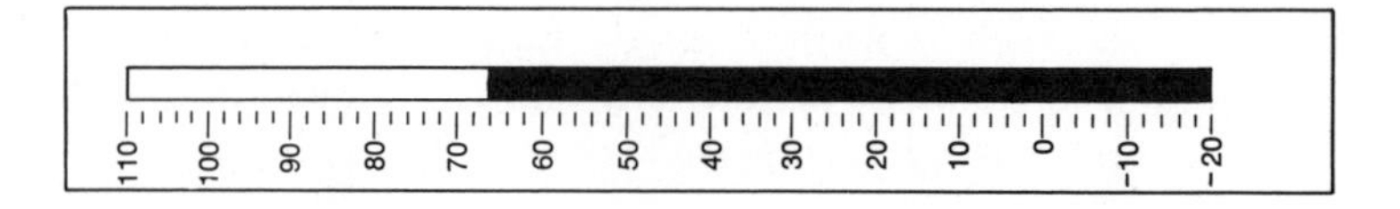


**Figure 14-20** A demonstration thermometer to help children learn how temperature is determined

thermometer, another should show the temperature on the demonstration thermometer so it can be read by the rest of the class.

## LEARNING TO TELL TIME

Children learn about two aspects of time: long-range—measured by a calendar—and the more immediate—measured by a clock or watch. Children should learn that calendars indicate the month of the year, the day of the week, and the number of the day. Many teachers like to incorporate the calendar into seasonal bulletin board arrangements. For the month of February, birthdays of famous Americans and Valentine's Day can be featured. Calendars such as the one in Figure 14-21, which you can make or buy, can be the basis for daily discussion; as a new date is posted, children should talk about such things as the year, month, day of the week, number of the day, number of days of the month that have already passed, and special days.

| March | | | | | | |
|---|---|---|---|---|---|---|
| Sun. | Mon. | Tues. | Wed. | Thur. | Fri. | Sat. |
| | 1 | 2 | 3 | 4 | 5 | |
| | | | | | | |
| | | | | | | |
| | | | | | | |
| | | | | | | |
| | | | | | | |

**Figure 14-21** Classroom calendar with removable tabs for recording dates

Later, children should learn to spell the names of the months and days and to use cardinal and ordinal interpretations of the numerals, such as "Today is October 30" and "This is the thirtieth day of October." In the intermediate grades, research projects about the history of our present calendar and ones that have been used in various cultures provide further information about this topic.

The ability to tell time by the clock is a skill every person needs to develop. Children learn to tell time by engaging in frequent activities with clocks that have movable hands. Large demonstration clocks with wooden or plastic gears synchronizing movements of the hands will cause fewer errors than those with hands that move independently. Smaller clocks with plastic gears are also available for individual children to use. Digital clocks and watches reveal the time directly with numerals. The increasing use of these instruments means that children should have opportunities to learn to read them as well as conventional clocks and watches.

The steps in learning to tell time are usually ordered in the following sequence:

1. Children learn to tell time to the hour. The *little hand* (*short* or *small* might also be used) indicates the hour. If you can, use a clock with only a little hand to introduce time to the hour. If such a demonstration clock is not available, cover the big hand so that only the little hand shows. After children have named the times as the hand is moved from numeral to numeral, write times on the chalkboard. Two ways of writing times to the hour are used: *9:00* and *9 o'clock*. In addition to naming times shown on the clock, children should also place the little hand so it points to numerals for times given by you.
2. Introduce the *big hand* (*long* or *large* can also be used). Show how the big hand always points to 12 when one of the hours is indicated. Again, provide practice in both reading time to the hour and placing the hands in their proper positions to indicate the hour.
3. Practice reading time to the half hour. Emphasize the fact that the big hand always points to 6 and the little hand is half-way between two numerals. Activities like those mentioned for telling time to the hour should be used.
4. Introduce reading time at 5-minute intervals. The ability to count by fives is a useful skill that will make it easier for children to learn to tell time to these intervals. Also, point out that 5 minutes pass as the big hand moves from 12 to 1, 1 to 2, and so on. To help children realize this and to grasp the meaning of 5 minutes, have them watch the classroom clock. Remind them often to keep their eyes on the big hand as it moves from one numeral to the next as 5 minutes pass. Tell children how to read the clock when the big hand points to a numeral other than 12 or 6, and show them how to write the time, for example, 6:20 and 8:45. To simplify the time-telling process, eliminate expressions like "a quarter past 5," "half past 9," and "a quarter to 10." These times can be expressed as "5:15," "9:30," and "10:45," and read as "five-fifteen," "nine-thirty," and "ten-forty-five," respectively. Time is expressed in figures like these in timetables, TV guides, and other written materials that mention times.
5. Introduce times to the minute, using procedures like the ones already mentioned.[10]

[10] For further information about instruction in telling time, see Beatrice Bachrach, "No Time On Their Hands," *The Arithmetic Teacher*, XX, No. 2 (February 1973), pp. 102–108. Some of the foregoing activities are adapted from this article.

Work sheets such as the one illustrated in Figure 14-22 should accompany lessons during which children learn to tell and write times.

In addition to learning to tell times by the clock and write them properly, children must learn to associate particular events of the day with the times at which they occur. For example, they should learn that school begins at 9:00, they go to lunch at 11:45, and the TV show *Wonderful World of Disney* begins at 7:30 in the evening. Activities that help children learn to figure elapsed time should be given, also. "If it's 6:30 now, what time will it be in 4 hours?" "Billy arrived at his grandmother's house at 7:00. He said he left home 3 hours earlier. What time was it then?"

Children in the middle grades should continue to practice until they learn to tell time accurately. Some will need to use demonstration clocks and work sheets until they can tell time correctly; others will need no special help. Class activities in the middle grades should also include such topics as the establishment of uniform time zones around the world, the history of telling time, and precision instruments, such as electronic timers, that record time.

Games are useful for helping children to tell time accurately once they have learned the basic skills. Games are particularly useful for older children whose skills have not been fully developed. "Calendar bingo"[11] helps develop understanding of days, weeks, months, years, A.D., B.C.,

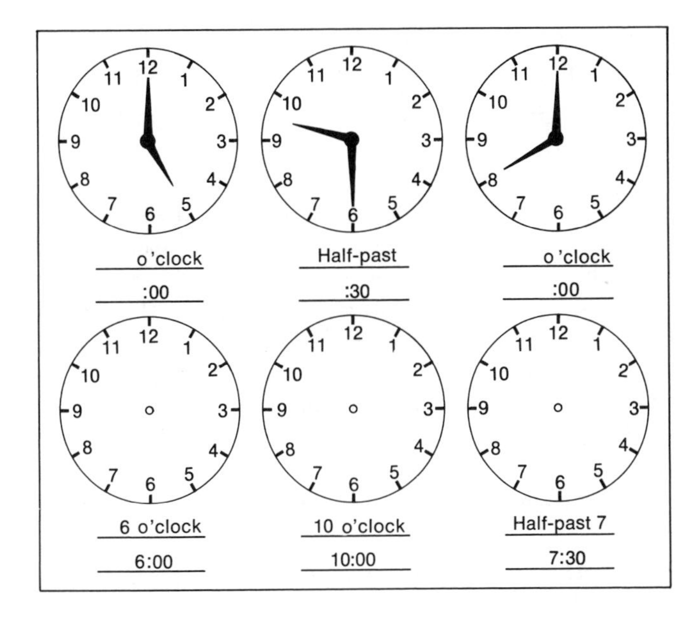


**Figure 14-22** Work sheet for primary-grade children as they learn to tell time to the hour and half hour

[11] Nikki Bryson Schriener, *Games & Aids for Teaching Math* (Hayward, Calif.: Activity Resources Company, 1972), pp. 37–58.

and significant and less well-known holidays and birthdates of famous people. Clyde Corle's book *Skill Games for Mathematics*[12] contains four games, and *Games for Individualizing Mathematics Learning*[13] has two.

## LEARNING TO USE MONEY

Children who can mentally reverse actions (Piaget's reversibility) and can think in terms of the whole rather than parts are ready to learn about money. In a sense, instruction about money is the reverse of that for telling time. Instead of beginning with the largest unit, instruction about money begins with the smallest. Children learn about the penny, then the nickel, dime, quarter, half-dollar, and dollar. Experiences with real coins and bills are essential to teach the value of different pieces.

1. Begin by introducing the *penny.* Give individuals small numbers of pennies — up to ten — to count. Then show them such things as packs of gum, candy bars, and pencils that they can buy with these pennies so they can associate certain numbers of pennies with the items' costs.
2. Introduce the *nickel* and *dime.* Help children associate five pennies with the nickel, ten pennies with one dime, and two nickels with the dime. These are many-to-one matchings that must be carefully developed so that children will understand their meanings. Continue work with the values of the coins by having children match amounts of money with things they can buy with them.
3. Introduce the quarter, half-dollar, and dollar through activities similar to the foregoing ones.

Children in the primary grades should extend their understanding of money through dramatic activities. Small toys and games or pictures cut from magazines can be displayed in a "store" with teacher-prepared price tags; these items are to be sold in the "store." Real or play money can be used by children alternating as clerks and customers. The games books mentioned in footnotes 11, 12, and 13 also contain games for helping children learn about money and its value.

Older children might assist with the collection of money in the cafeteria, in a student-operated school supply store, or for Red Cross and Community Chest campaigns. Some schools use bank savings accounts to

[12] Clyde C. Corle, *Skill Games for Mathematics* (Danville, N.Y.: The Instructor Publications, 1968), pp. 35–39.

[13] Leonard M. Kennedy and Ruth L. Michon, *Games for Individualizing Mathematics Learning* (Columbus, Ohio: Charles E. Merrill Books, Inc., 1973), pp. 109–112.

encourage children to save money. The educational value of such activities is limited unless children can handle and account for the money under adult supervision.

## Self-check of Objective 17

Describe at least one activity each for helping children learn about time and money.

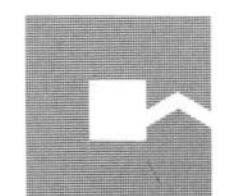

## SUMMARY OF MEASUREMENT CONCEPTS

As children progress in their study of measurement, they should periodically summarize the concepts they have learned. By the time children have completed the elementary school, most will have mastered these concepts:

1. The process of measuring means selecting a unit of measure and applying it to whatever is being measured. The result gives a comparison between the unit and the item being measured. The number of times the unit can be laid end to end with no overlapping or gaps (in the case of linear measure) is its measure. When the unit of measure is named, for example, 6 centimeters, the item's measurement is stated.
2. Measurements are determined for quantities that are continuous rather than discrete. Because the quantities are continuous, all measurements are *approximate* rather than exact.
3. Units of measure must be of the same nature as what is being measured. For example, line segments are units used for determining linear measurements, and square regions are units for determining area measurements.
4. Units of measure must be uniform wherever they are used. A government establishes standards that guarantee uniformity throughout the country.
5. A measuring instrument is selected on the basis of the job to be done. Generally, the smaller the object to be measured, the more precise the instrument. A meter stick or trundle wheel is suitable for measuring a school yard, whereas a centimeter ruler is used to determine the dimensions of a desk top.

## Self-check of Objective 18

Give an oral or written summary of the important measurement concepts included in the elementary school program.

## Common Pitfalls and Trouble Spots

Two common pitfalls associated with teaching about measures and measuring processes are discussed.

The first is that care must be taken to be certain that children are ready for any measuring activities in which they are expected to participate. The premature introduction of activities dealing with standard units of measure and measuring instruments does not give children opportunities to experiment with nonstandard units in both free-play and structured situations to learn underlying concepts and skills. The result is that many children do not develop an understanding of the system of measures or become knowledgeable users of various measuring instruments. The scope-and-sequence chart in this chapter, which is based on Piaget's research dealing with conservation, offers the recommendation that no standard units of measure be introduced before the middle of the second grade. In view of present-day knowledge of how children learn mathematics, this is a reasonable recommendation.

A standard activity associated with measures has been conversion of units from the customary to the metric system and vice versa. Thus, children have converted a given number of feet to the equivalent measure in meters, or a number of liters to quarts. Such activities have not proven to be effective either as a means of helping children understand the systems of measure or to learn a useful skill. Rather than have children convert from one system to another, impress upon them the importance of choosing the appropriate unit each time they must measure, applying the proper measuring instrument accurately, and then reporting the measurement in the correct way. As indicated earlier, until the metric system becomes predominant, children should learn to use common units in both the metric and customary systems.

Many children have trouble using measuring instruments in everyday situations and some confuse perimeter and area measures. These two trouble spots are discussed.

First, children who seemingly have a good understanding of a measuring system and its units of measure quite often cannot use measuring instruments properly and purposefully. This difficulty is likely to arise when measuring activities are confined to the mathematics period. In order for children to transfer skills and knowledge learned during this period, you must plan activities that require that they use measuring instruments in other situations. For example, during a plant growth experiment in science, they can use scales to weigh soil mixtures, volume containers to measure water and nutrients, thermometers to measure temperature, and rulers to measure daily growth.

Second, the distinction between area and perimeter measures will be clear if square units are used to measure area, and string and rulers are used to measure perimeters. Even so, some children confuse the names of the two measures, and apply one when they should use the other. A discussion about the word *perimeter* will usually remove the confusion. Most children are familiar with two words that will make the meaning clear—*periscope* and *meter.* A periscope is used for looking around the surface of the water when a submarine is submerged. *Peri-* is from the Greek language and means "around." *Meter* is from the Latin language and means "a device for measuring." Thus, perimeter means "the measure of the distance around the boundary of a region."

---

## Self-check of Objectives 19 and 20

Identify two pitfalls commonly associated with teaching about measures and the processes of measuring. Describe orally or in writing how each pitfall can be avoided.

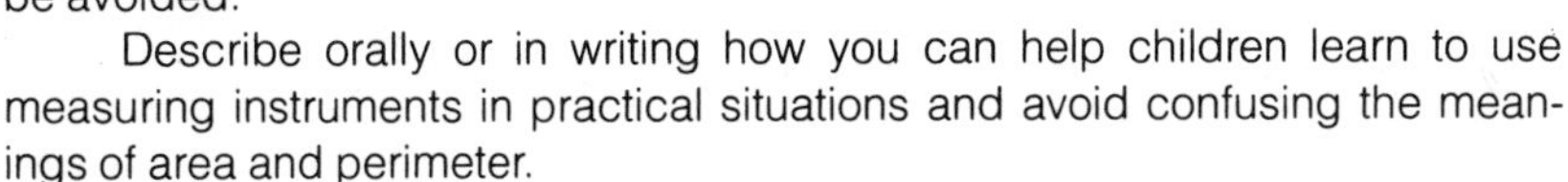

Describe orally or in writing how you can help children learn to use measuring instruments in practical situations and avoid confusing the meanings of area and perimeter.

---

## SUMMARY

The signing of the Metric Conversion Act of 1975 by President Ford set in motion the process of converting the United States from the English (customary) system to the metric system of measures. Even though the conversion process is slow, the metric system is receiving increasing emphasis in school mathematics programs. The metric system has several advantages over the English system. Chief among these are the systematic way it is organized around a decimal base, and the logical relationships among its various parts.

Remember that children must be conservers of quantity in its various forms before they can fully understand the meaning of measures. Until they become conservers, their activities should be exploratory—they should pour water, sand, and rice from one container to another, feel temperature differences, put objects on balance scales, and order objects by length, height, and size—to build background for understanding linear, weight, capacity, area, and temperature concepts. Nonstandard units are used to introduce the meanings of different measures—linear, weight, capacity, volume, area, time, temperature, and angle. Later, various commonly used metric and customary units are introduced, and children learn the apparatus and processes for using each one. By the time children complete the elementary school, they should be able to name commonly used metric and customary units, describe the structure of the metric system, and use measuring instruments accurately.

There are five measurement concepts that are developed over time during the elementary school years:

1. A unit of measure is selected and applied to the object being measured.
2. All measurement is approximate rather than exact.
3. Units of measure that are applied directly must be of the same nature as the object being measured.
4. Units of measure must be uniform wherever they are used.
5. The basis for selecting a measuring instrument is the nature of the object to be measured.

Teachers should avoid introducing concepts and units of measure before children are cognitively ready for them and having children convert measurements from one system of measure to the other.

Children need opportunities to use measuring instruments in a variety of settings once the instruments are introduced. Some children need special help in avoiding confusion of area and perimeter measures.

## STUDY QUESTIONS AND ACTIVITIES

1. Add to your collection of teaching-learning aids by including games and puzzles dealing with measures and processes of measuring, and devices for a metric laboratory. Include apparatus for linear, capacity, area, volume, and angle measure.
2. Many children's books about the metric system have been published. Review some and select several you would like to have in your classroom. Identify reasons why you accept or reject a particular book.
3. Examine an elementary school textbook series. Compare the sequence of measurement activities in it with the sequence recommended in this chapter. Are there any significant differences? If so, identify them.
4. Take the metric quiz in Sak's article (see this chapter's reading list). If you do not do well, schedule time to read appropriate sections of this chapter again, as well as the articles reviewed in the reading list.

## FOR FURTHER READING

*The Arithmetic Teacher*, XX, No. 4 (April 1973). The six major articles in this issue deal with the metric system. There are articles about its history, Britain's and Hawaii's experiences with metrication, and methods and materials for teaching it. The May issue of the same year also has four articles dealing with the metric system.

Bachrach, Beatrice. "No Time on Their Hands," *The Arithmetic Teacher*, XX, No. 2 (February 1973), 102–108. Presents a five-phase program for helping children learn to tell time. Describes and illustrates activities and sample work sheets from the program.

Brougher, Janet Jean. "Discovery Activities with Area and Perimeter," *The Arithmetic Teacher*, XX, No. 5 (May 1973), 382–385. Presents discovery activities dealing with geometric figures having constant perimeters and varying areas and having constant areas and varying perimeters. These activities can serve as the basis for teacher-made problem cards dealing with area and perimeter.

Chalupsky, Albert B., and Jack J. Crawford. "Preparing the Educator to Go Metric," *Phi Delta Kappan*, LVII, No. 4 (December 1975), 262–265. The authors studied the experiences of five nations in transition to metrics. Their study showed that attention must be paid to these issues: establishing and maintaining a schedule for transition, a need for continuing communication and coordination, development and selection of appropriate classroom devices, teacher training, development of classroom practices and strategies, and resistance to a temptation for unneeded precision.

Ginaites, Stephen J. "Sense with Cents," *The Arithmetic Teacher*, XXV, No. 4 (January 1978), 43. Describes a procedure using a 10 by 10 hundreds board and smaller regions to represent dollars and coins. Activities with the materials aid children in learning coin values and change-making skills.

Lindquist, Mary M., and Marcia E. Dana. "The Neglected Decimeter," *The Arithmetic Teacher*, XXV, No. 1 (October 1977), 10–17. The decimeter is more suitable than either the centimeter or meter for introducing children to standard units of linear measure, claim these writers. Five types of activities, with many examples of each, are presented.

May, Lola. "Introduce Metrics . . . Now," *Early Years*, II, No. 3 (November 1971), 75–76. The brief introduction offers reasons for teaching the metric system. The eleven activities discussed are suitable for primary-grade children.

Reisman, Fredricka K. "Children's Errors in Telling Time and a Recommended Sequence," *The Arithmetic Teacher*, XVIII, No. 6 (October 1971), 152–155. Children experience greatest difficulty learning the shorter units of time. Some specific types of errors, their causes, and a teaching sequence to avoid them are discussed.

Ropa, Adrienne. "Roll out the Meters," *Instructor*, LXXXVI, No. 9 (May 1977), 78–79. Self-adhesive meter and dekameter strips, each a different color, serve as a means for building children's understanding of meter, dekameter, hectometer, and kilometer. Both indoor and outdoor activities are described.

Sak, Theresa. "Metric Quiz—Test Your Metric Savvy," *Instructor*, LXXXVI, No. 2 (October 1976), 58–59. This semi-humorous programmed article is a good test of a person's knowlege of the basic units of the metric system.

Steffe, Leslie P. "Thinking about Measurement," *The Arithmetic Teacher*, XVIII, No. 5 (May 1971), 332–338. Piaget's theories of cognitive development and some of their applications to learning the meaning of measurement are discussed in this thought provoking article.

Strangman, Kathryn B. "Grids, Tiles, and Area," *The Arithmetic Teacher*, XV, No. 8 (December 1968), 668–672. Instructing children about area-measuring processes using grids and irregularly shaped regions is discussed. The process enables children to formulate generalizations of the important concepts involved.

Walter, Marion. "A Common Misconception about Area," *The Arithmetic Teacher,* XVII, No. 4 (April 1970), 286–289. The common misconception is that figures with common perimeters will have the same area, regardless of the measures of their sides. The examples given here can serve as the basis for several problem cards to challenge mathematically mature fifth and sixth graders.

Williams, David E., and Brian Wolfson. "Play Metric—Games to Help Kids Think Metric," *Instructor,* LXXXVI, No. 8 (April 1977), 62–63, 66. Three games—"Shuffleboard," "Roll a Meter," and "Metric Bet"—are described and illustrated. All are teacher-made.

Yvon, Bernard R. "Metrics with Marcel and Marcette," *The Arithmetic Teacher,* XXV, No. 1 (October 1977), 26–27. Marcel and Marcette are meter-long cloth snakes. Instructions are given for making them and some of the ways they can be used by young children to learn about the meter.

Zalewski, Donald. "Some Dos and Don'ts for Teaching the Metric System," *The Arithmetic Teacher,* XXVI, No. 4 (December 1978), 17. Briefly discusses a list of six "dos" and two "don'ts" to consider while teaching the metric system.

# 15 Nonmetric Geometry

Upon completion of Chapter 15, you will be able to:

1. Distinguish between a topological and a Euclidean view of space.
2. List four important topological relations included in the primary program, and describe children's activities for learning about them.
3. Explain why young children's early work with geometry in the Euclidean sense should begin with space (three-dimensional) figures rather than lines and planes.
4. Discuss a sequence of activities that might be used to develop children's understanding of solid figures.
5. Demonstrate activities at both the primary and the intermediate levels for learning about plane figures.
6. Identify geometric figures associated with lines and line segments.
7. Describe and demonstrate activities dealing with points and lines.
8. Distinguish between congruent and similar figures, and describe materials and activities children can use to develop their understanding of both kinds of figures.
9. Give a definition of symmetry, and describe materials and activities children can use to learn about it.
10. Illustrate with appropriate models three transformations: translations (slides), rotations, and reflections.
11. Describe two pitfalls commonly associated with teaching geometry, and describe ways to avoid them.
12. Identify a trouble spot children encounter in geometry, and explain how it can be minimized.

Key Terms you will encounter in Chapter 15:

| | | | |
|---|---|---|---|
| topology | apex | tetrahedron | quadrilateral |
| proximity | polygon | octahedron | closed curve |
| separation | rectangle | dodecahedron | open curve |
| order | right prism | icosahedron | congruent |
| enclosure | sphere | net | isosceles triangle |
| space figure | cylinder | parallelogram | scalene triangle |
| prism | cone | Euler's formula | right triangle |
| pyramid | polyhedron | geoboard | obtuse triangle |
| vertex | Platonic solid | concave | acute triangle |
| edge | cube | convex | pentagon |

| | | | |
|---|---|---|---|
| hexagon | right angle | radius | flip motion |
| heptagon | obtuse angle | similar | reflection |
| octagon | acute angle | symmetry | point symmetry |
| point | circle | transformation | line symmetry |
| line | tangent | translation | perpendicular |
| ray | diameter | slide motion | |
| line segment | chord | parallel | |
| angle | arc | turn motion | |

"There is no agreement on what aspects of geometry should be taught in elementary school. Recommendations vary widely. Until further research, including classroom testing of teaching procedures, provides a more conclusive answer, teachers will have to use their own judgment." This quotation from the 1970 edition of this book is still true. Agreement has not yet been reached about geometric content and its organization or procedures for helping children learn geometry.

The areas of nonmetric geometry in this chapter and the procedures discussed are based on the research of Piaget and others who have studied children's learning and successful practices in both British and United States schools. In Chapter 2 Piaget's four stages of mental growth are listed and discussed. They are repeated here:

1. The sensorimotor stage (zero to two years).
2. The preoperational stage (two to seven years).
3. The concrete operations stage (seven to eleven years).
4. The formal operations stage (eleven years and older).

These stages have been taken into account throughout the preceding chapters where materials and procedures for helping children learn about numbers and operations are discussed. Their implications for teaching geometry are considered now because they provide guidelines for selecting and organizing learning experiences.

A comprehensive discussion of content, materials, and procedures covering all aspects of elementary school geometry is beyond the scope of this book. The topics that are included are representative of important and useful ones for children to understand. You are encouraged to read one or more books dealing specifically with elementary school geometry for additional information. The titles of some of these books are listed in Appendix B.

## TOPOLOGICAL VERSUS EUCLIDEAN CONCEPTS OF SPACE

A topological view of space does not require that figures maintain a fixed shape, as they must in Euclidean geometry. Topology is sometimes called "rubber-sheet geometry" because shapes may be altered so that they assume new configurations, much as a figure drawn on a sheet of rubber will if the sheet is stretched and pulled in different ways. For example, a square may be altered to assume the shape of a circle or some other simple closed curve; or an open figure, such as one shaped like the letter C, might be changed to look like the letter S. Even though the shapes' configurations change, certain characteristics remain unaltered. The shape that was once a square remains a simple closed curve, while the open figure remains open with only two end points. In Euclidean geometry, all figures, whether line, plane, or space, always remain rigid and unchanged.

According to Piaget, children's first concepts about the world (space) around them are topological rather than Euclidean. He says that very young children do not view people and objects as being rigid and unchanging, but rather they view them in the topological sense. To children in the sensorimotor stage, the appearances of people and objects change as the positions from which they are viewed change. This topological way of viewing things continues on into the preoperational stage. But during this second stage, children change their views of people and things as they develop their understanding of four important topological (spatial) relations: proximity, separation, order, and enclosure.

### 1. Proximity

Proximity refers to the nearness of one object to another. Naturally, very young children are interested in things near to them because they can touch and manipulate them. Things that are out of their reach are usually of little interest, unless the child sees something that is eye-catching, such as a shiny part of a swinging mobile. Objects that are out of sight do not exist in the mind of the sensorimotor child.

Gradually, children engage in activities that help them recognize that out-of-sight objects exist, and to clarify the distinction between near and far and the relationships of objects in terms of their nearness to each other. The prenumber activities discussed in Chapter 5, particularly the ones dealing with classification and spatial relationships, provide experiences that help develop children's understanding of proximity. Once children have classified a set of materials or arranged a set of beads to form a pattern on a string, you can ask such questions as: "Which red car is nearest to the green car?" "Which black bead is farther from the blue bead?"

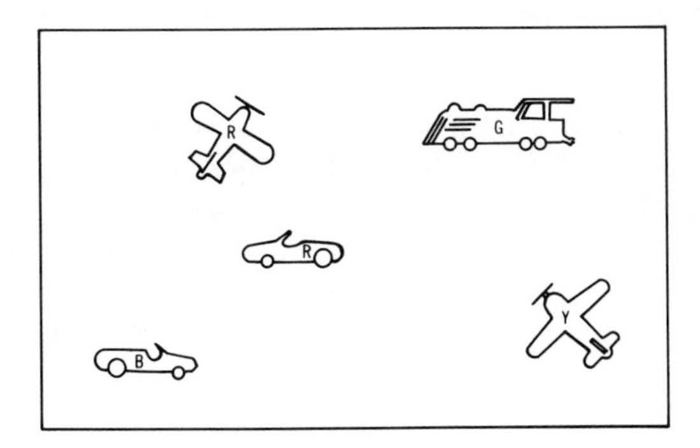


**Figure 15-1** Objects on magnetic board for discussion of proximity

You can also set up magnetic or flannel board displays and question children about the nearness of various objects on it (Figure 15-1). Magnetic plastic animals, flowers, vehicles of different kinds and colors, or pictures cut from magazines and fixed for either magnetic or flannel boards can be used. "Which is nearer to the red airplane, the red race car or the green railroad engine?" "Which object is farthest from the green railroad engine?" "I want Sally to move it so it is nearer to the railroad engine than it is to the yellow airplane."

## 2. Separation

Until children achieve an understanding of separation, they cannot clearly visualize an object as having separate parts, or a collection as being made up of separate objects. All of the parts run together in the child's mind. Children's drawings provide a good indication of their understanding of separation. Drawings of human figures are particularly enlightening. A child who understands separation will put the arms and legs in place with the body properly separating them, and the eyes will be in their proper position on the face with the nose separating them.

You should have small groups of children meet at a magnetic or flannel board to further their understanding of separation. Poker chips and strips of magnetic ribbon or felt disks and lengths of yarn can be used for the following activities:

1. Put two chips about 6 inches apart on the board, with a piece of magnetic ribbon between them. Have the children describe what

they see. During the discussion, say that the ribbon separates one chip from the other.

2. Move the shapes about, putting them closer together or farther apart, with the ribbon still between them. After each move, ask if the ribbon still separates the chips.

3. Move both chips to the same side of the ribbon. "Does the ribbon still separate the two chips?"

Use the classroom setting and pictures cut from magazines or ones you draw to extend children's understanding of separation. In the classroom, for example, a pillar may separate two bookcases, or a ceiling beam may separate two rows of lights, and so on. Pictures showing a river separating two parts of a town, a railroad track separating two buildings, and a highway separating two sides of a canyon can be used.

### 3. Order

Activities that deal with continuous and discrete objects, classification, and patterns, such as the ones described in Chapter 5, contribute toward children's understanding of order. Children will see order in the way attribute blocks are organized, and the way beads or pegs are put into patterns on cords or pegboards. During early work, children may be able to copy patterns directly from models, but not be able to reverse the order. They will eventually be able to do this if they are given enough unhurried pattern-card activities. When they can reverse a pattern's order, they demonstrate that they have a clear understanding of order.

*Think and Color*[1] contains a number of activities that contribute to children's understanding of order, as well as other topological relationships. It offers an intermediate step between activities with manipulative materials and the more abstract concepts that come later in the program.

### 4. Enclosure

Enclosure includes the positioning of one point between two others on a line, a point within a closed curve on a plane, and a point within a closed space figure. To help children in the preoperational stage understand enclosure, use activities such as these:

**a. Enclosure on a Line.** Put three different-colored beads on a string. Have children describe what they see. If they have already considered

[1]Evelyn Neufeld and James Lucas, *Think and Color* (San Leandro, Calif.: Educational Science Consultants, 1971).

separation, they may mention that the center bead separates the other two. You can point out that the two outside enclose the bead in the middle. Help them see that the middle bead cannot be taken off without removing one of the two outside beads. In addition to beads on a cord, have three children stand so a boy is enclosed between two girls or a girl is enclosed by two boys; use a play clothesline and discuss items of clothing that are enclosed by other clothing on the line; and discuss how some chairs are enclosed by others in a row of chairs. Enclosure on a line is an important relation because it is one children encounter frequently in mathematics. For example, they will talk about a number being between two other numbers. (A number that is between two numbers can also be said to be *enclosed* by the numbers.)

**b. Enclosure on a Plane.** Toys and pictures can be used to develop children's understanding of enclosure on a plane. A rural scene that uses toy plastic fences, horses and cows, barns, and so on, can be set up to show animals and buildings enclosed by a fence. Put some animals and buildings outside the fence, too, so children can talk about the things that are enclosed by the fence and those that are not. A similar scene can be set up on a flannel board with a yarn fence enclosing felt animals and buildings. Young children's playgrounds are frequently separated from other parts of the school by a fence connected to the preschool or kindergarten classroom. This playground provides a good setting for a discussion of enclosure. Some play equipment is permanently placed inside the fence, while wagons and other wheeled toys are not. Discuss these pieces of equipment in terms of whether or not they are permanently enclosed by the fence.

**c. Enclosure in Space.** Boxes with lids, both large enough for children to get inside, and small enough for them to put objects inside, will help them understand enclosure by space figures. One teacher collected shoe boxes from stores so that her children could build a "brick" playhouse. A father made the framework for the roof, which the children "shingled" with pieces of cardboard. Not only did the house provide a setting for discussion of enclosure as children played inside and outside of it, but the process of building it offered opportunities for them to consider other mathematical concepts, such as measurement ("How long and wide will the house need to be so that two or three children can comfortably sit inside, along with several pieces of small furniture?"), counting ("How many bricks long will the house be?"), dealing with preliminary angle ideas ("How can we be sure our corners are square?"), and handling area concepts ("How many carpet squares do we need for the floor?").

## Self-check of Objectives 1 and 2

Give an oral or written statement in which you distinguish between a topological and a Euclidean view of space. List four topological relations included in the primary program, and describe at least one activity dealing with each relation.

## STUDY OF SPACE FIGURES

We live in a three-dimensional world, and children's experiences inside and outside of the home are with objects that have three dimensions. It is recommended that once children understand the prerequisite concepts of proximity, separation, order, and enclosure you should have them begin to look at space in the Euclidean sense by examining three-dimensional figures. Most primary-grade workbooks deal with line and plane figures first, so these books will need to be set aside, at least temporarily, if the recommendations given here are followed.

### 1. Play Activities

Commercial materials, such as Geo Blocks[2] and attribute materials mentioned in Chapter 5, as well as the colored wooden or plastic blocks, both large and small, commonly found in preschool and kindergarten classrooms, and boxes and cardboard tubes of all shapes and sizes should be available for children's play. Initial activities should give children freedom to construct whatever they like. They will design and create many imaginative structures; some will show a surprising amount of creativity. Many children have a natural inclination to make patterns and structures that balance, or are symmetrical. While you do not need to develop the notion of symmetry during early play activities, you can comment on the way two sides of a pattern look almost alike, or are the same.

Another activity that interests children is that of making their own structures to take home. Give children a collection of various-sized boxes and tubes, some glue, and tempera paint, and allow them freedom to make a truck, railroad engine, steamship, or whatever strikes their fancy. Again call attention to the way two sides look alike (if they do), and any other interesting aspects of their creations.

[2] A set of Geo Blocks contains more than 100 hardwood pieces cut into a wide variety of shapes with edges up to 4 inches long. In addition to the blocks, there are a *Teacher's Guide for Geo Blocks* and *Problem Cards for Geo Blocks*. All are produced by the Webster Division of the McGraw-Hill Book Company, Manchester Road, Manchester, Mo. 63011.

## 2. Examination and Discussion of Solid Figures

During play activities, children gradually learn some of the terminology for parts of space figures—corner, edge, square, triangle, circle. You should encourage them to use these words, but do not force the specialized names and definitions too soon. As children indicate their interest by asking for names of a figure's parts, organize them into small groups to examine and talk about their models. Work with each group until children are making the proper associations of terminology and parts. Later, you can make problem cards to guide their work in a mathematics laboratory or classroom learning center. Use plastic or tagboard models, if possible, because children can conceive of the figure itself and its inside and outside more easily with hollow and, especially, transparent models, than with wooden or other solid figures.[3]

For the present, help children recognize the common plane figures associated with the models' bases and faces—squares, rectangles, triangles, and circles—the models' edges and corners, and the fact that they have an inside and outside. (Vertex is the proper mathematical term for a corner, but reserve its use for older children. Also, the idea of square and other plane figures will not be a mature one, but will be adequate for the time being. Later, children will learn more precise definitions of the plane figures and the proper names for parts of solid figures.)

There are other activities for children to engage in during exploratory lessons.

1. Count the shapes of a particular kind in a model. "How many squares are there in your model? Are there any other shapes in it? What are they? How many are there? Are they all the same size?"
2. "How many edges and corners are there in each model?"
3. "Which models roll easily? Which don't roll? Which slide when they're pushed? Which roll when placed one way, but slide when placed another way?"
4. "Can you find objects in and around the school that look like your models?"
5. Have the children complete work sheets by matching or drawing lines between pictures of similar objects (Figure 15-2).
6. Have children make a bulletin board or individual booklets with pictures of objects that look like their models.

[3]See Leonard M. Kennedy, *Models for Mathematics in the Elementary School* (Belmont, Calif.: Wadsworth Publishing Company, 1967), pp. 158–173, for patterns and directions for making fourteen inexpensive clear plastic models.

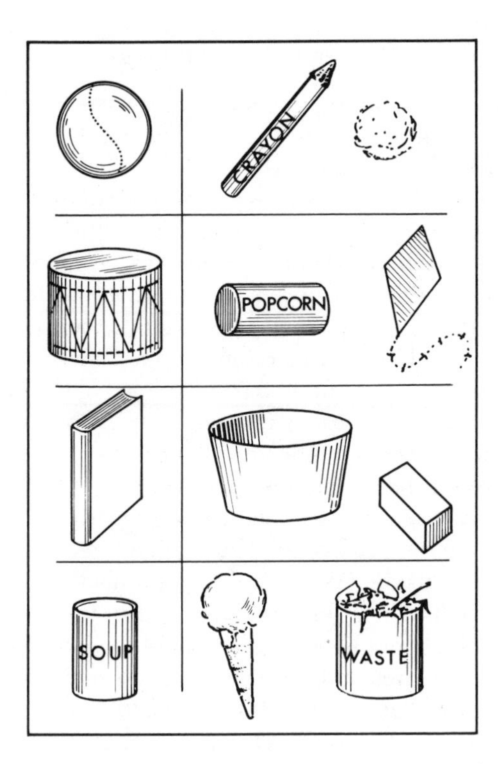


**Figure 15-2** Sample work page for shape recognition by primary-grade children

### 3. Formal Study of Solid Figures

Children in higher grades who have completed activities such as the ones already discussed should be involved in activities that deal with the figures in more detail and bring out the Euclidean view of thinking about them. Now they can begin using a greater variety of models and learn more precise terminology. Their earlier work should be extended to include study of prisms (Figure 15-3) and pyramids (Figure 15-4).

They should learn that a right prism has two parallel bases that are congruent polygons, and sides (faces) that are rectangles. (By the time children reach this stage of study, they will have learned about polygons and their interiors. These figures are discussed in a later section of this

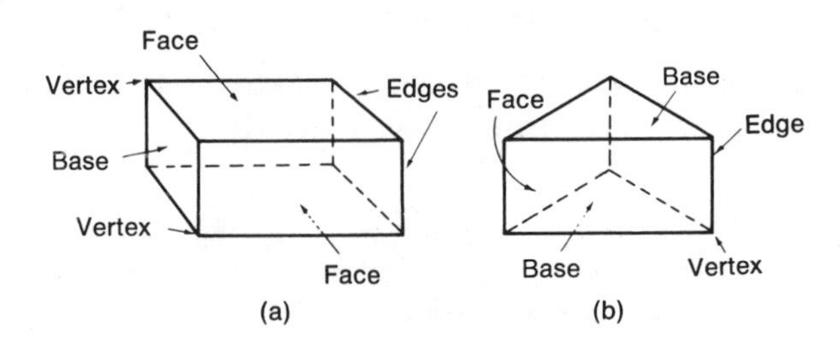


**Figure 15-3** Identification of the parts of a rectangular prism (a), and a triangular prism (b)

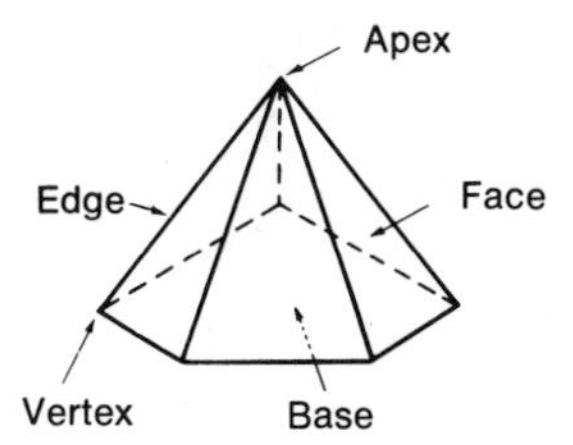


**Figure 15-4** Identification of the parts of a pentagonal pyramid

chapter.) If the bases are regular polygons, such as squares or equilateral triangles, the faces will be congruent rectangles. If the bases are not regular, such as a rectangle or an isosceles triangle, not all of the rectangular faces will be congruent. In a right prism the faces are perpendicular to the bases, so that they form right angles where they meet.

Pyramids have but one base and triangular faces that meet at a point at the top, or apex. The pyramids studied in the elementary school usually have regular polygons as bases, although polygons that are not regular are also found in pyramids. Both prisms and pyramids are named according to the polygon that forms their bases. In Figure 15-3 there are (a) rectangular and (b) triangular prisms, while in Figure 15-4 there is a pentagonal pyramid. Naturally, polygons with any number of sides can form the bases of these figures.

During their study, children should learn the proper names for each figure and its parts. The parts named in Figures 15-3 and 15-4 are ones children should learn. They should also learn that these figures are polyhedrons. A polyhedron is a closed figure formed by the union of four or more polygons and their interiors.

The solid figures associated with the circle—sphere, cylinder, and cone—should also be studied (Figure 15-5).

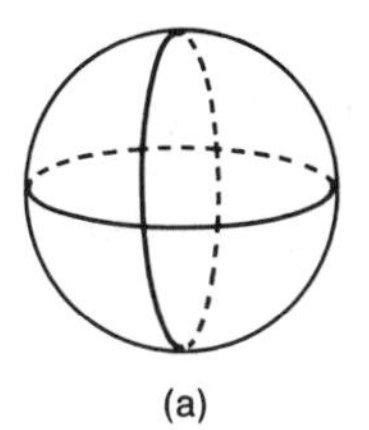
(a)

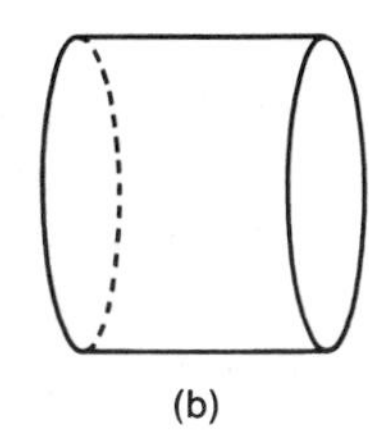
(b)

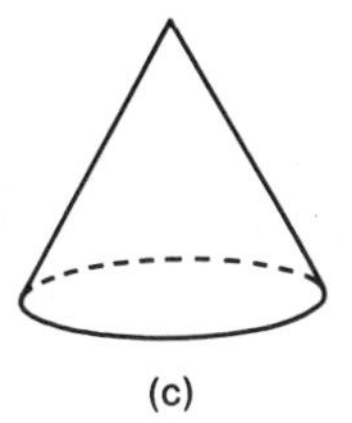
(c)

**Figure 15-5** Common geometric solids associated with the circle

Commercial or teacher-made plastic models of these figures should be used rather than solid ones, if possible, so that children can visualize points on, inside, and outside each model (Figure 15-6). They can grasp the idea that polyhedrons or figures associated with the circle separate space into three distinct sets of points: the set of points that are a part of the figure itself; the set of points that occupy the interior of the figure; and the set of points found outside of the figure.

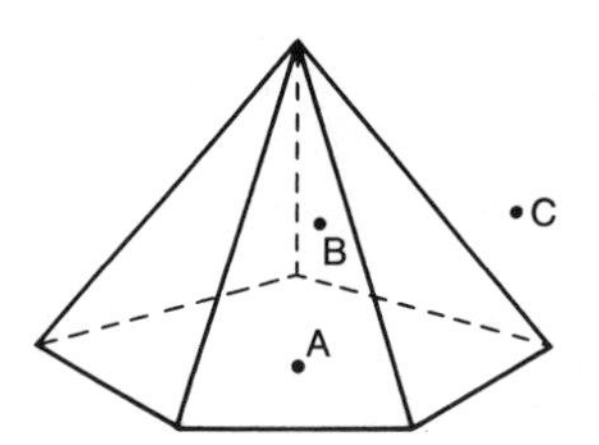


**Figure 15-6** A polyhedron separates space into three distinct sets of points. Point *A* is on the pyramid itself, point *B* is inside the pyramid, and point *C* is outside the pyramid

Here are other investigations with solid figures for older elementary school children. These make good problem card activities.

*One.* There are five figures called Platonic solids. They are formed by the union of four or more congruent regular polygons and their interiors. These are the regular tetrahedron, cube (or regular hexahedron), regular octahedron, regular dodecahedron, and regular icosahedron (Figure 15-7). Give children models of these and some nonregular solids and the above definition. Challenge them to identify the Platonic solids and to match each one with its name. (The names of many polyhedrons are combinations of the Greek word for the number of faces and *hedron*, which means "closed figure." *Polyhedron*, for example, means "many-sided closed figure.") Have children name the plane figures associated with each Platonic solid.

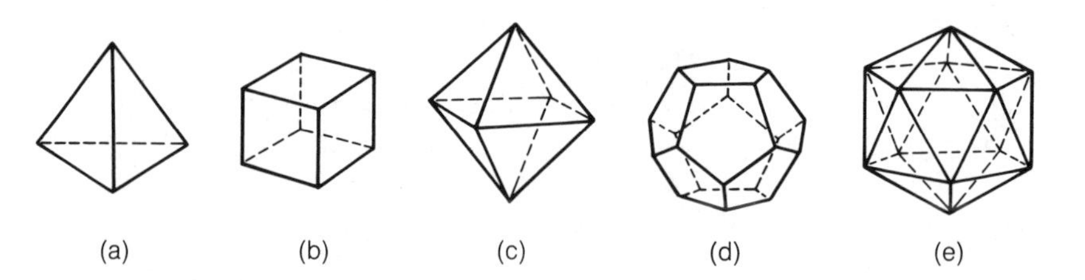


**Figure 15-7** The five Platonic solids: (a) regular tetrahedron, (b) cube (regular hexahedron), (c) regular octahedron, (d) regular dodecahedron, and (e) regular icosahedron

*Two.* Children are often confused by textbook illustrations of solid figures. Help them clear up their confusion by having them draw pictures of their models viewed from different positions. Two different types of illustrations are used in books, so children should draw some of each kind. Figure 15-8 shows a rectangular prism drawn both ways.

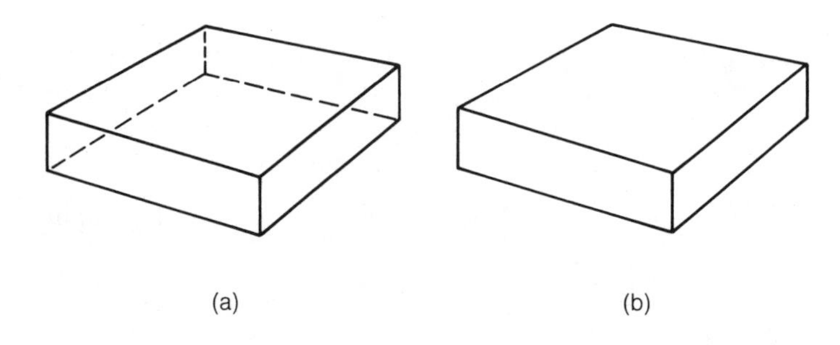


**Figure 15-8** Pictures of a geometric solid. In (a) unseen edges are illustrated by broken lines; in (b) unseen edges are omitted

*Three.* Many different types of models should be available for children to study. Children can make some themselves. Plasticine clay or Play-Doh can be used for nonhollow figures. These models are especially good to have on hand when children investigate symmetry of solid figures. (Symmetry is considered later in this chapter.) Soda straws connected with pipe cleaners and glue make good skeleton models (Figure 15-9). Skeleton models can also be made by joining lengths of pipe cleaners. Tagboard figures can be made from printed patterns which are cut and assembled by children themselves. As children make their own figures, they learn about edges, corners, the interiors of plane figures, and other parts of solid figures.

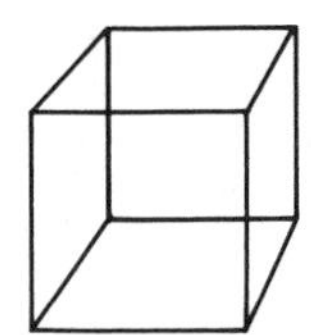
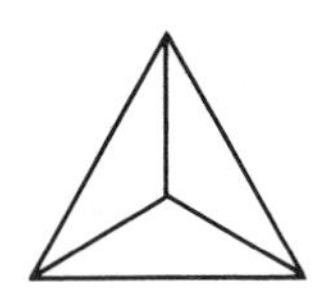
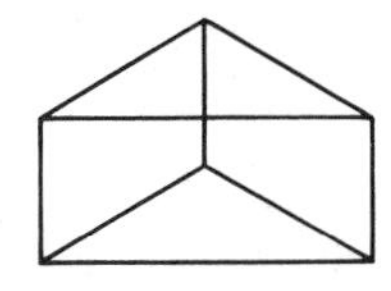

**Figure 15-9** Soda straw geometric solids

An interesting aspect of work with tagboard models is the study of nets of solid figures. A net is made by cutting some of the edges of a figure so that it will lie flat; or by drawing the outline of a figure and cutting it out. Two nets for a cube are shown in Figure 15-10. To begin, give children cardboard models to cut. Cut-down milk cartons, either with a part of one side left to form a sixth face, or cut down so there is an open top and only a base and four sides, give a plentiful supply of cubes. The nets in Figure 15-10 are for complete cubes. Challenge children to see how many ways they can cut the cartons to form different nets. A related activity is to give children a work sheet that contains nets so they can identify the ones that can be cut out and assembled to form a cube (Figure 15-11).[4] After children have marked their papers, they can cut out the nets and try to assemble them to make boxes to check their answers.

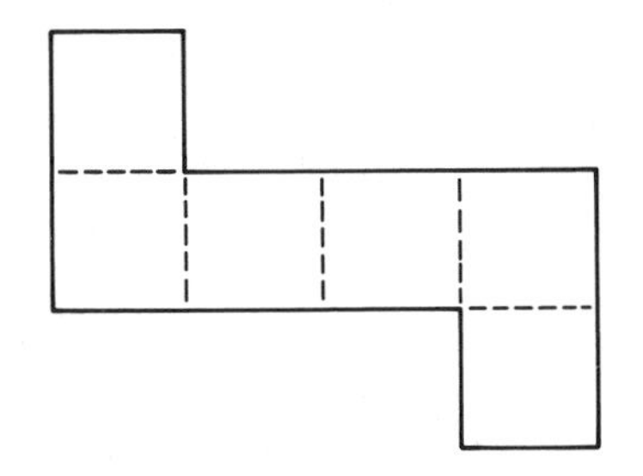
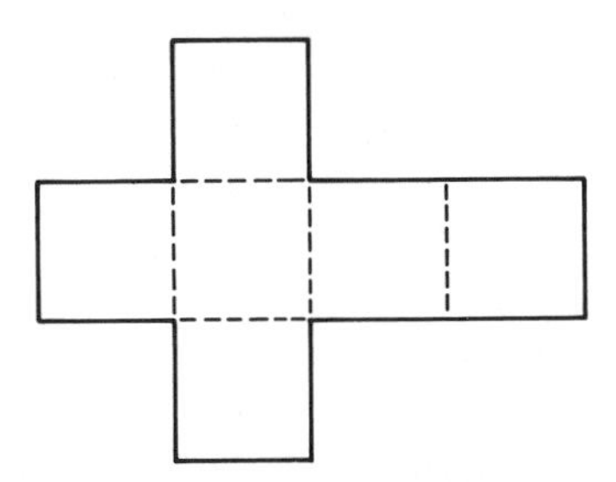

**Figure 15-10** Two nets for a cube

[4]This work sheet is adapted from Marion I. Walter, *Boxes, Squares and Other Things* (Washington, D.C.: The National Council of Teachers of Mathematics, 1970), p. 9. (Used by permission.) This book contains a number of investigations dealing with solid figures and their patterns, symmetry, tessalation, and geometric transformations.

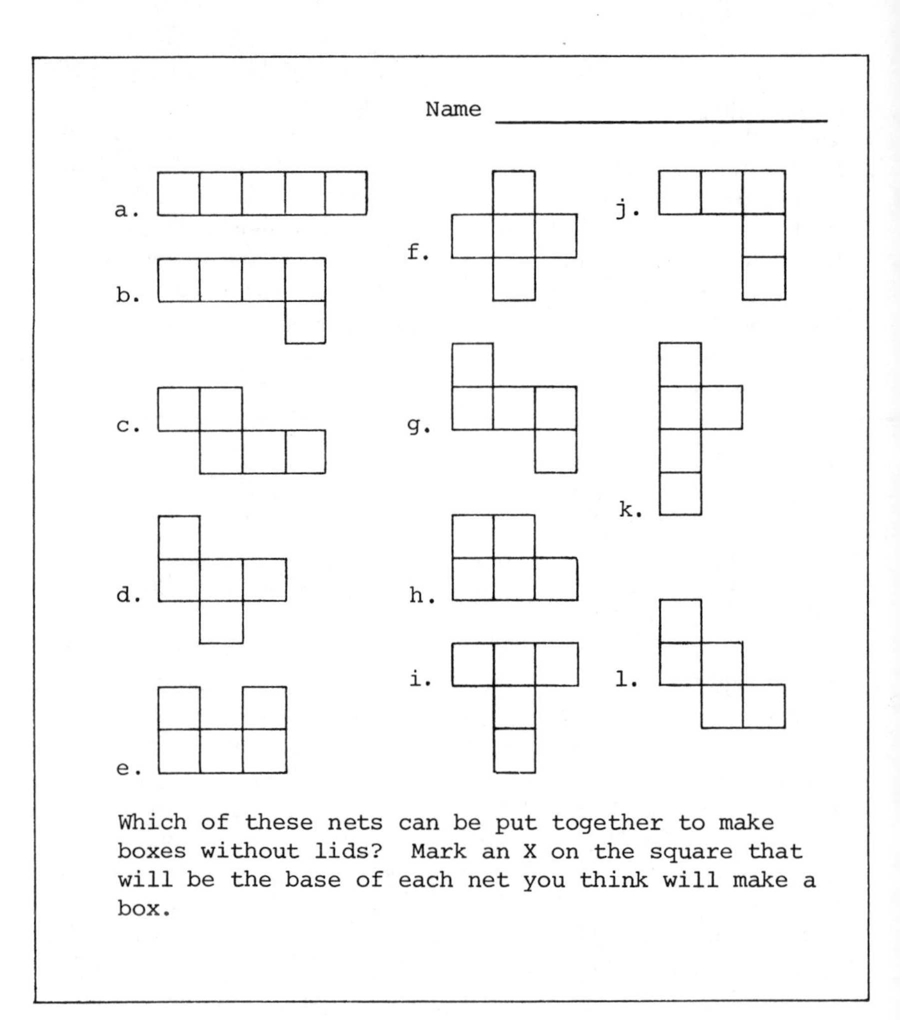

Name
a.
b.
c.
d.
e.
f.
g.
h.
i.
j.
k.
l.
Which of these nets can be put together to make boxes without lids? Mark an X on the square that will be the base of each net you think will make a box.


**Figure 15-11** Work sheet for making boxes without lids from five-square nets

Nets for other figures—pyramids, the five Platonic solids, and other polyhedrons; and models of spheres, cylinders, and cones—should be included in the children's investigations, too. Children will find that a sphere cannot be flattened and that if the top and bottom of a cylinder are removed and the side is cut straight down from one edge to the other, a rectangle is formed when the figure is flattened [Figure 15-12(a)], but if the cut is an oblique one, the flattened figure is a parallelogram (b).

*Four.* The novelty of Euler's formula will interest many children. Have them count the faces, vertices, and edges of each polyhedron model and list the numbers in a chart (Figure 15-13). When they add the number of faces and vertices and then subtract the number of edges from the sum, the answer is always 2. This can be summarized as $F + V - E = 2$. This formula is named for Leonhard Euler, a Swiss mathematician (1707–1783).

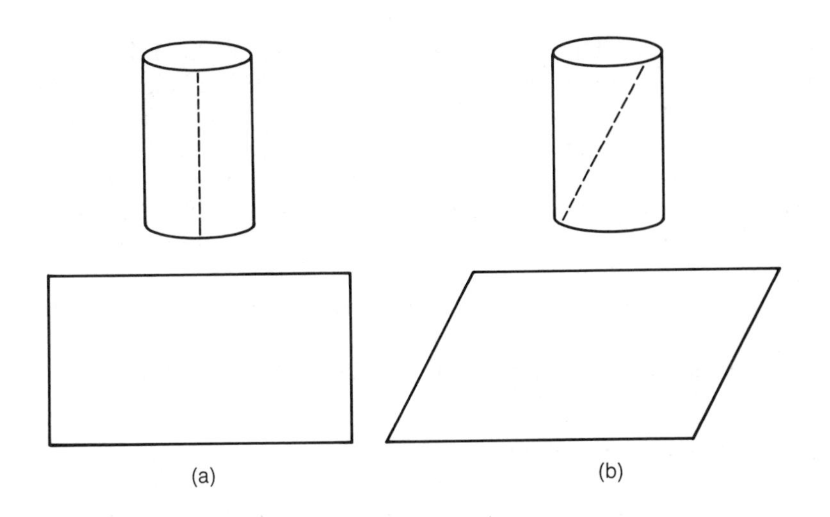


**Figure 15-12** A cylinder cut square with the edges gives a rectangle when flattened (a); it gives a parallelogram when the cut is oblique (b)

## Self-check of Objectives 3 and 4

Give an oral or written statement containing reasons for beginning the study of the Euclidean concept of geometry with space figures rather than with lines and plane figures.

Three levels of activities with space figures are discussed. Identify each level, and describe at least two children's activities for each one.

## ACTIVITIES WITH PLANE FIGURES

Growing out of investigations with solid figures are those that deal with plane figures. The activities already described will not be completed before children begin their study of plane figures and lines, which are described in the next section. Activities with the three types of figures will run simultaneously once all of them have been introduced.

The study of plane figures is a natural outgrowth of work with solid figures. As mentioned earlier, children will begin using terms such as *square, rectangle, triangle,* and *circle* as they use models of solid figures.

| Geometric Solid | Faces | Vertices | Edges | Invariant |
|---|---|---|---|---|
| Triangular Pyramid | 4 | 4 | 6 | 2 |
| Octagonal Prism | 10 | 16 | 24 | 2 |
| Square Prism | 6 | 8 | 12 | 2 |

**Figure 15-13** Part of a chart used to organize data about prisms and Euler's formula

### 1. Shapes and Posting Holes

Shapes cut from plywood, masonite, or heavy art paper, such as colored railroad board, are easily made. If they are cut carefully, you will have not only the shapes, as shown in Figure 15-14(a), but also their outlines in the pieces of material from which they were cut (b). These holes are called posting holes. Children use these models to get the "feel" of each shape by running their fingers around the outside of the shapes and the inside of their posting holes. Have children count the sides (edges) and corners of each shape, and talk about the way corners feel: they will feel "sort of flat," "sharp," or "square." Children can also talk about the differences in the feel of the edges of polygons compared with a circle.

Once they are familiar with some shapes, primary-grade children should search their environment for examples of the ones they know. Activities such as the following are interesting to most children:

1. Children can look around the classroom to find things that have shapes similar to the geometric figures they have been studying. Many of the figures will be represented by objects already in the room. You should prearrange the room to contain representative objects of certain shapes not usually present, particularly spheres and circles.

2. During their study of safety, children can be guided to recognize the different shapes of signs used along streets and highways (Figure 15-15). Replicas or pictures of them should be available.

3. Children can look in magazines for pictures of objects of different geometric shapes. The pictures can be displayed on bulletin boards.

Later, middle graders can use the cut-out models and the posting holes to investigate corners and other geometric ideas. For example, children will see that a rectangle's corners fit into each corner of a square-shaped posting hole; they will see that there is only one type of triangle that has a corner that fits each corner of a square posting hole.

You can then ask, "How many ways will a shape fit into its own posting hole?" This question leads to investigations about lengths of sides and types of angles, and to the identification of regular and irregular polygons. Mark the corners of each shape with numerals or colors so children can keep track of the different ways the shapes have been put into their posting holes. Have the children rotate a shape to see how many ways it fits, but make sure they don't turn it over. Figure 15-16 shows that a square fits into its posting hole four different ways, while a parallelogram fits in only two ways. If a four-sided figure (quadrilateral) fits in four ways, it is a regular polygon, while one that fits in only two ways will have opposite sides of the same length and opposite angles of the same size. Investigations with other quadrilaterals and polygons and circles will enable children to make conclusions about them, too.

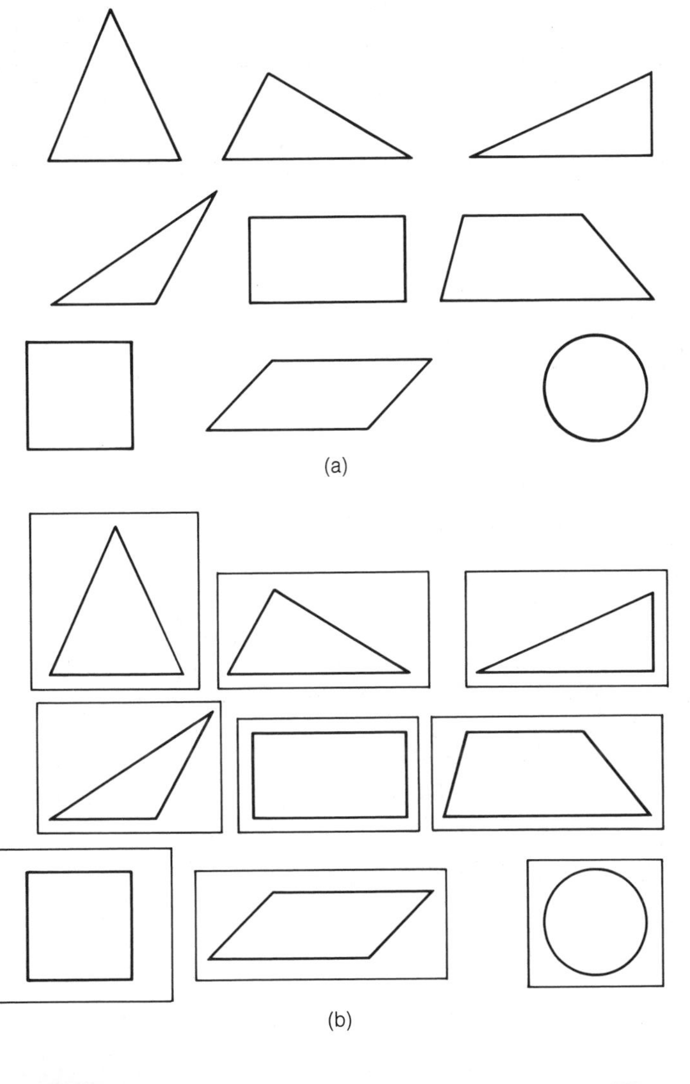


**Figure 15-14** Common geometric shapes (a) and their posting holes (b)

**Figure 15-15** Common highway signs that children should become acquainted with by shapes and colors

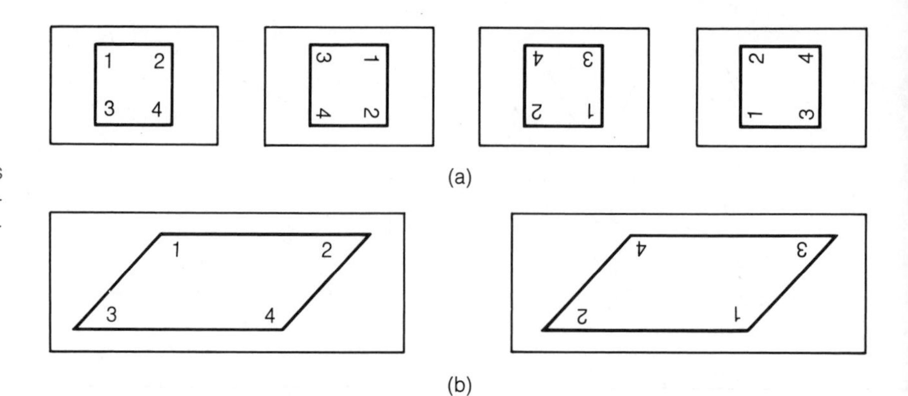


Figure 15-16 A square fits into its posting hole four different ways (a), while a parallelogram fits in only two ways (b)

## 2. Geoboards

A geoboard is a square piece of wood or plastic in which pegs are arranged in some orderly fashion. Arrangements in common use have nine, sixteen, twenty-five, or thirty-six pegs ordered in 3 by 3, 4 by 4, 5 by 5, and 6 by 6 patterns, while some have circular patterns (Figure 15-17). Boards with 100 pegs arranged in a 10 by 10 pattern are also useful because they have room for the outlines of several plane figures at one time. Rubber bands are used to make figures on the boards. Paper covered with rows and columns of evenly spaced dots should be available so children can copy

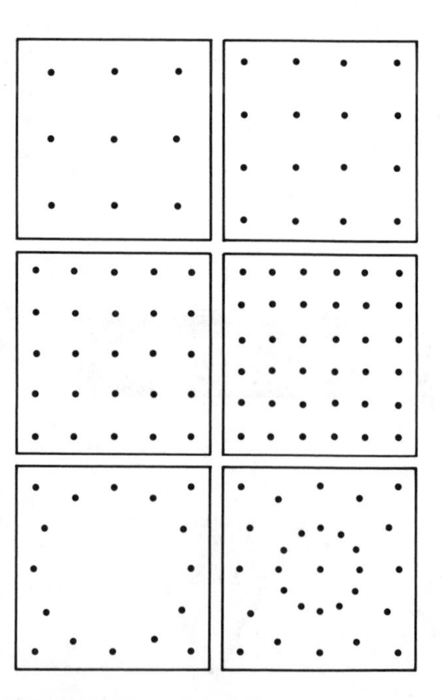

Figure 15-17 Examples of commonly used geoboards

figures from their geoboards. Geoboards and dot paper are produced commercially by a number of companies. Many teachers make geoboards with plywood and small finishing nails, and dot paper by the copy-machine process.

Geoboards make possible a wide range of investigations. In the primary grades, children can use them to investigate and gain an intuitive understanding of plane figures, both open and closed, and of points and line segments. Older children can use them to study different types of triangles, quadrilaterals, and other figures; and to learn about regular and irregular figures, concave and convex figures, and other more mature concepts.

The directions for the activities that follow are suitable for primary-grade children and can be written on problem cards or given orally.

*One.* Children can make closed figures with three, four, five, or more sides. At first, have them make only three-sided or four-sided figures. After their boards are covered with figures, have the children check to see that they made their figures with the designated number of sides. Then have the children study their figures to see how many different kinds they made. (At this stage, children will probably say that two congruent figures are different if they are in different positions on their boards. Don't try to get them to see that these figures are the same at this time. Later, children learn about congruence and will see that models of the same figure can be arranged on a geoboard in different ways.) Have the children count to see how many different figures they made. Then have them compare their boards to see the different figures made by their classmates. Have children talk about the different types of angles, but don't use such terms as *right*, *acute*, and *obtuse*. Let children say "square" for right angles, and say that an acute angle is "smaller" and an obtuse angle is "larger" than a square angle.

*Two.* Children can make a series of figures, each succeeding one having one more side than its predecessor. (Ten by ten boards are good for this activity. If smaller boards are used, children may need to copy each figure on dot paper before they make a new one on their boards. Be sure your children are mature enough to make accurate copies before you ask them to do this.) Have them check to be certain they have a figure of each kind, beginning with one that has three sides and continuing to the figure with the most sides, whatever it is on each child's board. Have them compare their figures with their neighbor's to see the differences, if any, in each kind. Talk about the changes in the figure's shapes as more sides are used. Ask, "Does the size of the angle increase as the number of sides increases?" Some children will probably make concave, as well as convex figures (Figure 15-18). Accept these as long as children recognize that they are closed figures and have the number of sides claimed by their makers. (The study of convex and concave figures is included in activities for older children.)

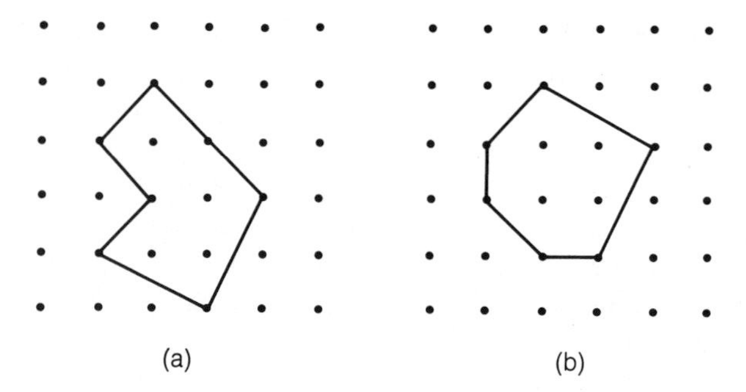


**Figure 15-18** Two six-sided figures. The polygon in (a) is concave; the one in (b) is convex

*Three.* Children can make open and closed curves. Open curves can be made by stretching a rubber band between two or more pegs, either in a straight line or going around a corner. Have children make models of both open and closed curves (Figure 15-19) so they can talk about the differences in the two figures. This is a good time to extend their understanding of enclosure. Surely, some children's closed curves will enclose one or more pegs (a), and some of the open figures will have pegs inside them (b). Discuss the fact that a rubber band cannot be stretched between one of the pegs inside the figure in (a) and a peg outside of it without crossing the rubber band that forms the closed curve. In (b), they can connect a peg inside with one outside without crossing the figure itself.

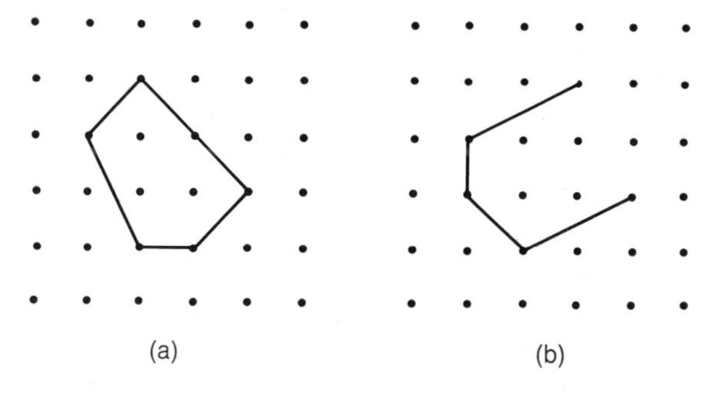


**Figure 15-19** The closed curve in (a) encloses three pegs, while the open curve in (b) has several pegs inside it

There are geoboard activities to help older children extend their understanding of plane figures. The ones that follow are examples of some you can give orally or present through problem cards.

*Four.* There are different kinds of four-sided figures (quadrilaterals). The quadrilaterals illustrated in Figure 15-20 are examples of several children can make. In addition to these, children should make other models of each type of figure. Each is named according to the types of sides and angles it has and the relationships among its sides and angles. Children should discuss each figure, noting similarities and differences with the others, and between models of the same figure. Information can be organized in a table, as in Figure 15-20.

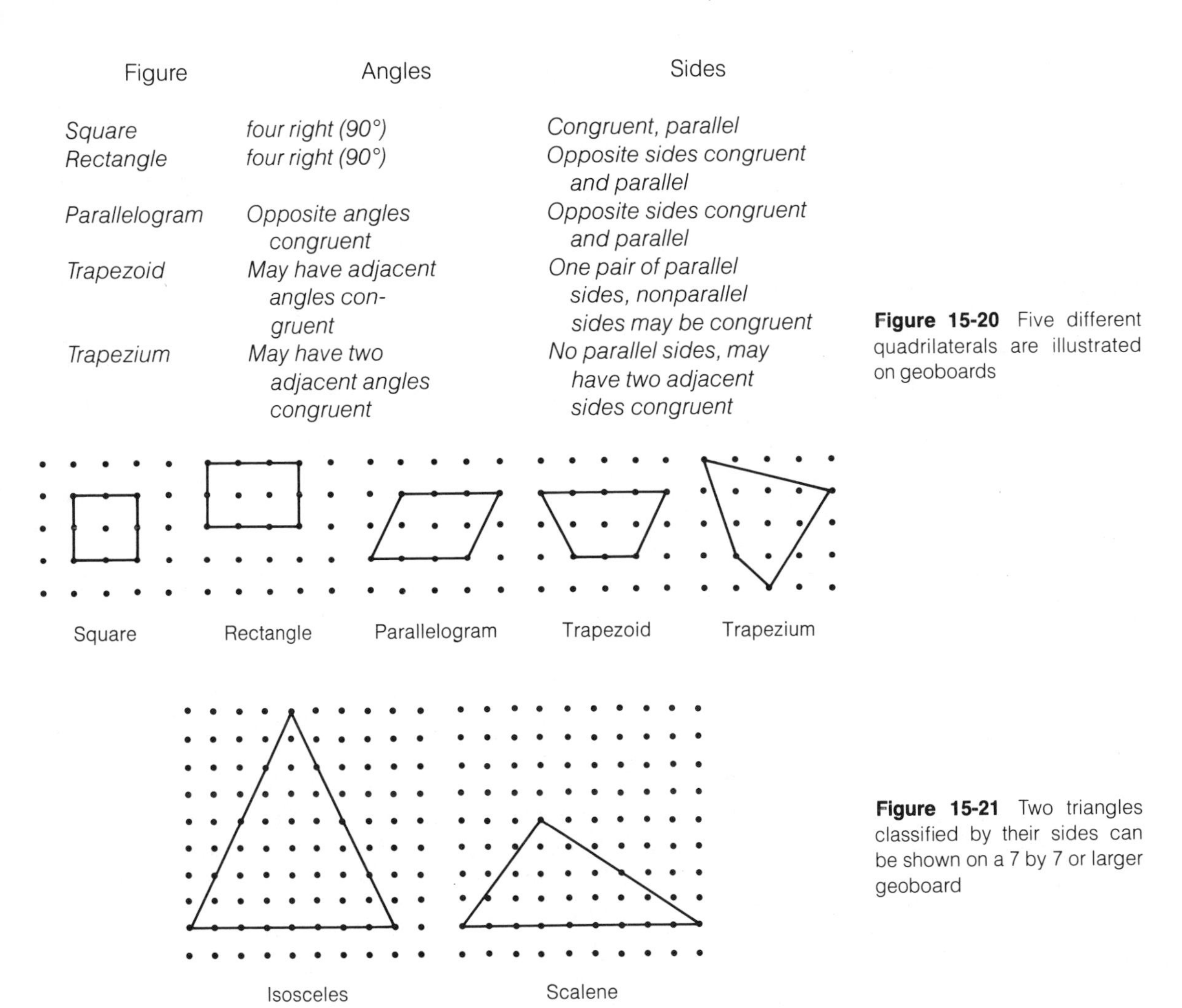

| Figure | Angles | Sides |
|---|---|---|
| *Square* | *four right (90°)* | *Congruent, parallel* |
| *Rectangle* | *four right (90°)* | *Opposite sides congruent and parallel* |
| *Parallelogram* | *Opposite angles congruent* | *Opposite sides congruent and parallel* |
| *Trapezoid* | *May have adjacent angles congruent* | *One pair of parallel sides, nonparallel sides may be congruent* |
| *Trapezium* | *May have two adjacent angles congruent* | *No parallel sides, may have two adjacent sides congruent* |



**Figure 15-20** Five different quadrilaterals are illustrated on geoboards

**Figure 15-21** Two triangles classified by their sides can be shown on a 7 by 7 or larger geoboard

Models of some triangles classified by their sides can be made on larger geoboards, such as 10 by 10 boards (Figure 15-21). During a study of these models, children will learn to classify triangles as isosceles, which have two congruent sides, and scalene, which have no congruent sides. Children can also learn that triangles are classified according to the types of angles they have, as shown in Figure 15-22. In (a), the three types of triangles classified by angles are illustrated. A right triangle has one right angle; an acute triangle has three acute angles; an obtuse triangle has one obtuse angle, which is an angle that is greater than 90°. These models can also be used to illustrate the meaning of the triangles' altitudes (b). For the right and acute triangles, a rubber band (shown by the broken line and representing the altitude) is inside of the triangle, while it is outside of the figure when the triangle is obtuse. (If the obtuse triangle is placed with

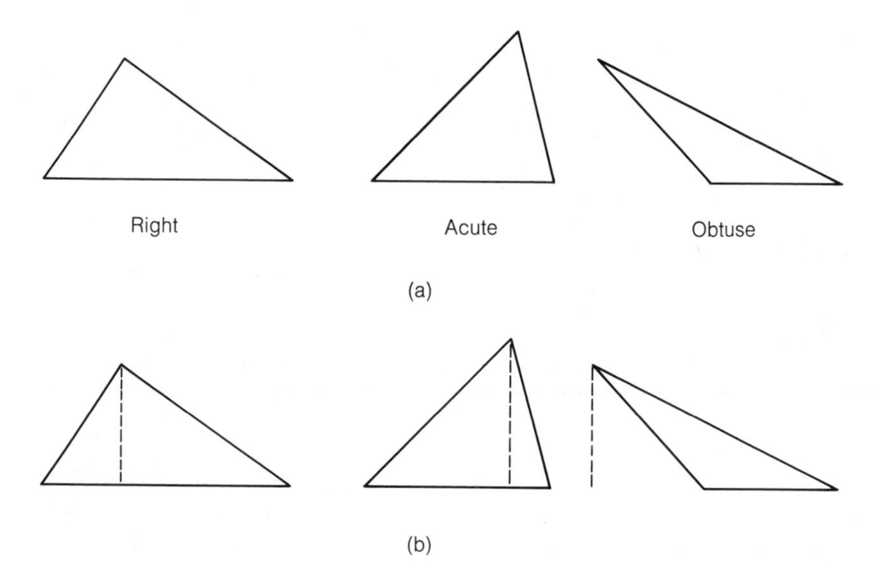


**Figure 15-22** Three triangles classified by angles (a) and their altitudes (b) can be represented on geoboards

the longest side at the bottom, the altitude can be shown by a broken line in its interior.)

*Five.* Children can study polygons and their diagonals. Have them make models of polygons with three, four, five, and more sides. Then have them show the diagonals in each of the figures. Again a table helps organize the information. From this table children should note the pattern that develops as the number of diagonals increases. This pattern's sequence should enable more mature children to tell the number of diagonals in polygons that have more sides than any of the ones they have shown on their geoboards.

| *Figure* | *Angles* | *Diagonals* |
|---|---|---|
| Triangle | 3 | 0 |
| Quadrilateral | 4 | 2 |
| Pentagon | 5 | 5 |
| Hexagon | 6 | 9 |
| Heptagon | 7 | 14 |

During their study of these polygons, children should compare their names with the names of polyhedrons. The similarity of some names will be noted as they compare *hexagon* and *hexahedron*, *octagon* and *octahedron*, and others. The first part of *hexagon* tells the number of sides, while the *-gon* tells that it is a plane figure.

*Six.* Children can study concave and convex polygons. Models such as the ones pictured in Figure 15-18 help them understand the difference between concave and convex polygons. A polygon, or any other closed curve, is concave when a line segment that connects two points on or inside the figure does not necessarily lie entirely within the figure. Segments connecting points on the left side may lie partly outside of the hexagon in Figure 15-18(a). The hexagon in (b) will have all of its segments inside the figure, so it is a convex figure.

---

## Self-check of Objective 5

Demonstrate one activity for work with plane figures at the primary level and one at the intermediate level using these devices: shapes and posting holes and geoboards.

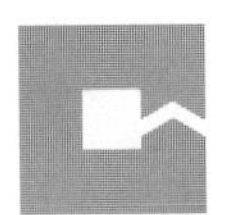

---

## ACTIVITIES WITH POINTS AND LINES

Many current programs include the study of lines, and the individual points of which they and all other geometric figures are made, as a part of the program for children in the early primary grades. However, according to Piaget, it is better to postpone the study of these figures until children are eight or nine years old. By this time most children will have a background of earlier experiences upon which to build their understanding of points and lines. Once points and lines are introduced, children will study them and plane and solid figures in such a way that work with one type of figure will extend and reinforce their understanding of the others.

### 1. Introducing Points

During earlier activities, the corners of solid and plane figures are sometimes called points. Now children should learn to think of points as locations in space. They can do this by locating points in relation to identifiable objects in the classroom. For example, a child whose desk has rounded corners can locate the point in space where two lines that represent the edges of the desk would intersect if extended from the desk (Figure 15-23). Another way to locate points is in relation to different

**Figure 15-23** A point can be located at the corner of a desk with rounded corners

positions in a classroom. "Go to the southeast corner of the room. Locate the point that is 2 meters from the east wall, 3 meters from the south wall, and 1 meter above the floor."

## 2. Introducing Lines and Parts of Lines

Once they understand the abstract nature of points as locations in space, children are ready to use representations of points to develop their understanding of lines and figures associated with them. The simplest of these figures is a line segment, which is composed of two points, called end points, and the points between them (Figure 15-24). Segments can be represented by a piece of taut string, a piece of straight wire, or a line drawn with a pencil and straightedge. The study of geometric figures should include the symbolism that represents them. The symbol $\overline{AB}$ is used to represent the line segment illustrated in Figure 15-24. A line

**Figure 15-24** A representation of a line segment

segment is only a part of a line; a line extends infinitely in opposite directions as it passes through two points. It may be represented by a picture of a line with arrowheads at each end, as in Figure 15-25(a); its symbol is $\overleftrightarrow{AB}$. Part of a line that consists of one end point and all the points extending from it infinitely in one direction is called a ray (b); its symbol is $\overrightarrow{AB}$.

**Figure 15-25** Representations of (a) a line and (b) a ray

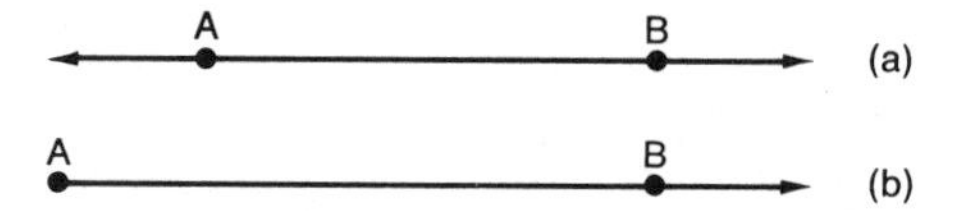


During the examination of representations of line segments, lines, and rays, children should learn about points that lie between, to the right of, and to the left of other points on lines or parts of lines. They should also learn that any point on a line separates the line into three distinct sets of points. One set is the point itself, another is the set of points on one side of it, and the third is a set of points on the other side of it (Figure 15-26). The parts of the line on either side of the separating point are referred to as half-lines. (Since a line is considered straight, a line or any part of it can be represented by a model drawn with a straightedge.)

**Figure 15-26** Point *A* separates the line into three distinct sets of points

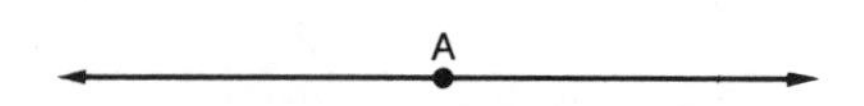

Curved lines can be represented with the same kinds of physical models as straight lines. (Models of curved lines are not drawn with a straightedge.)

### 3. Introducing Angles

An angle is formed by joining two rays that have the same end point but are not a part of the same line. In Figure 15-27, the angle formed by joining rays *BA* and *BC* is shown. The symbol for this angle may be written as $\angle ABC$ or $\angle CBA$.

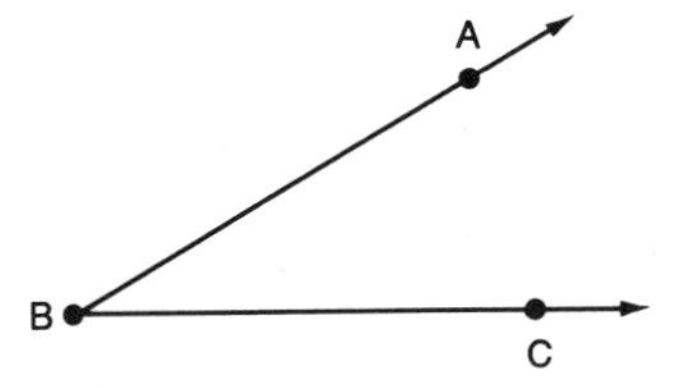


**Figure 15-27** Representation of an angle

Wooden slats with holes drilled in their ends and joined with small nuts and bolts make good models of angles. Slats are used in Figure 15-28 to show obtuse, acute, and right angles.

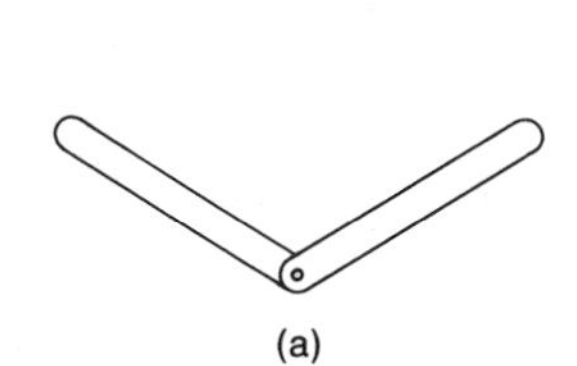
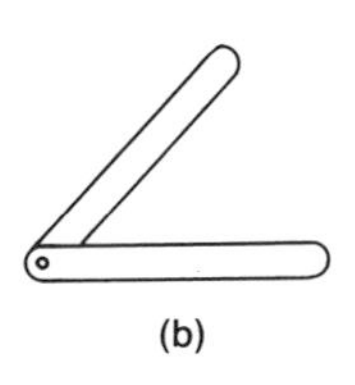
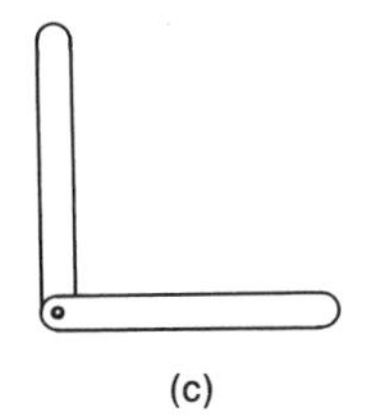


**Figure 15-28** Wooden slats representing (a) an obtuse angle, (b) an acute angle, and (c) a right angle

### 4. Points and Lines Associated with a Circle

A circle, shown in Figure 15-29(a), is a special simple closed curve. It consists of a set of points that are equidistant from a given point, the center. Lines or segments are associated with a circle in several ways. Each has a special name. A segment with one end point at the center and the other end point on the circle is a radius. A segment that has both end points on the circle is a chord. A diameter is a chord that passes through the center. A part of a circle is an arc. A line that contains one and only one point on the circle is a tangent. Each of these is illustrated in (b).

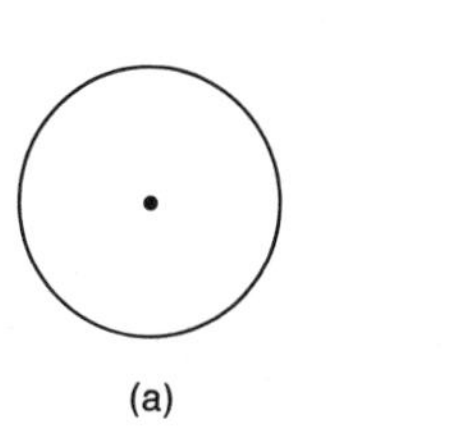
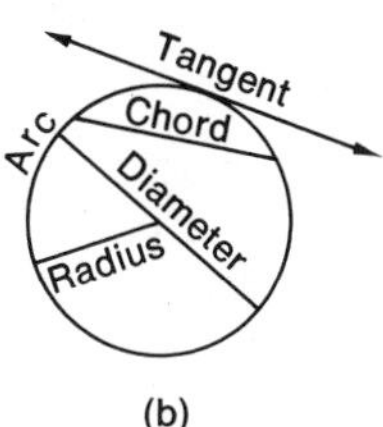


**Figure 15-29** (a) A representation of a circle and (b) representations of line segments, lines, and curved lines associated with a circle

## 5. Other Topics Dealing with Lines

Once children understand the nature of points and lines, they are ready for more abstract work with them and the figures they form. The following ideas are among the ones you might introduce.

*One.* There are simple closed curves and curves that are not simple (Figure 15-30). A simple closed curve can be represented by drawing a curved line that begins and ends at the same point without crossing itself (a). A closed curve that is not simple crosses itself at least once or does not have the same beginning and ending point, or is a combination of both conditions. See (b) and (c).

*Two.* An infinite number of points pass through any given point. One way to illustrate this idea is with a series of transparent overlays (Figure 15-31). Begin with a single point with one line passing through it (a). With each successive overlay, show more lines passing through the original point.

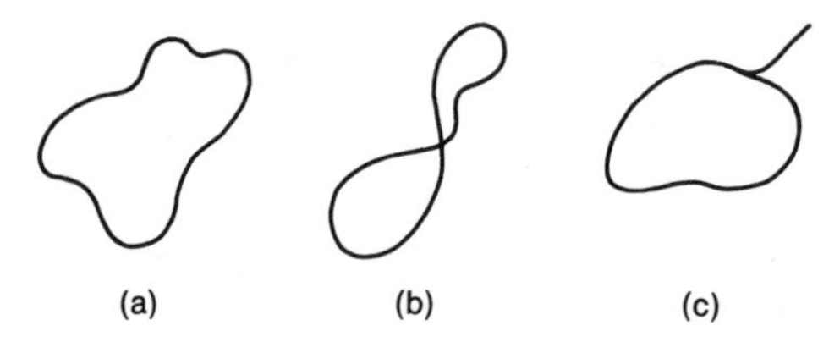


**Figure 15-30** Representations of (a) a simple closed curve, (b) a closed curve that crosses itself, and (c) a closed curve that does not have the same beginning and ending point

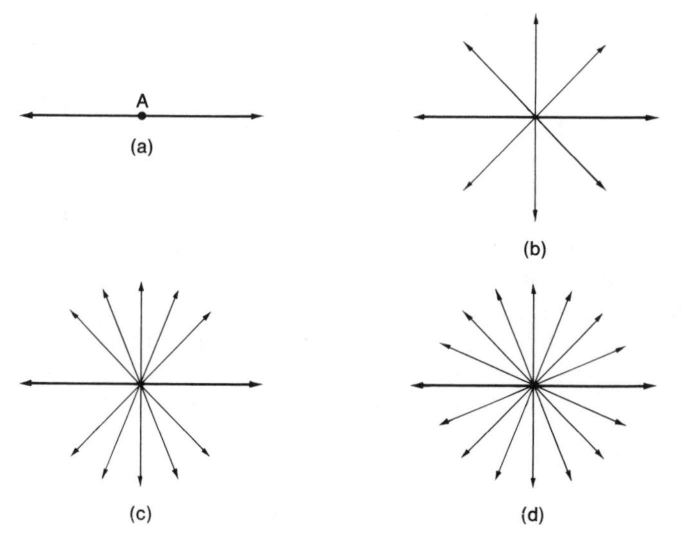


**Figure 15-31** Transparency and overlays showing that many lines pass through a point

*Three.* A line relates to a plane in one of three ways. It may pass through the plane at a single point, as seen in Figure 15-32(a); it may be a part of the plane itself (b); or it may be parallel to the plane (c).

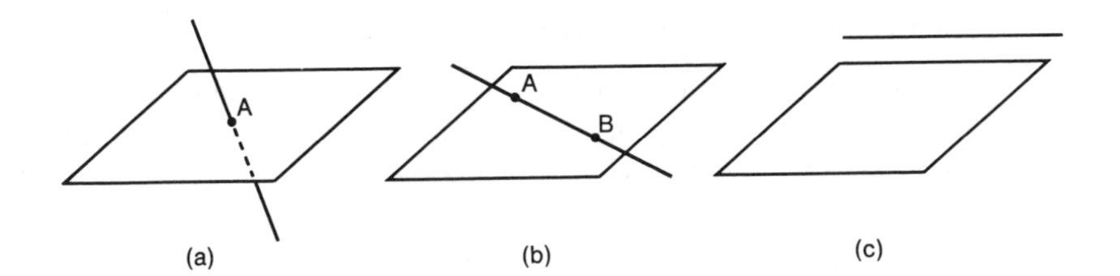


**Figure 15-32** Piece of wire and sheet of clear plastic used to represent the relationships of a line to a plane: in (a) the line passes through the plane; in (b) it is part of the plane; and in (c) it is parallel to the plane

*Four.* An abstract definition of a polygon can be introduced. A polygon is formed by the union of three or more line segments joined so that they form a simple closed curve. As children examine models of polygons to determine their characteristics and learn that polygons are classified by name according to the number of their sides, they should

| Polygon | Name of Polygon | Number of sides | Name of Region |
|---|---|---|---|
| | Triangle | 3 | Triangular |
| | Quadrilateral | 4 | Quadrilateral |
| | Pentagon | 5 | Pentagonal |
| | Hexagon | 6 | Hexagonal |
| | Heptagon | 7 | Heptagonal |
| | Octagon | 8 | Octagonal |

**Figure 15-33** A chart to classify polygons

develop a chart similar to the one in Figure 15-33 to organize their data. The name of the region that each polygon encloses should also be identified. This reflects the children's understanding that a polygon separates a plane into three distinct sets of points—the set that is the polygon itself, the set within the polygon, and the set outside the polygon. When children find the area of a region, they actually determine the measure of that part of the plane bounded by a simple closed curve. When they find a perimeter, they determine the measure of the line segments or curved lines that comprise a simple closed curve.

## Self-check of Objectives 6 and 7

Define and illustrate models of each of these figures: line, line segment, ray, angle, radius, diameter, tangent, and chord.

Describe an activity that can be used by children to learn the meaning of each term above.

## CONGRUENCE AND SIMILARITY

Two line segments are congruent when all points on one segment coincide exactly with all points on the other. Similarly, two plane figures or two space figures are congruent when all points of one figure coincide exactly with all points on the other. This is the concept on which standardized measurement, with its units for linear, area, and volume measures, is based.

Two figures are similar when they have the same shape and all of their corresponding parts are proportional. Thus all squares, except those that are congruent, are similar, while not all rectangles are. Other similar shapes are equilateral triangles, regular pentagons, hexagons, and circles, except when two like shapes are congruent. Two drawings of a room's floor plan are also similar if they are drawn to different scales.

### 1. Congruence

You can use several types of activities to introduce congruence or extend children's understanding of it. Problem cards can be prepared for many of them.

**a. Pattern Blocks.** Give pairs of children sets of pattern blocks. Have one child put together a pattern of his or her own choosing for the other child to copy.

**b. Attribute Blocks.** Have children separate a set of attribute blocks into groups of congruent blocks.

**c. Geoboards.** You can set up a geoboard with line segments, angles, and open and closed figures for children to copy. Leave enough space next to each figure for a child's copy of it, or give individual children their own boards.

**d. Tracings.** Cut out line-drawing figures of animals, automobiles, other objects and plane figures and glue them on cards, or draw simple line figures on cards for children to trace. Use either lightweight paper or clear plastic for the tracings. Figure 15-34 shows a triangle *ABC* and its tracing *MNO*. Tracings can also be used to show that two figures are not congruent. In Figure 15-35, the tracing of *XYZ*, represented by the broken lines on *DEF*, shows that the two angles are not congruent.

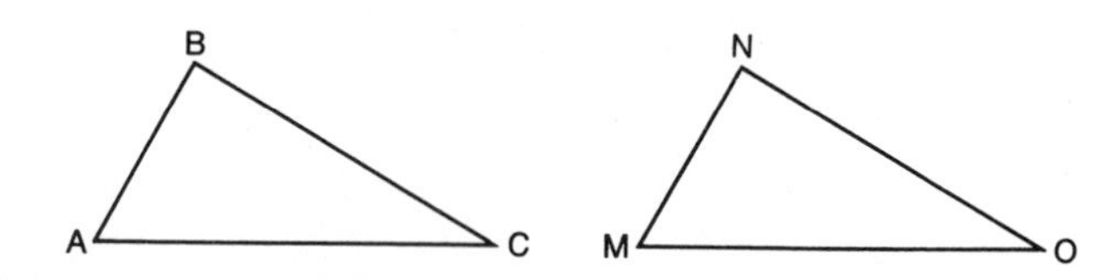


**Figure 15-34** A triangle *ABC* and its tracing *MNO*

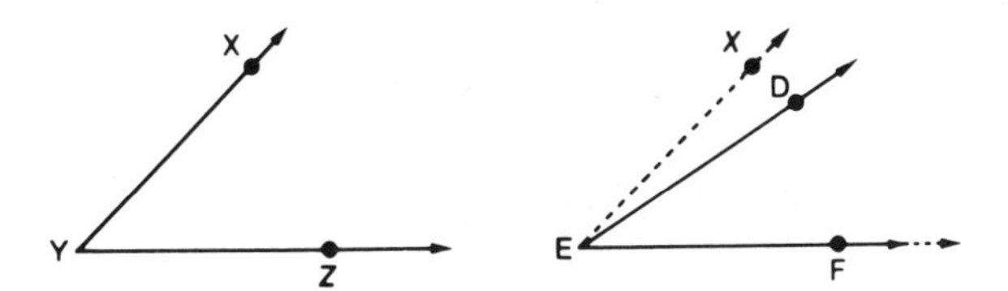


**Figure 15-35** The tracing of $\angle XYZ$ placed over $\angle DEF$ shows that the two angles are not congruent

e. **Square Paper.** Make copies of figures, including plane figures and outlines of letters of the alphabet, on squared paper for children to copy on their own sheets of paper. Or, have one child make a design of his or her own choosing on squared paper for a classmate to copy.

When activities have been completed, children should have opportunities to discuss their work and check each other's patterns and figures. Children who are used to working in pairs or small groups can discuss their work without your direct guidance once the concept of congruence is understood.

## 2. Similarity

Pattern blocks and sets of attribute materials with shapes of two different sizes, geoboards, and squared paper are good materials for activities dealing with similarity. Pattern block and squared-paper problem cards can be made for activities with these materials, while you can set up geoboard figures for activities with them. A 100-peg geoboard is good to use because more than one figure can usually be put on a single board at one time. For example, the triangle illustrated in the upper part of the first geoboard in Figure 15-36 can be given to a child with any of these directions: "Make a figure on your geoboard that is like this one, but smaller." "Make a figure on your geoboard that is like this one, but larger." "Make a figure on your geoboard that is like this one, but with sides that are half as long." "Make a triangle like this one, but make its sides one and a half times as long." The last two directions are for children who have a mature understanding of

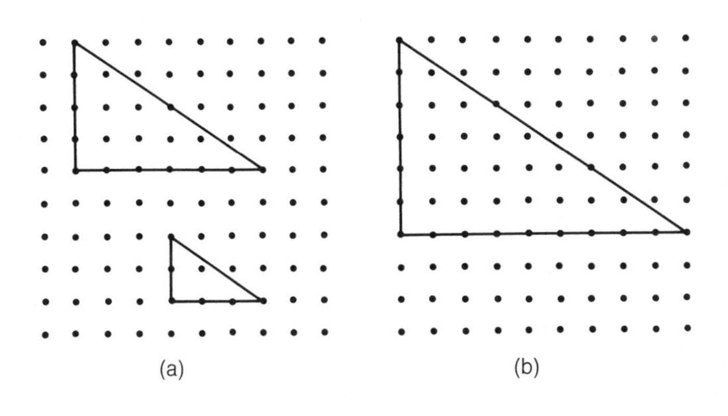


**Figure 15-36** Three similar triangles can be represented on 100-peg geoboards

similarity and the meaning of the fractions involved. Again completed work should be checked by children meeting in small groups. During a discussion, each child's work should be considered to see if he or she has followed directions properly. Activities with similarity also offer a good opportunity for children to consider the effect of a figure's area when its sides are increased or decreased by a given amount.

## Self-check of Objective 8

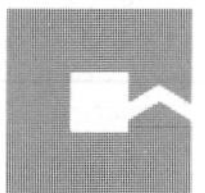

Give an oral or written statement distinguishing between congruent and similar figures. Describe at least three different investigations for developing children's understanding of congruence and similarity.

### SYMMETRY

Symmetry is found in both natural and man-made objects. Many geometric figures are symmetrical. Investigations in this area of geometry refine children's understanding of concepts such as congruence and the nature of plane figures.

During a visit to a British primary school, I observed a series of activities that introduced a group of seven-year-olds to symmetry. To begin, the teacher asked two children who were the same size to stand on a table at the front of the room. She told them to face each other and join hands. Then she gave them a series of directions: "Raise your hands high above your heads [Figure 15-37(a)]. Keep your hands high above your heads and bend your knees so they touch (b). Now let go of your hands and turn around so your backs are touching. Put your hands straight out in front of you, then push against each other and slowly move your feet forward so you look as though you are going to sit down (c)." After the children had modeled these and other positions, the class discussed what they had noticed as positions were changed. Among their comments, these were made: "Each one looked the same after they moved." "They always did the same thing." "It looked like one was in front of a mirror and we saw her image in it."

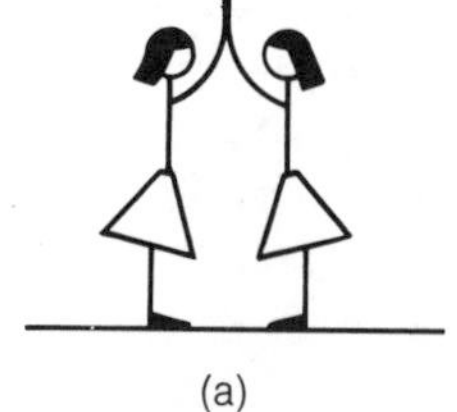

(a)

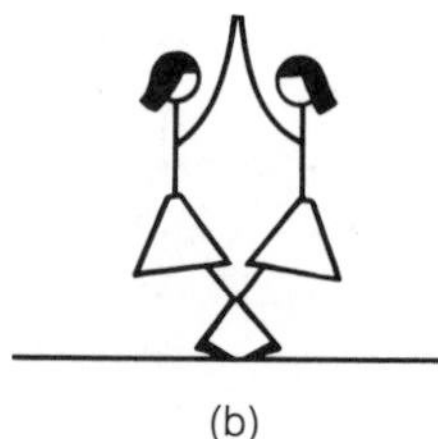

(b)

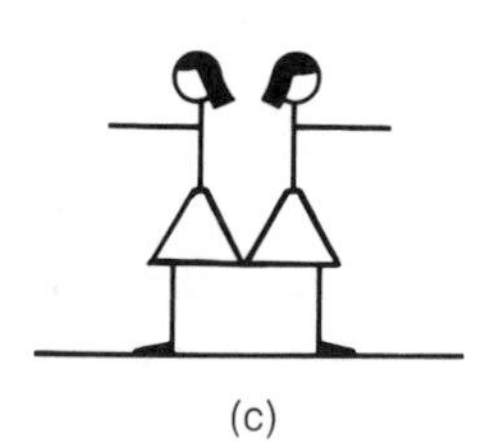

(c)

**Figure 15-37** Two children can model situations depicting mirror images to introduce symmetry

Next, the teacher had each of the children take a sheet of art paper from a pile in the center of the table and fold it in half. Each child put blobs of different-colored tempera paint on one side of the paper, and folded the clean half over it. Then the two halves were smoothed together so the paint was spread over them, and the paper was opened to reveal a unique symmetrical design.

At this time our group had to leave the room to see work being done in other classes. About an hour and a half later, one of the seven-year-olds found us and said, "Sir, we'd like you to come back to see the rest of our work." We followed him back to his room and found the children still eagerly at work. After they had finished the paint activity, the teacher had given each child a piece of white and a smaller piece of colored construction paper. She told them to fold the colored paper in half, cut out a design, open it, and paste the design on the white paper to make a pretty pattern. Some children were still working on this activity, but some had begun a third activity. They had folded a sheet of centimeter-squared paper in half and were making crayon designs on one half. When they finished the design, they put its mirror image on the second half. The walls of the room were gradually being covered with the children's work.

As lunch time neared, the teacher called a halt to the activities, with assurance to those with unfinished work that they would have time to complete it later. The children talked about their designs and noted that the two halves of each one looked alike, except that one was the opposite of the other, just as the two children had looked earlier. Not until this time did the teacher mention that the designs were symmetrical and ask the children to tell what they thought the word means.

Before we left the room, the teacher said to us, "You know, we spent all morning on these activities, and I don't usually spend so much time on one thing. But, I'm sure the children have a pretty good idea of what symmetry means now. They got so interested, I hated to stop them. We may not do any 'maths' for several days now, but I'd rather do it this way as long as the boys and girls are interested and are learning."

Our group left the room convinced that we had seen a good teacher at work. She knew what she wanted to accomplish and had taken the time to plan and organize activities to get the job done. She had materials ready for each activity, and she accounted for individual differences by letting each child work at his or her own pace.

There are many other symmetry activities for children. Instructions for younger children can be given orally, while those for older ones can be put on problem cards.

*One.* Paper-folding activities can show the number of lines of symmetry in different plane figures. Ask, "How many ways can you fold a square so that one half completely covers the other half?" Have the children try folding rectangles, different kinds of triangles, regular pentagons, other polygons, and circles. (A plane figure has symmetry if it has two matching halves and if one half is a mirror image of the other. If a line or fold can be made to separate the two halves so that each point on one half has a matching point on the other half and each point is equidistant from the line, the figure has line symmetry.)

*Two.* Three-dimensional objects, such as milk cartons, shoe boxes, and other small containers can be easily cut to see how many planes of symmetry each box (space figure) has. (A space figure has plane symmetry if the two parts form a mirror image after the figure has been cut.) Plasticine clay models are also good for this activity because the solid interior gives two faces that are clearly seen after a cut has been made. Ask such questions as, "How many ways can a cube be cut to give two parts that form mirror images? What about other prisms? pyramids? a sphere? a cylinder?"

*Three.* Structured materials, such as Cuisenaire rods and pattern blocks, can be used to make symmetrical patterns.

*Four.* Boxes with lids can be used to investigate rotational symmetry. (A figure has rotational symmetry if it can be rotated around a point at its center and then appear not to have been moved.) A lid for a square box can be put on four ways: a square has fourfold rotational symmetry. Compare this with the number of ways a square-shaped plane figure can be folded to show different lines of symmetry and the number of times a square-shaped figure can be put in its posting hole. Ask, "How many ways can the lid of a rectangular shoe box be put on? Is this number the same as the number of lines of symmetry on a rectangular-shaped piece of paper? What about circular boxes? triangular boxes?"

*Five.* Pictures of natural or man-made objects that have symmetry can be collected and organized on a bulletin board or in a class book.

*Six.* Mirror Cards are commercial materials for symmetry study. The copyrighted materials consist of a number of cards and small mirrors. Children determine whether it is possible to place a mirror on a card in such a way that the printed pattern and its reflection in the mirror will match the printed pattern on a second card. Some patterns are symmetrical and can be matched; others lack symmetry and cannot. Examples are shown in Figure 15-38.[5]

*Seven.* Symmetry is also found in number patterns. The symmetry of the addition table (Figure 7-14) and the multiplication table (Figure 9-12) can be shown by drawing a real or imaginary diagonal from the upper-left to the lower-right corner. It is this symmetry that makes the tables useful for illustrating the commutative property of addition and multiplication. Mature children might investigate Pascal's triangle to note the symmetry in it (Figure 15-39).

[5] These patterns are from Marion Walter, "An Example of Informal Geometry: Mirror Cards," *The Arithmetic Teacher*, XIII, No. 6 (October 1966), 448–452. Mirror Cards are available from Educational Services Incorporated, P.O. Box 415, Waterton, Mass. 02172.

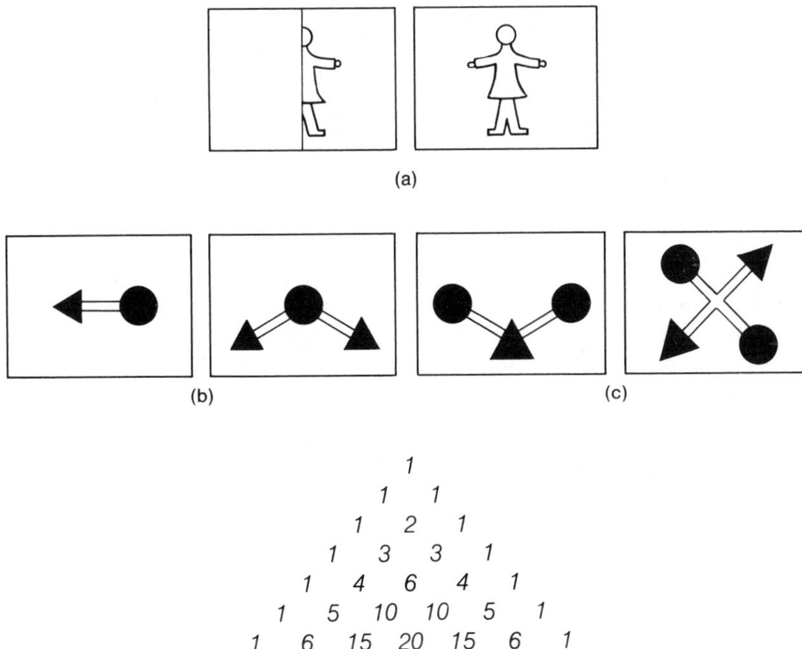


**Figure 15-38** Examples of Mirror Cards. A child is to determine which patterns at the right can be duplicated by placing a mirror on the cards at the left

**Figure 15-39** Pascal's triangle

## Self-check of Objective 9

Define symmetry. Describe several investigations that will help children understand symmetry.

## TRANSFORMATIONAL GEOMETRY

Transformational, or "motion," geometry is an intriguing area that can be studied by some children as early as primary grades, although it is more commonly studied in the intermediate grades. It deals with three basic motions that can be applied to geometric (and other) figures: translations, or sliding motions; rotations, or turning motions; and reflections, or flipping motions. Each motion is defined as a *rigid* motion because it follows certain rules. Only when a motion adheres to these rules does a transformation occur. The figure obtained as the result of a transformation is congruent with the original figure, and is frequently called a *congruency transformation*. The figure that comes from a transformation is also an *image* of the original figure. (There are other transformations than the three mentioned here, but they are usually reserved for study in the higher grades, and are not discussed in this book.)

First experiences should be exploratory, with investigations used to help refine understanding of concepts of congruency, symmetry, parallelism, and the nature of plane figures. The activities can include observations of transformations projected by an overhead projector. Children can also manipulate wire models of geometric figures to perform transformations at their desks. A few examples of the three types of motions follow.

### 1. Slide Motions

A slide motion occurs when a geometric figure moves along a plane so that the distance between all corresponding points on the two figures remains the same. The transformation from $\triangle ABC$ to $\triangle A'B'C'$ can be represented on an overhead projector. First, use a transparency to project models of the two triangles [Figure 15-40(a)]. Then place a copy of $\triangle ABC$ that is on an overlay (b) atop the triangle. Slide the copy across the transparency until it covers $\triangle A'B'C'$. The dotted lines on the overlay now join the vertices of the two triangles and indicate that the vertices on $\triangle A'B'C'$ are equidistant from the corresponding vertices on $\triangle ABC$ (c).

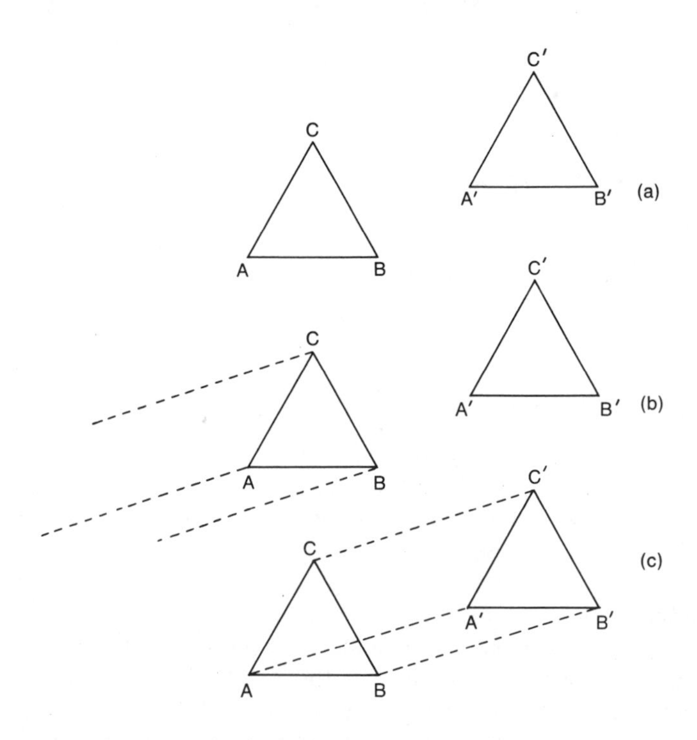


**Figure 15-40** A transparency (a) and an overlay (b) can be used to show the transformation of $\triangle ABC$ to $\triangle A'B'C'$ (c)

A slide can occur in any direction on a plane. An arrow indicates the direction and distance of a slide, or "programs" the transformation. In Figure 15-41 the arrow in (a) indicates that the square is to be moved up and to the left a distance equal to the length of the arrow. In (b), the completed transformation is represented.

The concept of parallelism can be discovered as children investigate line transformations. Two lines on a plane are defined as parallel if one is the image of the other. One figure is the image of another only when a slide motion has occurred. In Figure 15-42 the lines $l$ and $l'$ in (a) are parallel because a slide motion has occurred and the line $l'$ is the image of $l$. In (b), the lines are not parallel because one is not the image of the other.

## 2. Turn Motions

A turn motion occurs when a figure is rotated around a point at the center of the turn as in Figure 15-43. The turn motions easiest for children to understand are those with the point on the figure, as in (a), or outside the figure, as in (b), where the turn is around point *P*. These transformations can be demonstrated readily by using projected transparencies.

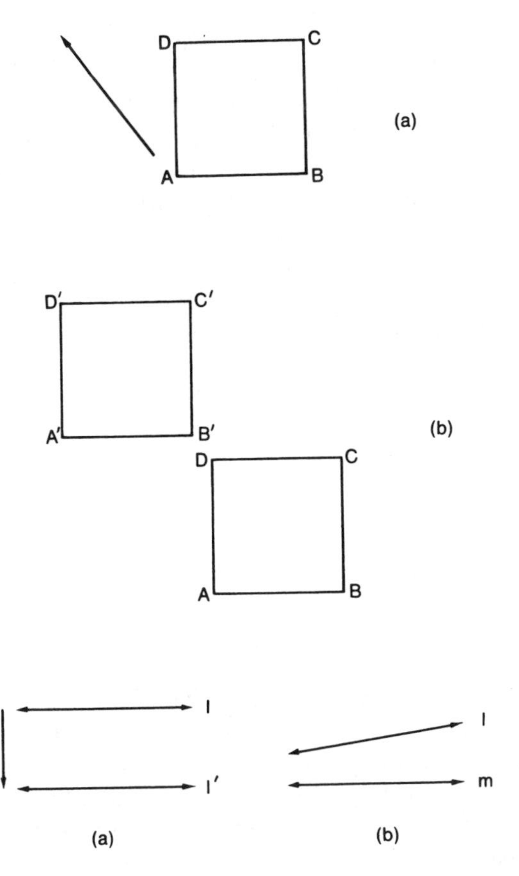


**Figure 15-41** An arrow can be used to "program" a transformation

**Figure 15-42** In (a), lines $l$ and $l$ are parallel because $l'$ is the image of $l$. In (b), lines $l$ and $m$ are not parallel because one is not the image of the other

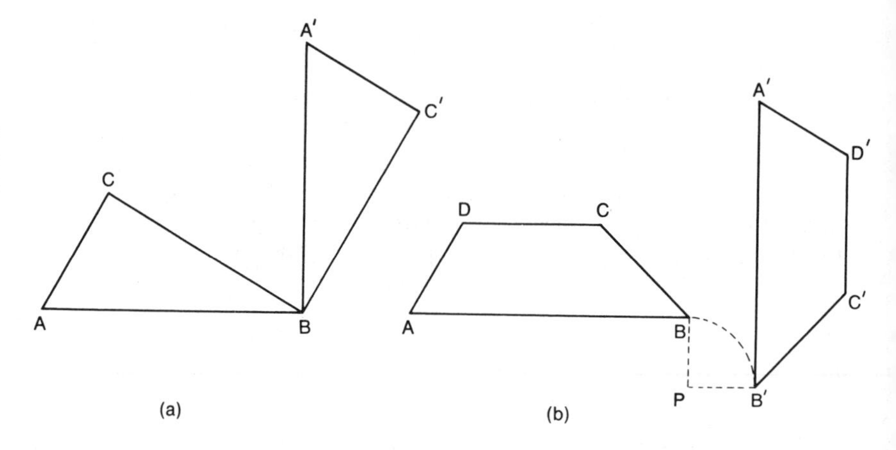


**Figure 15-43** A turn may be made around a point (a) on the figure or (b) away from the figure

## 3. Flip Motions

Flip motions create a reflection of the original figure. During their examination of these motions, children will extend their understanding of symmetry and further their understanding of regular plane figures. A flip action may take place in relation to a point or a line. In Figure 15-44(a) parallelogram $A'B'C'D'$ is the reflection of parallelogram $ABCD$. The line between the two parallelograms is equidistant from each and is the line of symmetry, or *flip axis*. In (b), the reflection is in respect to a point at the center of the figure. Children should be guided to observe that symmetry of the type considered here is found frequently in nature. The two halves of a butterfly have line symmetry; the parts of many flowers have point symmetry. Children should use mirrors to determine symmetry of figures. When a mirror is placed along the line in (a) and between halves of the drawing in (b), it will reflect figures that duplicate the originals.

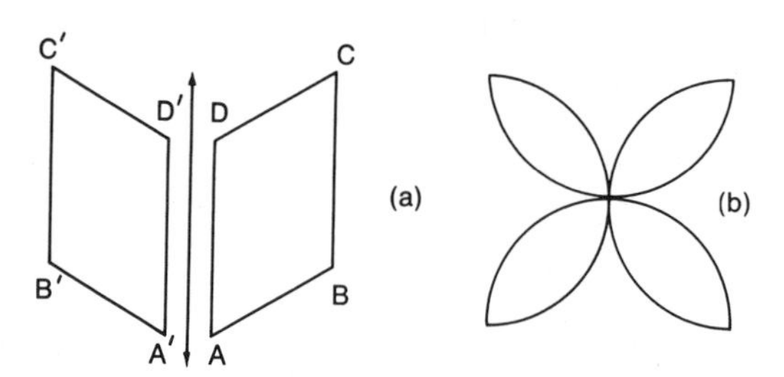


**Figure 15-44** The flip may take place in relation to (a) a line or (b) a point

Some figures, for example, the equilateral triangle in Figure 15-45, have both point and line symmetry. Each point on the left half of the triangle has a corresponding point on the right half that is equidistant from the point at the center. Each point on one half also has a corresponding point that is equidistant from the line that is the perpendicular bisector of the triangle's base.

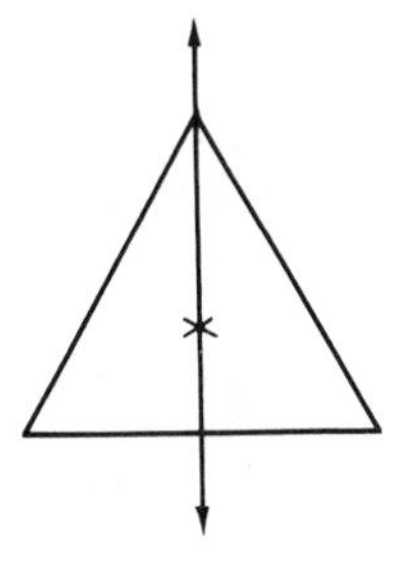

**Figure 15-45** A equilateral triangle has both point and line symmetry

A flip action will help children further their understanding of the meaning of perpendicular lines. Two lines are perpendicular to each other when the reflection of one remains unchanged after it has been flipped along the axis of the other line (Figure 15-46). Children can demonstrate this with pieces of wire bent in the shape of angles (Figure 15-47). The

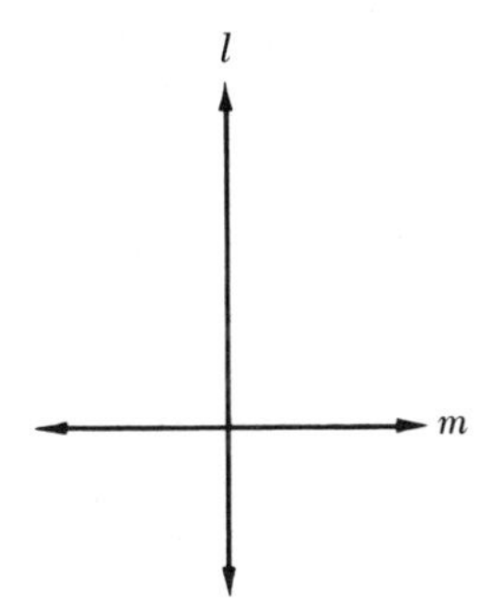


**Figure 15-46** Two lines are perpendicular if the line $m$ is invariant when a flip is made on the line $l$

wire should be placed on a piece of paper so a tracing can be made of it. Then it should be flipped along one axis and another tracing made. When the line that is not the axis continues as a part of the original nonaxis line, the two lines are perpendicular (a). When the line does not continue as a part of the same line, they are not perpendicular (b).

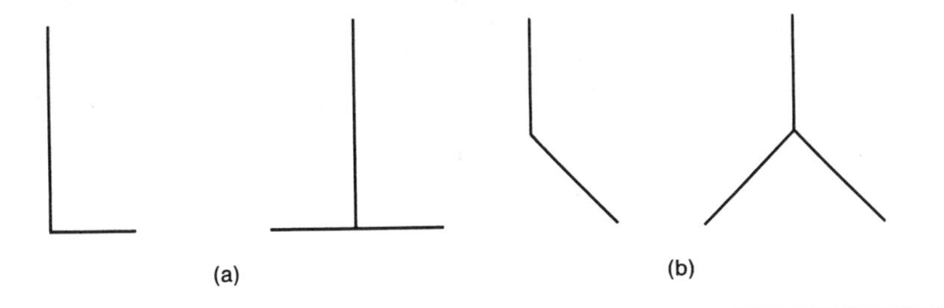


**Figure 15-47** Pieces of wire can be used to determine whether two lines are perpendicular. In (a) the tracing indicates that the lines are perpendicular; in (b) they are not

---

## Self-check of Objective 10

Demonstrate each of these geometric transformations with appropriate materials: translation (slide), rotation, and reflection.

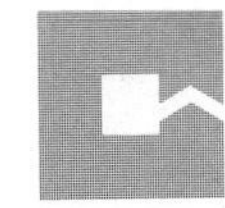

---

## Common Pitfalls and Trouble Spots

Two pitfalls associated with teaching geometry are discussed:

First, maintaining an orderly sequence of topics is frequently a problem. As indicated earlier, there is not full agreement about what geometric concepts and topics should be included in an elementary school program. As a result, there is great variation in what is presented from school to school and even grade to grade within a single school. Many schools and/or teachers exclude it entirely. Others include it, but select their topics and modes of presentation without regard to a planned program. This means that you cannot take it for granted that your children will have attained a particular level of understanding of geometry just because they are in a particular grade. You must diagnose their backgrounds and use the findings to help you plan suitably sequenced activities.

Second, the use of models is a potential pitfall. Many teachers use pictures (often line drawings of geometric figures) in a "show-and-tell" fashion unrelated to real-life situations. Piaget's research indicates that children need opportunities to manipulate models, such as regions and their posting holes, to construct models on geoboards and with paper and other materials, and to draw pictures of them, as well as to talk about them. Your geometry lessons should be more than "show-and-tell" activities by including provisions for these kinds of activities, regardless of the grade level at which they are presented.

A potential trouble spot for children new to the study of geometry is the confusion caused by unfamiliar words and terminology. Unless care is taken, they will have trouble understanding the words' meanings and uses. Keep these points in mind as you deal with geometry:

1. Include only essential terminology.
2. Keep the terminology as simple as possible. However, do be accurate in using it. It is all right to say "square corner" with primary children, but the terms "right angle" and "90° angle" should be used with older children.
3. Be sure parts of words are understood. For example, when children study quadrilaterals, they should learn that *quadri-* means "four," while *lateral* means "side" (both are from the Latin language); so, *quadrilateral* means "four-sided figure." (See Appendix A for a list of prefixes and suffixes used in geometry.)
4. Once a word is introduced, use it frequently in discussions and written materials.
5. Do not emphasize terminology at the expense of understanding of the concepts they represent.

---

## Self-check of Objectives 11 and 12

Describe two pitfalls associated with teaching geometry, and explain orally or in writing how each one can be avoided.

Children frequently have trouble with the vocabulary associated with geometry. List orally or in writing at least four things you can do to help them understand and use geometric terms properly.

---

## SUMMARY

While complete agreement on the content of nonmetric geometry has not been achieved, there is substantial agreement that children do profit from work with it. Piaget's research indicates that children first have a topological view of their world rather than a Euclidean view. Topologically oriented activities are recommended for preoperational-level children. Four topological concepts are discussed: proximity, separation, order, and enclosure.

There are three levels of geometric activities for elementary school children. Models of space, plane, and line figures should be available at all levels. Children's exposure to geometry should begin through play or gamelike activities, with models of solid, or space, figures rather than plane and line figures the subject of their attention at first. Once children have been exposed to the various common figures and recognize some of their more general characteristics, they are ready for the next level of study. Now they examine figures to note their likenesses and differences, begin to sort them according to common attributes, and informally discuss them and learn their names and other words associated with them. Finally, they move to the formal level of study that leads to an understanding of geometric figures and their places in natural and man-made objects. By the time children complete the elementary school, most of them will recognize and know the names, characteristics, and terms associated with common space, plane, and line figures.

Other geometric concepts introduced are congruence and similarity, symmetry, and transformational, or motion, geometry. You must use care to see that geometric concepts are sequenced properly and that you use a laboratory approach to teaching geometry. Care must also be taken to help children learn geometric terms and learn to use them properly.

## STUDY QUESTIONS AND ACTIVITIES

1. How is the geometry included in today's elementary school mathematics textbooks different from that included prior to 1960? Which topics were included before 1960? Examine modern textbooks to determine what geometry is included in them and at what grade it is introduced.

2. Complete the work sheet shown in Figure 15-11. Use cut-down milk cartons as models, if necessary.

3. Continue the table begun on page 450. What interesting pattern develops as the table is continued? How can one determine the number of diagonals in an $n$-sided figure?

4. Make a collection of models to help children develop an intuitive understanding of points, lines, planes, and space figures and concepts related to them. Why are models important during children's early study of nonmetric geometry?

5. Examine a copy of *Mirror Magic*, by Janet S. Abbott (Pasadena, Calif.: Franklin Publications, 1968) or a set of Mirror Cards and the manual for their use (St. Louis: McGraw-Hill Book Company, Webster Division, 1968). What geometric concepts should children acquire by using materials such as these?

6. Make a collection of pictures that illustrate symmetrical figures. Include illustrations of both plane and space figures having (a) point symmetry, (b) line symmetry, and (c) plane symmetry.

## FOR FURTHER READING

*The Arithmetic Teacher*, XX, No. 6 (October 1973). The editor's introduction and five major articles in this issue deal with content and procedures for teaching elementary school geometry. Materials and teaching strategies for teachers at all levels are included in the articles' discussions.

Black, Janet M. "Geometry Alive in Primary Grades," *The Arithmetic Teacher*, XIV, No. 2 (February 1967), 90–94. Discusses the use of work stations at which small groups of children manipulate materials to discover geometric concepts. Describes the materials and some children's uses of them.

Bright, George W. "Using Tables to Solve Some Geometry Problems," *The Arithmetic Teacher*, XXV, No. 8 (May 1978), 39–43. Describes and illustrates five problems dealing with shapes contained in square and triangular grids. Each one uses a table to organize data and reveal patterns that can be extended beyond the materials themselves.

Copeland, Richard W. *Math Activities for Children*. Columbus, Ohio: Charles E. Merrill Publishing Company, 1979. Activities in Chapter 3 deal with children's perceptions of space, perspective, and other geometric concepts.

Dana, Marcia E., and Mary M. Lindquist. "Let's Try Triangles," *The Arithmetic Teacher*, XXVI, No. 1 (September 1978), 2–9. Triangle puzzles provide the basis for a variety of geometry activities. Besides triangles, children deal with rhombuses, parallelograms, hexagons, trapezoids, symmetry, and even metric measurement.

———. "The Surprising Circle," *The Arithmetic Teacher*, XXV, No. 4 (January 1978), 4–10. Eight ideas, one for each grade from first through eighth, show how children can investigate circles. Each idea involves children in manipulative activities.

Inskeep, James E., Jr. "Primary-Grade Instruction in Geometry," *The Arithmetic Teacher*, XV, No. 5 (May 1968), 422–426. Inskeep presents information about geometry projects, the worth of geometry to young children, how to implement a geometry program into the mathematics curriculum, children's ability to understand geometry, and teaching suggestions and techniques.

Sanok, Gloria. "Living in a World of Transformations," *The Arithmetic Teacher*, XXV, No. 7 (April 1978), 36–40. This profusely illustrated article gives many examples of translations, rotations, and reflections, as well as point and line symmetry. In addition, there is a list of fourteen teaching suggestions.

Sensiba, Daniel E. "Geometry and Transformations," in *Enrichment Mathematics for the Grades*, 27th Yearbook, National Council of Teachers of Mathematics. Washington, D.C.: The Council, 1963. Examples of the three types of transformations are illustrated and described in Chapter 23.

Smith, Lewis B. "Geometry, Yes—But How?" *The Arithmetic Teacher*, XIV, No. 2 (February 1967), 84–89. Smith suggests how teachers can use discovery procedures with children as they study geometry. Three examples of discovery techniques are included.

———. "Pegboard Geometry," *The Arithmetic Teacher*, XII, No. 4 (April 1965), 271–272. Discusses pegboard geometry, which permits children to make models of many geometric figures. Line drawings show some of the discoveries children can make using pegboards.

*Teachers' Guide for Mirror Cards*. St. Louis: McGraw-Hill Book Company, Webster Division, 1968. The introduction explains and describes Mirror Cards, tells how children can learn from them, and presents ways of working with them.

Walter, Marion. "An Example of Informal Geometry: Mirror Cards," *The Arithmetic Teacher*, XIII, No. 6 (October 1966), 448–452. Mirror Cards, produced by Educational Services, Inc., are discussed by their originator. Examples of the cards are illustrated.

———. "Some Mathematical Ideas Involved in the Mirror Cards," *The Arithmetic Teacher*, XIV, No. 2 (February 1967), 115–125. Discusses symmetry in general and the use of Mirror Cards in particular. Suggestions of ways of presenting concepts of symmetry with Mirror Cards are described and illustrated.

Wenninger, Magnus J. *Polyhedron Models for the Classroom*. Washington, D.C.: National Council of Teachers of Mathematics, 1966. Presents illustrated directions for making polyhedron models. Includes many line drawings and photographs.

Williford, Harold. "What Does Research Say about Geometry in the Elementary School?" *The Arithmetic Teacher*, XIX, No. 2 (February 1972), 97–104. Williford reviews research dealing with elementary school geometry. He discusses the capabilities of children to learn geometry, and the instructional variables related to achievement, transfer effects, teachers, and feasibility studies in geometry. He concludes with a summary of his findings.

# 16 Elementary Number Theory, Probability, and Statistics and Graphs

Upon completion of Chapter 16, you will be able to:

1. Present orally or in writing three reasons for including some number theory in the elementary mathematics program.

2. Define odd and even numbers, and describe at least one activity with these numbers for both primary and intermediate levels of the elementary school.

3. Distinguish between prime and composite numbers.

4. Demonstrate an activity that can be used by children to discover the meanings of prime and composite numbers.

5. Describe the modernized sieve of Eratosthenes, and explain how it is used to determine prime numbers.

6. Use at least three processes for changing composite numbers so they are represented in prime factor form.

7. Demonstrate a process for finding all of the factors of a given composite number.

8. Give a definition of the greatest common factor for two or more numbers, and determine the GCF for two numbers in at least two different ways.

9. Demonstrate at least two procedures for finding the least common multiple for two or more numbers.

10. Use tests of divisibility to determine numbers that are divisible by 2, 3, 4, 5, 6, 7, 8, 9, 10, or 11.

11. Define commonly used probability terms, give appropriate examples of each one, and cite examples of ways these terms can be introduced to children.

12. Demonstrate at least two probability experiments for children.

13. List sources of data children can collect and organize in tables and graphs.

14. Gather data and organize them in a table and make a graph from the table.

15. Describe five activities for the minicalculator that extend children's understanding of number theory, probability, and statistics, and demonstrate the calculator's application for each one.

16. Identify two macropitfalls associated with teaching mathematics at any level.

17. Describe three ways you can keep abreast of developments in elementary school mathematics.

Key Terms you will encounter in Chapter 16:

number theory
odd number
even number
prime number
composite number
sieve of Eratosthenes
factoring
factor tree
factor
greatest common factor (GCF)
Venn diagram
least common multiple (LCM)
multiple
test of divisibility
probability
impossible event
likely event
certain event
equally likely event
statistics
graph
line graph
circle graph
Pascal's triangle
outcome

Number theory, probability, and statistics and graphs are relatively new topics in the elementary mathematics program. Reasons for including these topics are given in Chapter 1. If you have forgotten them, read pages 14 to 17 again.

## ELEMENTARY NUMBER THEORY

According to Dantzig, the theory of number was developed before the theory of arithmetic.[1] Long before efficient ways to use algorithms were worked out, people began to study the relationships among numbers. Number lore, the study of prime and composite numbers, tests of divisibility, and other aspects of number theory have a long history. Even so, ". . . the theory of number is the branch of mathematics which has found the least number of applications. Not only has it so far remained without

[1]Tobias Dantzig, *Number, the Language of Science*, 4th ed. (New York: Doubleday & Company, Inc., 1954), p. 38.

influence on technical progress, but even in the domain of pure mathematics it has always occupied an isolated position only loosely connected with the general body of the science."[2]

Today number theory has achieved a new position in the elementary school program. Where formerly study of number theory was incidental and haphazard, if there was any at all, now it is planned and organized to lead children to discover certain elementary generalizations about numbers. There are at least three reasons for its present position:

1. The study helps children understand that number theory can be useful in other areas of mathematics. Children who know about prime and composite numbers can use them to factor composite numbers and find the *greatest common factor* (GCF) and the *least common multiple* (LCM) of two or more numbers. Later, they can use this knowledge to find common denominators and simplify, or reduce *to lowest terms*, common fractions that represent fractional numbers.
2. Many children find that work with number theory is interesting and fun. Some find that work with prime and composite numbers leads to many interesting areas of investigation.
3. During the study of different aspects of number theory, many opportunities arise for work with the basic facts of the four operations, to practice computational skills, and to further develop a mathematics vocabulary. For example, the process of factoring numbers involves factors, or divisors, and multiples of numbers.

Many topics dealing with number theory are well suited for learning centers, with directions for activities on problem cards or tapes for playback on cassette recorders. Activities and materials, some in a problem card format, are suggested in this chapter.

---

## Self-check of Objective 1

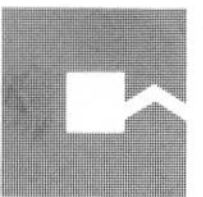

Give three reasons for including some number theory topics in today's mathematics program.

---

### 1. Odd and Even Numbers

Investigations into the nature of the whole numbers that give children an intuitive understanding of odd and even numbers can begin in the pri-

[2] Dantzig, p. 38.

mary grades. Children can use disks or structured materials, such as Cuisenaire rods, as manipulative materials while they follow directions on problem cards similar to the ones in Figure 16-1.

Have children keep a record of their work by drawing pictures of the sets and writing answers on response sheets. Children should also copy and complete the underlined sentences on the problem cards on their answer sheets. (You may record instructions for these activities on a cassette tape for children who are ready for the work but who might have difficulty reading the cards. Draw pictures of the chips that are shown on the problem cards to go along with the tapes.)

Follow this activity with investigations on the number line. Have children locate odd and even numbers on the line and then count by twos beginning at an even or an odd number and going forward or backward. The forward count can go to the end of the line, or to a designated number, such as 14 or 19. The backward count can go to 0 or 1, or to any designated number, such as 3 or 6.

A hundreds chart offers many opportunities for children to extend their experiences with odd and even numbers (see, for example, Figure 16-6). Counting activities similar to those done with the number line can be done with the chart. Children can also examine the chart to make these observations:

1. Beginning at 2, every other number is an even number; beginning at 1, every other number is odd.
2. The column on the left side of the chart contains all odd numbers ending in 1. Each alternate column contains odd numbers, ending in 3, 5, 7, and 9 across the chart.
3. The second column from the left contains all even numbers ending in 2. Each alternate column contains even numbers, ending in 4, 6, 8, and 0 across the chart.
4. The even numbers are multiples of 2. (Not all children will make this observation, especially if the concept of multiples has not yet been introduced. Do not force it upon those who don't make the observation.)

Intermediate-grade children can study odd and even numbers at a more abstract level. These generalizations can arise from their investigations:

1. If an even number is designated by $2n$, then its successor is $2n + 1$ and is an odd number. ($2n$ is used to designate an even number because every even number has 2 as a factor; therefore 2 times any number must be even.)
2. Zero is an even number.

ODD AND EVEN (1)

1. Get a bag of poker chips.

2. Take 8 chips from the bag and put them on your table so they look like this:

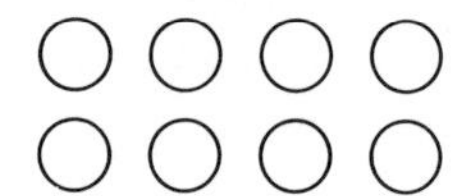

Count the chips in each row. There are ___ chips in each row.

3. Use a set of 10 chips this time. Put them in 2 rows with the same number of chips in each row. Draw a picture of your two rows of chips. There are ___ chips in each row.

**Figure 16-1** Problem cards for work with odd and even numbers

(2)

4. Put a set of 6 chips in 2 rows with the same number of chips in each row. Draw a picture of your two rows of chips. There are ___ chips in each row.

5. Do the same things with each of these sets of chips. Draw a picture of each set of chips after you have put them in two rows. Tell how many chips are in each row by writing the answer on your paper.

   a. 4 chips d. 20 chips
   b. 12 chips e. 18 chips
   c. 16 chips f. 30 chips

(3)

6. Make some other sets that have the same number of chips in each row when you make 2 rows. Draw a picture of each of your sets. Tell how many chips are in each of your sets.

7. Make a set of 7 chips and put them in 2 rows like this:

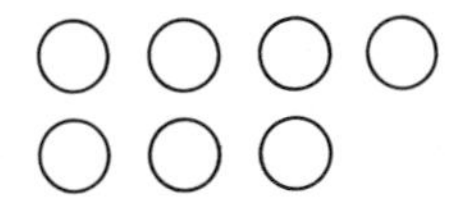

Does each row have the same number of chips in it? There are ___ chips in one row and ___ chips in the other row.

(4)

8. Make 2 rows with 5 chips. Draw a picture of your two rows. Did you put the same number of chips in each row? There are ___ chips in one row and ___ chips in the other row.

9. Make 2 rows with these sets of chips. Do it so there is always one more chip in one row than in the other row. Draw a picture of each of your sets after you have put them in 2 rows.
   a. 3 chips
   b. 9 chips
   c. 15 chips
   d. 11 chips
   e. 19 chips
   f. 23 chips

(5)

10. You have counted chips and have put each set into 2 rows. Some sets were put into rows with the same number of chips in each row. Tell which sets these are. Some sets cannot be put into 2 rows with the same number of chips in each row. Tell which sets these are.

11. Sets that can be put into rows with the same number of chips in each row are sets with an EVEN number of chips in them. Write the numerals for some even numbers on your paper.

    Sets that cannot be put into rows with the same number of chips in each row are sets with an ODD number of chips in them. Write the numerals for some odd numbers on your paper.

(6)

12. Make some different sets with an even number of chips in them. Tell how many chips are in each of your sets.

13. Make some sets with an odd number of chips in each one. Tell how many chips are in each of your sets.

3. When two even numbers are added, their sum is an even number: $2n + 2n = 4n$. When two odd numbers are added, their sum is an even number: $(2n + 1) + (2n + 1) = 4n + 2$. When an even number and an odd number are added, their sum is an odd number: $2n + (2n + 1) = 4n + 1$.

4. When two even numbers are multiplied, their product is an even number: $2n \times 2n = 4n^2$. When two odd numbers are multiplied, their product is an odd number: $(2n + 1) \times (2n + 1) = 4n^2 + 4n + 1$. When an even and an odd number are multiplied, their product is an even number: $2n \times (2n + 1) = 4n^2 + 2n$.

---

## Self-check of Objective 2

Define an even and an odd number. Demonstrate with materials an activity suitable for introducing odd and even numbers to primary-grade children. Describe an activity with these numbers for intermediate-grade children. Tell how these activities will help these children understand addition and multiplication better.

---

### 2. Prime and Composite Numbers

Intermediate-grade children can investigate the nature of prime and composite numbers. Activities with poker chips or other disks allow children to investigate different array patterns. Problem cards featuring the following activities can guide children's work, or you can present the work in a series of teacher-directed lessons.

1. Arrange sets of 4, 6, 8, 9, 10, and 12 chips into all of their possible arrays. The arrays for 6 chips are shown in Figure 16-2.

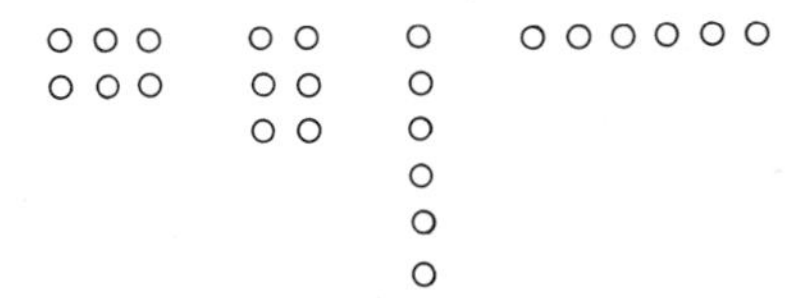

**Figure 16-2** The four arrays possible with six markers

2. Keep a record of the arrays for each set of chips. (Figure 16-3 shows a chart for doing this.)

3. Make arrays for sets of 2, 3, 5, 7, 11, 13, and 17 chips and record the results. The arrays for 7 chips are shown in Figure 16-4.

| Whole Number | Arrays | Number of Different Arrays |
|---|---|---|
| 6 | 2 by 3, 3 by 2, 6 by 1, 1 by 6 | 4 |

**Figure 16-3** Example of table to be developed by children to keep track of arrays made with different numbers of markers

**Figure 16-4** The two arrays that can be made with seven markers

4. Determine the array for 1 chip and record it.
5. Examine the chart for generalizations about the different arrays and factors for the numbers. (The completed table for sets of 1 through 10 chips is shown in Figure 16-5.) Children will note these generalizations:

   a. Some whole numbers—2, 3, 5, and 7—have only two arrays; these form a straight line rather than a rectangular pattern. The numerical description always contain a 1 and the numeral that tells the total number of markers in the array.
   b. Some whole numbers—4, 6, 8, 9, and 10—have more than two arrays. They can be arranged as straight-line arrays and in one or

| Whole Number | Arrays | Number of Different Arrays |
|---|---|---|
| 1 | 1 by 1 | 1 |
| 2 | 1 by 2, 2 by 1 | 2 |
| 3 | 1 by 3, 3 by 1 | 2 |
| 4 | 2 by 2, 1 by 4, 4 by 1 | 3 |
| 5 | 1 by 5, 5 by 1 | 2 |
| 6 | 2 by 3, 3 by 2, 6 by 1, 1 by 6 | 4 |
| 7 | 1 by 7, 7 by 1 | 2 |
| 8 | 2 by 4, 4 by 2, 8 by 1, 1 by 8 | 4 |
| 9 | 3 by 3, 9 by 1, 1 by 9 | 3 |
| 10 | 2 by 5, 5 by 2, 10 by 1, 1 by 10 | 4 |

**Figure 16-5** Table after the arrays for the numbers 1 through 10 have been developed

more rectangular patterns (square in the cases of 4 and 9). Numerals other than 1 and the numeral that names the number of markers in the set are used in the numerical expressions for these arrays.

c. Only one whole number—1—has exactly one array.

These generalizations open the way to further work with prime and composite numbers. Leave room on the chart to record other arrays, perhaps for all sets up to and including thirty chips. Children can then list all the prime numbers between 1 and 31. By the time children complete this activity, they should be able to give a suitable definition of prime and composite numbers: "A prime number has only two different whole-number factors: 1 and the number itself." "A composite number has three or more different whole-number factors." The special characteristics of the number 1 should be stressed: it has only one factor—itself—and it is a factor of every other number. Make it clear that it is a special number that is neither prime nor composite. It is treated as a *unit*.

Work with prime and composite numbers extends understanding of factors, divisors, and multiples. Children will already have encountered these terms in connection with their earlier study of multiplication and division. Now they should learn that the terms *factor* and *divisor* can be used interchangeably. When two whole numbers are multiplied, they yield a product; they can be called either factors or divisors of their product. (An exception to this statement is that 0 can be a factor but not a divisor.) The product of a pair of numbers can also be called a multiple of the two numbers. Thus 5 and 7 are factors or divisors of 35; 35 is a multiple of both 5 and 7.

a. **The Sieve of Eratosthenes.** Another activity for intermediate-grade children is an investigation of the sieve of Eratosthenes. Eratosthenes was a Greek astronomer-geographer who lived in the third century B.C. He devised a scheme for separating any set of consecutive whole numbers larger than 1 into prime and composite numbers. The set most commonly used in classrooms is the set 1 through 100. These numbers are usually arranged in a hundreds chart to give an orderly means for developing the sieve (Figure 16-6). Instructions such as the following can be presented on problem cards or during a directed lesson.

1. Put a ring around 1.
2. Two is the smallest prime number. Beginning with 4, mark out all of the multiples of 2 by putting a line through them.
3. The next prime number is 3. Some of the multiples of 3 have already been marked out. Which ones are these? Mark out the remaining multiples of 3.

**Figure 16-6** A hundreds chart can be used for developing the sieve of Eratosthenes

| | | | | | | | | | |
|---|---|---|---|---|---|---|---|---|---|
| 1 | 2 | 3 | 4 | 5 | 6 | 7 | 8 | 9 | 10 |
| 11 | 12 | 13 | 14 | 15 | 16 | 17 | 18 | 19 | 20 |
| 21 | 22 | 23 | 24 | 25 | 26 | 27 | 28 | 29 | 30 |
| 31 | 32 | 33 | 34 | 35 | 36 | 37 | 38 | 39 | 40 |
| 41 | 42 | 43 | 44 | 45 | 46 | 47 | 48 | 49 | 50 |
| 51 | 52 | 53 | 54 | 55 | 56 | 57 | 58 | 59 | 60 |
| 61 | 62 | 63 | 64 | 65 | 66 | 67 | 68 | 69 | 70 |
| 71 | 72 | 73 | 74 | 75 | 76 | 77 | 78 | 79 | 80 |
| 81 | 82 | 83 | 84 | 85 | 86 | 87 | 88 | 89 | 90 |
| 91 | 92 | 93 | 94 | 95 | 96 | 97 | 98 | 99 | 100 |

4. Four is marked out. The next prime number is 5. Mark out the multiples of 5 that are not already crossed out.
5. What is the smallest multiple of 7 that has not been marked out? Mark it out along with any other multiples of 7 that still remain.
6. Eleven is the next prime number. What multiples of 11 are smaller than 100? Are any of these left to be marked out? Why do you suppose no multiples of 11 are left?
7. The numbers that have not been marked out are all prime numbers (Figure 16-7). Make a list of these numbers and compare your list with a classmate's.

Boys and girls who are interested in further study of prime and composite numbers can complete these activities:

1. Determine the prime numbers between 100 and 200, 100 and 300, or within any other limits, using the sieve to find the multiples of increasingly larger prime numbers — 11, 13, 17, 19, 23, . . . .

**Figure 16-7** The sieve of Eratosthenes showing prime and composite numbers from 1 to 100

| | | | | | | | | | |
|---|---|---|---|---|---|---|---|---|---|
| ① | 2 | 3 | ~~4~~ | 5 | ~~6~~ | 7 | ~~8~~ | ~~9~~ | ~~10~~ |
| 11 | ~~12~~ | 13 | ~~14~~ | ~~15~~ | ~~16~~ | 17 | ~~18~~ | 19 | ~~20~~ |
| ~~21~~ | ~~22~~ | 23 | ~~24~~ | ~~25~~ | ~~26~~ | ~~27~~ | ~~28~~ | 29 | ~~30~~ |
| 31 | ~~32~~ | ~~33~~ | ~~34~~ | ~~35~~ | ~~36~~ | 37 | ~~38~~ | ~~39~~ | ~~40~~ |
| 41 | ~~42~~ | 43 | ~~44~~ | ~~45~~ | ~~46~~ | 47 | ~~48~~ | ~~49~~ | ~~50~~ |
| ~~51~~ | ~~52~~ | 53 | ~~54~~ | ~~55~~ | ~~56~~ | ~~57~~ | ~~58~~ | 59 | ~~60~~ |
| 61 | ~~62~~ | ~~63~~ | ~~64~~ | ~~65~~ | ~~66~~ | 67 | ~~68~~ | ~~69~~ | ~~70~~ |
| 71 | ~~72~~ | 73 | ~~74~~ | ~~75~~ | ~~76~~ | ~~77~~ | ~~78~~ | 79 | ~~80~~ |
| ~~81~~ | ~~82~~ | 83 | ~~84~~ | ~~85~~ | ~~86~~ | ~~87~~ | ~~88~~ | 89 | ~~90~~ |
| ~~91~~ | ~~92~~ | ~~93~~ | ~~94~~ | ~~95~~ | ~~96~~ | 97 | ~~98~~ | ~~99~~ | ~~100~~ |

2. Look for *twin primes*. Twin primes are two prime numbers that have exactly one composite number between them—3 and 5, 5 and 7, 11 and 13.

3. Test Goldbach's conjecture. Goldbach was a Russian mathematician who made a conjecture in 1742 that every even number greater than 2 can be written as the sum of a pair of prime numbers: $4 = 2 + 2$; $6 = 3 + 3$; $8 = 5 + 3$; $10 = 7 + 3$ or $5 + 5$. This conjecture has never been proven to be either true or false for all even numbers.

4. Play a game such as "prime drag,"[3] "prime-o," or "prime factor."[4]

---

## Self-check of Objectives 3, 4, and 5

Give an oral or written statement that distinguishes between prime and composite numbers.

Demonstrate how markers, such as poker chips, can be used by children to discover the difference between prime and composite numbers.

Prepare a modern sieve of Eratosthenes for the numbers 1 through 100, and demonstrate how to use it to determine the prime numbers smaller than 100.

---

### 3. Factoring Composite Numbers

Children's proficiency in finding the greatest common factor of a pair of numbers and using this factor to simplify common fractions, or in finding the least common multiple for two or more numbers and using this multiple to name common denominators for fractions, depends upon their ability to factor composite numbers quickly and accurately. A composite number is said to be factored completely when it is represented as a product expression that consists of two or more prime numbers. When the number 18 is factored completely, it is expressed as $2 \times 3 \times 3$; 36 is expressed as $2 \times 2 \times 3 \times 3$. Several procedures for factoring composite numbers will be discussed below.

**a. Use Basic Multiplication Facts.** Children can begin work with factoring by using small composite numbers, such as 4, 6, 9, 10, 15, and 21. These numbers have only a pair of prime number factors, so their prime

[3]"Prime Drag" is available from Creative Publications, P.O. Box 10328, Palo Alto, Calif. 94303.

[4]"Prime-o" and "Prime Factor" are available from Creative Teaching Associates, P.O. Box 7766, Fresno, Calif. 93727.

factorization is complete as soon as the two factors are named. You can help children use basic facts to factor these numbers: "Which two prime numbers are multiplied to give a product of 21?" The basic multiplication fact $3 \times 7 = 21$ gives the answer.

**b. Use Factor Trees.** Factor trees give a systematic way to factor numbers that are reasonably small but are the product of more than two prime numbers. Factor trees are created by expressing numbers in terms of successively smaller factors until all factors are prime numbers. The factor trees for 24 are shown in Figure 16-8. In (a), three different ways of beginning are shown; each begins with a different pair of factors. (The number pair 1 and 24 is not used.) In (b), the process is continued with each composite factor expressed by a pair of smaller factors. Note that 12 is factored in two ways in the two trees on the right and that the prime factors in any of the trees are represented by a repetition of their numerals. The tree on the left in (b) is completed because in its bottom line 24 is expressed as the product of prime numbers. The completed trees for the other factorizations are shown in (c), where each bottom line expresses 24

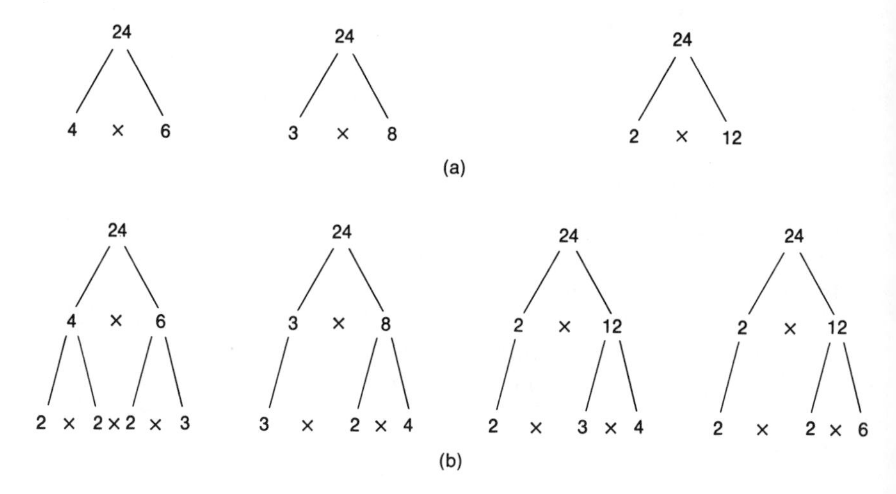


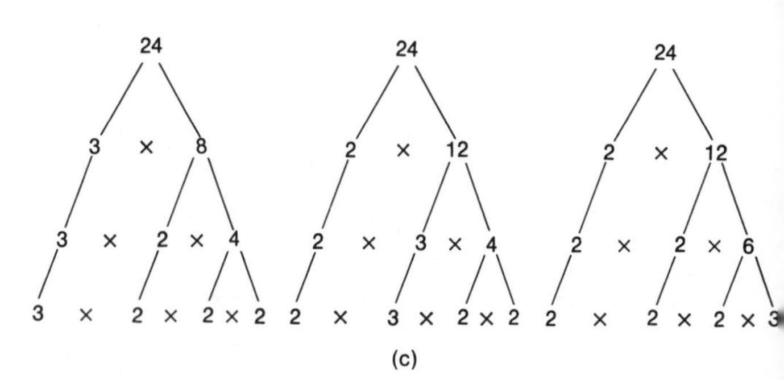


**Figure 16-8** Different factor trees for the number 24

as the product of prime numbers. The numerals at the bottom of each completed tree in (c) are 2, 2, 2, and 3.

Other numbers with two or more factor trees that can be used for practice are 16, 18, 28, 30, 36, and 40. After children have completed the factor trees for several small composite numbers, they should observe that the same prime numbers are in the bottom line of each number's factorization. For 24, there are three 2s and one 3; for 16 there are four 2s, and so on. Even though children do not have mathematical proof, they can observe that each composite number has only one prime factorization, if order of factors is disregarded. This property of numbers is called the *fundamental theorem of arithmetic*.

**c. Use an Algorithm.** Factor trees are useful for numbers small enough for children to easily see at least one pair of factors. For larger numbers, an algorithm for factoring numbers is useful. In the example in the margin an algorithm is used to factor 124. First, 124 is divided by 2 and the numeral for the quotient, "62," is written beneath the "124." Sixty-two is also divisible by 2. This division is completed, and the quotient's numeral is written beneath the "62." The number 31 is a prime number, so it is divided by itself. The three divisors show the prime factorization of 124 to be $2 \times 2 \times 31$. A second example is shown in the next algorithm in the margin. It shows that the prime factorization of 525 is $3 \times 5 \times 5 \times 7$.

| | |
|---|---|
| 2 | 124 |
| 2 | 62 |
| 31 | 31 |

| | |
|---|---|
| 3 | 525 |
| 5 | 175 |
| 5 | 35 |
| 7 | 7 |

In these two examples, the division by primes began with the smallest possible prime, which is 2 in the first example and 3 in the second. Each successive division was done with the smallest possible prime, and in both cases a prime number was repeated as a divisor. When the prime factorizations of the numbers are given, the prime numbers are written in order from smallest to largest. While it is not necessary to take the steps in this particular order, this order is recommended for two reasons:

1. Children can recognize the divisibility by 2, 3, and 5 more easily than they can by the larger prime numbers.
2. When numbers have been completely factored, the factors are already ordered according to size, and the writing of prime product expressions is simplified. One of the advantages of writing product expressions in this manner becomes obvious when the process of changing common fractions to simpler forms is done by factoring the numerators and denominators.

**d. Finding All the Factors of a Composite Number.** Another skill is that of finding *all* of the factors of a given composite number. For some numbers children can name all the factors by inspection. All of the factors of 6 are 1, 2, 3, and 6, while all the factors of 12 are 1, 2, 3, 4, 6, and 12.

Children use basic facts and the knowledge that 1 and the number itself are always factors as they list all the factors for small numbers like these. However, there are many other numbers that children cannot factor easily by inspection, so they need to learn a systematic procedure for finding all of the factors for these numbers. One procedure follows, using 42 as an example:

1. Begin with the prime factorization of 42: $2 \times 3 \times 7$.
2. List all of the obvious factors: 1, 2, 3, 7, and 42. (Six might also be listed now since children know that $6 \times 7 = 42$.)
3. Use the prime factors to name any other factors. If 2 and 3 are factors of 42, there must be a number to pair with each of them to give a product of 42. Divide 42 by 2 to find the factor to pair with 2. This factor is 21. Divide 42 by 3 to find the factor to go with 3; it is 14. (An alternate way to find these factors is to use the prime factorization in its $2 \times 3 \times 7$ form. The associative property of multiplication gives $2 \times (3 \times 7)$, which can be rewritten as $2 \times 21$, to show that 21 is the factor that goes with 2. The associative and commutative properties give $3 \times (2 \times 7)$, which can be rewritten as $3 \times 14$, to show that 14 is the factor that goes with 3.)

$$\begin{array}{r|r} 2 & 210 \\ 3 & 105 \\ 5 & 35 \\ 7 & 7 \\ \hline \end{array}$$

Once children understand this procedure, they can use it along with the algorithm to find all the factors for larger composite numbers. For example, the algorithm in the margin shows the prime factorization of 210. Altogether, the algorithm shows seven different factors: 2, 3, 5, 7, 35, 105, and 210. In addition, 1 is an obvious factor, giving a total of eight different factors found by using the algorithm. The four prime factors can be used to find the remaining factors. There must be a factor to pair with each prime number. Since 2 was used as the original divisor of 210, the number to pair with it—105—is already known. To determine the number to pair with 3, multiply the product expression $2 \times 5 \times 7$ to get 70. For the one to pair with 5, multiply the expression $2 \times 3 \times 7$ to get 42; for 7, multiply the expression $2 \times 3 \times 5$ to get 30. Next, use pairs of primes. When 2 and 3 are paired, they indicate that 6 is a divisor of 210. The 5 and 7 give a product of 35, which has already been identified as a factor by the algorithm. Thirty-five appears in the algorithm because dividing by 2 and then by 3 is the same as dividing by 6. The other prime numbers should be paired to give the remaining factors: $2 \times 5 = 10$, $3 \times 7 = 21$, $2 \times 7 = 14$, and $3 \times 5 = 15$.

A list makes it easy to see that no factors have been overlooked. As divisors are determined, children should list them, then order them from smallest to largest. Figure 16-9 shows one way to do this. Each pair of factors has been joined by lines; all the numbers are paired except when the factored number is the square of some number; the square root will not be paired with another number.

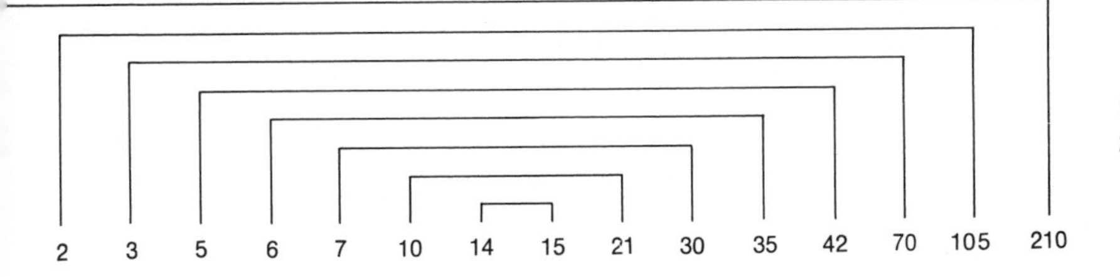


Figure 16-9 The factors of 210 listed and paired

## Self-check of Objectives 6 and 7

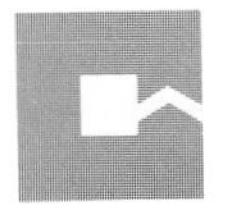

Children can use basic multiplication facts, factor trees, and an algorithm to change composite numbers to their prime factor forms. Tell when each of these procedures should be used, using appropriate examples to illustrate.

Every composite number has three or more factors. Demonstrate a way of finding all of the factors of 156.

### 4. Finding the Greatest Common Factor of Numbers

The process of finding the greatest common factor (GCF) of a pair of numbers is a useful skill when it is used to simplify (reduce to lowest terms) common fractions. Therefore children's work with GCFs should be coupled with simplifying common fractions. The meaning of the GCF and processes for finding greatest common factors are given in this chapter. The way GCFs are used to simplify common fractions is discussed in Chapter 11.

As the words *greatest common factor* suggest, for every pair of whole numbers, there is only one largest whole number by which both may be divided. The greatest common factor of the pair 2 and 4 is 2; for the pair 8 and 12, it is 4; for 5 and 7, it is 1.

Children can learn the meaning of these generalizations by using Venn diagrams and algorithms.

1. When two numbers are prime, their GCF is 1, since a pair of prime numbers, by definition, cannot have a common factor other than 1.

2. When a pair of numbers consists of one prime number and one composite number, the GCF will be 1, except when the composite number is a multiple of the prime number. One is the GCF of 7 and 12 because 12 is not a multiple of 7. Seven and 14 have 7 as the GCF because 14 is a multiple of 7.

3. When two numbers are composite, the GCF may or may not be greater than 1. If the GCF is 1, the composite numbers are *relatively prime*. Examples of relatively prime numbers are 9 and

28, where 1 is the GCF. Composite numbers that are not relatively prime have a GCF that is greater than 1. The numbers 8 and 12 have 4 as their GCF; they are not relatively prime.

a. **Using Venn Diagrams.** A Venn diagram is a good way to show the GCF of a pair of numbers. Children who have used Venn diagrams earlier, perhaps as they sorted attribute materials, will recognize the value of these diagrams in the new situation. If your children have not used Venn diagrams before, you must use care to see that the diagrams' uses are clear to them now. Begin by using diagrams for pairs of small composite numbers. Figure 16-10 shows a Venn diagram that illustrates the greatest common factor for 4 and 6. In circle A the prime factorization of 4 is shown, while in circle B the prime factorization of 6 is given. The "2" in the overlapping (shaded) parts of the two circles is common to both factorizations and is the GCF of 4 and 6.

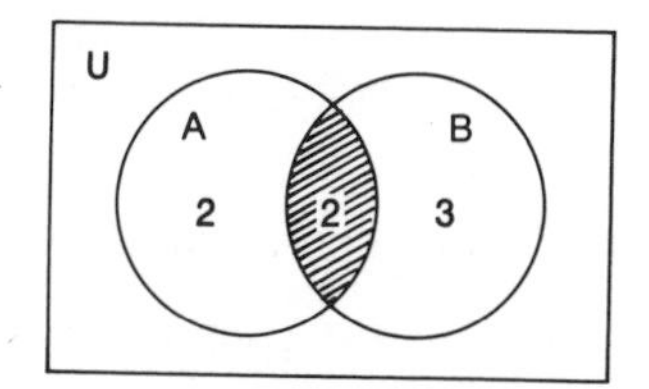


**Figure 16-10** A Venn diagram for indicating the GCF of 4 and 6

Venn diagrams can be used to find the GCF for any pairs of numbers. After two numbers have been factored, the numerals for prime factors of one will be placed inside one circle; the numerals for the prime factors of the other will be placed inside the second circle. Any factors common to both sets will be represented by numerals written in the overlapping parts of the two circles. These numbers form the intersection of the two sets; in other words, they are found in both sets. These numbers are multiplied to give the GCF of a pair of numbers.

Figure 16-11 shows the Venn diagram for finding the GCF of 36 and 42. The numerals that represent the numbers in the prime factorization of 36 are shown in circle A; those in the prime factorization of 42 are shown in circle B. In the overlapping part, or intersection, of the two circles a 2 and a 3 appear because in the prime factorization of both 36 and 42 there is at least one 2 and one 3. The shading indicates which parts of the circles show the GCF, which is 2 times 3, or 6. Children should be reminded that

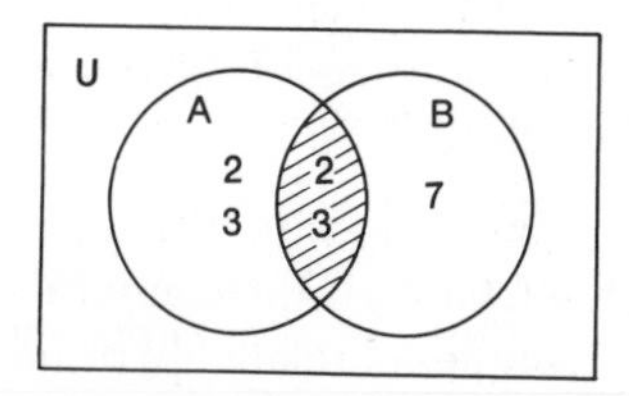


**Figure 16-11** A Venn diagram used to illustrate the process of finding the GCF for 36 and 42; the factors in the overlapping parts of the circles are multiplied to find it

when no prime numbers appear in the intersection of prime factorizations, the GCF is 1.

**b. Using an Algorithm.** A systematic procedure involving an algorithm can be taught to mature fifth- and sixth-grade children to help them find the GCF of two or more numbers. The procedure involves a series of divisions by common prime factors, beginning with the smallest prime factor for a pair of numbers. In the example in the margin the GCF of 12 and 18 is determined. The 12 and 18 are first divided by 2, the smallest prime number common to both. Next, the two quotients are divided by 3, which is the next smallest prime number common to both of them. The division by 3 results in two prime numbers, so the division is complete. The product of 2 and 3, the common divisors, gives 6 as the GCF for 12 and 18.

| | | |
|---|---|---|
| 2 | 12 | 18 |
| 3 | 6 | 9 |
| | 2 | 3 |

The GCF of 120, 140, and 210 is determined in the algorithm in the margin. The product of 2 and 5 yields 10 as the GCF of these three numbers. In this example the factoring ends with division by 5 because 12, 14, and 21 have no common factor greater than 1.

| | | | |
|---|---|---|---|
| 2 | 120 | 140 | 210 |
| 5 | 60 | 70 | 105 |
| | 12 | 14 | 21 |

## Self-check of Objective 8

Define the GCF for two or more numbers; then use the following two procedures to find the GCF of the numbers 24 and 56: Venn diagrams, the algorithm.

### 5. Finding the Least Common Multiple of Two or More Numbers

The process of finding the least common multiple (LCM) for two or more whole numbers is applied when the common denominator for two or more fractional numbers is determined. In the past the process of finding the LCM was commonly taught as a rote process, with the meaning of each step left unclear to the learner. Today systematic procedures can be developed with meaning so that children can apply them with confidence when they need to use LCMs to add or subtract with common fractions having unlike denominators. The three processes that follow can be developed over a period of time. First, children can use the listing process; later, mature children can learn the factorization and algorithm processes.

**a. Listing Multiples.** When children are finding the LCM for small numbers, they can use inspection of lists of multiples to identify the LCM. For example, for 2 and 3, ask, "What is the smallest multiple of both 2 and 3?" Some children will recognize immediately that it is 6. To verify this, have the children list the first several multiples of each number, as in Figure 16-12. The first twelve multiples of 2 are listed in (a), while the first eleven multiples of 3 are in (b). The common multiples are ringed, as in (c). Now children can see that 6 is the smallest multiple that has been ringed; it is the LCM of 2 and 3. The LCM of other pairs of small numbers should be identified in a similar way until the concept is clear.

2, 4, 6, 8, 10, 12, 14, 16, 18, 20, 22, 24,...

(a)

3, 6, 9, 12, 15, 18, 21, 24, 27, 30, 33,...

(b)

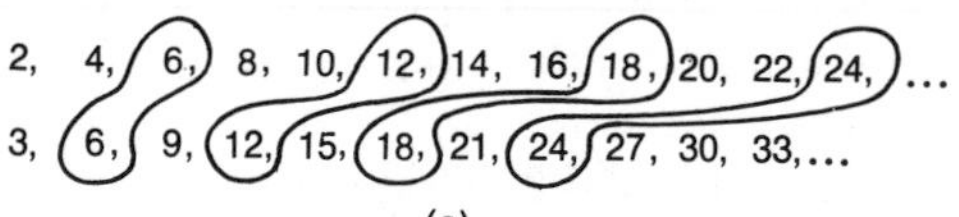

2, 4, 6, 8, 10, 12, 14, 16, 18, 20, 22, 24,...

3, 6, 9, 12, 15, 18, 21, 24, 27, 30, 33,...

(c)

**Figure 16-12** The listing of the multiples of (a) 2 and (b) 3, and (c) the first four common multiples of 2 and 3

The process of listing multiples is useful for finding the LCM of three or more numbers as well. The listing in Figure 16-13 shows that 40 is the LCM for 4, 5, and 8.

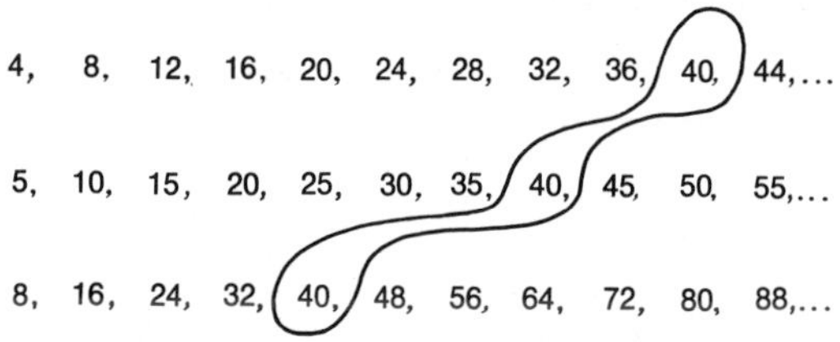

4, 8, 12, 16, 20, 24, 28, 32, 36, 40, 44,...

5, 10, 15, 20, 25, 30, 35, 40, 45, 50, 55,...

8, 16, 24, 32, 40, 48, 56, 64, 72, 80, 88,...

**Figure 16-13** The listing of multiples of 4, 5, and 8 show that 40 is their LCM

**b. Using a Factorization Process.** With larger numbers, the process of finding the LCM for two or more numbers by inspection is difficult, and a listing of multiples is tedious. A process that uses prime numbers is helpful for determining LCMs for larger numbers. To find the least common multiple for the fraction in the margin, proceed this way: Determine the prime factorization for each denominator—$12 = 2 \times 2 \times 3$ and $15 = 3 \times 5$. Every common multiple of 12 and 15, including the LCM, will include these prime factors. However, the LCM of 12 and 15 is not $2 \times 2 \times 3 \times 3 \times 5$, or 180. This is so because 3 is a factor of both numbers. Therefore it needs to be used only once in determining the LCM. The LCM—60—is the product of $2 \times 2 \times 3 \times 5$, which includes the prime factorization of 12 ($2 \times 2 \times 3$) and of 15 ($3 \times 5$). The LCM of 9 and 12 is determined this way: $9 = 3 \times 3$; $12 = 2 \times 2 \times 3$; $2 \times 2 \times 3 \times 3 = 36$.

$$\begin{array}{r} 1/12 \\ +1/15 \\ \hline \end{array}$$

A Venn diagram will help make the process clear to children who are familiar with its uses. In Figure 16-14 the prime factorization of 6—

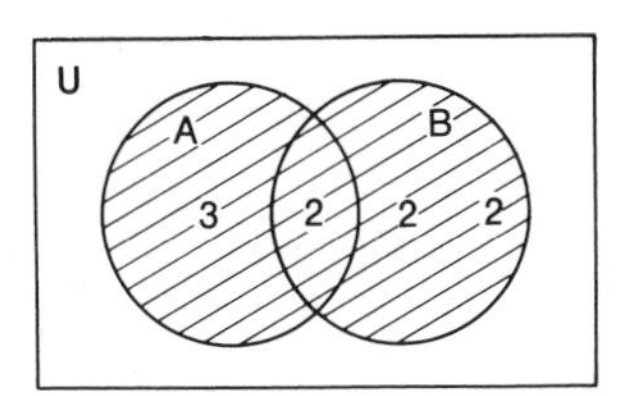


**Figure 16-14** Venn diagram used to illustrate the process of finding the LCM for 6 and 8 by factorization and multiplication of the factors common to the two sets of factors

$2 \times 3$—is shown in the A part of the diagram, while the prime factorization of 8—$2 \times 2 \times 2$—is in the B part. Since there is at least one 2 in each factorization, there is a 2 in the overlapping part of the diagram. The LCM of 6 and 8 is determined by finding the product of $3 \times 2 \times 2 \times 2$, which is 24. Students who are familiar with the concept of set union will recognize that the LCM for two or more numbers is determined by multiplying the factors that are in the union of factors in the prime factorizations of the numbers.

c. **Using an Algorithm.** The algorithm for finding the GCF for numbers can also be used for finding the LCM of two or more numbers. In the algorithm in the margin, the LCM for 12, 18, and 24 is determined. First, the three numbers are divided by 2, the smallest common prime divisor. The quotients of this division are divided by 3, which is a common divisor for all of them. There is no prime number divisor that is common to the three new quotients, so the algorithm continues with the division of 2 and 4 by 2. The 3 is brought down. All the quotients are now prime or are 1, so the division is complete. The LCM of 12, 18, and 24 is 72, which is the product of the expression $2 \times 3 \times 2 \times 3 \times 2$. The Venn diagram in Figure 16-15 is related to this algorithm. First, relate the prime numbers in the intersection of all three circles to the algorithm's first two divisors. Next, relate the prime number in the intersection of 12 and 24 to the third divisor. Finally, the numbers in the circles for 18 and 24 are the fourth and fifth divisors.

| | | | |
|---|---|---|---|
| 2 | 12 | 18 | 24 |
| 3 | 6 | 9 | 12 |
| 2 | 2 | 3 | 4 |
| | 1 | 3 | 2 |

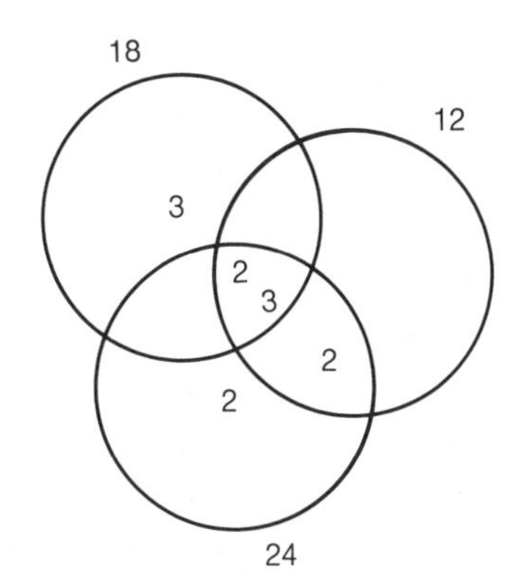


**Figure 16-15** Venn diagram showing the factorizations of 12, 18, and 24. The numbers' LCM is $2 \times 2 \times 2 \times 3 \times 3$, or 72

## Self-check of Objective 9

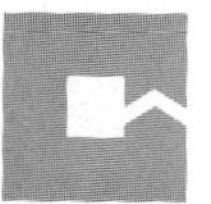

Use at least two different procedures to find the LCMs of these sets of numbers: 4, 6, and 8; 3, 12, and 15.

### TESTS OF DIVISIBILITY

Tests of divisibility are used to determine whether a given number is divisible by another given number. "Divisible by" implies division with a remainder of zero. Divisibility tests are useful as children determine the factors of numbers. The divisibility tests for 2, 3, 5, 7, and 11 are particularly valuable because these numbers are frequently used to determine prime factorizations of numbers. The tests of divisibility for 2, 5, and 10 can be developed as children learn division and relate that operation to repeated subtraction by 2, 5, and 10. The other tests require more direct instruction, and some of them, such as the tests for 7 and 11, should be taught only to mature fifth- and sixth-grade children.

Mathematical proofs for each of these tests exist, but are not explained in this book.

1. *Divisibility by 2.* Any whole number that ends in 0, 2, 4, 6, or 8 is divisible by 2.
2. *Divisibility by 3.* The test for divisibility by 3 requires first that the sum of the digits in the number be determined. If this sum is divisible by 3, the original number is also divisible by 3. The sum of the digits in 462 is 12, and 12 is divisible by 3; therefore 462 is divisible by 3. The sum of the digits in 496 is 19; therefore 496 is not divisible by 3.
3. *Divisibility by 4.* The test of divisibility by 4 involves the tens and ones places of the number to be tested. If the tens and ones are divisible by 4, the entire number is divisible by 4. The number 1344 can be renamed as 1300 + 44; 44 is divisible by 4, so the number 1344 is divisible by 4. In the number 1634, the number in the tens and ones places, 34, is not divisible by 4; neither is 1634.
4. *Divisibility by 5.* Every whole number that ends in 0 or 5 is divisible by 5.
5. *Divisibility by 6.* Any even whole number that is divisible by 3 is divisible by 6. To test a whole number for divisibility by 6, see, first, if it is even, and then apply the test for divisibility by 3. If the test shows that the number is divisible by 3, it will also be divisible by 6.

6. *Divisibility by 7*. There are several tests for divisibility by 7. A useful one for reasonably small whole numbers is: Multiply the digit in the ones place by 2. Subtract the product from the number named by the remaining digits. If the difference is divisible by 7, the entire number is divisible by 7. This test shows that for the number 672, $2 \times 2 = 4$, $67 - 4 = 63$, and $63 \div 7 = 9$. Therefore 672 is divisible by 7. For 1001, the test gives $2 \times 1 = 2$, $100 - 2 = 98$, and $98 \div 7 = 14$. The number 1001 is divisible by 7.

7. *Divisibility by 8*. A whole number is divisible by 8 if the part of the number that begins in the hundreds place is divisible by 8. The number 11,384 is divisible by 8 because 384 is divisible by 8. The number 16,692 is not divisible by 8 because 692 is not divisible by 8.

8. *Divisibility by 9*. A whole number is divisible by 9 if the sum of its digits is divisible by 9. The sum of the digits in the number 2331 is $2 + 3 + 3 + 1$, or 9; $9 \div 9 = 1$. The number 2331 is divisible by 9.

9. *Divisibility by 10*. Any whole number that has a 0 in the ones place is divisible by 10.

10. *Divisibility by 11*. There are several tests for divisibility by 11. A simple one to use with whole numbers involves subtracting the digit in the ones place of a number from the number named by the remaining digits. When the remainder is divisible by 11, the entire number is divisible by 11. If one subtraction does not make determination of divisibility by 11 obvious, repeat the process. For the number 121, apply the test as follows: $12 - 1 = 11$, $11 \div 11 = 1$. For the number 1452, $145 - 2 = 143$, $14 - 3 = 11$, $11 \div 11 = 1$. The number 1452 is divisible by 11.[5]

---

## Self-check of Objective 10

Explain how each test of divisibility shows which numbers are or are not divisible by 2, 3, 4, 5, 6, 7, 8, 9, 10, and 11.

---

## PROBABILITY

You should not attempt to take children too far into the study of probability. Children's intuitive understanding will enable them to grasp the simpler concepts of the topic, but the thin dividing line between the simpler

[5] The author is indebted to Leo Allison, mathematics instructor at Oakmont High School, Roseville, Calif., for information about this divisibility test.

and more complex topics should not be crossed.[6] Simple activities can be begun by children in the primary grades.

### 1. Probability Terms

By the time they complete elementary school, many children will understand the terms *probability, impossible event, likely event, certain event,* and *equally likely event.* Every event, or happening, can be assigned a number that states the likelihood of it occurring. The more likely an event is to occur, the higher the number assigned to it. The numerical measure that refers to the likelihood of an event occurring is that event's probability. An impossible event, for example, Christmas will come in the middle of summer in the northern hemisphere, is assigned a probability of 0. A certain event, for example, that an airplane will cross the Atlantic Ocean when flying from New York to London in an easterly direction, is assigned a probability of 1. If an event can happen, but is *unlikely* to, the value of its probability is close to 0; while a likely event has a value that is close to 1. Two equally likely events are each assigned the value of ½.

### 2. Probability Experiments

One way to introduce children to the concept of an event is to discuss the likelihood of certain things occurring. The School Mathematics Study Group publication *Probability for Primary Grades, Teacher's Commentary* suggests these as good beginning topics.[7]

"Which is more likely on the fourth of July (here in our town), rain or snow?"

"Billy is a very good student. Which is he more likely to do in tomorrow's test, pass or fail?"

"A new boy has joined the class. Mary is to guess his birthday. Alice is to guess how many brothers and sisters he has. Who is more likely to guess right?"

"When Mary guesses the new boy's birthday, which (in her guess) is the more likely to be right, the entire birthday, or the month alone?"

In the first example, the more likely event (in the United States) is rain, and this should be obvious to children. It is possible, of course, that neither event will occur. This gives you a chance to discuss the idea that even though *neither* event *may* happen, one is still *more likely* than the

[6] John D. Wilkinson and Owen Nelson, "Probability and Statistics—Trial Teaching in Sixth Grade," *The Arithmetic Teacher,* XIII, No. 2 (February 1966), 105.

[7] David W. Blakeslee, et al., *Probability for Primary Grades, Teacher's Commentary,* rev. ed. (Stanford, Calif.: School Mathematics Study Group, Leland Stanford Junior University, 1966), pp. 7–9. The same topics appear in SMSG's companion book, *Probability for Intermediate Grades, Teacher's Commentary,* pp. 8–10.

other. The second example is relatively simple, while the third and fourth ones will require careful consideration by children. In the final example, an important principle is brought out. It is more likely that the birth month will be named than the birthday. However, it is impossible for the birthday (less likely event) to be given correctly if the birth month (more likely event) is given incorrectly. "In this case, the comparison in likelihood is derived not from past experience or numerical considerations, but from logical necessity."[8] You can adjust these topics to suit your children, or use others of your own choosing.

A suitable activity for introducing probability is the removal of beads, without looking, from a bag. The beads should all be the same size and weight and should feel the same when touched. If a bag contains ten differently colored beads, each bead has an equal chance of being removed, so the probability of any given bead being removed is 1 in 10, or $1/10$.

To develop an understanding of this concept, a small number of beads should be used first. With only two beads, one red and one black, it will not take children long to conclude that the probability of drawing the red bead is equal to the probability of drawing the black bead from the bag; the probability for each event is 1 in 2, or ½. *Event* refers to each different combination of beads that can be drawn from a container under certain stated conditions. In the case of two beads in a bag, an event is the removal of either one of them.

To verify their conclusion, the children should make a record of many events. To simplify this, pairs of children might work as teams, with each pair having a bag and two beads. One child should remove the bead, while the other records the results on a simple chart (Figure 16-16). When everyone has made and tabulated 100 drawings, the children can compile a master chart with the accumulated data. The master chart's evidence will substantiate the conclusion that the probability for each event is 1 in 2.

Children should use other numbers of beads of two colors to determine the probabilities for other events. "What is the probability of drawing a red bead when we have two black and one red bead in the bag?"

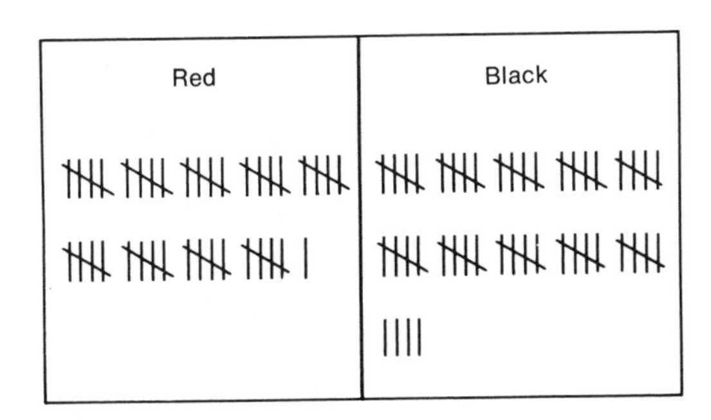


**Figure 16-16** Chart showing the results of drawing a red or black bead from a bag 100 times

[8] Blakeslee, et al., p. 9.

"What is the probability of drawing a black bead when we have six red beads and six black beads in the bag?"

These activities with readily available materials will reinforce children's understanding of the probability of simple events:

1. Use poker chips. If there are one red and one blue chip in a bag, what is the probability of drawing a red chip? a blue chip? Put other combinations of chips in the bag.
2. Use checkers. What is the probability of a checker stopping with crown face up after you spin it?
3. Use spinners that are half red and half blue.[9] What is the probability that the spinner will stop on red? on blue? Also, use spinners that are a quarter red and three-quarters green; or one-third red, one-third blue, and one-third yellow; or other combinations. (Until children have a good understanding of probability, do not use common fractions to express probabilities. Instead, use expressions such as 1 out of 2 and 3 out of 4.)

Later, children should use larger numbers of beads to discover how to predict probabilities when more than one bead at a time is removed from a container. Necessary equipment includes a container with an equal number of red and black beads—perhaps 150 of each—and a device for removing a given number of beads. An easily made device is a spoon fashioned from two pieces of masonite board. The top piece of masonite contains holes, or bowls, slightly larger than the beads. This piece is glued to the solid piece to complete the spoon (Figure 16-17). Make several spoons, each with a different number of bowls.

Be sure children understand how to draw beads and record their results before they begin independent or small group investigations. When an equal number of red and black beads are used with a three-bowl spoon, four events are possible. They can draw three red; two red and one black; one red and two black; or three black beads during any one draw-

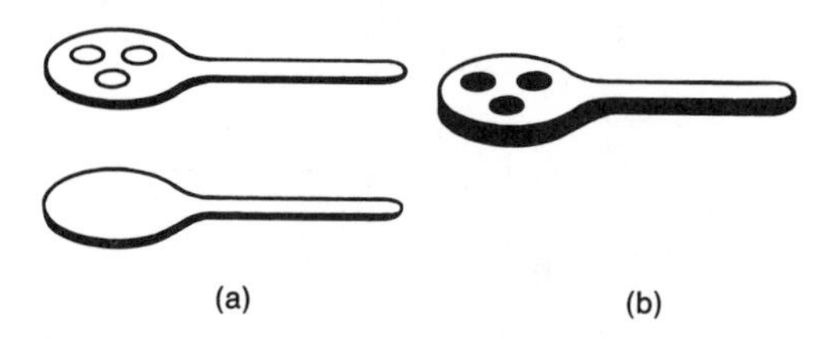


**Figure 16-17** Two parts of a "probability spoon" (a). The assembled "spoon" (b)

[9] You can make a spinner out of a margarine tub. You will need one tub with a lid, a straight pin, two small beads, a small cork, and a pointer cut from another lid. Assemble the spinner in this order: (1) Put a bead on the pin, (2) put the pin through the pointer, (3) put another bead on the pin, (4) put the pin through the tub's lid, (5) push the pin into the cork, and (6) put the lid on the tub.

ing. These events can be symbolized as (3R, 0B), (2R, 1B), (1R, 2B), and (0R, 3B), or, more simply, as ordered pairs in which the first numeral refers to the number of red beads and the second numeral refers to the number of black beads: (3,0), (2,1), (1,2), and (0,3).

So children can determine the probability of a given event, the bowls of each spoon should be lettered. Then children can record the different arrangements of beads in each event in a table. The table in Figure 16-18 shows eight possible outcomes that can occur if three beads are drawn. Three red or three black beads can be drawn in only one way. There are three ways each to draw two red and one black bead and one red and two black beads. The probability for each event is ⅛, ⅜, ⅜, and ⅛, respectively, as listed on the chart.

Comparing tables of events in which different spoons were used will yield an interesting observation concerning the number of possible outcomes when two, three, four, and so on beads are drawn. There are four possible outcomes when two beads are drawn; the number increases to eight when three beads are drawn; there are sixteen possible outcomes when four beads are drawn. "Can you see a pattern that will enable us to predict the number of possible outcomes when we draw five beads?" ($2^5$, or 32.) "Six beads?" ($2^6$ or 64.) "Any number of beads?" ($2^n$.)

Other probability activities for children to follow:

1. Give each child a familiar object—a thumb tack, small paper plate, or tetrahedron model with numbers or letters on its faces—and ask him or her to determine the probabilities for each object to land a given way after it is tossed or spun. Some of the objects will lead to events for which probabilities can be easily established (tetrahedron model, assuming it is not weighted, but rather is *fair*) while others do not (paper plate).

2. Have children investigate the probabilities of multiple events occurring. For example, "When we use our three-bowl spoon, what

| Events | Bowls | | |
|---|---|---|---|
| | A | B | C |
| (3,0) | R | R | R |
| (2,1) | R | R | B |
| | B | R | R |
| | R | B | B |
| (1,2) | B | B | R |
| | R | B | B |
| | B | R | B |
| (0,3) | B | B | B |

**Figure 16-18** Example of a table filled in as children determined the probabilities of different events

is the probability of drawing either three red or three black beads from the container?" "What is the probability that we will draw either all black beads or two black and one red bead from the container?"

3. Ask children to bring to school games that use dice or spinners for determining moves. Have them determine the probabilities for the occurrence of different events (moves).

---

## Self-check of Objectives 11 and 12

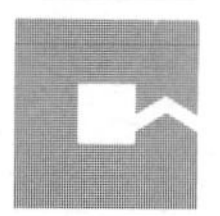

Define and give an example of each of these probability terms: *event, impossible event, certain event, unlikely event, likely event,* and *equally likely event.* Cite examples of situations that can serve as a means of introducing each term to children.

Demonstrate with appropriate materials probability investigations for children.

---

## STATISTICS AND GRAPHS

Statistics are used by planners in many fields. For example, an architect needs information about the ages and numbers of children scheduled to occupy a new school so adequate restroom facilities can be planned. Architects also need information about the number of left-handed children in a typical school population, the number of male and female teachers, and the adult activities planned for the building and the number of people likely to participate so they can plan a building and furnishings that are functional. An insurance actuary uses data about the number of deaths in a given population to prepare actuarial tables that list the number of deaths in different age groups. These tables are used to establish the insurance rates charged by the company.

### 1. Sources of Data

The statistical data collected and used by children are, of course, simpler than those needed in these examples. The National Council of Teachers of Mathematics lists examples of data that can be collected in the classroom.[10]

1. Shoe sizes of the pupils in the classroom.

[10] National Council of Teachers of Mathematics, "Collecting, Organizing, and Interpreting Data," in *More Topics in Mathematics* (Washington, D.C.: The Council, 1969), pp. 450–451. (Used by permission.)

2. Heights of the children.
3. Weights of the children.
4. Color of eyes or hair of the children.
5. Enrollment in classes, clubs, or events.
6. The favorite colors of the children.
7. Cost of the same item in different stores, as determined from newspaper advertisements.
8. TV programs seen the night before class.
9. Record of the temperatures for a week in a particular place in the room at three different times each day.
10. The number of automobiles that pass the classroom window during a five-minute period at the same time each day.
11. High and low temperatures of cities, found in a newspaper.
12. First names of fifteen people.
13. Birthdays of the children.
14. Weekly growth of a plant from seed to maturity.
15. Number of blocks each child lives from school.
16. Opinion polls.
17. The time taken for each classroom activity during the day.
18. Each pupil's favorite kind of meat.
19. Where each child went on his or her vacation.
20. Kinds of books read by the children.

Each of these examples offers children the opportunity to collect data from firsthand sources—themselves, their classmates, children in other classes, and the adults in their school. The use of firsthand data is preferred over those taken from almanacs, encyclopedias, or textbooks because they have greater meaning to children. Also, children get valuable experience in collecting, organizing, and interpreting data when they collect their own. Later, they can use their knowledge to read and interpret ready-made tables and graphs.

## 2. Tables and Graphs

Once children have gathered their data, they need to organize them so they can interpret them easily. Tables and graphs are useful for doing this. The table and graph in Figure 16-19 show how one group of children organized the data they collected about the colors of shoes their classmates wore to school over a period of time. They used a table (a) to record the daily shoe count. Then they put the information on the bar graph (b) so they could compare the number of shoes for each day and from day to day.

| Days / Shoe Colors | Mon. | Tues. | Wed. | Thurs. | Fri. | Mon. | Tues. | Wed. | Thurs. | Fri. | Mon. | Tues. | Wed. | Thurs. | Fri. | Mon. | Tues. | Wed. | Thurs. |
|---|---|---|---|---|---|---|---|---|---|---|---|---|---|---|---|---|---|---|---|
| White | 5 | 6 | 7 | | | | | | | | | | | | | | | | |
| Black | 11 | 10 | 8 | | | | | | | | | | | | | | | | |
| Brown | 8 | 9 | 9 | | | | | | | | | | | | | | | | |
| Comb. | 3 | 2 | 3 | | | | | | | | | | | | | | | | |
| Red | | | 1 | | | | | | | | | | | | | | | | |
| Totals | 27 | 27 | 27 | | | | | | | | | | | | | | | | |

(a)

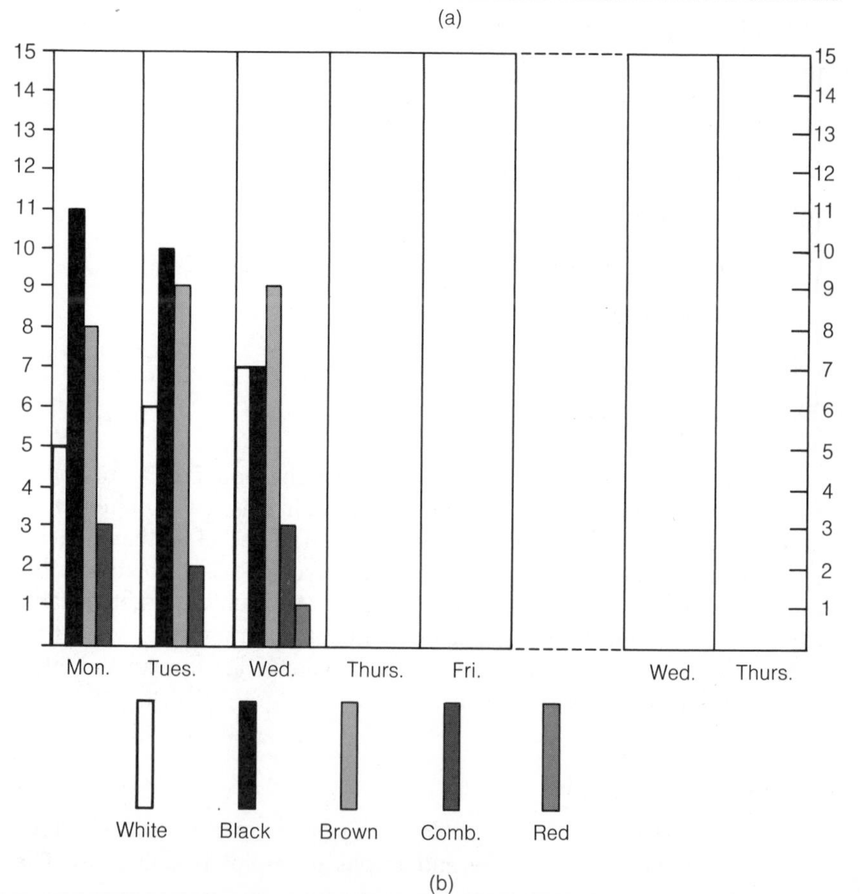


(b)

**Figure 16-19** Table (a) and graph (b) with data concerning children's shoe colors

Ms. Luna, the children's teacher, helped them learn to use the bar graph gradually. Several weeks before the record of shoe colors was begun, she had a one-day activity to introduce the idea of bar graphs. When she took the lunch count, she gave a red block to each child who was going home to eat, a blue block to each one who was buying a hot lunch, and a yellow block to each one who had brought lunch from home. The children stacked their blocks by color on a table and talked about the meaning of the number of blocks in each stack. A few days later Ms. Luna gave the children a similar activity. First, she marked a base line on a large sheet of paper and wrote the names of the days of the week beneath the line. Then she taped the paper to the wall. When she took the lunch count, she gave each child a square of colored gummed paper. Each child put his or her square on the paper to make a graph with columns to show the number of children going home, the number having hot lunch, and the number eating lunch brought from home. Ms. Luna helped each child stick his or her square on so that there were no gaps between squares and no overlapping squares. The children talked about the meaning of each column and why it was important to use care when they put the squares on the graph. This activity was repeated each day for the rest of the week. By the time the children began the shoe-color activity, they each had a good background for using tables and the more sophisticated bar graph.

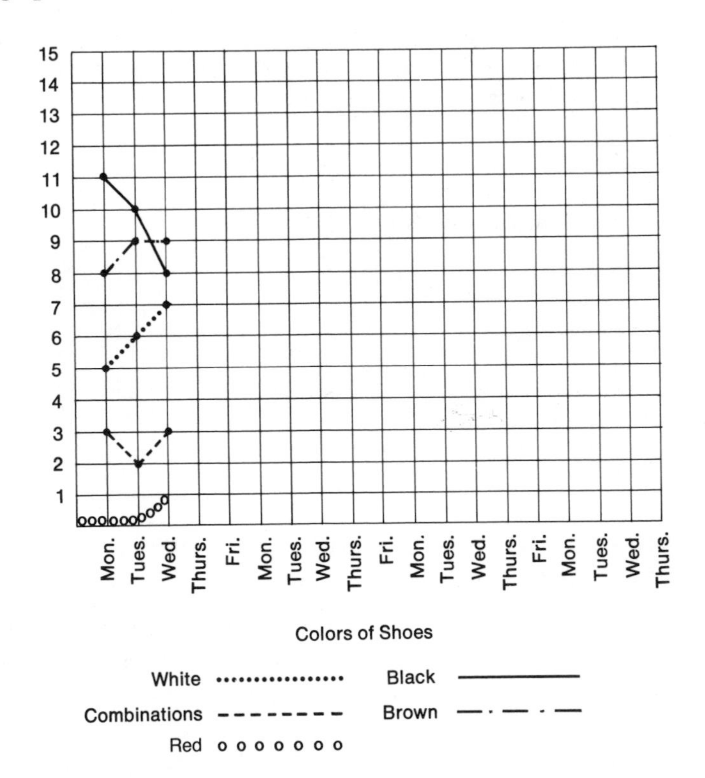


**Figure 16-20** Line graph with data concerning color of children's shoes

A line graph can be used to record the data from the shoe-color activity (Figure 16-20). Children can also use a picture graph. Each child can draw, color, and cut out a small shoe to paste on a sheet of paper that has a base line and names of the days on it.

Children should also learn to make and interpret circle graphs. These graphs usually show percents, so children should not use them before they have an understanding of percent and how to compute it. They also need to know how to measure degrees in a circle. The following topics, which are from the NCTM list, are sources of data for circle graphs:

1. The favorite colors of the children.
2. TV programs seen the night before class.
3. Opinion polls.
4. Each child's favorite kind of meat.
5. Kinds of books read by children.

The children who are sampled comprise the total population (100%) for a survey of the kind children might make of any of these topics. The number of children liking a certain color, for example, will be some percent of the total. Once the percent for each color has been determined, the children will need to figure out the portion of the circle's interior that will represent it. A completed graph showing the favorite ice cream flavors of one group of children is shown in Figure 16-21.

---

## Self-check of Objectives 13 and 14

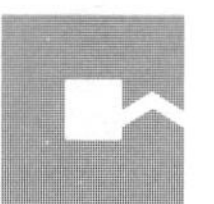

Twenty readily available sources of data for statistical study by children are listed. Name at least five of these, and add two more examples of your own.

Choose a statistical topic about which you can gather data from a group of children or classmates. Then gather and organize the data in a table and prepare two different graphs of it. (Use a hypothetical situation if real data cannot be collected.)

---

## THE MINICALCULATOR AND NUMBER THEORY, PROBABILITY, AND STATISTICS

These activities extend children's understanding of number theory, probability, and statistics. The minicalculator provides the means for simplifying the computation needed to complete each activity.

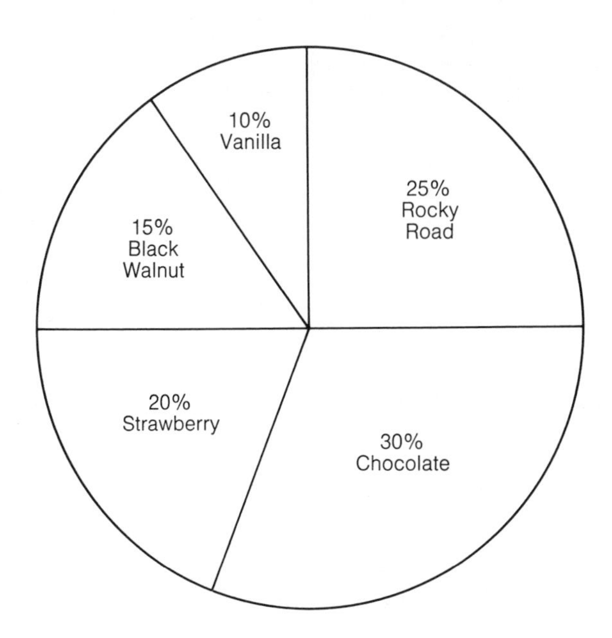


**Figure 16-21** Circle graph showing the favorite ice cream flavors of one group of children

## 1. Number Theory

**a. Finding Prime and Composite Numbers.** Which of these numbers are prime? Which are composite?

491 221 161 391 131 247 539 457

**b. Testing Formulas.** Test these formulas for finding prime numbers: $n^2 - n + 41 = p$ and $n^2 - 79n + 1601 = p$. The first formula yields prime numbers for all whole-number replacements for $n$ up to 41; the latter works for all whole-number replacements for $n$ through 79.

**c. Studying Prime and Composite Numbers.**[11] This investigation extends the activity with prime and composite numbers discussed on pages 477 to 479. Have the children complete a chart similar to the one in Figure 16-22 for the numbers 1 through 50. When the chart is finished, direct the children's attention to these questions:

1. Which numbers have two and only two factors? What kinds of numbers are these?

[11] This activity is adapted from William M. Fitzgerald, et al., *Laboratory Manual for Elementary Mathematics* (Boston: Prindle, Weber & Schmidt, Inc., 1973), pp. 49–50. Adaptation made with permission of the publisher.

| Number | Factors | Sum of Factors |
|---|---|---|
| 1 | 1 | 1 |
| 2 | 1,2 | 3 |
| 3 | 1,3 | 4 |
| 4 | 1, 2, 4 | 7 |
| 5 | 1, 5 | 6 |
| 6 | 1, 2, 3, 6 | 12 |
| 7 | 1, 7 | 8 |
| 8 | 1, 2, 4, 8 | 15 |
| 9 | 1, 3, 9 | 13 |
| 10 | 1, 2, 5, 10 | 18 |
| 11 | 1,11 | 12 |
| 12 | 1, 2, 3, 4, 6, 12 | 28 |
| 13 | 1, 13 | 14 |
| 14 | 1, 2, 7, 14 | 24 |

**Figure 16-22** The first fourteen counting numbers, their factors, and the sums of their factors

2. Which numbers have three and only three factors? What will be the next number that has exactly three factors?
3. Which numbers have four and only four factors? The numbers 8 and 27 have exactly four factors. How do they differ from the other numbers that have exactly four factors? Can you predict the next number of this kind? What number with exactly four factors follows 46?
4. Which numbers have five and only five factors? Can you describe these numbers? What is the next number that has exactly five factors?
5. Can you write a mathematical sentence that tells how to determine the sum of the factors when a number is prime? Can you write a sentence for finding the sum when a number has exactly three factors?

## 2. Probability and Pascal's Triangle

When a single coin is tossed, there are only two possible ways it can land; the coin can land either heads up or tails up. When two coins are tossed, there are four different outcomes: H-H, H-T, T-H, T-T. When three coins are tossed, these are the possible outcomes: H-H-H, H-H-T, H-T-H, T-H-H, H-T-T, T-H-T, T-T-H, and T-T-T. The number of ways one, two, and three coins can land are shown in this triangular arrangement of numbers:

| | |
|---|---|
| | 1 |
| 1 coin | 1—1 |
| 2 coins | 1—2—1 |
| 3 coins | 1—3—3—1 |

List all of the possible outcomes when four coins are tossed. What numbers indicate the ways four coins can land? Add these numbers to the triangle. Does the triangle give you enough information to predict the number of ways five coins can land? If not, list all of the possible outcomes for five coins. Extend the triangle to include complete rows for ten coins.

This arrangement of numbers is called Pascal's Triangle, after the 17th century mathematician, Blaise Pascal, who studied and wrote about it. With it, many interesting questions can be considered:

1. What is the sum of each row of the triangle? Can you describe the sequence of numbers that includes each sum?
2. When six coins are tossed, what is the probability of getting three heads and three tails? What percent of the possible outcomes is this? What is the probability of getting four heads and four tails when eight coins are tossed? What percent of the possible outcomes is this? Will the percent of possible outcomes for five heads and five tails when ten coins are tossed be greater or less than the percent for four heads and four tails when eight coins are tossed? Why do you think this is so?
3. When you toss seven coins, what is the probability of getting three heads and four tails? What is the probability of getting four heads and three tails? Is each of these probabilities the same percent of the total outcomes? How do you explain this?
4. When you toss nine coins, what is the probability that you will get four heads? Can you give the probability of getting at least four heads?

### 3. Statistics — Target Average[12]

**FINDING THE AVERAGE** PARTNERS ____________
____________

***Target Average***

1. Players take turns picking a number.
2. Each player picks as many numbers as indicated for the game.
3. As you pick numbers, enter them in your calculator and add them.
4. After the last number is picked, divide by the number of numbers.
5. The player whose average is closest to the target is the winner.

[12] This activity is from George Immerzeel, *'77 Ideas for Using the Rockwell 18R in the Classroom* (Foxboro, Mass.: New Impressions, Inc., 1976), p. 93. (Used with permission.)

| 12 | 9 | 3 | 19 | 26 | 44 | 49 | 25 | 65 | 16 |
|---|---|---|---|---|---|---|---|---|---|
| 15 | 96 | 20 | 51 | 78 | 18 | 15 | 19 | 30 | 0 |
| 47 | 13 | 45 | 29 | 14 | 45 | 21 | 41 | 14 | 39 |
| 16 | 76 | 33 | 12 | 48 | 30 | 45 | 6 | 40 | 19 |
| 2 | 10 | 21 | 20 | 17 | 22 | 35 | 28 | 10 | 34 |
| 17 | 49 | 37 | 17 | 33 | 60 | 39 | 13 | 40 | 35 |
| 0 | 87 | 14 | 55 | 65 | 31 | 93 | 24 | 18 | 47 |
| 18 | 11 | 87 | 32 | 15 | 50 | 15 | 36 | 2 | 13 |
| 31 | 62 | 23 | 75 | 18 | 89 | 57 | 10 | 35 | 91 |
| 27 | 43 | 18 | 31 | 11 | 27 | 61 | 28 | 37 | 43 |

Game 1
Pick 3 numbers
Target 15

Game 2
Pick 5 numbers
Target 12

Game 3
Pick 5 numbers
Target 20

Game 4
Pick 4 numbers
Target 30

Game 5
Pick 5 numbers
Target 40

Game 6
Pick 6 numbers
Target 50

## Self-check of Objective 15

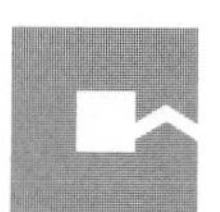

Five activities for using the handheld calculator to extend children's understanding of number theory, probability, and statistics are described. Explain each one, and demonstrate how the calculator is used for each.

## Common Pitfalls and Trouble Spots

Each chapter beginning with Chapter 4 is closed with a discussion of one or more pitfalls that inhibit effective teaching unless properly accounted for, and with trouble spots children face as they learn mathematics. These pitfalls and trouble spots are presented to make you aware of the kinds of problems mathematics teachers face. An examination of pitfalls in these chapters indicates that most of them have their genesis in two macropitfalls that encompass all mathematics topics and grades of the elementary school:

*Presenting material before children are ready for it.* You must constantly guard against presenting concepts that are too advanced for a child's level of cognitive development and for which he or she does not have the necessary mathematical background. You will do this if you (1) are aware of Piaget's levels of cognitive development, the characteristics of children at each level, and ways to assess each child's levels of growth, (2) know the hierarchy of concepts and skills for each topic, and use diagnostic procedures to determine each child's level of understanding, and (3) use appropriate performance objectives for each topic.

*Presenting material in inappropriate ways.* Generally, each topic or skill should be introduced at the concrete-manipulative level so children can investigate the new work at their own level of understanding. Manipulative materials are followed by pictorial-representational materials that lead children to a more abstract view of the topic. Finally, they deal with the concept or skill at an abstract level. Teachers who violate this sequence in the belief that their children do not need the concrete and/or semiconcrete experiences or that time can be gained for other activities usually find that their children do not understand the material as well as they had hoped and that any time saved must be spent reteaching the material.

The trouble spots at the close of chapters are examples of the problems children have while learning mathematics. While these examples do not exhaust the problems and errors, they should be sufficient to make you aware of the fact that children do have misconceptions about the mathematics they are learning and that some do establish faulty ways of dealing with operations. You should not blame either yourself or a child when mistakes are made, nor become discouraged with your teaching. When difficulties arise, step back to examine the reasons and think of ways to encourage the child to approach the work with renewed vigor and enthusiasm, or set it aside for the present while other work is undertaken.

---

## Self-check of Objective 16

Identify orally or in writing two macropitfalls associated with teaching mathematics at any level, and discuss briefly how each one can be avoided.

---

## CONCLUDING STATEMENT

One of the essential goals of any mathematics program is for children *to learn how to learn,* so they can continue to learn and understand mathematics after their formal education ends. In the same sense, every elementary school teacher must continue to learn about mathematics and how to teach it. After you have finished your studies, you can keep abreast of developments in the field of elementary school mathematics in several ways.

## 1. Reading

Professional books and journals provide up-to-date information concerning mathematics. No teacher should be content with the ideas from only one book about teaching mathematics. Even though the content of each book devoted to teaching mathematics in elementary school is similar, each author presents a different point of view. Also, each author usually discusses some topics not included in other books.

*The Arithmetic Teacher* is a journal devoted entirely to elementary school mathematics. Everyone who teaches mathematics to children should read it regularly. A one-year subscription to this journal is a benefit of membership in the National Council of Teachers of Mathematics. One can get information about the organization and all of its publications by writing the National Council of Teachers of Mathematics, 1906 Association Drive, Reston, Va. 22091.

Other publications—*School Science and Mathematics, Learning, Teacher,* and *Instructor,* for example—include articles dealing with elementary school mathematics.

## 2. Professional Meetings

The National Council of Teachers of Mathematics presents a number of mathematics conferences each year. In addition to an annual meeting each spring, nearly a dozen smaller meetings are held in various cities throughout the country. Each conference features speeches and workshops by authorities on teaching mathematics and displays of professional materials, as well as pupil-prepared exhibits. All teachers, whether or not they are NCTM members, are invited to these meetings. There is a fee for each one.

Throughout the country a number of state and local mathematics councils, clubs, or associations hold meetings in their areas. The names of these and their meeting dates are listed in *The Arithmetic Teacher* from time to time.

## 3. College and University Courses and Workshops

Many colleges and universities, as well as school districts, provide opportunities for continuing one's study of mathematics. Teachers can refine their teaching skills through courses and workshops of various types.

## 4. Miscellaneous Activities

Teachers may participate in other types of activities to improve their understanding of elementary school mathematics. In many districts, teachers select textbooks and other teaching materials. You can learn much about newer developments in the field by examining and evaluating new textbooks and learning aids. District and county offices frequently provide assistance through mathematics consultants or supervisors. Teachers who use these services usually find they learn much that helps them become better teachers. In every district, a few teachers are especially effective in teaching mathematics to children. If you can, arrange for the opportunity to see these expert teachers working with children.

Each teacher must select from among the opportunities available to him or her those that will be most helpful to keep abreast of developments in elementary school mathematics. The teacher who conscientiously strives to keep up with the subject may soon discover how little he or she actually knows about it, and will recognize more than ever the need for continual study.

---

## Self-check of Objective 17

Describe orally or in writing at least three ways you can keep abreast of developments in elementary school mathematics.

---

## SUMMARY

Work with elementary number theory is largely investigative and exploratory, with children developing an intuitive understanding of number theory topics rather than a formalized set of definitions and rules. Topics include odd and even numbers, prime and composite numbers, and tests of divisibility. Primary-grade children use markers in arrays to distinguish between odd and even numbers. Older children learn about prime and composite numbers in the same way. Children who understand prime and composite numbers and who know how to find the prime factorizations of composite numbers have the background they need for understanding greatest common factors and least common multiples. Knowledge of GCFs and LCMs is used while working with fractional numbers represented as common fractions.

Tests of divisibility, especially those for prime numbers such as 2, 3, 5, 7, and 11, are useful when finding the prime factorizations of composite numbers.

Children who have an understanding of probability and statistics and graphs have a means of collecting, organizing, and interpreting information from the world around them. There are simple probability experiments for primary-grade children, and more sophisticated ones for older children. Children themselves provide ample sources of statistical information to organize and then to prepare tables and graphs to report it. The macropitfalls associated with teaching mathematics at any level of the elementary school are (1) presenting topics for which children are not cognitively and/or mathematically ready and (2) presenting information in inappropriate ways. Teachers can keep abreast of developments in elementary school mathematics by reading professional books and journals such as *The Arithmetic Teacher,* attending national, state, and local conferences, participating in district and college or university courses and workshops, and in miscellaneous ways.

## STUDY QUESTIONS AND ACTIVITIES

1. What are the twin primes less than 100 after 11 and 13? What are twin primes between 100 and 200?
2. Test Goldbach's conjecture for even numbers between 2 and 50.
3. Make factor trees for 36, 42, and 54. Show various trees for each number.
4. Use the algorithm described in this chapter to factor 315, 462, and 735.
5. Use the factoring process to determine the GCF of 18 and 24, 54 and 96, and 210 and 462. Use a Venn diagram to illustrate the factorization and determination of the GCF for each pair.
6. Use the tests of divisibility for 3, 6, 7, 9, and 11 to determine by which of these numbers each of the following numbers can be divided.
   (a) 357 (b) 390 (c) 4,537 (d) 62,196
7. Put three red and three black checkers in a paper bag. Draw three checkers at a time for a total of 100 drawings. Record the events, using a large version of the table in Figure 16-18. How close to the $\frac{1}{8}$, $\frac{3}{8}$, $\frac{3}{8}$, and $\frac{1}{8}$ probabilities do your results come?
8. Experiment with different familiar objects, as indicated in the first activity suggested on page 495. Which objects lead to predictable events? Which do not? What are the probabilities for the predictable events? Compare the results of your activities with those of several classmates. Compile a list of materials that might be used when this activity is done with children.
9. Examine the "Ideas for Teachers" section of several copies of *The Arithmetic Teacher* (the feature began in 1971). Make copies of the "Ideas" you believe you will use in the future.

## FOR FURTHER READING

Avital, Shmuel. "The Plight and Might of Number Seven," *The Arithmetic Teacher,* XXV, No. 5 (February 1978), 22–24. The number seven comes alive to visit the author and tell all about its special characteristics. Many interesting possibilities for student investigations of the number seven (and others) arise from this article.

Dantzig, Tobias. *Number: The Language of Science,* 4th ed. Garden City, N.Y.: Doubleday & Company, Inc., 1954. Chapter 3 is of particular interest because of its discussion of number theory.

Frame, Maxine R. "Hamann's Conjecture," *The Arithmetic Teacher,* XXIII, No. 1 (January 1976), 34–35. "Hamann's Conjecture" was made by a seventh-grade student when he discovered that every even number from 2 through 250 can be expressed as the difference of two prime numbers. (Hamann discovered later that his conjecture was not the first expression of this idea when he did some research about it.)

Gilbert, Robert K. "Hey Mister! It's Upside Down," *The Arithmetic Teacher,* XXV, No. 3 (December 1977), 18–19. Three goals of mathematics instruction are to free students to think for themselves; to provide opportunities for children to discover order, pattern, and relations; and to train students in necessary skills. Activities related to making a graph of children's hair length with colored yarn serve to help further these goals.

Hewitt, Frances. "Pattern for Discovery: Prime and Composite Numbers," *The Arithmetic Teacher,* XIII, No. 2 (February 1966), 136–138. Children's discovery of prime and composite numbers through a study of arrays and tables is recommended.

Huff, Sara C. "Odds and Evens," *The Arithmetic Teacher,* XXVII, No. 5 (January 1979), 48–52. Colored pieces of cardboard, called "odds and evens" by the teacher and "magic counters" by second graders, served as a basis for investigations into the nature of odd and even numbers, addition of such numbers, and a game.

Lichtenberg, Betty P. "Zero Is an Even Number," *The Arithmetic Teacher,* XIX, No. 7 (November 1972), 535–538. Presents disk and number line patterns suitable for discoveries about odd and even numbers. The activities develop in an intuitive way the idea that zero is an even number.

Meyer, Jerome S. *Fun with Mathematics.* New York: Fawcett World Library, 1962. This book contains many different topics dealing with number theory. Particularly of interest is "The Magic of Numbers, " pp. 41–56.

Niman, John, and Robert D. Postman. "Probability on the Geoboard," *The Arithmetic Teacher,* XX, No. 3 (March 1973), 167–170. Gives an imaginative and interesting approach to probability. Regions are constructed with rubber bands; then the board is imagined to be a field. Children are asked, "What are the probabilities that a parachutist will land in the region, if the chances are equal that he will land in any part?" The article is a source of problem card activities.

Paige, Donald D. "Primes and Factoring," *The Arithmetic Teacher,* IX, No. 8 (December 1962), 449–452. Discusses arrays, sieve of Eratosthenes, factorization, and a process for finding the LCD.

Parkerson, Elsa. "Patterns of Divisibility," *The Arithmetic Teacher,* XXV, No. 4 (January 1978), 58. The patterns for divisibility by 3, 7, and 11 developed by a class of sixth graders are described and illustrated. The pattern for 11 is different from the one described in this chapter.

Pincus, Morris, and Frances Morgenstern. "Graphs in the Primary Grades," *The Arithmetic Teacher,* XVII, No. 6 (October 1970), 449–501. Pincus and Morgenstern illustrate a number of graphs that are suitable for kindergarten and primary-grade children. They conclude with a list of eleven additional topics about which data can be collected and graphed.

Rasof, Elvin. "Prime (Candy Bar) Numbers," *The Arithmetic Teacher,* XV, No. 1 (January 1968), 67–69. Presents a process to help children discover prime and composite numbers by using arrays and tables.

Reid, Constance. *From Zero to Infinity.* New York: Thomas Y. Crowell Company, 1960. Discusses each number—0 through 9—in a separate chapter. Discusses each number's origin, its lore, and its interesting characteristics. A separate chapter treats infinity. This book contains much number theory.

Roy, Sneh P. "LCM and GCF in the Hundred Chart," *The Arithmetic Teacher,* XXVI, No. 4 (December 1978), 53. Counting on the hundred chart and marking certain numerals with circles and triangles serves as a means of determining the LCM and GCF for pairs of numbers.

Scheur, Donald W., Jr. "All-Star GCF," *The Arithmetic Teacher,* XXVI, No. 3 (November 1978), 34–35. "All-Star GCF" is a game based on football. Teams composed of one or two children each determine moves based on the greatest common factors for a number. A board layout, rules, and a series of sample plays are included.

Smith, Rolland R. "Probability in the Elementary School," in *Enrichment Mathematics for the Grades,* 27th Yearbook, National Council of Teachers of Mathematics. Washington, D.C.: The Council, 1963, chap. 8. Smith discusses briefly the place of probability in elementary school, some background information for teachers, and a few probability topics that can be presented to children.

Souviney, Randall J. "Probability and Statistics," *Learning,* V, No. 4 (December 1976), 51–52. Beginning with letters from children's names as a basis for analysis, a series of statistical and probability activities are possible. The activities could form a unit for grades four through six.

———. "Quantifying Chance," *The Arithmetic Teacher,* XXV, No. 3 (December 1977), 24–26. Three activities using nonstandard dice, social studies book pages, and bags of colored marbles provide experiences for developing children's understanding of probability and their skill in making predictions about the occurrences of probability events.

Swafford, Jane, and Robert McGinty. "Story Numbers," *The Arithmetic Teacher,* XXVI, No. 2 (October 1978), 16–17. Another way of visualizing prime and composite numbers. Rectangular and triangular pieces of colored paper provide the models for constructing a city's skyline showing combinations of factors. All prime numbers form two-story buildings.

Tinnappel, Harold. "On Divisibility Rules," in *Enrichment Mathematics for the Grades,* 27th Yearbook, National Council of Teachers of Mathematics. Washington, D.C.: The Council, 1963, chap. 16. This chapter presents the rules for divisibility by 2, 3, 4, 5, 7, 9, and 11, and their proofs.

Webb, Leland F., and James D. McKay. "Making Inferences from Marbles and Coffee Cans," *The Arithmetic Teacher,* XXVI, No. 1 (September 1978), 33–35. A probability device is constructed from a coffee can, a funnel, and plastic tubing. Marbles come out at random through holes in the tubing. Several activities are described.

Yee, Albert H. "Mathematics, Probability and Decision Making," *The Arithmetic Teacher,* XIII, No. 5 (May 1966), 385–387. Yee presents some of the reasons for teaching probability in elementary school. He discusses some of the immediate values to children, but concludes that the long-range value of improving children's decision-making processes may be the primary reason for helping them understand probability.

# Appendixes

# Prefixes and Suffixes of Mathematical Terms

## PREFIXES

bi-, *L.*—twice, two
centi-, *L.*—hundred
deci-, *L.*—ten, tenth
deka-, *Gr.*—ten
dia-, *Gr.*—through
dodeca-, *Gr.*—twelve
geo-, *Gr.*—earth
hecto-, *Gr.*—hundred
hepta-, *Gr.*—seven
hexa-, *Gr.*—four
icosa-, *Gr.*—twenty
isos-, *Gr.*—equal
kilo-, *Gr.*—thousand
milli-, *L.*—thousand
octa-, *Gr.*—eight
para-, *Gr.*—side by side
penta-, *Gr.*—five
peri-, *Gr.*—around
poly-, *Gr.*—many
quadri-, *L.*—four
rect-, *L.*—right
tetra-, *Gr.*—four
tri-, *Gr.* and *L.*—three

## SUFFIXES

-angle, *L.*—angle
-hedron, *Gr.*—side, face
-gon, *Gr.*—angle
-lateral, *L.*—side
-meter, *Gr.*—measure

Note: *L.* indicates Latin, while *Gr.* indicates Greek.

# B Bibliography of Teachers' Books

The books listed in Appendix B are teachers' books. They cover a wide variety of topics of interest to teachers of elementary school mathematics.

**These books contain interpretations of Piaget's research and its implications for the teaching of mathematics.**

Brearley, Molly, and Elizabeth Hitchfield. *A Guide to Reading Piaget.* New York: Shocken Books, Inc., 1966.

Copeland, Richard W. *How Children Learn Mathematics,* 3rd ed. New York: The Macmillan Company, 1979.

———. *Math Activities for Children.* Columbus, Ohio: Charles E. Merrill Publishing Company, 1979.

Hyde, D. M. G. *Piaget and Conceptual Development.* New York: Holt, Rinehart and Winston, Inc., 1970.

Isaacs, Nathan. *New Light on Children's Ideas of Numbers.* London, Eng.: Ward Lock Educational Company, Limited, 1960.

National Council of Teachers of Mathematics. *Piagetian Cognitive-Development Research and Mathematical Education.* Washington, D.C.: The Council, 1971.

**These books present information about mathematics laboratories and classroom learning centers.**

Biggs, Edith E., and James R. MacLean. *Freedom to Learn.* Ontario, Canada: Addison-Wesley (Canada) Ltd., 1969.

Fitzgerald, William M., et al. *Laboratory Manual for Elementary Mathematics.* Boston: Prindle, Weber & Schmidt, Incorporated, 1973.

Greenes, Carole E., Robert E. Willcutt, and Mark A. Spikell. *Problem Solving in the Mathematics Laboratory: How to Do It.* Boston: Prindle, Weber & Schmidt, Incorporated, 1972.

Hooten, Joseph R., Jr., and Michael L. Mahaffey. *Elementary Mathematics Laboratory Experiences.* Columbus, Ohio: Charles E. Merrill Books, Inc., 1973.

Kaplan, Sandra Nina, et al. *Change for Children.* Pacific Palisades, Calif.: Goodyear Publishing Company, Inc., 1973.

Kennedy, Leonard M. *Experiences for Teaching Children Mathematics.* Belmont, Calif.: Wadsworth Publishing Company, 1973.

Kidd, Kenneth P., Shirley S. Myers, and David M. Cilley. *The Laboratory Approach to Mathematics.* Chicago: Science Research Associates, Inc., 1970.

Moore, Carolyn C. *Why Can't We Do Something Different?* Boston: Prindle, Weber & Schmidt, Incorporated, 1973.

Reys, Robert E., and Thomas R. Post. *The Mathematics Laboratory: Theory to Practice.* Boston: Prindle, Weber & Schmidt, Incorporated, 1973.

Schall, William E., ed. *Activity-Oriented Mathematics, Readings for Elementary Teachers.* Boston: Prindle, Weber & Schmidt, Incorporated, 1976.

**These books are of particular interest to teachers in preschool, kindergarten, and primary-grade classrooms.**

Baratta-Lorton, Mary. *Workjobs.* Menlo Park, Calif.: Addison-Wesley Publishing Company, 1972.

———. *Workjobs II.* Menlo Park, Calif.: Addison-Wesley Publishing Company, 1978.

———. *Mathematics Their Way.* Menlo Park, Calif.: Addison-Wesley Publishing Company, 1977.

Gardner, K. L., J. A. Glenn, and A. I. G. Renton. *Children Using Mathematics.* Oxford, Eng.: Oxford University Press, 1973.

Lovell, Kenneth. *The Growth of Understanding in Mathematics: Kindergarten through Grade Three.* New York: Holt, Rinehart and Winston, Inc., 1971.

Nuffield Mathematics Project. *I Do, And I Understand.* New York: John Wiley & Sons, Inc., no date.

———. *Mathematics: The First 3 Years.* New York: John Wiley & Sons, Inc., 1970.

Schools Council. *Mathematics in Primary Schools.* London, Eng.: Her Majesty's Stationery Office, 1969.

Thyer, Dennis, and John Maggs. *Teaching Mathematics to Young Children.* New York: Holt, Rinehart and Winston, Inc., 1971.

**These books contain games dealing with various mathematics topics. All of the games can be easily made by teachers and their children.**

Becker, Jan, Mary Laycock, and Genevieve Waring. *Enhance Chance.* Hayward, Calif.: Activity Resource Company, Inc., 1973.

Corle, Clyde G. *Skill Games for Mathematics.* Dansville, N.Y.: The Instructor Publications, Inc., 1968.

Henderson, George L. *Let's Play Games in Metrics.* Skokie, Ill.: National Textbook Company, 1974.

Holt, Michael, and Zoltan Dienes. *Let's Play Math.* New York: Walker and Company, 1973.

Kennedy, Leonard M., and Ruth L. Michon. *Games for Individualizing Mathematics Learning.* Columbus, Ohio: Charles E. Merrill Books, Inc., 1973.

Platts, Mary, ed. *Plus,* rev. ed. Stevensville, Mich.: Educational Service, Inc., 1975.

Schreiner, Nikki B. *Activities with Squares for Well-Rounded Mathematics.* Palos Verdes Estates, Calif.: Touch and See Educational Resources, 1973.

———. *Games & Aids for Teaching Math.* Hayward, Calif.: Activity Resources Company, Inc., 1972.

———. *More Games & Aids for Teaching Math.* Palos Verdes Estates, Calif.: Touch and See Educational Resources, 1973.

**These books contain information about assessing children's mathematical background and level of skills development, and Piagetian-type tests.**

Ashlock, Robert B. *Error Patterns in Computation,* 2nd ed. Columbus, Ohio: Charles E. Merrill Books, Inc., 1977.
Nuffield Mathematics Project. *Checking Up I.* London, Eng.: John Murray, Ltd., 1970.
———. *Checking Up II.* London, Eng.: John Murray, Ltd., 1972.
Reisman, Fredricka K. *A Guide to the Diagnostic Teaching of Arithmetic,* 2nd ed. Columbus, Ohio: Charles E. Merrill Books, Inc., 1978.
Suydam, Marilyn N. *Evaluation in the Mathematics Classroom.* Columbus, Ohio: ERIC Information Analysis Center for Science, Mathematics and Environmental Education, Ohio State University, 1974.

**These books are all yearbooks of the National Council of Teachers of Mathematics. They are listed in the order of their publication and contain titles only since they are all available from the same source.**

*Enrichment Mathematics for the Grades,* 1963.
*More Topics in Mathematics for Elementary School Teachers,* 1969.
*Historical Topics for the Mathematics Classroom,* 1969.
*The Slow Learner in Mathematics,* 1972.
*Instructional Aids in Mathematics,* 1973.
*Geometry in the Mathematics Curriculum, 1973.*
*Mathematics Learning in Early Childhood,* 1975.
*Measurement in School Mathematics,* 1976.
*Organizing for Mathematics Instruction,* 1977.
*Developing Computational Skills,* 1978.

**There are two journals published by the National Council of Teachers of Mathematics that contain valuable information for elementary school teachers:**

*The Arithmetic Teacher* (eight issues per year).
*Journal for Research in Mathematics Education* (four issues per year).

**These books deal with specific areas of mathematics and provide more details about teaching these areas.**

1. Calculators

Caravella, Joseph R. *Minicalculators in the Classroom.* Reston, Va.: The National Education Association, 1977.
Chinn, William G., Richard A. Dean, and Theodore N. Tracewell. *Arithmetic and Calculators: How to Deal with Arithmetic in the Calculator Age.* San Francisco: W. H. Freeman & Company, 1978.
Immerzeel, George. *Using Calculators in the Classroom.* Ormond Beach, Fla.: Camelot Publishing Company, 1976.

2. Geometry

Elliott, H. A., James R. MacLean, and Janet M. Jorden. *Geometry in the Classroom: New Concepts and Methods.* Toronto: Holt, Rinehart and Winston of Canada, Limited, 1968.
Hartung, Maurice L., and Ray Walch. *Geometry for Elementary Teachers.* Glenview, Ill.: Scott, Foresman and Company, 1970.
Johnson, Paul B., and Carol H. Kipps. *Geometry for Teachers.* Belmont, Calif.: Brooks/Cole Publishing Company, 1970.
Nuffield Mathematics Project. *Shape and Size.* New York: John Wiley & Sons, Inc., 1968.
Walter, Marion I. *Boxes, Squares, and Other Things.* Washington, D.C.: The National Council of Teachers of Mathematics, 1970.

3. Logic

Dienes, Z. P., and E. W. Golding. *Learning Logic, Logical Games.* New York: Herder and Herder, Inc., 1966.
Nuffield Mathematics Project. *Logic.* New York: John Wiley & Sons, Inc., 1972.

4. Measurement

Bitter, Gary G., Jerald L. Mikesell, and Kathryn Maurdeff. *Activities Handbook for Teaching the Metric System.* Boston: Allyn & Bacon, Inc., 1976.
Hanson, Susan G., comp. *Make Your Own Metric Measuring Aids.* Palo Alto, Calif.: American Institute for Research, Metric Studies Center, (no date).
Henry, Boyd. *Teaching the Metric System.* Chicago: Weber Costello, (no date).
Mathematics Education Task Force. *Inservice Guide to Teaching Measurement, An Introduction to the Metric System.* Sacramento, Calif.: California State Department of Education, 1975.
Michigan Council of Teachers of Mathematics. *Activities in Metric Measurement.* Lansing, Mich.: The Council, 1976.
Michigan Department of Education. *Teachers Resource Guide for Metric Education.* Lansing, Mich.: The Department, (no date).
Smart, James R. *Metric Math: The Modernized Metric System (SI).* Belmont, Calif.: Brooks/Cole Publishing Company, 1974.
Whitman, Nancy C., and Frederick G. Braun. *The Metric System, A Laboratory Approach for Teachers.* New York: John Wiley & Sons, 1978.

5. Probability and Statistics

Blakeslee, David W., et al. *Probability for Intermediate Grades, Teacher's Commentary.* Stanford, Calif.: School Mathematics Study Group, Leland Stanford Junior University, 1966.
———. *Probability for Primary Grades, Teacher's Commentary.* Stanford, Calif.: School Mathematics Study Group, Leland Stanford Junior University, 1966.
Nuffield Mathematics Project. *Probability and Statistics.* New York: John Wiley & Sons, Inc., 1969.

**Miscellaneous books are listed below. These cover a variety of topics not classifiable under any of the preceding headings.**

Conference Board of the Mathematical Sciences, National Advisory Committee on Mathematical Education. *Overview and Analysis of School Mathematics—Grades K–12.* Washington, D.C.: The Conference Board, 1975.

Ginsburg, Herbert. *Children's Arithmetic, the Learning Process.* New York: D. Van Nostrand Company, 1977.

Rising, Gerald R., and Joseph B. Harkin. *The Third "R"—Mathematics Teaching for Grades K–8.* Belmont, Calif.: Wadsworth Publishing Company, Inc., 1978.

Romberg, Thomas A. *Individually Guided Mathematics.* Reading, Mass.: Addison-Wesley Publishing Company, 1976.

Suydam, Marilyn N., and J. Fred Weaver. *Using Research: A Key to Elementary School Mathematics.* Columbus, Ohio: ERIC Information Analysis Center for Science, Mathematics and Environmental Education, Ohio State University, 1975.

# C Suppliers of Metric and Minicalculator Materials

This appendix contains information about sources of materials for teaching about the metric system and using handheld calculators in elementary school classrooms. The sources are representative of those from which materials are available; no attempt was made to list all possible sources. Neither names of specific items nor prices are given. Persons who are interested in knowing about a company's materials should write for a catalog.

## Part I—Metric Materials

*Activity Cards*

B.H.U.
23358 Hartland Street
Canoga Park, CA 91304

Creative Teaching Associates
P.O. Box 7766
Fresno, CA 93727

Creative Teaching Press, Inc.
5305 Production Drive
Huntington Beach, CA 92649

Mutual Aids
1953½ Hillhurst Avenue
Los Angeles, CA 90027

Prentice-Hall Learning Systems, Inc.
P.O. Box 47X
Englewood Cliffs, NJ 07632

Teachers
P.O. Box 393
Manhattan Beach, CA 90266

*Books for Children*—Workbooks, books of Ditto masters

Addison-Wesley Publishing Company
2725 Sand Hill Road
Menlo Park, CA 94025

The Dairy Queen System
P.O. Box 14312
Dayton, OH 45414

Harcourt Brace Jovanovich, Inc.
3800 Lakeville Highway
Petaluma, CA 94952

Hayes School Publishing Company, Inc.
321 Pennwood Avenue
Wilkinsburg, PA 15221

Laidlaw Brothers
Thatcher and Madison
River Forest, IL 60305

Scott, Foresman and Company
1900 East Lake Avenue
Glenview, IL 60025

Silver Burdett Company
Morristown, NJ 07960

*Converters*—Some of these are for converting within the metric system; others are for converting from customary to metric and vice versa.

Jaydee Specialties
P.O. Box 536
Wilmette, IL 60091

Kelm Manufacturing Company
3149 U.S. 33 North
Benton Harbor, MI 49022

Metric Genie Company
P.O. Box 305
Corte Madera, CA 94925

National Microfilm Association
8728 Colesville Road
Silver Springs, MD 20910

Perrygraf
2215 Colby Avenue
Los Angeles, CA 90064

Sterling Plastics
Division of Borden Chemicals
Mountainside, NJ 07092

Vari-Vue International, Inc.
650 South Columbus Avenue
Mount Vernon, NY 10550

*Films*—16mm, film loops, and filmstrips

Barr Films
P.O. Box 5667
Pasadena, CA 91107

Coronet Media
65 E. South Water Street
Chicago, IL 60601

Creative Learning
P.O. Box 324
Warren, RI 02885

Denoyer-Geppert Audio-Visuals
5235 Ravenswood Avenue
Chicago, IL 60640

Educational Activities, Inc.
1937 Grand Avenue
Baldwin, NY 11510

Educational Projections Company
1911 Pickwick Lane
Glenview, IL 60025

Encyclopedia Britannica
Educational Corp.
425 North Michigan Avenue
Chicago, IL 60611

Eye Gate Media
146-01 Archer Avenue
Jamaica, NY 11435

Houghton Mifflin Company
777 California Avenue
Palo Alto, CA 94304

Pathescope Educational Media, Inc.
71 Weyman Avenue
New Rochelle, NY 10802

Perennial Education, Inc.
P.O. Box 226
Northfield, IL 60093

SVE
1345 Diversey Parkway
Chicago, IL 60614

Walt Disney Educational Media
Company
500 South Buena Vista Street
Burbank, CA 91521

Xerox Educational Productions
245 Long Hill Road
Middletown, CT 06457

*Games*

Creative Teaching Associates
P.O. Box 7766
Fresno, CA 93727

Education Plus
18584 Carlwyn Drive
Castro Valley, CA 94546

Ideal School Supply Company
11000 South Lavergne Avenue
Oak Lawn, IL 60453

Kent Educational Services
P.O. Box 903
Ovieda, FL 32765

Metrix Corporation
P.O. Box 19101
Orlando, FL 32814

Milton Bradley Company
Springfield, MA 01101

*Kits*—Self-contained collections of measuring devices, problem cards, workbooks, films, cassettes, and so on.

Charles E. Merrill Publishing Company
1300 Alum Creek Road
Columbus, OH 43216

Cuisenaire Company of America
12 Church Street
New Rochelle, NY 10805

Imperial International Learning Corp.
Box 548
Kankakee, IL 60901

Leicestershire Learning Systems
Hill Mill
Chestnut Street
Lewiston, ME 04240

Modern Math Materials
1658 Albemarle Way
Burlingame, CA 94010

Prentice-Hall Media, Inc.
150 White Plains Road
Tarrytown, NY 10591

Redco Science, Inc.
181 Main Street
Danbury, CT 06810

Science Research Associates, Inc.
259 East Erie Street
Chicago, IL 60611

Silver Burdett Company
Morristown, NJ 07960

SVE
1345 Diversey Parkway
Chicago, IL 60614

Zweig
20800 Beach Boulevard
Huntington Beach, CA 92648

*Posters*

Encyclopedia Britannica Educational Corp.
425 North Michigan Avenue
Chicago, IL 60611

The Math Group
396 East 79th Street
Minneapolis, MN 55420

Milton Bradley Company
Springfield, MA 01101

*Teaching Materials*—Classroom devices such as metersticks, centimeter rulers, liter containers, gram and kilogram weight (mass) sets, bathroom scales, and so on.

Activity Resources Company, Inc.
P.O. Box 4876
Hayward, CA 94540

The Atlantic and Pacific Commerce Company
4061 Port Chicago Highway
Concord, CA 94520

Creative Publications, Inc.
P.O. Box 10328
Palo Alto, CA 94303

Cuisenaire Company of America
12 Church Street
New Rochelle, NY 10805

Dick Blick
P.O. Box 1267
Galesburg, IL 61401

Educational Teaching Aids
159 W. Kinzie Street
Chicago, IL 60610

Enrich
760 Kifer Road
Sunnyvale, CA 94086

GW School Supply
P.O. Box 14
Fresno, CA 93707

Ideal School Supply Company
11000 South Lavergne Avenue
Oak Lawn, IL 60453

Math-Master
P.O. Box 1911
Big Springs, TX 79720

*Miscellaneous*

Lily
Owens-Illinois
P.O. Box 1035
Toledo, OH 43666
(Manufactures drinking cups with a metric motif)

U.S. Department of Commerce
National Bureau of Standards
Washington, D.C. 20234
(Publishes bulletins, posters, booklets, and other information items about the metric system)

### Part II—Calculator Materials

*Activity Cards*

Addison-Wesley Publishing Company
2725 Sand Hill Road
Menlo Park, CA 94025

*Books for Children*

Immerzeel, George. *'77 Ideas for Using the Rockwell 18R in the Classroom.* Foxboro, Mass.: New Impressions, Inc., 1976.

Judd, Wallace P. *Using Your Calculator.* Chicago: Science Research Associates, Inc.

Prigge, Glenn, and Jane D. Gawronski. *Calculator Activities.* Big Springs, Texas: Math-Master, 1978.

Rade, Lennert, and Burt A. Kaufman. *Adventures with Your Hand Calculator.* St. Louis: CEMREL, Inc., 1977.

*Calculators*

Educational Calculator Devices, Inc.
P.O. Box 974
Laguna Beach, CA 92652

Ju-Rav Equipment
P.O. Box 1145
Pleasanton, CA 94566

Stokes Publishing Company
P.O. Box 415
Palo Alto, CA 94302

Telesensory Systems, Inc.
P.O. Box 10099
Palo Alto, CA 94304

Texas Instruments, Inc.
P.O. Box 5012, M/S54
Dallas, TX 75222

*Films*—16mm and filmstrips

BFA Educational Media
P.O. Box 1795
Santa Monica, CA 90406

Encyclopedia Britannica
Educational Corp.
425 North Michigan Avenue
Chicago, IL 60611

*Kits*

Ju-Rav Equipment
P.O. Box 1145
Pleasanton, CA 94566

Science Research Associates, Inc.
259 East Erie Street
Chicago, IL 60611

Texas Instruments
P.O. Box 5012 M/S54
Dallas, TX 75222

*Miscellaneous*

Calculator Information Center
1200 Chambers Road
Columbus, OH 43212
(Calculator Information Center publishes periodic bulletins about classroom uses of handheld calculators.)

# GLOSSARY/INDEX